Handbook of Engaged Sustainability

Satinder Dhiman
Editor-in-Chief

Joan Marques
Editor

Handbook of Engaged Sustainability

Volume 2

With 99 Figures and 54 Tables

Editor-in-Chief
Satinder Dhiman
School of Business
Woodbury University
Burbank, CA, USA

Editor
Joan Marques
School of Business
Woodbury University
Burbank, CA, USA

ISBN 978-3-319-71311-3   ISBN 978-3-319-71312-0 (eBook)
ISBN 978-3-319-71313-7 (print and electronic bundle)
https://doi.org/10.1007/978-3-319-71312-0

Library of Congress Control Number: 2018941852

© Springer International Publishing AG, part of Springer Nature 2018
This work is subject to copyright. All rights are reserved by the Publisher, whether the whole or part of the material is concerned, specifically the rights of translation, reprinting, reuse of illustrations, recitation, broadcasting, reproduction on microfilms or in any other physical way, and transmission or information storage and retrieval, electronic adaptation, computer software, or by similar or dissimilar methodology now known or hereafter developed.
The use of general descriptive names, registered names, trademarks, service marks, etc. in this publication does not imply, even in the absence of a specific statement, that such names are exempt from the relevant protective laws and regulations and therefore free for general use.
The publisher, the authors and the editors are safe to assume that the advice and information in this book are believed to be true and accurate at the date of publication. Neither the publisher nor the authors or the editors give a warranty, express or implied, with respect to the material contained herein or for any errors or omissions that may have been made. The publisher remains neutral with regard to jurisdictional claims in published maps and institutional affiliations.

Printed on acid-free paper

This Springer imprint is published by the registered company Springer International Publishing AG part of Springer Nature.
The registered company address is: Gewerbestrasse 11, 6330 Cham, Switzerland

*Humbly Dedicated to the
Well-Being of All Beings,
Our Shared Destiny, and
the Well-Being of Our Mother Nature!*

Preface

Sustainability *Matters*: Treading *Lightly* on Our Planet

> *The work an unknown good person has done is like a vein of water flowing hidden underground, secretly making the ground green* – Thomas Carlyle

As we sail through the precarious first decades of the twenty-first century, a new vision is emerging to end poverty, protect the planet, and ensure well-being for all as part of our shared destiny. Achieving these goals will require the concerted efforts of governments, society, educational institutions, the business sector, and informed citizens. To help achieve these goals, we propose that organizations and B-schools highlight *sustainability* so that they can become willing contributors to the vision of cultivating a harmonious society and a sustainable planet.

The uniqueness of this *Handbook* lies in its emphasis on the "engaged" dimension of sustainability.

The questions that we explore in this august volume are:

1. What can *I* do to change the future of the planet?
2. What can *organizations* do to improve the well-being of our society/planet?
3. What can we *all* do that is good for us, good for society, and good for the planet?

Within this *rethink* of sustainability, the ethico-spiritual basis of sustainability is never out of the frame. By definition, the topic of sustainability requires a broad interdisciplinary approach – as in "our total footprint on the planet," not just "our carbon footprint." Our work proposes to bring together the two allied areas of sustainability and spirituality in a dialectical manner, with ethics acting as a balancing force and spirituality playing the role of the proverbial invisible hand guiding our quest for sustainability.

Why Sustainability Matters?

The Exorbitant Economic Costs of Unsustainability

Hurricanes Harvey, Irma, and Maria are not the only disasters to ravage the world, says *US News and World Report*: natural disasters have hit nearly every continent in 2017, claiming the lives and livelihoods of hundreds of thousands of people in South Asia, North America, Central America, and Africa.[1] Commenting on the state of affairs of recent devastating disasters in North America, meteorologists and environmental risk experts predict that our planet faces a continuing grave risk from natural disasters.[2]

AccuWeather predicts that Hurricane Harvey paired with Hurricane Irma collectively cost the United States nearly $300 billion.[3] This is just one example of the exorbitant economic costs of our unsustainable footprint. The foregoing context underscores even more that *sustainability matters*.

Experts agree that the issues of climate change, dwindling biodiversity, and land degradation need to be urgently addressed. Responsible public policy changes at the macro level and mindful consumption at the individual level can go a long way to lessen the damage done to our planet. Business also needs to share the responsibility of protecting the environment in a major way. These observations have far-reaching implications for developing a sustainability mindset and also indicate the great potential for sustainability studies in management education.

United Nations Environment Programme (UNEP) offers the following overall recommendations:

1. Enhance sustainable consumption and production to reduce environmental pressures by addressing drivers associated with manufacturing processes and consumer demand.
2. Implement measures to reduce pollution and other environmental pressures.
3. Reduce dependence on fossil fuels and diversify energy sources.
4. Enhance international cooperation on climate, air quality, and other environmental issues.

---

[1] See: 10 of the Deadliest Natural Disasters of 2017. US News and World Report, Sept 20, 2017. Retrieved Sept 27, 2017: https://www.usnews.com/news/best-countries/slideshows/10-of-the-deadliest-natural-disasters-of-2017

[2] Annie Sneed, Hurricane Irma: Florida's Overdevelopment Has Created a Ticking Time Bomb, *Scientific American*, September 12, 2017. Retrieved September 27, 2017: https://www.scientificamerican.com/article/hurricane-irma-floridas-overdevelopment-has-created-a-ticking-time-bomb/

[3] Andrew Soergel, Forecast Predicts Heavy Economic Damage from Harvey, Irma. US News and World Report, Sept 11, 2017. Retrieved Sept 27, 2017: https://www.usnews.com/news/articles/2017-09-11/irma-harvey-forecast-to-do-290-billion-in-economic-damage

5. Low-carbon, climate-resilient choices in infrastructure, energy and food production coupled with effective and sustainable natural resource governance are key to protecting the ecological assets that underpin a healthy society.[4]

Cultivating a Sustainability Mind-Set

The contributors in this volume examine myriad sustainability issues objectively and offer diverse perspectives for reflection. A vibrant vision is presented to the readers in the form of a series of engaging *affirmations*:

1. We believe that it is impossible to separate economic developmental issues from environmental issues. In its most practical aspect, sustainability is about understanding close interconnections among environment, society, and economy.
2. We believe that the way to achieve sustainable, harmonious living in all spheres is through lived morality and spirituality at the personal level, team level, and societal level. We call it *engaged sustainability*.
3. We believe that excessive desire, anger, and greed are subtle forms of violence against oneself, others, and the planet.
4. We are not only unaware of these mental pollutants; we are often unaware how unaware we are.
5. We believe that a focus on engaged sustainability will help us harness what is good for us, good for society, and good for the planet.
6. We believe that achieving this goal will require a shift from being a consumer to being a contributor.
7. Only an individual life rooted in the continuous harmony with nature – a life based on moral and spiritual awareness – can be sustainable for the entire creation.

In this *Handbook*, our overarching goal has been to explore the application of sustainability to a wide variety of contemporary contexts – from economics of consumption and growth to government policy to achieving a sustainable planet. These three perspectives – ecology, equity, and economics – can serve as guiding principles. *This framework has far-reaching implications for developing a sustainability mindset and also indicate the great potential for garnering a higher awareness about sustainability.* As a point of departure, we can start with and build upon the World Commission on Environment and Development's 1987 definition of sustainability as economic development activity that "meets the needs of the present without compromising the ability of future generations to meet their own needs."

---

[4]Rate of Environmental Damage Increasing Across Planet but Still Time to Reverse Worst Impacts. Retrieved September 27, 2017: http://www.un.org/sustainabledevelopment/blog/2016/05/rate-of-environmental-damage-increasing-across-planet-but-still-time-to-reverse-worst-impacts/

In the web of life, everything is linked with everything else: you cannot pluck a flower without disturbing a star, as the poet Francis Thompson observed. It is, therefore, incumbent on tomorrow's leaders to examine these issues holistically and objectively and to seek out diverse perspectives for reflection.

Our economic system is committed to maximizing productivity and profits. This new credo for sustainability asks for an additional commitment: examine existing belief systems in light of the evidence presented, rather than scrutinizing the evidence in the light of preexisting notions. Believe nothing; research everything. This expectation is at the heart of every scientific endeavor. It requires us to be aware of confirmation-bias and premature cognitive commitment. There is a difference between being on the side of the evidence and insisting that the evidence be on your side.

This is the most important key to understanding all profound questions of life. Aristotle is reported to have said the following of his teacher, Plato: "Plato is dear; still dearer is the truth."

Human Undertakings: Prime Driver of Climate Change

When the real enemy is within, why fight an external war? – Gandhi

Calling the post 1950s the *Anthroposene* (literally, the "era of humans"), Ricard Matthieu tells us that this is the first era in the history of the world when human activities are profoundly modifying and degrading the entire system that maintains life on earth. He states that the wealthy nations are the greatest culprits: An Afghan produces 2500 times less CO_2 than a Qatari and a thousand times less than an American.[5] With a note of urgency, Ricard rightly observes:

> If we continue to be obsessed with achieving growth, with consumption of natural resources increasing at its current exponential rate, we will need three planets by 2050. We do not have them. In order to remain within the environmental safety zone in which humanity can continue to prosper, we need to curb our endless desire for "more."[6]

Likewise, observations throughout the world make it clear that climate change is occurring, and rigorous scientific research demonstrates that the greenhouse gases emitted by human activities are the primary driver.[7]

These conclusions are based on multiple independent lines of evidence, and contrary assertions are inconsistent with an objective assessment of the vast body of peer-reviewed science. Moreover, there is strong evidence that ongoing climate

---

[5]Matthieu Ricard, *Altruism: The Power of Compassion to Change Yourself and the World*, trans. by Charlotte Mandell and Sam Gordon (New York: Little, Brown and Company, 2015), 8.
[6]Ibid.
[7]See: Climate Change 2007: Synthesis Report. Retrieved August 19, 2015: http://www.ipcc.ch/publications_and_data/ar4/syr/en/spm.html

change will have broad impacts on society, including the global economy and on the environment. For the United States, climate change impacts include sea level rise for coastal states, greater threats of extreme weather events, and increased risk of regional water scarcity, urban heat waves, western wildfires, and the disturbance of biological systems throughout the country. The severity of climate change impacts is expected to increase substantially in the coming decades.[8]

If we are to avoid the most severe impacts of climate change, emissions of greenhouse gases must be dramatically reduced. In addition, adaptation will be necessary to address those impacts that are already unavoidable. Adaptation efforts include improved infrastructure design, more sustainable management of water and other natural resources, modified agricultural practices, and improved emergency responses to storms, floods, fires, and heat waves.[9]

Some Positive Initiatives About Climate Change Adaptability

While climate change is a global issue, it is felt on a local scale. In the absence of national or international climate policy direction, cities and local communities around the world have been focusing on solving their own climate problems. They are working to build flood defenses, plan for heat waves and higher temperatures, install water-permeable pavements to better deal with floods and storm water, and improve water storage and use.

According to the 2014 report on Climate Change Impacts, Adaptation and Vulnerability from the United Nations Intergovernmental Panel on Climate Change, governments at various levels are also getting better at adaptation. Climate change is starting to be factored into a variety of development plans: how to manage the increasingly extreme disasters we are seeing and their associated risks, how to protect coastlines and deal with sea-level encroachment, how to best manage land and forests, how to deal with and plan for reduced water availability, how to develop resilient crop varieties, and how to protect energy and public infrastructure.[10]

On June 18, 2015, Pope Francis officially issued 184-page encyclical *Laudato si'*, Italian for *"Praise Be to You."* Subtitled as *On Care for Our Common Home*, it is a new appeal from Pope Francis addressed to "every person living on this planet" for

---

[8]The conclusions in this paragraph reflect the scientific consensus represented by, for example, the Intergovernmental Panel on Climate Change and US Global Change Research Program. Many scientific societies have endorsed these findings in their own statements, including the American Association for the Advancement of Science, American Chemical Society, American Geophysical Union, American Meteorological Society, and American Statistical Association. See: Statement on climate change from 18 scientific associations (2009).

[9]Statement on climate change from 18 scientific associations (2009). Retrieved February 25, 2018: http://www.aaas.org/sites/default/files/migrate/uploads/1021climate_letter1.pdf

[10]NASA: *Global Climate Change: Vital Signs of the Planet*. Retrieved August 19, 2015: http://climate.nasa.gov/solutions/adaptation-mitigation/

an inclusive dialogue about how we are shaping the future of our planet. As Yardley and Goldstein review in the *New York Times*, the encyclical boldly calls for:

> ...a radical transformation of politics, economics and individual lifestyles to confront environmental degradation and climate change, blending a biting critique of consumerism and irresponsible development with a plea for swift and unified global action.[11]

Laudato si' has a wider appeal; it has found resonance with Buddhists, Hindus, with Jews, Muslims, Protestant and Orthodox Christians, as well as with atheists and agnostics.

Regarding the Entire World as One Family

When we see unity in diversity, it helps us develop universal outlook in life which is so essential sustaining the sanctity of our war ravaged planet. By developing universal pity and contentious compassion toward all and everything, one is then able to make peace with the world and feel at home in the universe. Let us seek and share the underlying truth of mutuality that does not lead to unnatural differences and disharmony. That is the truth of our identity behind diversity – the essential oneness of all that exists. By seeking the truth that is equally good to all existence, we will be able to revere all life and truly redeem our human existence. Only then can we ensure equally the happiness and welfare of all beings. That will be our true gift of sustainability to the universe.

Concluding Reflections

Our organizations and societies are human nature writ large; therefore, we believe that the solution to society's current chaos lies in the spiritual transformation of each one of us. For material development to be sustainable, spiritual advancement must be seen as an integral part of the human development algorithm. The choice we face is between conscious change and chaotic annihilation. The last century has highlighted both the creative and the destructive power of human ingenuity. Whereas humanity's greatest gains in this century came in the areas of science and technology, we also witnessed the horror of two world wars, the rise of international terrorism, and economic and financial meltdowns. Many believe that the greatest harm occurred in the erosion of moral and spiritual values.

In his splendid little book, *The Compassionate Universe*, Eknath Easwaran notes the *urgency* of the responsibility of humans to *heal* the environment – "the only creatures on Earth who have the power – and, it sometimes seems, the inclination –

---

[11] Jim Yardley and Laurie Goodstein "Pope Francis, in Sweeping Encyclical, Calls for Swift Action on Climate Change." *The New York Times*. 18 June 2015.

to bring life on this planet to an end."[12] As human beings, we are given the power to think and power to do. We are also given the power to choose, the free will. We can choose to live differently and create our own reality. This is perhaps the most unique gift we have that needs to be harnessed and realized if humanity has to have a shared future.

We have one planet to live. Let's cultivate it together.

Satinder Dhiman
Joan Marques

---

[12]Eknath Easwaran, *The Compassionate Universe: The Power of Individual to Heal the Environment* (Petaluma, California: Nilgiri Press, 1989), 7.

Contents

Volume 1

Part I Mapping the Engaged Sustainability Terrain 1

Selfishness, Greed, and Apathy 3
Satinder Dhiman

To Eat or Not to Eat Meat 37
Satinder Dhiman

Moving Forward with Social Responsibility 63
Joan Marques

Social Entrepreneurship 91
Joan Marques

Sustainable Decision-Making 113
Poonam Arora and Janet L. Rovenpor

Transformative Solutions for Sustainable Well-Being 139
Annick De Witt

The Spirit of Sustainability 169
Eugene Allevato

Part II Systemic Paradigm Shifts About Sustainability 199

Just Conservation 201
Helen Kopnina

Ethical Decision-Making Under Social Uncertainty 221
Julia M. Puaschunder

Oceans and Impasses of "Sustainable Development" 243
Will McConnell

Environmental Stewardship 273
Sachi Arakawa, Sonya Sachdeva, and Vivek Shandas

The Theology of Sustainability Practice 297
Peter Newman

Bio-economy at the Crossroads of Sustainable Development 309
José G. Vargas-Hernández, Karina Pallagst, and Patricia Hammer

Environmental Intrapreneurship for Engaged Sustainability 333
Manjula S. Salimath

Part III Training the Mind, Education is the Heart of Sustainability 357

The Sustainability Summit 359
Dennis Heaton

Empathy Driving Engaged Sustainability in Enterprises 383
Ritamoni Boro and K. Sankaran

Education in Human Values 405
Rohana Ulluwishewa

Utilizing Gamification to Promote Sustainable Practices 427
Kristen Schiele

Sustainable Higher Education Teaching Approaches 445
Naomi T. Krogman and Apryl Bergstrom

Expanding Sustainable Business Education Beyond Business Schools 471
Christopher G. Beehner

Business Youth for Engaged Sustainability to Achieve the United Nations 17 Sustainable Development Goals (SDGs) 499
George L. De Feis

Part IV Global Initiatives Toward Engaged Sustainability 525

Time Banks as Sustainable Alternatives for Refugee Social Integration in European Communities 527
Joachim Timlon and Mateus Possati Figueira

Supermarket and Green Wave 549
Josi Paz

Social License to Operate (SLO) 579
Michael O. Wood and Jason Thistlethwaite

Intercultural Business 603
Francisco J. Rosado-May, Valeria B. Cuevas-Albarrán, Francisco J.
Moo-Xix, Jorge Huchin Chan, and Judith Cavazos-Arroyo

Low-Carbon Economies (LCEs) 631
Elizabeth Gingerich

Volume 2

Part V Innovative Initiatives Toward Engaged Sustainability 651

Ecosystem Services for Wine Sustainability 653
Sukhbir Sandhu, Claudine Soosay, Howard Harris, Hans-Henrik Hvolby,
and Harpinder Sandhu

Gourmet Products from Food Waste 683
Inés Alegre and Jasmina Berbegal-Mirabent

**Collaboration for Regional Sustainable Circular Economy
Innovation** 703
Rajesh Buch, Dan O'Neill, Cassandra Lubenow, Mara DeFilippis, and
Michael Dalrymple

From Environmental Awareness to Sustainable Practices 729
Ragna Zeiss

Community Engagement in Energy Transition 755
Babak Zahraie and André M. Everett

**Relational Teams Turning the Cost of Waste into Sustainable
Benefits** 779
Branka V. Olson, Edward R. Straub, William Paolillo, and Paul A. Becks

Teaching Circular Economy 809
Helen Kopnina

Part VI Sustainable Living and Environmental Stewardship 835

Smart Cities 837
John E. Carroll

Sustainable Living in the City 869
Mine Üçok Hughes

Urban Green Spaces as a Component of an Ecosystem 885
José G. Vargas-Hernández, Karina Pallagst, and Justyna Zdunek-
Wielgołaska

Strategic Management Innovation of Urban Green Spaces for Sustainable Community Development 917
José G. Vargas-Hernández, Karina Pallagst, and Patricia Hammer

Application of Big Data to Smart Cities for a Sustainable Future 945
Anil K. Maheshwari

People, Planet, and Profit 969
Dolors Gil-Doménech and Jasmina Berbegal-Mirabent

Ecopreneurship for Sustainable Development 991
Parag Rastogi and Radha Sharma

Part VII Contemporary Trends and Future Prospects 1017

Responsible Investing and Environmental Economics 1019
Carol Pomare

Responsible Investing and Corporate Social Responsibility for Engaged Sustainability 1043
Raghavan 'Ram' Ramanan

The LOHAS Lifestyle and Marketplace Behavior 1069
Sooyeon Choi and Richard A. Feinberg

To Be or Not to Be (Green) 1087
Ebru Belkıs Güzeloğlu and Elif Üstündağlı Erten

Agent-Based Change in Facilitating Sustainability Transitions 1135
Katariina Koistinen, Satu Teerikangas, Mirja Mikkilä, and Lassi Linnanen

Designing Sustainability Reporting Systems to Maximize Dynamic Stakeholder Agility 1157
Stephanie Watts

Index 1185

About the Editor-in-Chief

Satinder Dhiman, Ph.D., Ed.D., M.B.A, M.Com., serves as Associate Dean, Chair, and Director of the MBA Program and Professor of Management at Woodbury University's School of Business. He holds a Ph.D. in Social Sciences from Tilburg University, Netherlands; a Doctorate in Organizational Leadership from Pepperdine University, Los Angeles; an M.B.A. from West Coast University, Los Angeles; and an M.Com. (with gold medal) from the Panjab University, India. *He has also completed advanced Executive Leadership programs at Harvard, Stanford, and Wharton.* In 2013, Dr. Dhiman was invited to be the opening speaker at the prestigious TEDx Conference @ College of the Canyons in Santa Clarita, California. He serves as the President of International Chamber of Service Industry (ICSI).

Professor Dhiman teaches courses pertaining to ethical leadership, sustainability, organizational behavior and strategy, and spirituality in the workplace in the MBA program. He has authored, co-authored, and co-edited over 17 management, leadership, and accounting related books and research monographs, including most recently authoring: *Holistic Leadership* (Palgrave 2017); *Gandhi and Leadership* (Palgrave 2015), and *Seven Habits of Highly Fulfilled People* (Personhood 2012); and co-editing and co-authoring, with Dr. Marques, *Spirituality and Sustainability* (Springer 2016) and *Leadership Today* (Springer 2016).

He is the *Editor-in-Chief* of two multi-author *Major Reference Works: Springer Handbook of Engaged Sustainability* and *Palgrave Handbook of Workplace Spirituality and Fulfillment* and *Editor-in-Chief* of *Palgrave Studies in Workplace Spirituality and Fulfillment* and editor of *Springer Series in Management, Change, Strategy* and *Positive Leadership*. Some of his forthcoming titles include: *Leading without Power: A New Model of Highly Fulfilled Leaders*; *Bhagavad Gītā: A Catalyst for Organizational Transformation* (both by Palgrave MacMillan); and *Conscious Consumption: Diet, Sustainability and Wellbeing* (Routledge, 2019).

Recipient of several national and international professional honors, Professor Dhiman is also the winner of Steve Allen Excellence in Education Award and the prestigious ACBSP International Teacher of the Year Award. He has presented at major international conferences and published research with his colleagues in *Journal of Values-Based Leadership, Organization Development Journal, Journal of Management Development, Journal of Social Change, Journal of Applied Business and Economics*, and *Performance Improvement*. He also serves as Accreditation Mentor and Site Visit Team Leader for the Accreditation Council for Business Schools and Programs (ACBSP) for various universities in America, Canada, Europe, and India.

Professor Dhiman is the Founder-Director of Forever Fulfilled, a Los Angeles-based Wellbeing Consultancy, that focuses on workplace wellness, sustainability, and self-leadership.

About the Editor

Joan Marques has reinvented herself from a successful media entrepreneur in Suriname, South America, to a groundbreaking "edupreneur" (educational entrepreneur) in California, USA. She currently serves as Dean at Woodbury University's School of Business, in Burbank, California, where she works on infusing and nurturing the concept of "Business with a Conscience" into internal and external stakeholders, using every reputable resource possible. She is also a Full Professor of Management and teaches business courses related to Leadership, Ethics, Creativity, and Organizational Behavior in graduate and undergraduate programs.

Joan holds a Ph.D. in Social Sciences from Tilburg University's Oldendorff Graduate School (2011); and an Ed.D. in Organizational Leadership from Pepperdine University's Graduate School of Education and Psychology (2004). She also holds an M.B.A. from Woodbury University (2000) and a B.Sc. equivalent degree (HEAO) in Economics from MOC, Suriname (1987). Additionally, she has completed post-doctoral work at Tulane University's Freeman School of Business (2010).

Dr. Marques is a frequent speaker and presenter at academic and professional venues. In 2016, she gave a TEDx-Talk at College of the Canyons in California, titled "An Ancient Path Towards a Better Future," in which she analyzed the Noble Eightfold Path, one of the foundational Buddhist practices, within the realm of contemporary business performance.

Joan's research interests pertain to Awakened Leadership, Buddhist Psychology in Management, and Workplace Spirituality. Her works have been widely published and cited in both academic and popular venues. She has written more than 150 scholarly articles and has (co)authored more than 20 books, among which are *Ethical Leadership, Progress with a Moral Compass* (Routledge, 2017); *Leadership, Finding Balance Between Acceptance and Ambition* (Routledge, 2016); *Leadership Today: Practices for Personal and Professional Performance* (with Satinder Dhiman - Springer, 2016); *Spirituality and Sustainability: New Horizons and Exemplary Approaches* (with Satinder Dhiman – Springer, 2016); *Business and Buddhism* (Routledge, 2015); and *Leadership and Mindful Behavior: Action, Wakefulness, and Business* (Palgrave MacMillan, 2014).

Contributors

Inés Alegre Managerial Decisions Sciences Department, IESE Business School – University of Navarra, Barcelona, Spain

Eugene Allevato Woodbury University, Burbank, CA, USA

Sachi Arakawa Portland State University, Portland, OR, USA

Poonam Arora Department of Management, School of Business, Manhattan College, Riverdale, NY, USA

Paul A. Becks Welty Building Company, Akron, OH, USA

Christopher G. Beehner Center for Business, Legal and Entrepreneurship, Seminole State College of Florida, Heathrow, FL, USA

Jasmina Berbegal-Mirabent Department of Economy and Business Organization, Universitat Internacional de Catalunya, Barcelona, Spain

Apryl Bergstrom Department of Resource Economics and Environmental Sociology, University of Alberta, Edmonton, AB, Canada

Ritamoni Boro Justice K. S. Hegde Institute of Management, Nitte University, Nitte, India

Rajesh Buch Walton Sustainability Solutions Initiatives, Arizona State University, Tempe, AZ, USA

John E. Carroll Department of Natural Resources and Environment, University of New Hampshire, Durham, NH, USA

Judith Cavazos-Arroyo Business, Universidad Popular Autónoma del Estado de Puebla, Puebla, Puebla, Mexico

Sooyeon Choi Department of Consumer Science, Purdue University, West Lafayette, IN, USA

Valeria B. Cuevas-Albarrán Centro Intercultural de Proyectos y Negocios, Universidad Intercultural Maya de Quintana Roo, José Ma. Morelos, Quintana Roo, Mexico

Michael Dalrymple Walton Sustainability Solutions Initiatives, Arizona State University, Tempe, AZ, USA

George L. De Feis Department of Management, Healthcare Management, and Business Administration, Iona College, School of Business, New Rochelle, NY, USA

Annick De Witt Copernicus Institute of Sustainable Development, Utrecht University, Utrecht, The Netherlands

Mara DeFilippis Walton Sustainability Solutions Initiatives, Arizona State University, Tempe, AZ, USA

Satinder Dhiman School of Business, Woodbury University, Burbank, CA, USA

Elif Üstündağlı Erten Faculty of Economics and Administrative Sciences, Department of Business Administration, Ege University, İzmir, Turkey

André M. Everett Department of Management, University of Otago, Dunedin, New Zealand

Richard A. Feinberg Department of Consumer Science, Purdue University, West Lafayette, IN, USA

Mateus Possati Figueira Arceburgo, MG, Brazil

Dolors Gil-Doménech Department of Economy and Business Organization, Universitat Internacional de Catalunya, Barcelona, Spain

Elizabeth Gingerich Valparaiso University, Valparaiso, IN, USA

Ebru Belkıs Güzeloğlu Faculty of Communication, Department of Public Relations and Publicity, Ege University, İzmir, Turkey

Patricia Hammer IPS Department International Planning Systems, Faculty of Spatial and Environmental Planning, Technische Universität Kaiserslautern, Kaiserslautern, Germany

Howard Harris School of Management, University of South Australia, Adelaide, SA, Australia

Dennis Heaton Maharishi University of Management, Fairfield, IA, USA

Jorge Huchin Chan Desarrollo Empresarial, Universidad Intercultural Maya de Quintana Roo, José Ma. Morelos, Quintana Roo, Mexico

Hans-Henrik Hvolby Department of Materials and Production, Centre for Logistics, Aalborg University, Aalborg, Denmark

Katariina Koistinen School of Energy Systems, Lappeenranta University of Technology, Lappeenranta, Finland

Helen Kopnina Faculty Social and Behavioural Sciences, Institute Cultural Anthropology and Development Sociology, Leiden University, Leiden, The Netherlands

Leiden University and The Hague University of Applied Science (HHS), Leiden, The Netherlands

Naomi T. Krogman Department of Resource Economics and Environmental Sociology, Faculty of Graduate Studies and Research, University of Alberta, Edmonton, AB, Canada

Lassi Linnanen School of Energy Systems, Lappeenranta University of Technology, Lappeenranta, Finland

Cassandra Lubenow Walton Sustainability Solutions Initiatives, Arizona State University, Tempe, AZ, USA

Anil K. Maheshwari Maharishi University of Management, Fairfield, IA, USA

Joan Marques School of Business, Woodbury University, Burbank, CA, USA

Will McConnell Woodbury University, Burbank, CA, USA

Mirja Mikkilä School of Energy Systems, Lappeenranta University of Technology, Lappeenranta, Finland

Francisco J. Moo-Xix Centro Intercultural de Proyectos y Negocios, Universidad Intercultural Maya de Quintana Roo, José Ma. Morelos, Quintana Roo, Mexico

Peter Newman Curtin University Sustainability Policy (CUSP) Institute, School of Design and Built Environment, Curtin University, Perth, Australia

Dan O'Neill Walton Sustainability Solutions Initiatives, Arizona State University, Tempe, AZ, USA

Branka V. Olson School of Architecture, Woodbury University, Burbank, CA, USA

Sindik Olson Associates, Los Angeles, CA, USA

Karina Pallagst IPS Department International Planning Systems, Faculty of Spatial and Environmental Planning, Technische Universität Kaiserslautern, Kaiserslautern, Germany

William Paolillo Welty Building Company, Akron, OH, USA

Josi Paz Consultant - Communication and International Cooperation, University of Brasília, Brasilia, DF, Brazil

Carol Pomare Ron Joyce Center for Business Studies, Mount Allison University, Sackville, NB, Canada

Julia M. Puaschunder The New School, Department of Economics, Schwartz Center for Economic Policy Analysis, New York, NY, USA

Graduate School of Arts and Sciences, Columbia University, New York, NY, USA

Princeton University, Princeton, NJ, USA

Schwartz Center for Economics Policy Analysis, New York, NY, USA

Raghavan 'Ram' Ramanan Desert Research Institute, Dallas, TX, USA

Parag Rastogi Management Development Institute, Gurgaon, India

Francisco J. Rosado-May Universidad Intercultural Maya de Quintana Roo, José Ma. Morelos, Quintana Roo, Mexico

Janet L. Rovenpor Department of Management, School of Business, Manhattan College, Riverdale, NY, USA

Sonya Sachdeva Northern Research Station, USDA Forest Service, Evanston, IL, USA

Manjula S. Salimath Department of Management, College of Business, University of North Texas, Denton, TX, USA

Harpinder Sandhu College of Science and Engineering, Flinders University, Adelaide, SA, Australia

Sukhbir Sandhu School of Management, University of South Australia, Adelaide, SA, Australia

K. Sankaran Justice K. S. Hegde Institute of Management, Nitte University, Nitte, India

Kristen Schiele California State Polytechnic University, Pomona, CA, USA

Vivek Shandas Portland State University, Portland, OR, USA

Radha Sharma Management Development Institute, Gurgaon, India

Claudine Soosay School of Management, University of South Australia, Adelaide, SA, Australia

Edward R. Straub Velouria Systems LLC, Brighton, MI, USA

Satu Teerikangas Turku School of Economics, Turku University, Turku, Finland

School of Construction and Project Management, Bartlett Faculty of the Built Environment, University College London, London, UK

Jason Thistlethwaite School of Environment, Enterprise and Development (SEED), University of Waterloo, Waterloo, ON, Canada

Joachim Timlon Ronneby, Sweden

Mine Üçok Hughes Department of Marketing, California State University, Los Angeles, Los Angeles, CA, USA

Rohana Ulluwishewa (Former) Massey University, Palmerston North, New Zealand

José G. Vargas-Hernández University Center for Economic and Managerial Sciences, University of Guadalajara, Guadalajara, Jalisco, Mexico

Núcleo Universitario Los Belenes, Zapopan, Jalisco, Mexico

Stephanie Watts Susilo Institute for Ethics in a Global Economy, Associate Professor of Information Systems, Questrom School of Business at Boston University, Boston, MA, USA

Michael O. Wood School of Environment, Enterprise and Development (SEED), University of Waterloo, Waterloo, ON, Canada

Babak Zahraie Research Management Office, Lincoln University, Lincoln, New Zealand

Justyna Zdunek-Wielgołaska Faculty of Architecture, University of Technology, Otwock, Poland

Ragna Zeiss Faculty of Arts and Social Sciences, Department of Technology and Society Studies, Maastricht University, Maastricht, The Netherlands

Part V

Innovative Initiatives Toward Engaged Sustainability

Ecosystem Services for Wine Sustainability

A Case in Point of Sustainable Food Systems

Sukhbir Sandhu, Claudine Soosay, Howard Harris, Hans-Henrik Hvolby, and Harpinder Sandhu

Contents

Introduction	654
Ecosystem Services Approach and Wine Industry	656
Engaged Sustainability at Grape and Wine Organization	660
Wine Company	660
Environmental Issues	661
Ecosystem-Based Risks and Opportunities	661
Ecosystem Service Approach in a Wine Company	662
Identification of Environmental Issues	664
Identification of Ecosystem-Based Risks and Opportunities	665
Management and Policy Outcomes	667
Management of Environmental Issues	667
Management of Ecosystem-Based Risks and Opportunities	668
Mechanisms and Policy Support for Managing Ecosystem Services	670
Reflection Questions	678
Cross-References	679
References	679

S. Sandhu (✉) · C. Soosay · H. Harris
School of Management, University of South Australia, Adelaide, SA, Australia
e-mail: Sukhbir.Sandhu@unisa.edu.au; Claudine.Soosay@unisa.edu.au; Howard.Harris@unisa.edu.au

H.-H. Hvolby
Department of Materials and Production, Centre for Logistics, Aalborg University, Aalborg, Denmark
e-mail: hhh@m-tech.aau.dk

H. Sandhu
College of Science and Engineering, Flinders University, Adelaide, SA, Australia

e-mail: Harpinder.Sandhu@flinders.edu.au

© Springer International Publishing AG, part of Springer Nature 2018
S. Dhiman, J. Marques (eds.), *Handbook of Engaged Sustainability*,
https://doi.org/10.1007/978-3-319-71312-0_43

Abstract

This study investigates the concept of ecosystem services in an Australian grape and wine company and explores risks and opportunities to achieve environmental sustainability in this organization. Ecosystem service approach is an emerging paradigm to address natural resource degradation and achieve sustainability in agribusiness organizations. A case study method is used to identify environmental issues at one of the premium wine organizations based in South Australia. This study conducts semi-structured interviews with multiple informants to analyze how this organization integrates ecosystem services approach in their management systems. These semi-structured interviews with multiple informants identified three categories of environmental issues: (1) primary (water use efficiency, soil health, carbon emissions), (2) secondary (energy, water availability), and (3) tertiary (waste water recycling, salinity in soil, loss of biodiversity, impacts due to climate change projections, winery waste management, soil carbon). We used Ecosystem Based Business Risk Analysis Tool (EBBRAT) and found freshwater availability as a major risk for this organization. This tool led to the identification of key areas, such as biological control of insect pests, maintaining biodiversity and management of soil, as an opportunity for the wine company to enhance sustainability. This study highlights ecosystem service approach to achieve sustainability in wine and other agribusiness organizations. This case study is followed by two interactive exercises to illustrate the application of the ideas discussed in the chapter. The chapter concludes with lessons learnt to develop sustainable food systems and some questions that reflect the ideas presented in the chapter and are aimed at shifting the focus toward food sustainability.

Keywords

Sustainable food systems · Environmental management · Ecosystem-based approach · Ecosystem-based business risks and opportunities · Ecosystem services · Natural capital

Introduction

Agriculture occupies one-third of the planet's land area, producing food for increasing human population. At the same time, clearing of land for agriculture and agricultural intensification has been damaging the global environment and human health (MEA 2005; IPCC 2007, 2012; Wratten et al. 2013). This has resulted in the decline of life-support functions and processes widely known as ecosystem services (MEA 2005). Ecosystem services are the benefits obtained either directly or indirectly, from natural and managed ecological systems (Daily 1997; MEA 2005; Pascual et al. 2017). Ecosystem services also provide natural capital for large number of industries such as agriculture, timber industry, power generation companies, etc. (TEEB 2010). The degradation of the natural capital, followed by the decline in global ecosystem services, has become an important issue for governments, societies, and businesses throughout the world (Stern Report 2006). In response to the increased environmental performance demands from community, governments, and other stakeholders, the agriculture industry is being forced to consider a range of environmental management programs (Céspedes-Lorente and Galdeano-Gómez 2004). A new approach to capture, measure, compare, and communicate sustainability in agriculture involves using the concept of ecosystem services (MEA 2005). The study investigates the concept of ecosystem services in an Australian agribusiness organization (henceforth called wine company) and explores ecosystem-based risks and opportunities to achieve environmental sustainability in this organization.

Agribusiness and ecosystems are intricately linked (Hanson et al. 2008). For example, the food industry depends upon land and water resources for the production of grains, meat, fiber, horticultural products, etc. Implications of loss of ecosystem services for the natural resource-based industry such as agribusinesses are far more direct than for manufacturing industries due to their direct reliance on ecosystems and their services (Grigg et al. 2009; TEEB 2010; Sandhu 2010). Current research has, however, largely constrained itself to the environmental practices of visibly polluting industries such as the chemical, automobile, mining, and energy industries (Bansal and Hunter 2003; Nishitani 2009). Consequently, there is limited knowledge about the environmental practices adopted by agribusinesses (Magdoff and Foster 2000; Jansen and Vellema 2004). Agribusinesses include organizations that are involved in primary production of agricultural products (e.g., grains, milk, meat, wool, fruit, and vegetables), processing (e. g., cereals, dairy, wine, sugar industry), manufacturing (e.g., seed, fertilizers, agrochemical industry), and retailing (e.g., processed food, chocolate etc.). Being a natural resource-dependent industry, the agribusiness industry is extremely vulnerable to the effects of environmental and climate changes (MEA 2005; TEEB 2015). While environmental degradation can undermine economic development for business in general, agribusinesses are particularly vulnerable, as healthy and functional ecosystems underpin the continuity of agribusinesses (Sandhu et al. 2012; Houdet et al. 2012; TEEB 2015). There is thus an urgent need for examining the consequences of changing natural environment, declining natural resources, and the resultant impacts on agribusinesses (Maloni and Brown 2006; Hoffman 2007; Heyder and Theuvsen 2010). So far, agribusinesses have been responding to the regulatory, voluntary, market pressures to address the challenges of changing environment (WBCSD 2010; Houdet et al. 2012). However, due to increasing stakeholder pressure, organizations which are more directly reliant on natural resources now need an explicit focus on incorporating ecosystem services into their business decisions (Hanson et al. 2008; Sandhu et al. 2012). Experts are now suggesting that given the scale of changes happening in the natural environment, it will be difficult, if not impossible, to carry on agribusinesses without integrating environmental concerns into decision making (Gladwin et al. 1995; Shrivastava 1995; Dunphy et al. 2007; Hanson et al. 2008; Sandhu et al. 2010).

The global agriculture industry is the largest industry on the planet, with 1.3 billion people dependent on agriculture for their livelihoods, and contributes about $4 trillion in global GDP (about 6% annually – World Bank 2017). In Australia, agriculture contributes about 3–4% annually to the national GDP and consumes 65% of the total natural resources including land, water, and biodiversity (Beeton et al.

2006; Hochman et al. 2013). The total value of agricultural export was $31 billion in the year 2015–2016 (DFAT 2016). The five dominant agricultural enterprises in Australia are livestock, grains, wine, dairy, and horticulture. The Australian wine industry is the sixth largest export industry in Australia (with a 5.2% share of agricultural exports and contributes about $2 billion annually to agricultural exports – DFAT 2016). It has grown considerably since the 1990s. Australia, with 145,000 ha area under vineyards, is among the leading wine grape-producing countries, with 2.3% of the total global vineyard area. It produces 4.3% of the total global wine production and is the fourth largest wine exporters in the world. South Australia has the largest wine-producing area in Australia with over 70,000 ha under vineyards. Wine exports contribute over $1 billion every year to the state economy, second only to wheat exports. The South Australian wine industry was established in the 1840s, after the arrival of European migrants. Over the last few years, the wine industry is experiencing uncertainties in global markets due to oversupply, impacts of climate change, etc. (Hayman et al. 2007; SAWIC 2010). It is increasingly being recognized that while there is a need to maintain balance between demand and supply in the short term, the future focus has to be on improving competitiveness in the long term (ABARE 2006).

The wine industry, in particular, is actively exploring alternative approaches to enhance competitive advantage. One way of thinking about these environmental issues and identifying opportunities for improving practice is to consider the concept of ecosystem services. This study examines an organization from the South Australian wine industry. The choice of business from the wine industry was dictated by the fact that it is the dominant wine region in Australia and contributes significantly to the state economy and national wine exports.

In the following section, an overview of the application of ecosystem services approach to the wine industry is provided. The study used the case of a premium grape- and wine-producing company based in South Australia as an example of engaged sustainability in integrating the challenges posed by ecosystems to achieve wine sustainability. It identifies environmental issues that this organization is dealing within the production of grapes and wine making. It then captures ecosystem-based risks and opportunities in this organization using a modeling tool: Ecosystem Based Business Risk Analysis Tool (EBBRAT – Sandhu et al. 2012). Insights are also provided on how this organization integrates ecosystem services approach into their management system to achieve sustainability. The chapter concludes by discussing how sound policy and management of environmental issues and ecosystem service approach can help develop sustainable food systems.

Ecosystem Services Approach and Wine Industry

Ecosystem services approach integrates the ecological, social, and economic dimensions of managing natural resources by the wine industry (Sandhu 2010). It includes identification and classification of ecosystem services that are relevant to wine industry. This approach explores sustainable management of natural resources by the wine industry for the production of grapes and wine. Enhancement of ecosystem services can help develop sustainable agriculture and food systems (Wratten et al. 2013). Key concepts around sustainable agriculture and food systems used in the chapter are described in Box 1.

> **Box 1 Key Concepts in Sustainable Agriculture and Food Systems**
> *Transformative agriculture and food systems.* Current agriculture and food systems are focused on one measure of success, i.e., production per unit area. However, it ignores social and environmental impacts and dependencies of agriculture and food systems. These impacts include loss of biodiversity, pollinators, pollution of water ways, pesticide poisoning, negative impacts on human health, social inequities (lower wages of farm workers), poor well-being of farmers and their families, loss of heritage and traditional knowledge, etc. Therefore, in order to transform agriculture and food systems, there is a need to understand and evaluate all impacts and dependencies (Sukhdev et al. 2016). This evaluation can help develop appropriate policy response for minimizing environmental and health impacts and improving long-term social, environmental, and economic sustainability of agriculture and food systems.
> *Natural capital.* Natural capital is described as the stock of natural resources that are extracted from nature for providing raw materials in manufacturing and other goods and services to sustain life (Costanza et al. 1997; UNU-IHDP and UNEP 2014). It is also known as the environmental or ecological capital. Examples include land, soil, water, air, biodiversity, etc.
> *Social capital.* Social capital is defined as the networks together with shared norms, values, and understandings that facilitate cooperation within or among groups (Keeley 2007, p. 103). It includes social equity, societal interactions, social rules, norms, community customs, culture, etc.
> *Human capital.* Human capital is defined as the knowledge, skills, competencies, and attributes embodied in individuals that facilitate the creation of personal, social, and economic well-being (Keeley 2007, p. 29).
> *True cost accounting.* The current farm accounts only include costs associated with inputs such as fertilizers, pesticides, irrigation, labor, capital costs, etc. They do not account for any depletion of natural resources, greenhouse gas emissions, and associated environmental costs including damage to air, water, and biodiversity, social inequities, and impacts on health and any positive environmental benefits. This exclusion of social and environmental costs and benefits in the economic system often results in perverse outcomes such as wrong policies that favors short-term gains over long-term sustainability. Therefore, true cost accounting (TCA) is a tool that is used in environmental management accounting to include all costs and benefits associated with the production, distribution, and consumption of food (Bebbington et al. 2001; TEEB 2015).
> *Biodiversity.* Biodiversity is the variety and variability of all living things on Earth. In agriculture, it includes above- and belowground genetic and species

(*continued*)

> **Box 1** (continued)
> diversity and all the ecological processes that contribute to the production of food. All the crops and livestocks have originated from the wide genetic resources. Food, forage, and fiber production depend on functioning biodiversity which is often known as functional agricultural biodiversity (Gurr et al. 2004).
> *Ecosystem services*. These are defined as the benefits obtained from natural and managed systems by human beings to support their livelihoods and well-being (Daily 1997). These include many categories of goods and services, for example, food, fish, timber, pollination, water regulation, waste treatment, soil erosion control, pest and disease suppression, cultural and recreational services, etc.
> *Payments for ecosystem services*. Payments for ecosystem services (PES) are defined as a voluntary transaction where an ecosystem service is bought by a buyer from a service provider (UNDP 2017). PES mechanisms include incentives to land managers in exchange of managing their land for the provision ecosystem services.
> *Well-being*. Human well-being is defined as the state of social and economic prosperity that leads to happiness and better health of individuals. In agriculture and food systems, it relates to better health of farm environment, animals, workers, and farming families. In the value chain, it includes better conditions for all workers in the processing and distribution of food and improved health of consumers.

The ecosystem services are currently in decline (MEA 2005), and it is of pressing concern for wine businesses because they rely on these ecosystem services for their continuity. However, since the wine industry can affect and be affected by ecosystem services, it is in a unique position to alleviate these challenges through a careful assessment of ecosystem service dependence and usage (Athanas et al. 2006; Genier et al. 2008; Houdet et al. 2012). It is therefore not surprising that stakeholder groups are increasingly pressurizing wine businesses to address these challenges.

Being a land-based industry, the wine industry is dependent on the services provided by healthy and functional ecosystems. The wine industry depends on four categories of ecosystem services: provisioning, regulating, supporting, and cultural services (De Groot et al. 2002; MEA 2005; Boyd and Banzhaf 2007; Wratten et al. 2013; CICES 2017). The provisioning, regulating, and supporting ecosystem services contribute to natural capital, while cultural ecosystem services contribute toward social and human capital. Natural capital includes well-functioning biodiversity and ecosystems, social capital constitutes societal interactions, relationships, formal and informal institutions, and human capital includes skills and knowledge (UNU-IHDP and UNEP 2014). The wine industry includes grape-growing and wine-making operations. The ecosystem services associated with these two operations in the wine industry are discussed briefly below (Table 1).

Table 1 List of ecosystem goods and services associated with agribusiness organizations and wine industry

	Agribusiness organizations	Wine industry
Provisioning services	1. Crops food crops (grains, oilseed, tea, coffee, vegetables, cocoa (chocolate products), coffee beans (coffee drink), seed (seed crops), horticultural products (fruits)) 2. Livestock (meat), milk (milk products) 3. Capture fisheries and aquaculture (fish, prawns, shrimps) 4. Wild foods (non-timber food products) 5. Timber and other wood fiber 6. Other fibers (cotton, hemp, silk) 7. Biomass fuel 8. Freshwater (irrigation, drinking water) 9. Genetic resources (plant breeding) 10. Biochemicals, natural medicines, and pharmaceuticals (medicinal plants)	1. Wine grapes 2. Wine 3. Vine varieties 4. Freshwater availability
Regulating services	11. Air quality regulation (greenhouse gas regulation: Emission reduction through supply chain management) 12. Global climate regulation (climate regulation: Carbon emission reduction) 13. Regional/local climate regulation 14. Water regulation (water regulation: Water use efficiency) 15. Erosion regulation (soil erosion control: Payments to upstream users for improving watershed services) 16. Water purification and waste treatment	5. Pesticides drift management 6. Carbon sequestration by above ground and below ground biomass 7. Water use efficiency 8. Maintaining soil cover with vegetation for managing soil erosion
Supporting services	17. Disease regulation (biological control) 18. Pest regulation (biological control: Natural enemies, pheromones, etc.) 19. Pollination (pollination: Horticultural products) 20. Mineralization of plant nutrients (nutrient cycling: Fertilizer) 21. Soil formation 22. Nitrogen fixation	9. Natural disease control 10. Natural pest control 11. Nutrient cycling with soil management
Cultural services	23. Recreation and ecotourism (ecotourism (nature's trail, camping in outback), recreation (visits to national parks, bird watching)) 24. Ethical values (social welfare schemes by business in local communities) 25. Aesthetic information 26. Cultural and artistic values 27. Historical values 28. Science and education values 29. Employment	12. Wine tourism 13. Ethical wine 14. Employment to locals

Provisioning services include goods and services produced by ecosystems. The wine industry is highly dependent on ecosystem goods (grapes) which are produced in vineyards. Wine grapes and wine are the final products under this category associated with the wine industry.

Regulating services include ecosystem services that regulate climate, air quality, soil erosion, etc. In the wine industry, regulating services support production of wine grapes. In vineyards, large amounts of agrochemicals are used to control pests/diseases and to supply nutrients to vines. Good management practices can minimize pesticide drifts and impacts on air quality and leaching of excessive nutrients into soil.

Supporting services include processes that support the production of provisioning services. In the wine industry, these are nutrient cycling, biological insect pest/disease control, etc. These ecosystem services support the provision of goods and services such as wine grapes. Nutrient cycling is managed in vineyards by better soil management using organic composts. Biological control of insect pests is enhanced by providing a habitat to the natural enemies of insect pests.

Cultural services include the aesthetics and recreational component of ecosystems. In wine industry these include recreation, wine tourism, ethical values, etc. The wine industry is largely dependent on wine tourism to earn revenue from the visitors, through wine sales, and also to improve sustainable wine image for marketing advantage.

A case study method (Yin 2003) is used to identify environmental issues at one of the premium wine organization based in South Australia. Ecosystem-based risks and opportunities are identified in this organization using the Ecosystem Based Business Risk Analysis Tool (EBBRAT; Sandhu et al. 2012). It also analyzes how this organization integrates the ecosystem service approach into their management system to achieve sustainability.

Engaged Sustainability at Grape and Wine Organization

Wine Company

The wine company in this study belongs to the exclusive group of *Australia's First Families of Wine* (a group of 12 family-owned Australian wineries – AFFW 2017). The case study organization established their first vineyard and wine-producing unit dating back to the early 1860s. The wine-producing company is one of the premium grape- and wine-producing company (hereafter, wine company) based in South Australia. A descriptive and in-depth case study approach is used to examine how ecosystem services approach can help in achieving environmental sustainability in this wine company. A single case study method was used to study environmental issues in this wine company as it allows in-depth examination of the general issues associated with the industry. Moreover, tools and analysis of ecosystem-based risks and opportunities are not being applied widely in the wine or agribusiness industry. This case is intended to develop context-dependent knowledge and management systems to improve environmental sustainability and apply ecosystem services approach in the industry.

Environmental Issues

Quality wine grape production depends upon adequate supply of natural resources such as water, soil, and sunshine. Apart from quality soil and water resources, healthy and functional ecosystems are also required for quality wine production. For example, grape growing requires nutrient supply by soil, moisture in the soil profile, and suppression of insect pests, weeds and diseases, etc. Degradation of these functions, loss of biodiversity and, declining natural resources (including impacts on water resources) pose a number of risks to the continuity of grape production in vineyards. Therefore, the tool used in this study (EBRAT) integrates ecosystem services into identifying risks and then developed strategies to improve environmental sustainability. An in-depth analysis of the environmental issues that the wine company has to deal with using semi-structured interviews is examined. This information was supplemented by information obtained from the company website, business publications, and brochures of the company. These in-depth interviews were conducted with the key staff (production manager, environmental manager, vineyard and winery staff) over the period December 2008 to December 2010. The initial interviews were conducted to understand and identify the key issues that this company has to face regarding environment and ecosystems. The discussion was around key topics: (1) What are the environmental issues at your organization? (2) Do you think there are environmental issues related with value chain in your organization? (3) Does your organization have impact/dependence on ecosystems? Subsequent interviews were primarily focused on understanding the processes to deal with the issues identified in the previous interview and included questions such as (1) how do you deal with these issues? and (2) why do you want to deal with these issues?

The analysis of the interviews led to the identification of the major themes relating to the vineyards and wine-making operations at the wine company. In accordance with the methodology suggested by Riemen (1986), meanings were then formulated from the significant statements. These formulated meanings were arrived at by reading, re-reading, and reflecting on the significant statements in the original transcription to get the meaning of the statements in the original context. The key issues were then divided into three categories – primary, secondary and tertiary issues with primary being the most pressing issues.

Ecosystem-Based Risks and Opportunities

Ecosystem-based risks and opportunities associated with the wine company using the Ecosystem Based Business Risk Analysis Tool are examined (EBBRAT; Sandhu et al. 2012). EBBRAT is a modeling tool which allows agribusinesses to assess their current reliance on ecosystem services. It helps organizations to identify ecosystem-based risks and opportunities and develop strategies for responding to the increasing demands to improve environmental performance (Sandhu et al. 2012). EBBRAT seeks to qualitatively and quantitatively explore and identify the risks and opportunities that agribusiness face because of their dependence on ecosystem services.

EBBRAT consists of a database of relevant ecosystem services with their description and examples. It uses Microsoft Excel spreadsheet for inputs and outputs. It comprises four categories of ecosystem services based on the MEA. These four categories are subdivided into 29 ecosystem services (Table 1).

A firm's dependence and impact on ecosystem services is entered into a data entry sheet. Dependence of a particular ecosystem service relevant to the firm's performance is measured on a scale, ranging from 0, which indicates no relevance, to 10, which is highly relevant. The firm's impact is quantified as either positive, negative, or no impact on a scale of -10 to $+10$. A positive impact indicates that the company's operations improve the ecosystem services, whereas a negative impact indicates that it results in the decline of ecosystem services. These responses are generated in the output worksheet, and data are automatically organized in the analysis sheet into different dependence categories and different impact categories with different color schemata. A graph is then generated showing the risks and opportunities based on impact dependence profiles of ecosystem services. Risks increase toward the bottom right side of the graph as company's operations are dependent on these ecosystem services and impact negatively on them, whereas opportunities increase toward the upper right corner of the graph as a company's operations are positively impacting on the ecosystem services involved. Overview of the steps involved in the use of EBBRAT are summarized in Fig. 1 (adapted from Ranganathan et al. 2008). In this case, these steps were carried out with the team consisting of an environmental manager, two operational managers of vineyards, and one staff from winery operations. Database of ecosystem services prepared by the United Nations Environment Program led project the Economics of Ecosystems and Biodiversity (TEEB 2010) and the Ecosystem Services Partnership (ESP, https://www.es-partnership.org/) was used in this study. These are publicly available databases. First, database of ecosystem services related to agriculture and natural systems was explored to identify relevant ones for wine industry especially those which are associated with grape growing at vineyard level. Several ecosystem services ($n = 29$) associated with grape growing and wine production were screened out of these global databases (Table 1). For each ecosystem service, the magnitude of dependence was estimated by the team relevant to the wine company on a scale of 1–10. Similarly, all negative and positive impacts of wine operations on each ecosystem service were estimated on a scale ranging from -10 to $+10$ by the team. This exercise provided an estimation of the impact and dependence on each ecosystem services. All the positive impacts provided opportunities for the wine company to further explorations, whereas the negative impacts pointed toward risks.

The next section presents the findings.

Ecosystem Service Approach in a Wine Company

Findings are provided across the two main themes that illustrate the case of ecosystem service approach to achieve sustainability in the wine company. First, key environmental issues in the wine company relating to vineyards and wine production

Ecosystem Services for Wine Sustainability

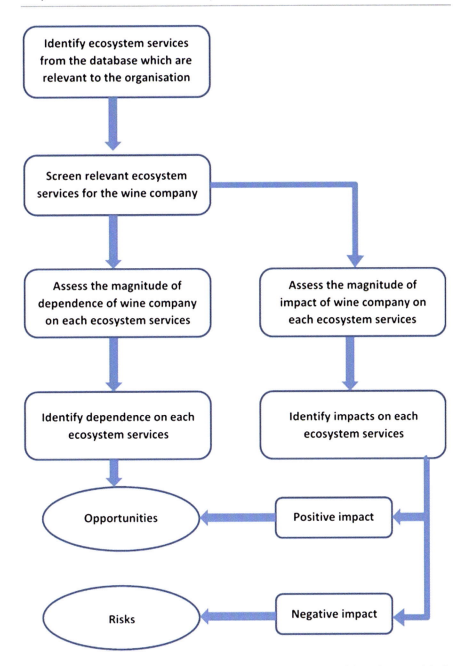

Fig. 1 Overview of the steps involved in the use of EBBRAT to assess risks and opportunities in the wine company

aspects are identified. These were classified into three categories: primary, secondary, and tertiary issues based on their importance. Second, ecosystem-based risks and opportunities were identified by the application of EBBRAT. These results were used to develop measures to adopt ecosystem services approach, which was then employed to improve sustainability at this wine company.

Identification of Environmental Issues

On the basis of in-depth interviews in this organization, environmental issues were classified into three categories – primary, secondary, and tertiary based on their importance (Fig. 2).

Water use efficiency, soil health, and carbon emissions were regarded as the primary issues because stakeholders such as community, regulations, and international markets (buyers and suppliers) considered these issues very important (Fig. 2). Secondary environment issues identified in this company were energy and water

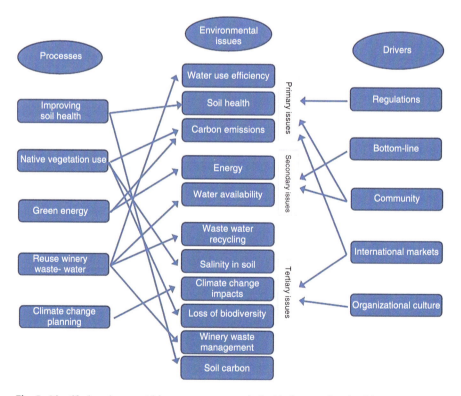

Fig. 2 Identified environmental issues, processes to deal with them, and major drivers of environmentalism at the wine company. On left side, each arrow indicates the process that deals with a particular environmental issue. On the right side, arrows indicate the type of driver that addresses each category of the environmental issues

availability (Fig. 2). There is a high impact on the business bottom line due to these environmental issues as they have significant costs involved. Tertiary issues identified were waste water recycling, salinity in soil, loss of biodiversity, impacts due to climate change projections, winery waste management, and soil carbon (Fig. 2). Five key drivers were also identified that drive environmentalism at the wine company. These were regulations, bottom-line impact, community pressure, international markets, and organizational culture. Regulations, community, and international markets help to address primary environmental issues. Bottom-line impact and community pressure addresses secondary environmental issues. International markets and organizational culture drive addressing of tertiary environmental issues. Five different processes were also identified that are being used to deal with the identified environmental issues. These were improving soil health, using native vegetation, using green energy, reusing winery waste water, and planning for climate change.

Identification of Ecosystem-Based Risks and Opportunities

Ecosystem services relevant to grape growing and wine making were analyzed using EBBRAT to identify risks and opportunities in this wine company (Fig. 3). The wine company was highly dependent on some key provisioning services (wine grapes, wine, vine varieties, freshwater availability), regulating services (pesticide drift

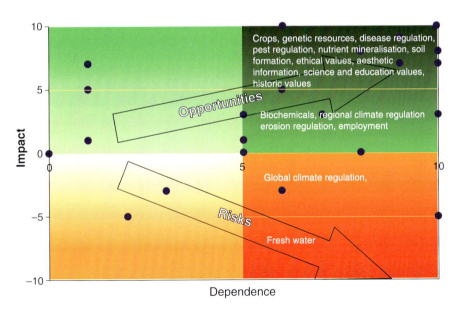

Fig. 3 Ecosystem-based risks and opportunities output by EBBRAT in the study organization. Arrows showing the risks and opportunities based on impact and dependence profiles. Risks increase toward the bottom right corner. Ecosystem services that are in this area are at risk due to company operations and also tend to impact negatively on the production. Opportunities increase toward the top right corner

management, carbon sequestration by vegetation, water use efficiency, maintaining soil cover), supporting services (disease/pests regulation, nutrient cycling), and cultural services (wine tourism, ethical wine, local employment). The impacts of grape growing and wine making on these ecosystem services were also identified which led to the identification of key risks through this tool.

Freshwater availability was the key high-level risk, identified in this study using EBBRAT. Grape growing and wine making involve large use of freshwater for irrigation and processing, respectively. Due to prolonged drought and low-rainfall region, freshwater availability was identified as one of the major risks to the continuity and sustainability of their operations. Global climate regulation (variability and change) pose great risks to the overall shift in seasons, rainfall patterns, pest pressure, and availability of other resources required to continue grape and wine production. The intensity of this risk was assessed as medium as the organization has opportunities to develop management action plan to mitigate this risk. Other risks identified in the medium category were greenhouse emissions and energy use. Low- to medium-intensity risks identified in the study were due to chemical pollution (use of agro-chemicals), soil erosion in vineyards, and soil health issues. These are considered as low-to-medium intensity as best practices to manage vineyards can be adopted in a short period of time to address these risks. Availability of workforce in vineyard and wine-making facility was considered low-risk areas as this organization has permanent employees and relatively stable access to seasonal labors during grape harvesting and wine-making time.

EBBRAT also led to the identification of some opportunities where this company can invest and develop management options to deal with those risks. For example, in the management of insect pests/disease through biological control, soil nutrition by using mulches to improve nutrient cycling, management of soil erosion by cover crops and mulching practices. Biodiversity in the vineyards can be enhanced by planting native plants and shrubs in the surrounding. This will not only enhance aesthetics but also help in bringing back biodiversity in the landscape. This wine company can invest in creating markets for biodiversity and develop ecotourism based on these measures.

Based on this assessment of risks and opportunities available, this wine company has undertaken strategic measures to adopt ecosystem services approach to improve vineyard sustainability (Table 2). This wine company has taken several measures to respond to these pressures due to their reliance on natural resources. This response not only helped them to manage ecosystem services but also improve their business bottom line and sustainability (including associated natural, social, and human capital). Management of freshwater by extending dams to store rainwater facility allows them to use water during the dry summers and save on costs associated with buying water and also reducing emissions related in transporting it to the site. Climate adaptation has been a major strategy to keep producing good wine grapes from more than century-old vineyards and maintain quality in the wines.

The wine company has also extended grape growing through establishing new vineyards in cooler area which are more resilient to climate impacts and erratic rainfall patterns. To enhance biodiversity and respond to the community needs,

Table 2 Ecosystem-based risks identified by using EBBRAT, their intensity and measures adopted by the wine company

Ecosystem-based risks	Intensity	Measures
Freshwater availability	High	Rainwater harvesting dam extended
Global climate regulation	Medium	Planning for adoption to climate variability
		Started operation in high rainfall zone
Greenhouse gas emissions	Medium	Capturing carbon through tree (biodiversity) plantings
Energy	Medium	Planning to include "green" energy sources for wine making
Chemical pollution (pesticides, herbicides)	Low to medium	Adoption of organic/biodynamic practices for pests and disease control
		Integrated pest management
		Ecosystem services strategies adopted in "the vineyard ecosystem management" project
Soil erosion	Low to medium	Inter-row vegetative cover throughout the year by native plants (Danthonia species and saltbush)
Soil health	Low to medium	Under-vine floor management using organic compost and mulches
Employment	Low	Employing locals for casual and permanent positions

native trees and plants are being planted near vineyards. Other responses include planning for "green" energy (solar and wind), less reliance on coal energy, managing insect pests using biological control methods, and incorporating organic/biodynamic principles in vineyard management.

Management and Policy Outcomes

There are three major outcomes of this study. First, environmental issues in wine company were identified. Second, ecosystem-based risks and opportunities were identified for the wine company. Based on these, this wine company has successfully adopted strategic measures to achieve sustainability. The third outcome is that the findings are potentially transferable to other organizations that operate in a similar context (i.e., agribusiness organizations).

Management of Environmental Issues

As governments and people start to respond to the urgency of the environmental issues, agribusinesses are being pressurized to adopt sustainable development principles (WBCSD 2009, 2010; IFC 2012; Houdet et al. 2012). The interviews with the staff at the wine company indicated that the primacy of the issues was dictated by stakeholder pressure. In this study, various environmental issues were identified (Fig. 2) because stakeholders such as community, regulators, and international markets (buyers and suppliers) considered these issues as being very important. The other issues identified in this study (such as improving water use efficiency, maintaining soil health and carbon emissions) are also strategic issues for the continuity of grape production at this organization. The primacy of these issues is driven by pressure to maintain the bottom line and pressures by community groups, markets, and culture at the organization (Fig. 2). For sustainable wine production, continuous supply of good-quality grapes is required. Their production depends on quality of soil, freshwater availability for irrigation, maintenance of pests in vineyards, and adaptation to changing climate. The issues identified in this study were aligned to the production objectives of the wine company in this study. Waste water, winery waste management, and biodiversity are important local issues as the rural community is very concerned about them. This organization works with local community and keeps other stakeholders informed about the decisions that may impact local biodiversity and water resources.

Processes to deal with the environmental issues are also identified (Fig. 2). Primary issues such as water use efficiency can be dealt with water recycling process. Soil health issues can be managed by improving soil fertility through compost and mulches. Carbon emissions from vineyard machinery use and wine-making operations can be managed by considering green energy options. Green energy options and recycling winery water were the two processes identified to address secondary issues. Tertiary issues were mostly operational issues in managing vineyards and in wine-making operations. The culture at this organization is inclusive and community participation (in field days, wine and food festivals, etc.) helps in identifying and providing solutions to such issues which involve workers and local community. Recycling of winery waste and water, native vegetation management and replanting, and climate adaptation planning were the major processes put in place to deal with these issues.

Management of Ecosystem-Based Risks and Opportunities

One of the key objectives of this study was to help wine companies understand the risks and opportunities that arise out of their dependence on ecosystem services to achieve sustainability. Ecosystem services approach helped this organization identify and manage risks and opportunities due to ecosystem change. Ecosystem services approach has thus enabled this wine company to better manage the natural resources on which they rely (Fig. 3). Some of these concepts are successfully used by the wine industry to enhance ecosystem services such as biological control of pests and improvement of soil health as demonstrated in the Box 2.

Box 2 Sustainable Wine Production by Enhancing Ecosystem Services in Vineyards
Below are two examples of successful improvement in two ecosystem services, biological pest control and soil health.

(*continued*)

Box 2 (continued)

Enhancing biological pest control. Light brown apple moth (LBAM, *Epiphyas postvittana*) is a serious pest in grapevines as its caterpillars roll shoots and feed on leaves and bunches causing reduction in grape yield and quality. This can cause significant economic damage. Conventional vineyards depend on pesticide to prevent losses. However, agrochemical use results in negative impacts such as loss of biodiversity, pesticide poisoning, and spray drift and is prohibited in certified organic vineyards. Therefore, management practices that can enhance natural pest control ecosystem services to manage such pests are required. A successful example is from the New Zealand wine industry, where growing strips of flowering buckwheat between vine rows decrease leafroller caterpillar population (Scarratt et al. 2008; Wratten et al. 2013). A small investment in growing buckwheat crop in vine rows results in increased number of beneficial parasitoid wasps that attack grape-feeding caterpillars and keep pest numbers under control. Buckwheat flowers provide nectar for the parasitic wasps and habitat for their survival.

Improving soil health. Soil heath is critical for sustainable agricultural production. Wine quality is also known to be affected by a number of environmental factors which are reflected in its *terroir*. Vineyards prefer deep soils with high microbial biodiversity and organic matter. High organic content leads to better microbiology of the soil, which leads to better nutrient cycling (Winkler et al. 2017). Soils with high organic matter have high water holding capacity. During dry period, this is extremely helpful in maintaining soil moisture. There are several plant diseases causing microbes that survive in soil during dormant season and transmit disease during ambient conditions. In vineyards, under-vine management can lead to suppression of the disease-causing microbes (Jacometti et al. 2007). Mulching with grass and compost can improve disease suppression ecosystem services (Wratten et al. 2013).

Mismanagement of ecosystem services can pose potential risks in the wine industry due to its high dependence and negative impacts. This case study developed risk and opportunity profile of a wine company and showed how strategic measures can be helpful in the governance, strategy, implementation, and reporting for long-term sustainability at the organization.

While several other tools are also available to assess ecosystem services such as ecosystem service review (ESR; Hanson et al. 2008) and Multiscale Integrated Models of Ecosystem Services (MIMES; Boumans and Costanza 2007), these are limited in their application and do not identify risks and opportunities as is by EBBRAT. EBBRAT identifies ecosystem-based risks that an organization has to deal with and also helps specify solutions to mitigate these through the adoption of ecosystem-based strategies (Houdet et al. 2012).

The ecosystem services approach, discussed in this paper, takes into account the interdependence between the natural capital and the wine industry and considers both risks and opportunities. The effects of changing environment on the wine company are tangible, so the issues can be best addressed by a strategic approach (Porter and Reinhardt 2007). As demonstrated in this study, the benefit of adopting the ecosystem services approach will lead to a more resilient wine industry that addresses the loss of biodiversity and declining natural resources and minimizes the impacts of climate change and other environmental issues (Table 2).

While our focus on this paper has been on the wine industry, this approach can also be applied to other agribusinesses who are seeking to direct organizational change through providing practices needed to improve social, environmental, and economic performance. This strategic approach offers potential to boost profits and create social good at regional scale (Sekine et al. 2010). It will also encourage business support for policies to protect and restore ecosystems. It can be further used to proactively develop strategies to identify and manage specific business risks and opportunities in value chains (Maloni and Brown 2006) arising from their company's dependence and impact on ecosystems and in reducing ecological footprints (Wackernagel and Rees 1996).

Mechanisms and Policy Support for Managing Ecosystem Services

Identification and acknowledgment of the role of ecosystem services in managing land-based agribusiness organizations can lead to the development of markets for ecosystem services (Kumar 2005). For example, an agribusiness organization involved in processing and marketing of breakfast cereals can also evolve into managing other valuable ecosystem services on the farmers' fields such as carbon credits, biodiversity credits, etc. (Narloch et al. 2011). Climate-induced risks to business organizations have prompted some of these organizations to identify and mitigate these risks through adoption of sustainable strategies (Bleda and Shackley 2008). At global scale, several studies have been conducted to understand the impacts of climate change and its socioeconomic consequences (Stern 2006; TEEB 2010). Markets have been created to mitigate the impacts of climate by reducing emissions, watershed restoration, biodiversity offsets, etc. (Waage and Stewart 2007). Some organizations are proactively forming partnerships (such as fair trade, rainforest alliance, roundtable for responsible soy, sustainable palm oil) for involving relevant stakeholders to manage natural landscapes and incorporating environment into daily operations. These initiatives are opening new streams of revenue for the businesses involved.

The wine company discussed in this study has adopted measures to respond to the risks identified by using ecosystem services approach (Table 2). This organization also explored opportunities identified for developing markets such as carbon credits by securing soil carbon through planting of native plants in and around the vineyards. Under the national carbon farming initiative, such efforts can result in opening of market for belowground and aboveground carbon maintained at their vineyards (CFI 2012).

Mechanisms such as market based instruments (MBIs; Whitten et al. 2003) and payments for environmental services (PES; Wunder 2005) that help to establish balance between the provision and consumption of ecosystem services have been developed around the world. While MBIs have already been in use for several decades especially in the context of environmental subsidies and taxes, PES is a relatively new concept which has originally emerged to denote a payment scheme. PES schemes such as water purification, biodiversity conservation, or carbon sequestration are increasingly becoming popular because of their perceived simplicity and cost-effectiveness (Pagiola and Platais 2007). As discussed in this case study, agribusinesses can use EBBRAT to help explore new markets from the identified opportunities as discussed in the case study (Table 2). To operationalize these tools, a greater effort is required to bring together the scientific knowhow through relevant policies, so that business can utilize them to ensure sustainability at both local and global level.

Increasingly, organizations, especially agribusinesses, are being pressurized into adopting more sustainable practices, by a wide range of stakeholders. It includes revealing their impacts and dependencies on natural, social, and human capital (Natural Capital Coalition 2016). This chapter explores ecosystem services approach to manage sustainability in a premium grape- and wine-producing company based in South Australia. Identifying environmental issues and risks helped this wine company to develop its response and adopt measures to improve sustainability through ecosystem services approach. The wine industry and other agribusiness organizations are increasingly being constrained by natural resources and climate risks. Therefore, there is high potential for the adoption of EBBRAT and similar tools to identify and minimize ecosystem-based risks and integrate ecosystem services into their management system to achieve long-term sustainability. This can help develop sustainable food systems.

Reflection Questions: Readers are invited to reflect on their experience of engaged sustainability transpired by the case study and in terms of the concept presented in the chapter.

1. In the case described in this chapter, a wine-producing organization identified its dependence on natural capital and ecosystem services. It demonstrated this concept on the production side (vineyards) of the wine operations. How can it improve its environmental sustainability through the entire value chain using the EBBRAT tool? List any limitations of the approach described in the above example.
2. Please comment on how sustainable agriculture and food systems can help improve environmental and public health. What are the opportunities to apply ecosystem services approach at national and global agriculture and food policy?
3. Multinational agribusiness companies have supply chains spanning many countries/regions across the globe and thus have high impact on the environment and natural resources. Explain how they can apply ecosystem services approach to manage their risks associated with the natural resources. Please discuss using examples.

4. Some global organizations recognize that to feed the growing population which is likely to be more than 9 billion by 2050 (and given that nearly 800 million people are currently malnourished), global agriculture should only focus on increasing per unit productivity. This is likely to intensify negative impacts on public and environmental health. Please comment whether this approach on *productivity only* is sufficient to achieve food security for all. Are their alternative approaches that can help reduce impact on natural resources and improve sustainable food production?
5. Consumer-driven sustainability is driving many organizations to modify their operations and become more socially responsible and sustainable. What is the role of consumers in driving local and global agriculture and food systems toward sustainability? Please discuss with examples.

Exercises in Practice: Based on the case study described above, two exercises are provided to illustrate how ecosystem-based approach can be applied to a dairy enterprise and wine business industry.

1. **Identification of ecosystem-based risks of a dairy enterprise**

Brief Description of a Business Organization

"Dairy enterprise" (a pseudonym) is a multinational organization based in Switzerland that processes and markets milk products across the globe. This organization has one unit based in India that sources milk from dairy farmers through its contract farming operations and pays fair prices, which are higher than the market prices. A part of its revenue is used for its social development program that focuses on developing women's health, children's education, and dairy farmers' well-being in rural areas in India. Milk is generally procured from farms and processed in the local chilling plants and then transported to a milk-processing factory located near Delhi. Milk is then processed to produce various milk products (including milk powder, cheese, yogurt and milk drinks, etc.), which are sold in domestic and international market through its retail outlets.

Task

To identify ecosystem-based risks in the entire value chain of "dairy enterprise."

Steps

Steps in the application of ecosystem-based approach (Fig. 4).
Aim: Use EBBRAT to identify ecosystem-based risks in the entire value chain of a dairy business (Fig. 5) **following the "guide for using ecosystem-based approach"** (Table 3 **and** Fig. 6).

Classification of ecosystem services

Ecosystem Services for Wine Sustainability

Fig. 4 Steps in the application of ecosystem-based approach

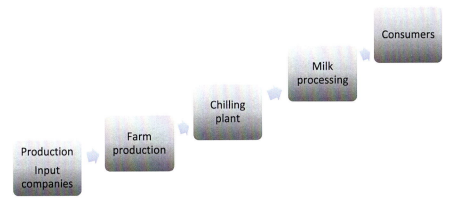

Fig. 5 Value chain of a dairy business

2. Assessing true cost of wine production

Brief description of the business organisation

A wine-producing organization based in Napa Valley, California, owns 400 ha of vineyards producing both chardonnay and merlot wines. The wine-making facility is located near the vineyards. The company employs local workers for both vineyard and wine-making facilities. Vineyards are irrigated by tapping into underground aquifers through bore well. There is a 50-ha area dedicated to native bush planting close to the vineyards. The wines are about 30 years old and have native vegetation around each vineyard as a biodiversity corridor to attract natural enemies of pests and disease. These native plants also help to sequester some of carbon dioxide from the atmosphere.

Task

What are the social, human, and natural capital associated with wine production?
The current business accounting system only recognizes costs associated with inputs, labor, capital costs, etc. It does not account for any depletion of natural resources, greenhouse gas emissions, and associated environmental costs including damage to air, water and biodiversity, social inequities, loss of natural, social and human capital, and any impacts on health. To develop sustainable agriculture and food systems, there is need to recognize the role of social, human, and natural capital (see Table 4).

Table 3 Classification and definition of four categories of ecosystem services and their types (Costanza et al. 1997; MEA 2005)

	Ecosystem services	Types	Definitions	Examples
	Regulating services: Ecosystems regulate essential ecological processes and life-support systems through biogeochemical cycles and other biospheric processes			
1		Gas regulation	Regulation of atmospheric chemical composition	Examples: CO_2/O_2 balance, O_2 for UVB, SOx levels
2		Climate regulation	Regulation of global temperature, precipitation, and other biologically mediated climatic processes at global or local levels	Example: Forests can impact regional rainfall levels
3		Disturbance regulation	Capacitance, damping, and integrity of ecosystem response to environmental fluctuations	Examples: Mangrove forests and coral reefs protect coastlines from storm surges; biological decomposition processes reduce potential fuel for wildfires
4		Water regulation	Influence ecosystems have on the timing and magnitude of water runoff, flooding, and aquifer recharge, particularly in terms of the water storage potential of the ecosystem or landscape	Examples: Permeable soil facilitates aquifer recharge; river floodplains and wetlands retain water – Which can decrease flooding during runoff peaks – Reducing the need for engineered flood control infrastructure
5		Water supply	Inland bodies of water, groundwater, rainwater, and surface waters for household, industrial, and agricultural uses	Examples: Freshwater for drinking, cleaning, cooling, industrial processes, electricity generation, or mode of transportation
6		Erosion control	Role vegetative cover plays in soil retention	Examples: Vegetation such as grass and trees prevents soil loss due to wind and rain; forests on slopes hold soil in place, thereby preventing landslides
	Supporting services: these are the services that are required to support the production of other ecosystem services			
7		Soil formation	Soil formation processes	Examples: Accumulation of organic material, weathering of rocks

(*continued*)

Table 3 (continued)

	Ecosystem services	Types	Definitions	Examples
8		Nutrient cycling	Storage, internal cycling, processing and acquisition of nutrients	Examples: Nitrogen fixation
9		Waste treatment	Recovery of mobile nutrients and removal or breakdown of excess or xenic nutrients and compounds	Examples: Waste treatment, pollution control detoxification
10		Pollination	Role ecosystems play in transferring pollen from male to female flower parts	Example: Bees from nearby forests pollinate crops
11		Biological control	Influence ecosystems have on the prevalence of crop and livestock pests and diseases	Example: Predators from nearby forests – Such as bats, toads, snakes – Consume crop pests
12		Habitat/ Refugia	Habitat for resident and transient production	Examples: Nurseries, habitat for migratory species, regional habitats for locally harvested species
	Provisioning services: These include food and services for human consumption, ranging from raw materials and fuel wood to the conservation of species and genetic material			
13		Food production	That portion of gross primary production extractable as food	Examples: Production of fish, crops, nuts, fruits
14		Raw materials	That portion of gross primary production extractable as raw material	Examples: Production of lumber, fuel, or fodder
15		Genetic resources	Genes and genetic information used for animal breeding, plant improvement, and biotechnology	Example: Genes used to increase crop resistance
	Cultural services: Cultural services contribute to the maintenance of human health and well-being by providing recreation, aesthetics, and education			
16		Recreation	Recreational pleasure people derive from natural or cultivated ecosystems	Examples: Hiking, camping, bird watching, scuba diving, going on safari
17		Cultural	Providing opportunities for noncommercial uses	Examples: Aesthetic, artistic, education spiritual, and/or scientific values

Background

Describe business operation in detail.

Identify various processes in the entire value chain: production, transportation, processing, marketing, waste etc.

Identify key products and their value chain.

Method

Apply ecosystem based approach throughout the entire value chain.

Create value chain map for the business.

Identify drivers of sustainability.

Identify ecosystem services dependencies through the entire value chain.

Identify impacts on ecosystems by business activities through the entire value chain.

Identify policy entry points for change.

Activity

The framework: Why should you conduct assessment of reliance and impacts on ecosystem services in the value chain?

Scope: Design the objective of the assessment?

Objective: What ecosystem services are of interest in this business organisation?

Assessment: Measure impact or dependencies using EBBRAT.

Trends: What are the changes in the state and trends of ecosystem services related to the business impacts and/or dependencies?

Interpret results: Discuss

Action: How will you apply the results for improving various processes?

Fig. 6 Guide for using ecosystem based approach in dairy enterprise

Table 4 Different types of social, human, and natural capitals (Costanza et al. 1997; Keeley 2007; UNU-IHDP and UNEP 2014)

Social capital	Market design, regulations, rules
	Civil and criminal laws; judicial systems
	Community rules, norms, customs, culture
	Constitutions; judiciaries; law and order; tax systems
	Social equity; communal harmony; cultural diversity
	Societal interactions, relationships
	Formal and informal institutions
Human capital	Health
	Education
	Skills and knowledge
	Traditional community knowledge
	Public databases
	Non-patent knowledge, intellectual outputs
	Motivation and capacity for relationships of the individual
Natural capital	Biodiversity, ecosystems
	Agriculture fields, forests
	Community forests, national parks
	Grazing commons

Aim: Use EBBRAT to identify and value social, human, and natural capital in the entire value chain of wine business following the "guide for using ecosystem-based approach" (Fig. 7).

Lessons Learned to Develop Sustainable Food Systems and Their Implications for Business Organizations

- Agriculture worldwide is under immense pressure to simultaneously increase production while minimizing impacts on environment and human health, which have begun to pose increasing risks to society.
- Natural resource-dependent business organizations face high risks due to changes in the ecosystems and their services.
- Organizations that source their raw material from nature such as agribusiness can play an important role to turn ecosystem-based risks into opportunities for the sustainability of business organization.
- Revealing all negative and positive impacts on social, human, and natural capital in the entire value chain can help improve transparency for shareholders, investors, and consumers.
- Utilizing ecosystem-based approaches can help conserve natural capital and protect environment while maintaining global supply chains for agribusiness companies.
- Sustainable food and agriculture systems can help eradicate hunger and poverty by providing access to healthy and nutritious food, thereby contributing to the achievement of globally agreed sustainable development goals of the United Nations.

Background

Describe business operation in detail.

Identify various processes in the entire value chain: production, transportation, processing, marketing, waste etc.

Method

Apply ecosystem based approach throughout the entire value chain.

Create value chain map for the business.

Identify natural, social and human capital.

Identify ecosystem services dependencies through the entire value chain.

Identify impacts on ecosystems by business activities through the entire value chain.

Identify policy entry points for change.

Activity

The framework: Why should you conduct assessment of reliance and impacts on ecosystem services in the value chain?

Scope: Design the objective of the assessment?

Objective: What are the impacts and dependcies on natural, social and human capital in this business organisation?

Assessment: Measure impact or dependencies using EBBRAT.

Trends: What are the changes in the state and trends of ecosystem services related to the business impacts and/or dependencies?

Interpret results: Discuss.

Action: How will you apply the results for improving various processes?

Fig. 7 Guide for using ecosystem based approach in the wine company

Reflection Questions

1. Sustainability paradigm is not limited to environment and nature, but there is a need to integrate social and human capital in holistic thinking. Reflect on your understanding about the role of social, human, and natural capital in a business organization.

2. What are the best ways to capture and communicate impacts and dependencies on natural, social, and human capital in business organizations?
3. How can business benefit from revealing their impacts on natural capital and ecosystem services?
4. What are the global frameworks that are developing knowledge bases to incorporate value of nature into business?
5. Business has a role to play for inclusive growth and development to achieve sustainable development goals (SDGs) as agreed by all member countries of the United Nations from 2016 to 2030. How can businesses best address this commitment?

Cross-References

▶ Bio-economy at the Crossroads of Sustainable Development
▶ Collaboration for Regional Sustainable Circular Economy Innovation
▶ Community Engagement in Energy Transition
▶ Responsible Investing and Corporate Social Responsibility for Engaged Sustainability

References

ABARE. (2006). Australian wine industry: Challenges for the future. ABARE Research Report.
AFFW. (2017). Australia's first families of wine. www.australiasfirstfamiliesofwine.com.au/. Accessed 20 Mar 2017.
Athanas, A., Bishop, J., Cassara, A., Donaubauer, P., Perceval, C., Rafiq, M., Ranganathan, J., & Risgaard, P.. (2006). Ecosystem challenges and business implications. Earthwatch Institute, World Resources Institute, WBCSD and World Conservation Union. Gland: IUCN.
Bansal, P., & Hunter, T. (2003). Strategic explanations for the early adoption of ISO 14001. *Journal of Business Ethics, 46*, 289–299.
Bebbington, J., Gray, R., Hibbitt, C., & Kirk, E. (2001). *Full cost accounting: An agenda for action*. London: Certified Accountants Educational Trust.
Beeton, R. J. S. (Bob), Buckley, K. I., Jones, G. J., Morgan, D., Reichelt, R. E., & Trewin, D. (2006). Australia state of the environment 2006. Independent report to the Australian government minister for the environment and heritage. Canberra: Department of the Environment and Heritage.
Bleda, M., & Shackley, S. (2008). The dynamics of belief in climate change and its risks in business organizations. *Ecological Economics, 66*, 517–532.
Boumans, R., & Costanza, R. (2007). The multiscale integrated earth systems model (MIMES): the dynamics, modeling and valuation of ecosystem services. GWSP issues in global water system research. Global assessments: Bridging scales and linking to policy. The Global Water System Project.
Boyd, J., & Banzhaf, S. (2007). What are ecosystem services? *Ecological Economics, 63*, 616–626.
Céspedes-Lorente, J., & Galdeano-Gómez, E. (2004). Environmental practices and the value added of horticultural firms. *Business Strategy and the Environment, 13*, 403–414.
CFI. (2012). The carbon farming initiative. Department of Climate Change and Energy Efficiency, Australian Government. http://climatechange.gov.au/cfi. Accessed 20 Mar 2017.

CICES. (2017). The common international classification of ecosystem services. https://cices.eu/. Accessed 20 Mar 2017.
Costanza, R., dArge, R., de Groot, R., Farber, S., Grasso, M., Hannon, B., Limburg, K., Naeem, S., Oneill, R. V., Paruelo, J., Raskin, R. G., Sutton, P., & van den Belt, M. (1997). The value of the world's ecosystem services and natural capital. *Nature, 387*, 253–260.
Daily, G. C. (Ed.). (1997). *Nature's services: Societal dependence on natural ecosystems*. Washington, DC: Island Press.
De Groot, R. S., Wilson, M., & Boumans, R. M. J. (2002). A typology for the classification, description and valuation of ecosystem functions, goods and services. *Ecological Economics, 41*, 393–408.
DFAT. (2016). *Trade at glance*. Canberra: Department of Foreign Affairs and Trade, Government of Australia.
Dunphy, D., Griffiths, A., & Benn, S. (2007). *Organizational change for corporate sustainability* (2nd ed.). London: Routledge.
Genier, C., Stamp, M., & Pfitzer, M. (2008). Corporate social responsibility in the agri-food sector: Harnessing innovation for sustainable development. FSG social impact advisors.
Gladwin, T. N., Kennelly, J. J., & Krause, T. S. (1995). Shifting paradigms for sustainable development: implications for management theory and research. *Academy of Management Review, 20*, 874–908.
Grigg, A., Cullen, Z., Foxall, J., & Strumpf, R. (2009). Linking shareholder and natural value. Managing biodiversity and ecosystem services risk in companies with an agricultural supply chain. Fauna & Flora International, United Nations Environment Programme Finance Initiative and Fundação Getulio Vargas. http://www.naturalvalueinitiative.org/download/documents/Publications/LSNVExecSummary.pdf.
Gurr, G. M., Wratten, S. D., & Altieri, M. A. (Eds.). (2004). *Ecological engineering for pest management: advances in habitat manipulation for arthropods*. Australia: CSIRO Publishing.
Hanson, C., Ranganathan, J., Iceland, C., & Finisdore, J. (2008). The corporate ecosystem services review: Guidelines for identifying business risks and opportunities arising from ecosystem change. World resources institute, World Business Council for Sustainable Development, Meridian Institute, p. 48.
Hayman, P., McCarthy, M. J., Sadras, V. O., & Soar, C. J. (2007). Can we identify dangerous climate change for Australian viticulture? In Proceedings of the 13th Australian wine industry technical conference, Adelaide.
Heyder, M., & Theuvsen, L. (2010). Corporate social responsibility in agribusiness: development and evaluation of a PLS model. EAAE seminar, Capri.
Hochman, Z., Carberry, P. S., Robertson, M. J., Gaydon, D. S., Bell, L. W., & McIntosh, P. C. (2013). Prospects for ecological intensification of Australian agriculture. *European Journal of Agronomy, 44*, 109–123.
Hoffman, A. J. (2007). If you're not at the table you're on the menu. *Harvard Business Review, 85*, 34–38.
Houdet, J., Trommetter, M., & Weber, J. (2012). Understanding changes in business strategies regarding biodiversity and ecosystem services. *Ecological Economics, 73*, 37–46.
IFC. (2012). *Policy and performance standards on environmental and social sustainability*. Washington, DC: International Finance Corporation, World Bank.
IPCC. (2007). Climate change 2007: The physical science basis. http://ipcc-wg1.ucar.edu/wg1/docs/WG1AR4_SPM_Approved_05Feb.pdf.
IPCC. (2012). Managing the risks of extreme events and disasters to advance climate change adaptation. A special report of working groups I and II of the intergovernmental panel on climate change. Cambridge, UK/New York: Cambridge University Press, 582 p.
Jacometti, M. A., Wratten, S. D., & Walter, M. (2007). Management to under storey to reduce inoculum of *Botrytis cinerea*: Enhancing ecosystem services in vineyards. *Biological Control, 40*, 57–64.
Jansen, K., & Vellema, S. (2004). *Agribusiness and society: Corporate responses to environmentalism, market opportunities and public regulation*. London: Zed Books.

Keeley, B. (2007). *Human capital: How what you know shapes your life*. Paris: OECD Publishing. https://doi.org/10.1787/9789264029095-en.

Kumar, P. (2005). *Market for ecosystem services*. Winnipeg: International Institute for Sustainable Development (IISD). http://www.iisd.org/.

Magdoff, F., & Foster, J. B. (2000). *Hungry for profit: The agribusiness threat to farmers, food and the environment*. New York: Monthly Review Press.

Maloni, M. J., & Brown, M. E. (2006). Corporate social responsibility in the supply chain: An application in the food industry. *Journal of Business Ethics, 68*, 35–52.

MEA. (2005). *Ecosystem and human well-being: Synthesis. Millennium ecosystem assessment*. Washington, DC: Island Press.

Narloch, U., Drucker, A. G., & Pascual, U. (2011). Payments for agrobiodiversity conservation services for sustained on-farm utilization of plant and animal genetic resources. *Ecological Economics, 70*, 1837–1845.

Natural Capital Coalition. (2016). Natural capital protocol – food and beverage sector guide. (Online) Available at: www.naturalcapitalcoalition.org/protocol.

Nishitani, K. (2009). An empirical study of the initial adoption of ISO 14001 in Japanese manufacturing firms. *Ecological Economics, 68*, 669–679.

Pagiola, S., & Platais, G. (2007). *Payments for environmental services: From theory to practice*. Washington, DC: World Bank.

Pascual, U., Balvanera, P., Díaz, S., Pataki, G., Roth, E., Stenseke, M., et al. (2017). Valuing nature's contributions to people: The IPBES approach. *Current Opinion Environmental Sustainability, 26*, 7–16.

Porter, M. E., & Reinhardt, F. L. (2007). A strategic approach to climate. *Harvard Business Review, 85*, 22–26.

Ranganathan, J., Raudsepp-Hearne, C., Lucas, N., Irwin, F., Zurek, M., Bennett, K., Ash, N., & West, P. (2008). *Ecosystem services: A guide for decision makers*. Washington, DC: World Resources Institute.

Riemen, D. J. (1986). The essential structure of a caring interaction: Doing phenomenology. In P. M. Munhall & C. J. Oiler (Eds.), *Nursing research: A qualitative perspective* (pp. 85–105). Norwalk: Appleton-Century Crofts.

Sandhu, S. (2010). Shifting paradigms in corporate environmentalism: From poachers to gamekeepers. *Business and Society Review, 115*, 285–310.

Sandhu, H., Nidumolu, U., & Sandhu, S. (2012). Assessing risks and opportunities arising from ecosystem change in primary industries using ecosystem-based business risk analysis tool. *Human and Ecological Risk Assessment, 18*, 47–68.

Sandhu, S. K., Ozanne, L., Smallman, C., & Cullen, R. (2010). Consumer Driven Corporate Environmentalism: Fact or Fiction?', *Business Strategy and the Environment, 19*, 356–366.

SAWIC. (2010). Wine: A partnership 2010–2015. South Australian Wine Industry Council. http://www.pir.sa.gov.au/wine.

Scarratt, S. L., Wratten, S. D., & Shishehbor, P. (2008). Measuring parasitoid movement from floral resources in a vineyard. *Biological Control, 46*, 107–113.

Sekine, K., Boutonnnet, J., & Hisano, S. (2010). Emerging standard complex and corporate social responsibility of agro-food businesses: A case study of dole food company. *The Kyoto Economic Review, 77*, 67–77.

Shrivastava, P. (1995). The role of corporations in achieving ecological sustainability. *Academy of Management Review, 20*, 936–960.

Stern Report. (2006). Stern review: The economics of climate change. http://www.hm-treasury.gov.uk/independent_reviews/stern_review_economics_climate_change/stern_review_report.cfm.

Sukhdev, P., May, P., & Müller, A. (2016). Fix food metrics. *Nature, 540*, 33–34.

TEEB. (2010). The economics of ecosystems and biodiversity report for business –executive summary. http://www.teebweb.org.

TEEB. (2015). *TEEB for agriculture & food: An interim report*. Geneva: United Nations Environment Programme.

UNDP. (2017). Payments for ecosystem services. Retrieved 8 Nov 2017, from http://www.undp.org/content/sdfinance/en/home/solutions/payments-for-ecosystem-services.html.
UNU-IHDP and UNEP. (2014). *Inclusive wealth report 2014. Measuring progress toward sustainability*. Cambridge: Cambridge University Press.
Waage, S., & Stewart, E. (2007). *The new markets for environmental services: A corporate manager's guide to trading in air, climate, water and biodiversity assets*. Business for Social Responsibility.
Wackernagel, M., & Rees, W. (1996). *Our ecological footprint: Reducing human impact on the earth*. Gabriola Island: New Society Publishers.
WBCSD. (2009). Corporate ecosystem valuation initiative. www.wbcsd.org/DocRoot/pdK9r5TpPijC1XXpx7QR/EcosystemsServices-ScopingReport_280509.pdf.
WBCSD. (2010). Vision 2050: The new agenda for business. The World Business Council for Sustainable Development.
Whitten, S., Van Bueren, M., & Collins, D. (2003). *An overview of market-based instruments and environmental policy in Australia*. In Proceedings of the sixth annual AARES national symposium. Canberra: Rural Industries Research and Development Corporation.
Winkler, K. J., Viers, J. H., & Nicholas, K. A. (2017). Assessing ecosystem services and multifunctionality for vineyard systems. *Frontiers in Environmental Science, 5*. https://doi.org/10.3389/fenvs.2017.00015.
World Bank. (2017). World Bank national accounts data. http://data.worldbank.org/indicator/NY.GDP.MKTP.CD. Accessed 20 Mar 2017.
Wratten, S., Sandhu, H., Cullen, R., & Costanza, R. (Eds.). (2013). *Ecosystem services in agricultural and urban landscapes*. Hoboken/Chichester: Wiley-Blackwell.
Wunder, S. (2005). Payments for environmental services: Some nuts and bolts CIFOR occasional paper no. 42. Jakarta: Center for International Forestry Research.
Yin, R. K. (2003). *Case study research: Designs and methods* (3rd ed.). Los Angeles: Sage.

Gourmet Products from Food Waste

Rethinking Food Management and Social Justice

Inés Alegre and Jasmina Berbegal-Mirabent

Contents

Introduction	684
Sustainability in the Food Industry	686
Analyzing The Food Supply Chain: Where Is Food Wasted?	688
Worldwide Initiatives That Fight against Food Waste	690
Hotels and Restaurants	691
Supermarkets and Retailers	693
Food for the Needy	694
Other Food Waste Related Initiatives	694
Case Study: Espigoladors	695
Reflection Questions	697
Exercises in Practice	698
Exercise 1: The Zero-Waste Challenge	698
Exercise 2: National Food Waste	699
Exercise 3: Healthy Eating	700
Exercise 4: Debate	700
Engaged Sustainability Lessons	700
Chapter-End Reflection Questions	701
Cross-References	701
References	701

I. Alegre (✉)
Managerial Decisions Sciences Department, IESE Business School – University of Navarra, Barcelona, Spain
e-mail: ialegre@iese.edu

J. Berbegal-Mirabent
Department of Economy and Business Organization, Universitat Internacional de Catalunya, Barcelona, Spain
e-mail: jberbegal@uic.es

© Springer International Publishing AG, part of Springer Nature 2018
S. Dhiman, J. Marques (eds.), *Handbook of Engaged Sustainability*,
https://doi.org/10.1007/978-3-319-71312-0_22

Abstract

Thousands of tons of food are wasted every day. Food can be discarded because it is out of date, it has lost some properties, excessive quantities are prepared at home, restaurants, or schools, etc. These causes are easily identifiable and are typically included in studies concerning food waste. However, there is one cause frequently ignored in food waste debates: the food that is left on the fields because its aesthetics do not conform to the market specifications in terms of size, weight, or color. This food is nutritious and perfectly suitable for consumption but it is excluded from the commercialization process. Espigoladors is a Spanish social enterprise that tackles precisely this issue. It collects the fruit and vegetables that are left on the fields and transforms them into high quality juices, creams, and jams. In addition, it creates more social value by employing people at social risk. This chapter presents the case of Espigoladors as a creative example of tackling an important societal issue and compels the readers to think about food management and social justice.

Keywords

Waste management · Sustainable food systems · Misshapen food · Social enterprise · Sustainability · Social justice · Agriculture · Espigoladors

Introduction

Sustainability is a broad term that, contrary to what one might expect, it is difficult to define in a single sentence. This term does not mean the same to each person. What is more, it might have different meanings depending on the context. To illustrate this point, we conducted an experiment. We wanted to know what people, aged between 18 and 24, think about sustainability. A survey was developed and responses from 112 undergraduate Business students were collected. We did not ask them for a definition, but instead, we requested them to indicate examples of sustainable habits they were already incorporating in their daily routines. Answers are shown in Table 1.

Results speak for themselves. Sustainability means reducing the consumption of resources, but it also refers to reutilizing things, or avoiding the use of harmful products or raw materials that pollute and contaminate. But it is not that simple. Sustainability embraces different facets that need to be simultaneously considered in order to have the full picture of what it really means. Thus, following the well-accepted "three pillar interpretation" or the "triple bottom line," we can define sustainability as the simultaneous pursuit of **environmental quality**, **social equity**, and **economic prosperity** (Klöpffer and Ciroth 2011; Balakrishnan et al. 2003). That is, sustainability encompasses ecological, personal, and economic interests, which, taken together, contribute to improving the planet.

The terms sustainability and sustainable development were first coined in 1983. It was not until 1987 when the Brundtland Report (Brundtland Commission 1987)

Table 1 List of sustainable habits of students

1.	Reduce water consumption: Turn off the tap while cleaning the teeth, soaping, or cooking; take a shower instead of a bath; do not use the dishwasher and the washing machine until they are at their full capacity.
2.	Reduce electricity consumption: Turn off the lights when leaving, take advantage of sunlight, turn off electronic devices when not used, make responsible use of air conditioning/heating, use led or low consumption bulbs.
3.	Avoid the use of private transportation. Instead, walk to places, use public transport, or use alternative means of transportation that cause less pollution (e.g., bicycle), share vehicles, etc.
4.	Do not throw food: Reuse it for another meal, give it another use (e.g., fertilizer).
5.	Do not buy products that have contaminating containers (e.g., sprays, plastic bags) that cannot or are not that easy of being recycled.
6.	Separate waste to be recycled.
7.	Recycle: Give another use to old things, use paper on both sides.
8.	Buy ecofriendly products.
9.	Do not use fertilizers.
10.	Do not smoke.
11.	Do not soil the forest.
12.	Do not cut trees.

settled the fundamentals and defined it as "meeting the needs of the present generation without compromising the ability of future generations to meet their own needs." According to this definition, the essence of sustainable development is pleasing fundamental human needs while preserving the life-support systems of planet. This principle implies using resources at rates that do not exceed the Earth's capacity to replace them (Godfray et al. 2010).

Recently, sustainability has also been considered as a form of **innovation** (e.g., Boons and Lüdeke-Freund 2013; Schaltegger et al. 2012; Schaltegger and Wagner 2011). Increasingly, firms are introducing practices in which waste is used as a raw material for new products or applications (Mirabella et al. 2014), leading to the emergence of concepts such as "**cradle to cradle**," "**circular economy**," and "**zero waste economy**." Sustainable resource management is thus grounded on the notion that "waste" can be turned out into a "resource." However, despite several initiatives and policies have been implemented, both academics and practitioners lament the fact that companies are introducing sustainability practices as a way to cut operating costs, rather than for a deeper commitment to the environment.

In this book chapter, we explore sustainability in the food industry. We first have an overview of all the actors that play a role in the food supply chain with a special focus on food waste, then we present initiatives that have taken place around the world to fight against food waste. Finally, we present the case of Espigoladors, a company that reduces food waste in an innovative and interesting manner. At the end of the chapter, several reflection questions and practical exercises are proposed.

Sustainability in the Food Industry

Food waste is generated throughout all the stages of production and consumption. About 1.3 billion tons of food is wasted every year (http://www.un.org/sustainable development/sustainable-consumption-production/). At the same time, almost 1 billion people go undernourished, and another 1 billion is starving. This is the era of paradoxes: while a growing number of countries are dealing with overconsumption of food, a significant proportion of the world's population is suffering from food scarcity. In this context, a key question emerges: How can food waste/surplus be managed more sustainably?

According to the report published by the IVL Swedish Environmental Research Institute (Stenmarck et al. 2016), countries in the EU-28 generate 88 million tons of food waste each year. This amount equates to 173 kilograms of food waste per person, being the household the sector that contributes the most to food waste (circa 50%). Said differently, **food management** has implications alongside the entire supply chain, starting with the farmer, moving through the food processing, and finishing with the final consumer (Papargyropoulou et al. 2014; Parfitt et al. 2010).

In order to better understand how food is managed, it is necessary to examine the different steps and stakeholders involved in the food supply chain and understand its main challenges and opportunities. Let's start analyzing the farmer, that is, the agricultural sector. Here, the main challenge lies on how to produce more food without damaging the environment. Indeed, the projected global food demand by 2050 is expected to increase by 70% (Godfray et al. 2010). In order to meet demand, food supply has to increase. Three strategies can help increase food supply: (i) either more food is produced in the same area of land, making the land more productive; (ii) same food is produced per area, but less energy is consumed to produce it; and (iii) less food is wasted at this stage.

To produce more food from the same area of land, alternative **farming** methods have to be developed. The new techniques can focus on the way fields are cultivated and propose new methods of organizing the crops or retaining and improving the soil; or can focus on the seeds, trying to develop high-production seeds, sometimes through natural methods, others through genetic modifications, a technique that has been largely criticized.

To produce the same with a lower energy consumption is another alternative to increase food supply, assuming that the energy saved is used to grow more food in other fields. Energy includes electricity and fuel but mainly water, which is the most prominent resource used in agriculture. **Sustainable agriculture** practices are being developed, most of them based on the use of renewable energy sources. In this respect, new policies and incentive schemes are required if we are to meet the demands without compromising environmental integrity or public health (Tilman et al. 2002).

Finally, reducing food waste at this stage is also a way to increase food supply; the objective would be 100% harvesting efficiency. This means that no edible food is left on the fields after harvesting, something that does not occur nowadays. In fields that are manually harvested, a relevant amount of food is currently left on the fields due to aesthetic reasons, the food is perfectly suitable for consumption but is misshapen and does not conform to market specifications in terms of size, color, or weight. In fields that are harvested with machines, a significant portion of food is left on the fields due to not having the appropriate technologies to harvest to the full potential. Technical developments can improve the latter, but not the former, which is precisely what the business case presented in this chapter is concerned with.

Sustainable and at the same time efficient agricultural techniques receive public institutional support in some countries, that put high taxes on inefficient food production and no taxes on efficient and sustainable food production (Goodland 1997).

From the side of food processing companies, that is, the intermediary between the farmer and the shops, two challenges arise. One is internal as is common in all industries: efficiency. That is, reducing costs, including food waste and energy consumption, while increasing production maintaining quality and ensuring safety. The other is external and comes from a societal claim: transparency. Food processing companies are more and more under the scrutiny of society that demands more information on the food they consume. Information about the ingredient, but also about the origin of the food and properties such as the pesticides used in its farming are expected to be clearly communicated. To that end, food processing companies might benefit from better labeling each type of product, including livestock, to reflect all its characteristics (Tilman et al. 2002). This way consumers would have available relevant information that would allow them choosing between alternative food products.

Food processing companies supply and distribute their products to grocery stores, supermarkets, restaurants, and other type of catering services like schools or hospitals. These organizations have the challenge to correctly predict demand in order to adapt their orders and avoid buying more food than necessary. And have also a tremendous responsibility of avoiding food waste. A recent study has shown that food waste in UK restaurants approximately costs 23% of their turnover (Papargyropoulou et al. 2014). In addition, this food surplus is left in landfills. Due to the natural decomposition process, methane and carbon dioxide are produced. Both gases contribute to climate change. It is estimated that in Europe, the food sector is causing approximately 22% of the global warming problem.

At the end of the food supply chain are consumers. Individual consumers have many responsibilities concerning food sustainability and in particular on reducing food waste. First, consumers are continuously taking decisions about what food to buy and to which supplier. By taking conscious decisions, consumers can influence the way food processing companies act. Second, consumers are responsible for what they buy, and they should buy only what is necessary and is going to be consumed. If not consuming it, consumers should find alternative ways of reusing surpluses. Tons of foods are thrown away because they are no longer wanted or have passed its best. This is **avoidable food waste**. Finally, another path to follow in order to be more protective and respectful of biodiversity and ecosystems is one of **sustainable diets**. According to Alsaffar (2016), sustainable diets promote a healthy life for present and future generations. They contribute to food and nutrition security while having a low environmental impact. The Mediterranean diet is an example of this. It has a high intake of plant-derived foods such as vegetables, fruits, nuts, and cereals. On the contrary, meat consumption is reduced. Plant- or grain-based diets are more sustainable than diets rich in animal products because they use fewer natural resources and are less taxing on the environment (Sabaté and Soret 2014).

An effective implementation of all these practices will require coordination among federal agencies, ministries, local administrations, and businesses, which often tend to have different objectives. "The goal is no longer simply to maximize productivity, but to optimize across a far more complex landscape of production, environmental, and social justice outcomes" (Godfray et al. 2010: 817). Indeed, food management is a complex process that has to be viewed in an integrated manner, taking into account all actors at all levels of the food supply chain. As we have outlined above, each of these stakeholders has its own interests and challenges derived from their distinct business activity; however, there is something in common to all of them, the ability to reduce food waste. Reducing food waste has enormous potential, not only because we are minimizing the resources employed to produce the food we eat and lower the environmental impact, but also because by being more efficient we can also save money. Being respectful with the environment is not detrimental to individual interests. Food waste is a poor use of current resources and is socially, ethically, economically, and environmentally detrimental.

Analyzing The Food Supply Chain: Where Is Food Wasted?

Quantifying global food waste is challenging. Information is scarce and reports assessing food waste have to rely on limited datasets obtained across different stages of the food value chain, and later, extrapolated to the larger picture (Parfitt et al. 2010). Despite these limitations, several estimates have been made. According to Grandhi and Singh (2016), approximately 1.3 billion tons of the total food produced for human consumption never reaches the human stomach. Where is thus all this amount of food lost?

Avoidable food waste takes place at the different stages of the food value chain (Bagherzadeh et al. 2014): during **production** or immediately after harvesting on the farm; after produce leaves the farm for **handling**, **storage**, and **transport**; during industrial or domestic **processing** and/or **packaging**; during **distribution** to markets, including losses at wholesale and **retail markets**; and finally at **consumption**, including the home or business of the consumer (including restaurants, caterers, etc.). Table 2 shows some examples of how food loss and waste can occur at each of these stages.

There are several **sources of inefficiency** conducive to food waste. These sources are mainly due to financial, managerial, and technical limitations in harvesting techniques, storage, and cooling facilities. Indeed, the lack of financial resources has been signaled as one of the main reasons leading to significant inefficiencies in food management and treatment. From the viewpoint of the stakeholders involved in the early stages of the food value chain, **financial resources** are critical in the sense that, in case of being limited, the logistics and infrastructures needed to preserve and

Table 2 Examples of food loss and waste at the different stages of the food value chain (Adapted from Lipinski et al. (2013))

Production	Handling and storage	Processing and packaging	Distribution and market	Consumption
Fruits bruised during picking or threshing Crops sorted out post-harvest for not meeting quality standards Crops left behind in fields due to poor mechanical harvesting or sharp drops in prices Fruit that is uneconomical to harvest Fish discarded during fishing operations Leave food in the field in response to either market forces or weather/pest-related damage	Edible food eaten by pests Edible produce degraded by fungus or disease Livestock death during transport to slaughter or not accepted for slaughter Fish that are spilled or degraded after landing	Milk spilled during pasteurization and processing Edible fruit or grains sorted out as not suitable for processing Livestock trimming during slaughtering and industrial processing Poor order forecasting and inefficient factory processes Fish spilled or damaged during canning/smoking	Edible produce sorted out due to quality (non-compliant with aesthetic) Edible products expired before being purchased ("best before" and "use-by" dates) Edible products spilled or damaged in market	Edible products sorted out due to quality Food purchased but not eaten Food cocked but not eaten

transport the raw material without altering the original attributes becomes difficult to manage.

There are, however, other sources of inefficiency. One of these relates to **information asymmetries** (Kouwenhoven et al. 2012). A fragmented supply base, the absence of proper communication channels among the different stakeholders and a misalignment of incentives, conduces to reactive value chains that, far from identifying common areas for improvement in order to develop a more sustainable food system, seek for individual benefits that very frequently end up in system inefficiencies. Another critical factor is the fast-**changing consumption patterns**. Producers and suppliers have little time to adjust their production to the demand. This higher demand uncertainty requires flexibility and innovation in the food value chain.

From farm to fork, almost one-third of all the food produced is lost (Göbel et al. 2015). Although the percentages might vary across countries, there is a common pattern. In developed countries, 24% of global food loss and waste occurs at the production, another 24% during handling and storage, and the largest percentage, approximately 35%, takes place at the final consumer level (Lipinski et al. 2013).

These figures mean that approximately 35–40% of the food that arrives at the consumption stage is thrown away (Nahman and de Lange 2013). Consumers seem, however, not to be completely aware of this situation. In a consumer-driven society with increased households and an illusion of abundance, buying decisions are influenced by aggressive discounts and special offers, which result in consumers' tendency to overbuy (Stuart 2009). Food is thrown away in small quantities, and there is a widespread sentiment that wastage is an inevitable part of consumption. In this respect, consumer education is essential to address this issue. Unfortunately, there are few opportunities for exactly knowing how much food is wasted and its associated cost. According to a study conducted by Gunders (2012), American families throw out 25% of the food and beverages they buy, with an estimated cost for an average family of four that ranges from $1,365 to $2,275 annually.

Implications of food waste are not only at the household level. From a social and environmental perspective, food waste has devastating consequences. First, all the edible food discarded signals that if properly managed it could have been used to feed those in need. Second, organic decomposition at landfills or incineration originates smokes and substances that pollute the planet. Third, efforts employed to produce, manipulate, and distribute the food that ends up going to waste, entails the consumption of natural resources (e.g., land, energy, water, etc.). Furthermore, a recent report from the Food and Agriculture Organization of the United Nations (FAO 2013) reported that all the food that is produced but not eaten is responsible for adding 3.3 billion tons of unnecessary greenhouse gases to the planet's atmosphere.

According to the OECD report (Bagherzadeh et al. 2014), practices to minimize food waste in developing countries should be directed toward improving the initial stages of the value chain, while in most developed regions or urban areas the strategy should be focused on stages "close to the fork."

In this setting, policymakers and nongovernmental agents should adopt an entrepreneurial mindset and, in order to minimize food waste, convert an altruistic effort into a profitable business (Grandhi and Singh 2016). Indeed entrepreneurs play a key role in this process (Kouwenhoven et al. 2012), bringing new inventions that fight against value chain inefficiencies, increase food availability, and alleviate poverty. Table 3 suggests some possible approaches for reducing food waste.

Next section describes some entrepreneurial initiatives that illustrate how to overcome some situations leading to food waste.

Worldwide Initiatives That Fight against Food Waste

Following the interest and concerns of the last decade about sustainability, initiatives to improve food management sustainability, and in particular, to reduce food waste, have flourished all around the world. Below we describe several innovative actions taken by different actors in the food supply chain and in different countries, all with the same objective of reducing food waste, therefore, improving **food management efficiency**.

Table 3 Approaches for reducing food waste at the different stages of the food value chain. Adapted from Lipinski et al. (2013)

Production	Handling and storage	Processing and packaging	Distribution and market	Consumption
Facilitate donation of unmarketable crops	Improve access to low-cost handling and storage technologies	Re-engineer manufacturing processes	Facilitate increased donation of unsold goods	Facilitate increased donation of unsold goods from restaurants and caterers
Improve availability of agricultural extension services	Improve ethylene and microbial management of food in store	Improve supply chain management	Change food date labeling practices	Conduct consumer education campaigns
Improve market access	Introduce low-carbon refrigeration	Improve packaging to keep food fresher for longer	Change in-store promotions	Reduce portion sizes
Improve harvesting techniques	Improve infrastructure	Packaging food into small packs and loose items wherever possible		Follow the recommendations on how to store fresh food

Hotels and Restaurants

People tend to leave food when they eat out. Hotels, restaurants, pubs, and bars generate thousands of tons of food waste per year. This food waste is made up of things like peelings and inedible by-products (e.g., bones, coffee grounds, tea leaves) but the majority is perfectly good food. Waste is also generated due to kitchen errors, or because of spoiled or out-of-date food. Likewise, when eating more than one course, the likelihood of leaving part of the main dish is high, as people also want to eat a starter or a dessert. Plate fillers like salads, vegetables, or chips are the most likely to remain uneaten.

What can be done to reduce food waste while keeping customers satisfied? Several initiatives include menus with greater flexibility and personalization. That is, menus with different portions sizes (consumers pay according to the size) or menus where customers can choose the plate filler they prefer. It is important being flexible on customers' request concerning some ingredients so that restaurants ensure consumers do not receive food they would not finish. Doggy bags are another alternative. Increasingly, restaurants are offering the possibility to ask for a container to take leftovers home. This is already quite common in some countries, but very rare and not culturally acceptable in others. By taking action in some of these directions, we can contribute to reduce food waste and have clean plates at the end of a meal out.

Action is well underway in the food and hospitality sector. Some organizations, such as Waste and Resources Action Program (WRAP) in the UK, have led successful initiatives by involving several societal actors including institutions, city authorities, and businesses. In particular, dozens of organizations signed up WRAP's voluntary agreement on food waste reduction in 2015. In 2017, a follow-up evaluation was done showing that, among the different actions undertaken, 555,000 tons of CO_2e have been saved from food and packaging waste. Another outstanding achievement is the one arising from the Hospitality and Food Service agreement, which has reported business savings of £67 million.

Similarly, large companies such as Unilever are also committed to reduce food waste. In this respect, Unilever launched an app that helps businesses operating in the food sector to track food waste generation (https://www.unilever.com/sustainable-living/reducing-environmental-impact/waste-and-packaging/reducing-food-waste/). By identifying what kinds of food are being wasted, companies are more likely to introduce mechanisms to reduce this waste. Information is a key point here as many hotels and restaurants do not really know how much food are they wasting and what is the real cost of that. This is the objective of some new companies that have developed applications to solve that issue. One example of this is *Winnow* (www.winnowsolutions.com/), a firm created in 2013 in the UK that uses cutting-edge technology to monitor and record exactly how food is being wasted. The technology consists of a smart meter technology attached to the food waste bin. Staff use a touchscreen that records all the steps. In just a few seconds it is possible to identify the type of food thrown away and at what stage. Daily reports are sent pinpointing key opportunities to cut waste, benchmark multiple sites, and track performance. Based on the statistics Winnow has recorded, customers can save up to 3–8% of their costs using this system, which translates in an expected ROI that ranges between to 2x to 10x. In 2015, Winnow opened a regional office in Singapore to lead the food waste revolution in the Asia Pacific region. A similar system is the one developed by *MintScraps* (www.mintscraps.com/). Starting at NYC's BigApps 2013 competition, MintScraps is an American firm that uses an online platform to empower restaurants to track and reduce their waste. As it is advertised in the website, "by implementing new waste management solutions, restaurants and food service businesses can have a better understanding of their waste, uncover cost savings, and support sustainability initiatives." Relying on analytics and online technologies the system developed by MintScraps graphically illustrates what is being wasted and the associated cost in a quick and easy manner. Thus, while getting an idea of how much money restaurants are losing MintScraps contributes to change bad business practices and increase productivity. This technology identifies and prevents **avoidable food waste**.

Although the two aforementioned examples help reduce food waste in hotels and restaurants, consumer leftovers are difficult to avoid. In that respect, *Resq Club* (https://resq-club.com/) is a Finnish startup that in 2016 launched its services in Sweden, enabling customers to "rescue" surplus food portions from restaurants in Stockholm, Göteborg, and Malmö. The system works as follows. Through an app, customers can make their orders and decide from which restaurant they want to pick up the leftovers. Portions are typically sold with a discount that ranges between 40% and 70%. The service has gained popularity and it has more than 200 active partners (including restaurants, bakeries, cafes, and hotels) and over 17,000 registered users.

Supermarkets and Retailers

At the end of a day, in supermarkets and groceries, there are many products that are not consumed and are about to expire but are still in good conditions. It is important to differentiate the sell-by date and the use-by date. While the use-by date specifies the recommended date by which the product should be used or consumed, the former indicates that although the sell-by date passes, the food is still safe to eat. However, many people still think that they have to toss it. Similarly, supermarkets and grocery stores might decide to throw it away because the sell-by date is approaching.

Aiming at fighting against food waste, there are several startups working to reclaim out of date food. One example is the German startup *FoodLoop* (www.foodloop.net/en/). A retailers' platform that ties grocer inventory system to consumer-facing mobile apps to provide real-time deals and personalized offers based on consumers' interests, purchase history, and location. Using the app, customers can receive discount offers on products that are about to expire in the supermarkets or small shops. The Italian app, *Last Minute Sotto Casa* (www.lastminutesottocasa.it/), has a similar purpose and functioning. When there is a product about to expire or when there is an excess of food that it is likely to remain unsold, the owner of the grocery or supermarket posts a discounted offer in the app. Customers receive a notification when a new offer is posted. If sound, they can buy it through the app and then go to the shop to pick up what they have purchased.

Some start-ups have expanded this service and gone a step further by allowing individuals and household to exchange for free and/or sell their leftovers. *Olio* (https://olioex.com/) is a free app developed in 2015 in the UK that allows users to snap photos of spare food with a brief description. The app incorporates an instant messenger that allows users to arrange a pick-up. According to their website, one of the things users value most of this app is the opportunity to meet neighbors and exchange food with them. In two years, the app has been downloaded 85,000 times, redistributing more than 125,000 items. Also from the UK, *The Real Junk Food Project* (http://therealjunkfoodproject.org/about/) collects food surpluses (either from individuals or supermarkets) that are in perfect conditions. Starting in 2013, their business model consists in taking all this food to a bar, cook it, and transform it into a delicious meal. Customers can go to the bar and pay what they believe this meal would cost. If someone cannot afford anything in exchange, they can help with the cleaning or collaborate with anything they can bring.

Other endeavors that fight against food waste are those programs where suppliers (supermarkets, households, restaurants etc.) offer discounts for products that are going to expire, do not meet standards of beauty, or have not been sold out and there are large amounts. Some Finish start-ups that fall within this category are *From Waste to Taste* (http://waste2taste.com/) and *Froodly* (http://froodly.com/). In the USA, *Food Cowboy* (www.foodcowboy.com/) and *Spolier Alert* (www.spoileralert.com/) also help food businesses manage their unsold inventory.

Table 4 Some examples of organizations whose aim is to donate food to the needy while reducing food waste

Country	Organisation	Description	Funding year
Australia	OzHarvest	Collects quality food surpluses from more than 2,000 commercial outlets (e.g., fruit and vegetable markets, supermarkets, hotels, wholesalers, farmers) and deliver it, directly and free of charge, to more than 900 charities.	2004
Australia	Second bite	Redistributes high-quality surplus fresh food to the homeless, women and families in crisis, youth at risk, indigenous communities, asylum seekers, and new arrivals. Food is donated by farmers, wholesalers, markets, supermarkets, caterers, and events.	2005
USA	Zero percent	An app that allows restaurants and shops make lists of leftover foods, which are sent to food pantries and social dining rooms. The organization is responsible for the collection and delivery of food.	2012
India	Feeding India	App in which users can post warnings whenever they have food to give. Volunteers go after it to bring it to poor families, orphanages, nursing homes, or shelters. Food is also collected from restaurants, hostels, and companies.	2014
USA	412 FoodRescue	Collects food that is not sold by retailers, wholesalers, restaurants, caterers, and universities among others, and is later delivered to organizations that need food surpluses.	2015

Food for the Needy

Within this category, we can include those organizations that allow individuals, restaurants, hotels, supermarkets, and any other type of donors to give surplus food to needy people such as homeless, social dining rooms, or new arrivals. This task is possible thanks to a group of volunteers, who go after this food and help in distributing it correctly. Table 4 summarizes some of these initiatives.

Other Food Waste Related Initiatives

What to do with large amounts of food surpluses? *Love Food Hate Waste* (www.lovefoodhatewaste.com/) can help you reduce your food waste. This initiative aims to raise food waste awareness by giving us some suggestion on how to start taking action. The website includes easy practical everyday suggestions, recipes, and articles that show users how to waste less food while saving money.

Do "ugly" products taste bad? The answer is no. Nevertheless, an odd physic appearance might prevent people to buy food that does not look as "perfect" as it should. Misshapen products are thus difficult to be sold out. *Imperfect* (www.imperfectproduce.com/) is a mission-driven start-up that operates in the Bay Area (San Francisco) and Los Angeles. It is a home produce delivery service focused on finding "a home" for these misshapen fruits and veggies. Discounts can reach 30–50% of the original price. The business model also generates extra revenue to California farmers. *Hungry Harvest* (www.hungryharvest.net/) is another solution that sells ugly fruit and veggies at a reduced price. In addition, for every harvest they deliver, 2 pounds are donated to help feed someone in need. It has only been two years since they started and they have already reduced 2 million pounds of food from going to the landfill and donated almost 450,000 lbs to people in need. They are currently operating in Maryland, DC, Northern Virginia, Philly, and South Jersey.

Is it possible to transform food surpluses into another product? For sure! Some food is thrown away because the offer exceeds the demand or because it cannot be sold out as it has expired. *WiSErg* (https://wiserg.com/) transforms food scraps and food surpluses into organic fertilizer (e.g., landfills, composters, and digesters). The process starts by placing WISErg Harvesters at food service facilities. This machine ingests and processes food scraps in an odorless, pest-free, self-contained system. During the transformation process, the Harvester captures and stabilizes valuable nutrients from the food scraps. The resulting material is transported to a nearby WISErg facility, where it is processed into liquid fertilizer. Similarly, *Re-Nuble* (www.re-nuble.com/), a social enterprise founded in 2011, creates an organic, nontoxic, liquid nutrient for hydroponic growers and traditional gardeners as an affordable and effective alternative to chemicals. This product can be later used to cultivate new products.

Case Study: Espigoladors

There are fruits and vegetables that grow imperfect and delicious. **Espigoladors** (http://www.espigoladors.cat) is a nonprofit organization located in **Barcelona, Spain**, whose objective is to fight against food waste and empower people at social risk in a transformative, participative, inclusive, and sustainable way. In particular, Espigoladors sends volunteers into the fields to pick leftover produce for distribution to the vulnerable and unemployed. Volunteers are individuals, but also companies and other type of organizations. Very frequently, the same vulnerable and unemployed that receive the fresh fruit and vegetables also go into the fields themselves to help in the picking. With all the fruit and vegetables picked, Espigoladors gives 90% to nonprofit organizations that distribute the fresh food among their beneficiaries, the other 10% is processed by the four full-time Espigoladors' employees that transform it into high-quality juices, sauces, creams, and jams, that are later sold under the brand *Imperfect*. The project has multiple benefits for all its stakeholders. Volunteers enjoy a day in the fields doing a different activity that, in addition to be enjoyable, is going to help people in need. On top of the social intrinsic reward involved in volunteering in such type of activity, volunteers also increase their awareness about food waste. Nonprofit organizations receive the fresh product from Espigoladors and later distribute them among their beneficiaries. For the nonprofit organization, to be supplied by Espigoladors is a great opportunity as obtaining fresh fruit and vegetables is not straightforward for this type of organizations that are more accustomed to deal with canned food as it can be stored for longer periods of time and is therefore less demand dependable. For people at risk of social exclusion, in addition to receiving fresh food through nonprofits, there is the possibility to participate as volunteers in the fruit picking. This is a great opportunity for them to mingle with other people, do a new activity, and feel useful and empowered which has a tremendous positive psychological impact on them.

Fruit and vegetables are left on the fields because of several reasons. In occasions, producers generate some surplus that they do not manage to sell, in other cases, fruit and vegetables do not have the adequate size, color, or shape to be accepted by sellers. Those misshapen products are perfectly healthy and tasty but their physical appearance works against them. Potatoes too big, broken carrots, bananas with black spots, or a curvy cucumber do not fit with grocery stores' strict cosmetic standards.

The word *espigoladors* means **gleaner** in Catalan, the mother tongue of Mireia Barba, the founder. Gleaners are the people in charge to glean the fields, that is, to collect what was left over in the fields once the main harvesting was finished. This activity was very frequent in the nineteenth century but had progressively disappeared until being nearly unknown nowadays.

Mireia Barba, Espigoladors' founder, was grown in Gelida, a 7,000 inhabitants' town on the outskirts of Barcelona, Spain. There she observed how the delicious misshapen lemons of his grandfathers' garden could not be found in the local supermarket. She obtained a double degree in business and social work. In 2013, a project started by her daughters' primary school teacher in order to raise awareness among children about the importance of not throwing away food, kept Mireia thinking about the issue. She visited several organizations that dealt with the issue of food waste and food needs including nonprofits, governmental initiatives, and community kitchens for the underprivileged. She observed that some supermarkets or grocery stores threw food that was about to expire, but still in good conditions, including ripen fruit and vegetables. At the same time, some people were picking what was left over on garbage containers around the city. With the idea that food in good conditions must not be spoiled in the field if it can be of profit, in 2014, Mireia Barba, together with two colleagues: Jordi Bruna, a financial analyst, and Marina Pons, an expert in communication, funded Espigoladors. Espigoladors philosophy is that of 0% waste.

Espigoladors had a very good social response from the beginning. The initiative has won many prizes and has received numerous support, both in financial terms through grants and prizes but also in social terms, being able to create a strong network of supporters and advisors. The advisory team now includes famous cookers as well as business school professors, government representatives, and leaders of nonprofit organizations.

Espigoladors project aims to respond to three social needs:

1. Food waste: According to the Food and Agriculture Organization of the United Nations (FAO), one-third of the food produced in the world for human

consumption gets lost or wasted. This accounts for approximately 1.3 billion tones. Fruit and vegetables have the highest wastage rate, with around 45% of losses. Losses occur all along the food value chain. In Europe and North America, per capita food losses are approximately 280 kg of food per year. In Catalonia region, which is the area of influence of Espigoladors, 700,000 tons per year are squandered.
2. Work opportunities for underprivileged people: The Statistics Bureau of the European Union, EUROSTAT, calculated that, in Europe, 23% of the population is at risk of poverty or social exclusion. In some countries, the unemployment rate is also very high, this is the case of Spain, where Espigoladors is located, a country with a 20% unemployment rate in 2014, the year in which the company was created.
3. Access to healthy and fresh food: 17% of the adult Spanish population is obese. This percentage is over 30% for the USA. Obesity is directly related to food consumption and particularly to low consumption of fresh fruits and vegetables that have been substituted by high calories processed aliments.

And does so with a business plan that relies on three main activities

1. Picking: Creating agreements with companies, grocery-stores, and producers. In the case of producers, Espigoladors guarantees that with a 24 to 48 h prior notice from the producer, a team of gleaners will be in the field ready to pick up what has been left. For supermarkets and grocery stores, Espigoladors commits to pick up whatever the store is going to throw away.
2. Donation: Espigoladors will donate 90% of the products picked to social organizations. Particularly to community kitchens and nonprofit organizations in that act in the area of food waste and food provision for the underprivileged.
3. Transformation: 10% of the fruit and vegetables will be processed under the imperfect brand and sold. This part of the business plan is critical, as is the one that provides revenues to economically sustain the rest of the operations. Although the vegetables are given to Espigoladors for free, the company has some associated costs like transportation, processing, and overhead.

Throughout its value chain of picking-donation-transformation, Espigoladors wants to raise social awareness about food waste and to, whenever possible, include people at risk of social exclusion to actively participate in the project either as volunteers in the picking or as employees in the transport and transformation part of the process.

Espigoladors is still a small company with a very local impact, on average, Espigoladors picks around 50,000 kg of fruit and vegetables annually with a team of 6 people full time and approximately 60 volunteers.

Reflection Questions

1. Draw the food value chain, that is, think about the different stages from food production to food consumption. Where is some risk of food waste?

a. Enumerate all the stages where the risk of food waste can occur. Find information to understand how much food is lost in each of the stages.
   Hint: This TED talk video can help: https://www.ted.com/talks/tristram_stuart_the_global_food_waste_scandal
b. Espigoladors effort to reduce food waste acts on one of the first stages of the food value chain: food that is left on the fields after harvesting. Where do you think most of the food waste happens, at the earlier stages of the food value chain or at the later stages? Would that depend on the development of the country under study? Find information that supports your opinion.
   Hint: The webpage of the FAO can help: http://www.fao.org/save-food/resources/keyfindings/en/
2. Currently, Espigoladors is only acting within its region, particularly in a ratio of 150 km from Barcelona, where its main offices are. How scalable is Espigoladors' business model? Can Espigoladors reach a higher radio of action? Can the same business model expand to other cities, areas, regions, countries? Find arguments that support your opinion.
3. Espigoladors aims to have a social, environmental, and economic impact. Besides the kg per year of fruits and vegetables recovered and given a second chance by Espigoladors, what other additional measures of impact can be used? Is kg/year the best impact measure for the type of activity Espigoladors runs?

Exercises in Practice

Exercise 1: The Zero-Waste Challenge

Although food waste happens at different stages of the food value chain, as consumers, we are especially responsible for the final part of this chain. The stage that includes shopping, cooking, and consumption also produce a certain amount of food waste that is entirely under our responsibility. The objective of this exercise is to become more aware of our behavior in terms of food squandering at our homes.

During a week, observe closely the amount of food you buy and the amount you consume. You can do that at an individual level if you are the responsible for your own grocery shopping, cooking, and consumption, or at a family level if it makes more sense to you. Or even at a school level if the exercise is done as part of a school project. Try to start in a week where your fridge is relatively empty!

1. List all the food you buy, specifying date, quantity, and time. For example:

Item	Date	Quantity	Total price
Lettuce	17 October	1 unit	1,75
Cucumber	17 October	4 units	2,31
Almonds	19 October	150 grams	4,20
...	...	...	...

2. Calculate all the food waste during cooking. You can do that either by observing how much is thrown when cooking or simply weighing the organic trash at the end of the day. Food waste at this stage will contain anything that can be either eatable or used for other purposes. For example, the peduncle and the seed of a cherry are not eaten by humans, but can be used for garden composting which would reduce waste and make the process much more sustainable and environmentally friendly. So you should count that as food waste.
3. Once food is cooked and ready to eat, calculate food waste after consumption. How much has been leftover? If leftovers can be consumed the following days, then they would not count as food waste, but if the leftovers are thrown away, then they are.
4. Add all food waste calculations to have an overall picture of your own (or your family/school) food waste during that period of time. Using the table elaborated in step one, calculate the waste in kilograms and also in monetary terms.
5. You have now your food waste for a week, multiply this number by 52 to have your food waste for a full year.
6. First individually and then in teams, reflect on which strategies could be applied to reduce food waste at this stage and list them. Present your team ideas to the rest of the group.
Idea!: Store potatoes with apples to keep them from sprouting, and keep them away from onions. Onions will make them go bad faster.
Do you know other tricks to better store your food?
7. Implement the ideas developed in step 6 during the following week and, following the same method outlined above, calculate what is your food waste level now. Have you improved? Have you been successful at the zero-waste challenge proposed? You can think of a prize for the winner.

Exercise 2: National Food Waste

1. Find statistics about the amount of food wasted in your particular country. Data can be in kg/person or in a total number of kg or tones.

Statistics about food waste are difficult to get and sometimes unreliable, so try to build your own. Think about the supply and demand of food in your country.

2. Food supply:
   a. Find how much food is produced in your country. Find data about fruit and vegetable production as well as livestock and fish.
   b. Find out your country food imports and exports. Together with the information gathered in question 2a, calculate the overall food production of your country. This would be the supply of food.
3. Food demand:
   a. What is the number of inhabitants in your country? Understand the proportion of children versus adults and also women versus men.

b. Find what is the average intake of an average adult male, adult female, and children in terms of calories. Translate these calories into kg of food by considering a standard healthy diet. Compute now the amount of food consumed by all the people in your country.
c. If you want to sophisticate your analysis, consider also other factors that can help you calculate the national intake such as the percentage of obese people.
4. Compute the difference between the number you have obtained in question two and the number you have obtained in question 3. Does this number coincide with the number found in question one? Why?

Exercise 3: Healthy Eating

Is food waste somehow related to healthy eating? How and why?

(a) In the text, the Mediterranean diet is mentioned as an example of healthy and sustainable diet. What are the characteristics of a healthy and sustainable diet? Design a healthy diet plan for the following week. Think about breakfast, lunch, and dinner for each of the days.
(b) Watch the TED video of Ellen Gustafson about obesity and hunger (https://www.ted.com/talks/ellen_gustafson_obesity_hunger_1_global_food_issue).

   Discuss in teams what are the main reasons for the issue presented in the video and propose some solutions.

Exercise 4: Debate

In the text, genetically modified seeds are mentioned as a technique to increase crops production.

(a) What are genetically modified seeds?
(b) Are genetically modified seeds healthy?
(c) What are the advantages and disadvantages of genetically modified crops?

Prepare arguments in favor and against genetically modified seeds. In class, you will be randomly assigned to a team, either in favor or against genetically modified seeds. Which group will better debate? Think a prize for the winner team.

Engaged Sustainability Lessons

– Food management is an important part of sustainable development. The food supply chain has many stakeholders, each of them facing different challenges and with different objectives. However, all actors have an active responsibility for managing food appropriately.

- Food waste is a global issue.
- Many initiatives exist to reduce food waste all along the food supply chain. All societal actors should be actively involved in these initiatives, consumers, businesses, politicians, and educators, all have an important role in contributing to food waste reduction.

Chapter-End Reflection Questions

1. Many online tools exist to calculate your environmental footprint, some of them break up the calculation into different types of footprints, one of them being the Food Footprint. Calculate your Food Footprint, for example using the tool offered online by the NGO Redefining Progress (http://rprogress.org/index.htm), or calculate your global Environmental Footprint using the tools offered by the World Wide Fund for Nature (http://footprint.wwf.org.uk) or the Global Footprint Network (http://www.footprintnetwork.org/resources/footprint-calculator/). Reflect on your results.
2. Unfortunately, not only food is wasted in our world. Also, some other precious goods like water or energy suffer from squandering. Can the lessons learned in this chapter be applied to other type of goods?
3. How could you fight against food waste in your family, in your school, in your community? Are there already some organizations taking care of the issue? Could you be actively involved in them? If there are not, is there anything you could do?

Cross-References

▶ Ecosystem Services for Wine Sustainability
▶ Supermarket and Green Wave
▶ To Eat or Not To Eat Meat

References

Alsaffar, A. A. (2016). Sustainable diets: The interaction between food industry, nutrition, health and the environment. *Food Science and Technology International, 22*(2), 102–111.

Bagherzadeh, M., Inamura, M., & Jeong, H. (2014). Food waste along the food chain. OECD Food, Agriculture and Fisheries Papers, 71. OECD Publishing.

Balakrishnan, U., Duvall, T., & Primeaux, P. (2003). Rewriting the bases of capitalism: Reflexive modernity and ecological sustainability as the foundations of a new normative framework. *Journal of Business Ethics, 47*(4), 299–314.

Boons, F., & Lüdeke-Freund, F. (2013). Business models for sustainable innovation: State-of-the-art and steps towards a research agenda. *Journal of Cleaner Production, 45*(April), 9–19.

Brundtland Commission. (1987). *Our common future. The United Nations World Commission on Environment and Development.* Oxford: Oxford University Press.

FAO. (2013). *Food wastage footprint: Impacts on natural resources.* Rome: FAO. Available at: www.fao.org/docrep/018/i3347e/i3347e.pdf. Retrieved 5 Nov.

Göbel, C., Langen, N., Blumenthal, A., Teitscheid, P., & Ritter, G. (2015). Cutting food waste through cooperation along the food supply chain. *Sustainability, 7*(2), 1429–1445.

Godfray, H. C. J., Beddington, J. R., Crute, I. R., Haddad, L., Lawrence, D., Muir, J. F., Pretty, J., Robinson, S., Thomas, S. M., & Toulmin, C. (2010). Food security: The challenge of feeding 9 billion people. *Science, 327*(5967), 812–818.

Goodland, R. (1997). Environmental sustainability in agriculture: Diet matters. *Ecological Economics, 23*(3), 189–200.

Grandhi, B., & Singh, J. A. (2016). What a waste! A study of food wastage behavior in Singapore. *Journal of Food Products Marketing, 22*(4), 471–485.

Gunders, D. (2012). Wasted: How America is losing up to 40 percent of its food from farm to fork to landfill. NRDC Issue Paper, IP 12-06-B. New York City: Natural Resources Defense Council.

Klöpffer, W., & Ciroth, A. (2011). Is LCC relevant in a sustainability assessment? *The International Journal of Life Cycle Assessment, 16*(2), 99–101.

Kouwenhoven, G., Nalla, V. R., & von Losoncz, T. L. (2012). Creating sustainable businesses by reducing food waste: A value chain framework for eliminating inefficiencies. *International Food and Agribusiness Management Review, 15*(3), 119–138.

Lipinski, B., Hanson, C., Lomax, J., Kitinoja, L., Waite, R., & Searchinger, T. (2013). Reducing food loss and waste. World Resources Institute Working Paper.

Mirabella, N., Castellani, V., & Sala, S. (2014). Current options for the valorization of food manufacturing waste: A review. *Journal of Cleaner Production, 65*(February), 28–41.

Nahman, A., & de Lange, W. (2013). Costs of food waste along the value chain: Evidence from South Africa. *Waste Management, 33*(11), 2493–2500.

Papargyropoulou, E., Lozano, R., Steinberger, J. K., Wright, N., & bin Ujang, Z. (2014). The food waste hierarchy as a framework for the management of food surplus and food waste. *Journal of Cleaner Production, 76*(August), 106–115.

Parfitt, J., Barthel, M., & Macnaughton, S. (2010). Food waste within food supply chains: Quantification and potential for change to 2050. *Philosophical Transactions of the Royal Society of London B: Biological Sciences, 365*(1554), 3065–3081.

Sabaté, J., & Soret, S. (2014). Sustainability of plant-based diets: Back to the future. *The American Journal of Clinical Nutrition, 100*(Supplement 1), 476S–482S.

Schaltegger, S., & Wagner, M. (2011). Sustainable entrepreneurship and sustainability innovation: Categories and interactions. *Business Strategy and the Environment, 20*(4), 222–237.

Schaltegger, S., Lüdeke-Freund, F., & Hansen, E. G. (2012). Business cases for sustainability: The role of business model innovation for corporate sustainability. *International Journal of Innovation and Sustainable Development, 6*(2), 95–119.

Stenmarck, A., Jensen, C., Quested, T., Moates, G., Buksti, M., Cseh, B., Juul, S., Parry, A., Politano, A., Redlingshofer, B., & Scherhaufer, S. (2016). *Estimates of European food waste levels*. IVL Swedish Environmental Research Institute.

Stuart, T. (2009). *Waste: Uncovering the global food scandal*. London: W.W. Norton.

Tilman, D., Cassman, K. G., Matson, P. A., Naylor, R., & Polasky, S. (2002). Agricultural sustainability and intensive production practices. *Nature, 418*(6898), 671–677.

Collaboration for Regional Sustainable Circular Economy Innovation

Rajesh Buch, Dan O'Neill, Cassandra Lubenow, Mara DeFilippis, and Michael Dalrymple

Contents

Introduction	704
Evolution of the Circular Economy	706
The Hannover Principles	707
Industrial Ecology and Symbiosis	707
Biomimicry	707
Cradle to Cradle™	708
The Natural Step	708
Ellen MacArthur Foundation	708
Circular Economy in Practice	709
Unsustainable Trajectory of City of Phoenix	711
Discovering Sustainable Solutions Through Collaboration	711
Sustainable Circular Economy in Phoenix	714
Leadership Buy-In and Funding Strategies	714
The Resource Innovation Campus	715
ASU and City of Phoenix Projects	716
Economic Impact Analysis	717
Regional Green Organics System Design	719
Sustainability Solutions Festival	720
Parks Turf Compost Study	721
RISN and Regional Partners	722
RISN and Global Partners	723
Public–Private–Academic Collaboration	724
Engagement Is Key to a Sustainable Circular Economy	725
Cross-References	727
References	727

R. Buch (✉) · D. O'Neill · C. Lubenow · M. DeFilippis · M. Dalrymple
Walton Sustainability Solutions Initiatives, Arizona State University, Tempe, AZ, USA
e-mail: rbuch@asu.edu; dan.oneill@asu.edu; clubenow@asu.edu; Mara.Defilippis@asu.edu; mick.dalrymple@asu.edu

© Springer International Publishing AG, part of Springer Nature 2018
S. Dhiman, J. Marques (eds.), *Handbook of Engaged Sustainability*,
https://doi.org/10.1007/978-3-319-71312-0_24

Abstract

Recognizing the looming, long-term sustainability challenges to the metropolitan Phoenix region's economy, environment, and community, City of Phoenix leadership shifted attention to a new paradigm that focuses on integrated resource management, the concept of "zero waste," and the circular economy. Understanding that this effort could not be achieved through current municipal efforts alone, Phoenix turned to Arizona State University, to establish local, regional, and global collaborative networks focused on creating value and economic development opportunities from solid waste streams.

Phoenix, Arizona State University, and our local collaborating partners in the public, nonprofit, and private sectors have developed a strategy to build a regional sustainable circular economy focusing on innovations and solutions that support a socially balanced, environmentally restorative, and resource regenerative economy. This strategy involves solutions development, community education and engagement, and the building of a Resource Innovation Campus and incubator focused on attracting resource innovators and entrepreneurs to Phoenix to build viable, market-driven closed-loop resource business models in the region.

Keywords

Sustainability · Circular economy · Stakeholder engagement · Systems modeling · Innovation · Outreach · Decision support · Sustainable development goals

Introduction

The **circular economy** has emerged as an alternative paradigm for the management of resources to drive sustainable development that contrasts with the linear materials economy model of extract–consume–dispose that has characterized economic activity since the industrial revolution (Senge et al. 2001). As defined in the United Nations' report, Our Common Future, **sustainable development** is defined as "development that meets the needs of the current generation without compromising the ability of future generations to meet their own needs" (UN WCED 1987). While the linear economy stresses the environment and risks the economic and social foundation of the global system, a circular economy is one that is restorative and regenerative by design and which aims to keep products, components, and materials at their highest utility and value at all times (MacArthur 2012).

A promising approach in the pursuit of sustainable development, the circular economy incorporates concepts and design principles from several schools of thought to encourage the creation of closed-loop resource systems and equitable, healthy, productive societies. While it has been gaining traction as a sustainability strategy, circular economy theory is not always easily translated into tangible systems and sometimes falls short of its societal goals. Examples of circular economy in practice do not always produce broad, socially desirable outcomes and sometimes also diverge from the principles of circular economy theory. A modified approach in circular economy theory and implementation is needed to reconcile the disparity between theory and practice.

The most recent revision of the United Nations **sustainable development goals** (SDGs) and agenda call for action "in areas of critical importance" for humanity and the planet:

- People – we are determined to end poverty and hunger, in all their forms and dimensions, and to ensure that all human beings can fulfill their potential in dignity and equality and in a healthy environment.
- Planet – we are determined to protect the planet from degradation, including through sustainable consumption and production, sustainably managing its natural resources and taking urgent action on climate change, so that it can support the needs of the present and future generations.
- Prosperity – we are determined to ensure that all human beings can enjoy prosperous and fulfilling lives and that economic, social, and technological progress occurs in harmony with nature.
- Peace – we are determined to foster peaceful, just, and inclusive societies which are free from fear and violence. There can be no sustainable development without peace and no peace without sustainable development.
- Partnership – we are determined to mobilize the means required to implement this agenda through a revitalized Global Partnership for Sustainable Development, based on a spirit of strengthened global solidarity, focused in particular on the needs of the poorest and most vulnerable and with the participation of all countries, all stakeholders, and all people (United Nations Statistical Commission 2016).

The 17 global goals (and 169 associated targets) emphasize the triple bottom line approach to development that requires the environment, the economy, and society to simultaneously benefit, and also demonstrate the inherently ethical nature of the pursuit of sustainability. Social issues including equity, empowerment, education, health, social protection, and job opportunities are highlighted as top priority areas along with environmental issues. In fact, more than half (8 out of 17) of the SDGs are focused directly on these social issues. Having been developed by the UN, the SDGs are the best representation of where the world needs to prioritize action for creating a prosperous future for all and provide a good framework by which to assess the effectiveness of a circular economy solution as being a desirable sustainable solution.

It is proposed that this modified approach to circular economy reconciles this disparity and works toward achieving the sustainable development goals and can be termed the **Sustainable Circular Economy**. It incorporates research, methods, and knowledge from social, environmental, and economic dimensions and has the goal of creating an overall benefit to society. The sustainable circular economy approach differs from other circular economy theories and methods by recognizing that current circular economy strategies lack effective integration of social processes and structures that must play an integral role in implementing circular strategies in order to deliver sustainable outcomes. Therefore, in order to achieve the sustainable goals of sustainable development, the social dimension needs to be more explicitly integrated into the development of circular economy solutions. Existing frameworks for circular economy intend to be inclusive of the social dimension in principal, but real-world implementation efforts have by-and-large focused on material and energy synergies in the design phase, rather than also impacting local social conditions and relationships. Many applications of circular economy that have been designed in this way have had limited success, when success is defined as improving overall well-being for today's generations and future generations.

The sustainable circular economy offers a strategy that allows human and cultural social interactions to drive the transition to circular economies, in addition to restoring environmentally responsible resource flows. When developed and applied in this way, the circular economy concept can be more effectively used to strive toward sustainable and prosperous futures for the world and its inhabitants.

Evolution of the Circular Economy

The circular economy offers a platform for societies to rethink their material and energy use in ways that align with natural cycles. The current materials economy is a "take–make–dispose" system that transforms valuable resources into waste and attempts to impose an unnatural and temporary linear process upon cyclic natural systems. This system is problematic simply because it is unrealistic – the planet and its systems operate in cycles, and the "linear" processes imposed on them by humans are, in reality, producing waste that is destructive and persistent to Earth's systems as it cycles. As a result, chemicals and materials with valuable properties accumulate in landfills while societies expend energy to extract virgin resources from finite supplies. Energetic and material value from discarded products gets allocated toward degradation instead of utility. A circular economy, in contrast, maximizes utility of valuable materials through closed-loop systems. Energy and resources provide functional utility in multiple or infinite life cycles when used as raw material for something else at the end of its current life. A transition to a circular economy can enable society to practice conscious development in a way that more intelligently couples economic growth with the use and reuse of finite resources. Circular economy supports and, in turn is supported by, frameworks for developing a resilient global human system that thrives in the longer term (MacArthur 2012). When coupled with a sustainable systems perspective, a circular economy can support synergistic growth with the natural systems of the finite planet.

Today's evolving concept of a circular economy has emerged from the integration of several schools of thought and academic disciplines, each emphasizing the necessity of working cyclically. The original contributions can be drawn back to Walter Stahel's work in the 1980s, and the development of the closed loop economy, as described in "The Product Life Factor" (MacArthur 2012; Stahel 1982). While the following list is not exhaustive of additional circular economy theory contributors, the most notable and significant contributions have come from the development of the Hannover Principles (McDonough and Braungart 1992), industrial ecology (Allenby and Graedel 1995), Cradle to Cradle^{TM} (McDonough and Braungart 2010), biomimicry, and The Natural Step (Robèrt and Anderson 2002; Nattrass and Altomare 2013), each described in more detail below.

The Hannover Principles

While planning for the 2000 World's Fair, the City of Hannover, Germany, decided to tackle the difficult issue of imagining and encouraging a sustainable future by theming EXPO 2000 as "Humanity, Nature and Technology." To ensure that all construction and preparation for the fair represented the city's commitment to sustainable development, the planners commissioned nine design principles to inform international design competitions for constructing the event. Known as "**The Hannover Principles**," these guidelines aimed to provide a platform for designers to adapt their work toward sustainable ends for the environment and for humanity. The principles emphasize the rights of humanity and nature to coexist, recognizing the relationship between spirit and matter, the responsible use of natural resources, and the importance of continuous improvement through sharing of knowledge (McDonough and Braungart 1992).

Industrial Ecology and Symbiosis

Many of the original concepts behind the circular economy first appeared in **Industrial Ecology**, by Allenby and Graedel (1995). Industrial ecology has been defined as a "systems-based, multidisciplinary discourse that seeks to understand emergent behavior of complex integrated human–natural systems" (Allenby 2006). Essentially, industrial ecology presents a way of looking at industrial systems as man-made versions of natural ecosystems. **Industrial symbiosis** has been integrated into systems where the waste products of one industry are used as raw materials for others. Just as "waste" products in natural systems are used as food or nutrients for other agents in a food web, industrial ecosystems cooperate in industrial symbiosis through exchange of byproducts, resources, and infrastructures that are able to achieve greater economic and environmental benefits than if they were acting alone (Allenby and Graedel 1995; Allenby 2006).

Biomimicry

Biomimicry is a new science that studies nature's models and then emulates these forms, processes, systems, and strategies to solve human problems (Benyus 1997). While one of the earliest examples of biomimicry is Leonardo da Vinci's designs for a flying machine, the scientific discipline was more recently popularized in the 1980s and applied to development of products like replaceable carpet tiles (modeled after forest leaves) and Velcro (modeled after burrs). More recently, biomimicry practitioners have been applying the approach at many scales, "from biota to biosphere," and positioning biomimicry as a critical tool for realizing the circular economy.

Cradle to Cradle™

Cradle to Cradle™ is a design philosophy (McDonough and Braungart 2010) that considers all material involved in industrial and commercial processes to be nutrients, which can therefore be utilized in a closed-loop system where there is no waste. Cradle to Cradle™ (C2C) models typically categorize materials used in manufacturing processes as biological or technical nutrients, where all materials can be returned to the earth as biological nutrients after use or re-enter the industrial system as a technical nutrient (C2C™ product standard). C2C™ emphasizes more than the elimination of waste and the improvement of resource efficiency. One of its three principles is to "celebrate diversity," which is described as using social fairness, encouraging stakeholder engagement, supporting local biodiversity, and cultivating creativity through technology diversity. C2C™ practitioners place a special emphasis on materials, especially with respect to chemistry, in an effort to eliminate harmful chemicals from buildings and products.

The Natural Step

The Natural Step is an international not-for-profit organization dedicated to education, advisory work, system change initiatives, innovation, and research in pursuit of sustainable development (Robèrt and Anderson 2002; Nattrass and Altomare 2013). The Natural Step works with diverse organizations to apply its Framework for Strategic Sustainable Development. In this framework, a sustainable society is defined as one in which nature is not subject to systematically increasing:

- Concentrations of substances extracted from the earth's crust
- Concentrations of substances produced by society
- Degradation by physical means, and, in that society
- Conditions that undermine peoples' capacity to meet their needs

These system conditions describe The Natural Step's principles for moving toward a sustainable society.

Ellen MacArthur Foundation

The United Kingdom-based **Ellen MacArthur Foundation** (EMF), established in 2010, mainstreamed the idea of circular economy with the aim of accelerating the transition to the circular economy. In the context of business and urban development practices, circular economy integrates sustainable product and package design and waste management with a more holistic sustainability approach that is focused on creating closed-loop systems throughout the entirety of supply chains and within geographic regions. EMF has advanced the concept of circular economy to the forefront of global sustainable development agendas and spurred action by many global corporations, other private organizations, and some municipalities to close loops in their materials and energy systems. EMF defines circular economy as one that is restorative and regenerative by design, one that aims to keep products, components, and materials at their highest utility and value at all times.

An important characteristic of circular economy systems is that they couple economic growth and development and the consumption of finite resources in a more intelligent, more sustainable way (Lieder and Rashid 2016). This synergistic approach promises to be mutually beneficial for the economy and the environment when successfully applied. The potential benefits of "going circular" include optimizing the use of materials, realizing new revenue streams, enhancing stakeholder relationships, and mitigating risk from future policy and industry shocks (Crane and Matten 2016). Research suggests that it has immense job creation and innovation potential, and an estimated savings of $1 trillion a year can be realized from global dematerialization (MacArthur 2012).

Circular Economy in Practice

The concept of circular economy has been applied in a variety of contexts as a means of transforming our current systems to be more sustainable. Systems can be circularized on different scales and with different types of boundaries. Two common ways that circular economy is applied is within "spatially dynamic" manufacturing systems, defined by supply chains, and within "spatially static" urban platforms, defined by geographic boundaries. In both cases, the end goal is to gain an economic benefit while simultaneously reducing the impact of production or development on the environment. These systems are often focused on using waste and energy as raw material inputs for other processes, while social impacts, capital, structures, and processes are often excluded. Additionally, these systems interact to create, use, and dispose of resources as humans carry out their daily lives.

In supply chains, circular economy has exploded as a business strategy in recent times. Prominent reports and case studies that promote the potential of the circular economy to deliver win–win situations for the environment and the economy have moved circular economy into the private sector mainstream. Large corporations are embracing the concept as a way to meet their environmental obligations without sacrificing shareholder value. In most examples, circular economy in supply chains is focused heavily on product stewardship and design-for-environment strategies. For instance, circular economy strategies in private industry might consist of takeback programs or leasing programs, in which the manufacturer creates products that are easy to disassemble, and maintain an internal responsibility for reutilizing material through successive life cycles. Because of the material focus of most circular economy strategies in supply chains, very few organizations tie social conditions into their circular economy strategies. The ones that do, however, are able to create higher levels of competitive advantage while creating social benefit, in addition to environmental and economic benefit. By integrating this third dimension, firms are able to use circular economy to create holistically sustainable supply chains and bring greater benefit to their business through increased competitive advantage, reputation, and legitimacy. For example, the mission of Interface is "to be the first company, that, by its deeds, shows the entire industrial world what sustainability is, in all its dimensions: people, process, product, place and profits" (Harel 2013). Wanting to hold true to its commitment, Interface officially adopted a social sustainability program in 2000 focused on the development of programs and processes that promote social interaction and cultural enrichment. Its emphasis is on protecting the vulnerable, respecting social diversity, and ensuring that we all put a priority on social capital (Interface USA 2015), and defined Interface's core values as: (1) human rights, (2) labor standards, (3) environment, and (4) ethical practices (Interface USA 2015). In including these social components, Interface's circular economy is enhanced to promote higher standards of living in its manufacturing companies around the world.

In urban platforms, circular economy can be applied at a variety of scales and are accordingly manifested in different forms, such as eco-industrial parks, eco-cities, or more general development guidelines. Masdar City in the United Arab Emirates (Cugurullo 2013) and Songdo International Business District in South Korea are two of the highest profile examples of such smart green cities (Shwayri 2013) that highlight different ways in which ignoring social factors can lead to circular economy failures. Masdar City claims to have officially adopted the triple-bottom-line definition of sustainability, placing equal importance on the well-being of environmental, economic, and social dimensions of the city (Cugurullo 2013). However, Masdar's sustainability performance has been heavily criticized for being weak in the social dimension. Some fear that it will become a luxury development for the rich and the technology-centric – a secluded metropolis that only furthers the division between wealthy and impoverished communities (Hodson and Simon 2010). If this is the case, Masdar City demonstrates how circular economy in practice can actively contribute to significant sustainability challenges – wealth disparity and social marginalization – rather than working toward an equitable and prosperous society. Similarly, Songdo International Business District, or New Songdo City, is a new smart, "ubiquitous" city being built on 1,500 acres of reclaimed land in Incheon, South Korea. New Songdo City's commitment to "encourage and foster sustainable design practices by incorporating the latest design standards and technologies that reduce energy consumption, increase energy efficiency, utilize recycled and natural materials and generate clean or renewable energy" reflects circular economy principles of designing out waste and using renewable energy sources (Shwayri 2013). However, New Songdo City's (Lobo 2015) struggle to attract permanent business tenants and residents highlights a different failing of a circular economy strategy that does not incorporate social factors than the failing found in Masdar City. In the case of New Songdo City, circular economy strategies were a failure because the lack of consideration given to social conditions led to a society that did not meet all of its residents' needs for a prosperous life.

Unsustainable Trajectory of City of Phoenix

The combination of existing markets, technologies, business models, governance institutions, and culturally embedded practices and policies are creating unsustainable cities, communities, and infrastructure. For the private sector, the linear model of extraction, production, consumption, and disposition is resulting in business risk founded on depleting resources and volatility of commodity prices and availability. For their part, public sector managers in urban regions respond to these dynamics, impacting business and consumption practices with waste management requiring perpetual landfilling.

Additionally, the population in the Sun Corridor, the mega-region in Arizona that includes Phoenix and Tucson and runs between the US–Mexico border in Nogales in the south to Prescott in the center of the state, is expected to double and reach a population of nine million people by 2050.

In 2013, the **City of Phoenix**'s recycling rate was 16% (compared to the national average of 34%), and the fleet of Phoenix Public Works Department trucks collected and disposed of more than one million tons of garbage, traveling more than seven million miles to local landfills. This is equal to traveling to the moon and back 14 times – every year.

The likely outcome of this scenario – our current linear economy practices, with the projected increase in population and consumption – is a decrease in the availability of finite resources such as raw materials, water, and energy. That decrease would in turn lead to an increase in the financial, social, and environmental cost of managing those resources, such as increasing costs of resources and raw materials, the potential loss of jobs, the potential for environmental pollution, an increase in the amount of waste produced, and ultimately a reduction in the overall standard of living for all citizens.

The challenge for Phoenix, the largest municipality of the region and fifth largest city in the USA, was how to optimize resources in a way that was economically feasible. The cost of infrastructure and additional employees would have to be offset by opportunities for profit in order to make waste management pay for itself in the long run.

Discovering Sustainable Solutions Through Collaboration

To meet this challenge, the Phoenix Public Works Department, led by then-Director **John Trujillo**, developed a strategic action plan with a goal of achieving 40% waste diversion by 2020, which was announced by **Mayor Greg Stanton** in 2013. The plan was branded as "**Reimagine Phoenix: Turning Trash into Resources**," and it set the course for a new sustainable waste management

Fig. 1 Representatives from the City of Phoenix, Arizona State University, the Arizona Diamondbacks, Telemundo, Mayo Clinic of Arizona, PetsMart, and other partners announce the launch of the city's Reimagine Phoenix initiative at Chase Field (Photo courtesy of ASU Walton Sustainability Solutions Initiatives)

trajectory for Phoenix and the entire Phoenix metropolitan area (Fig. 1). Under this plan, Phoenix established both short- and long-term goals that would improve the diversion rate while allowing sustainable development and economic development to occur simultaneously.

However, Phoenix was quick to realize that they could not accomplish these ambitious goals alone; collaboration with trusted partners was key to the success of the Reimagine Phoenix plan. The city established a partnership with the Global Sustainability Solutions Service, a sustainability consulting service within the Walton Sustainability Solutions Initiatives at **Arizona State University (ASU)**.

The centerpiece of this partnership became the **Resource Innovation and Solutions Network or RISN**. Recognizing that the world at large is becoming buried in waste, RISN was established to advance integrated resource management through a global network of public and private partners using collaboration, research, innovation, and application of technologies to create economic value, driving a sustainable circular economy.

Perhaps the greatest strength of RISN is the unique and powerful partnership between Phoenix and ASU. With a population of 1.5 million within a regional population of near 4.5 million, Phoenix is one of the fastest growing metropolitan areas in the country. The City of Phoenix was also named the Top Performing City overall by Governing and Living Cities in 2017.

ASU is the largest public university in the USA and has a current student enrollment of more than 98,000 students. For 3 years running, ASU was named America's Most Innovative University by US News and World Report and ASU's School of Sustainability, founded in 2006, is the first comprehensive degree-granting institution of its kind in the USA. Today, the **Julie Ann Wrigley Global Institute of Sustainability** at ASU is home to more than 425 Sustainability Scientists and Scholars.

Together, Phoenix and ASU have immense resources, knowledge, and the ability to bring global, regional, and local stakeholders together. Each partner has unique resources, abilities, and assets that contribute to success. It is only by working together that such a powerful impact and societal transition can take place.

The first long-term goal for Phoenix and ASU was to develop a Center of Excellence in waste management that would provide support and momentum beyond immediate programs. The center would also serve as a development and research hub focused on creating value and economic opportunity out of waste streams. In July 2014, Phoenix and ASU officially entered into an intergovernmental agreement to develop a strategic plan for the center.

With the agreement in place, a team from ASU facilitated nine collaborative regional stakeholder workshops on a variety of topics relevant to the design of a more sustainable waste management system. The workshops were developed to gain a clear understanding of the current waste system throughout the region, build a collaboratively developed future vision, and develop a strategy to reach that future. The workshops covered the following topics:

- Regional waste model
- Technology roadmap
- Communication and collaboration
- Sustainable urban metabolism
- Policy
- Research and funding
- Big data
- Facility design charrette
- Food scraps

Conversations, insights, and findings from these workshops were integrated into a business plan for evolving the Center of Excellence, including a transformation from a single-entity center to a network model, recognizing that to include all of the best knowledge, ideas, and practices for waste management system design would require collaboration between all experts and stakeholders. The ultimate goal was for the network, which was renamed RISN, to expand beyond the Phoenix metropolitan area and incorporate global information and knowledge from leaders and practitioners around the world. RISN would become a global network, originating and headquartered in Phoenix.

Sustainable Circular Economy in Phoenix

Critical to the evolution of RISN was the change in focus from integrated resource management and intelligence toward the implementation of a sustainable circular economy.

Recognizing the similarity between EMF, other circular economy agendas, the vision for RISN, and the future Phoenix-area waste management system, the RISN partnership refocused its initiative to pursue the realization of a circular economy in the Phoenix metropolitan area, with a focus on waste-to-resources as the ultimate goal.

To meet this goal, RISN has to work with Phoenix and its Reimagine Phoenix initiative to create a systemic, cultural, and behavioral shift among Phoenix residents and businesses. This focus widens the target set to surpass the city's waste diversion goal of 40% by 2020. In order to enable behavior change and a cultural shift in Phoenix, the city created a media campaign with a primary message that repositions the concept of trash as a valuable resource rather than material to be thrown away – trash is valuable raw material for new products, rather than waste.

Phoenix is motivated to position itself as a national and international leader in innovative and strategic solid waste management. Toward this end, Phoenix is investing in infrastructure for mixed waste and/or solid waste diversion technologies that will help to divert recoverable material in the municipal solid waste stream from the landfill and create a circular system focused on job creation, new revenue for Phoenix, and innovative and sustainable economic development.

Other RISN "hubs," centers based throughout the world, would have the same objective – to build regional sustainable circular economy platforms in their own locations and regions. RISN would serve as a capacity builder, convening platform and facilitator for realizing the circular economy at all of its hubs.

The Phoenix–ASU partnership is the first city–university partnership focused on sustainable circular economy in the world.

Leadership Buy-In and Funding Strategies

Finding and disseminating funding for such an aspirational goal can be a significant challenge in itself. In July 2014, Phoenix Public Works director John Trujillo successfully gained council approval for the funding of RISN, which included $2 million over a 4-year span to cover both operations and project implementation. ASU would also contribute $1 million over the same 4-year time span in addition to providing its sustainability resources and expertise.

To gain buy-in for such a large initiative, Trujillo and the Public Works Department were very strategic about the positioning of RISN and how it would benefit Phoenix. A strategy to inform and educate city council members and other important stakeholders about circular economy and the related economic opportunities played an important role in gaining acceptance. Information was disseminated over time through the Transportation and Infrastructure subcommittee and then to full council about the many opportunities that implementing a circular economy could bring to the Phoenix area.

Trujillo used these efforts as a way to position Phoenix as a national and international leader in innovation, solid waste management, and fiscally responsible sustainability. He emphasized the impact that the circular economy would have on future generations, in light of a rapidly increasing population throughout the region. Circular economy provides a way for Phoenix to prepare for a 100% increase in population projected by 2050. Trujillo educated the council on the opportunities for new business development and to attract innovative technologies to the city, which subsequently funded a position dedicated to RISN in the city's economic development department.

Phoenix's commitment to innovation for sustainable resource management is evidenced by its investment in the RISN partnership, funding research, and solutions development projects executed by the ASU Global Sustainability Solutions Service that advance the diversion of waste and create economic value through the enhancement and creation of new technologies, systems, and programs. Examples of such projects include development of a Resource Innovation Campus and other projects discussed below.

The Resource Innovation Campus

Because Phoenix had ideally located property and direct access to waste feed streams at one of its waste transfer stations, the city looked to create the **Resource Innovation Campus** (RIC – Fig. 2). The RIC will sit on a city-owned site that includes operating materials recovery facilities, a recently opened municipal composting facility, a closed landfill, and more than 100 acres of vacant land. This industrial space will be able to house fully developed companies that will build and operate their waste processing technologies on site, creating jobs and bringing new businesses to Phoenix. Businesses based at the RIC will use innovative waste processing technologies to transform the material resources from the municipal solid waste stream available onsite (Fig. 3) to create new products and processes.

A "call for innovators" and a series of request for proposals already have been issued by Phoenix to seek out these innovators and populate the campus with companies that will foster job growth and economic development for the Phoenix region through circular economy technologies. The RIC will also be the headquarters for RISN and house an on-site incubator space, where experts in entrepreneurship will assist young companies in developing marketable circular economy technology solutions.

The plan is for this industrial park to become an internationally recognized hub for circular economy technological innovation that will attract similar companies who desire to benefit from being involved in the growing circular economy industry. These activities will accelerate the availability of resources and awareness and foster innovation through spontaneous collaboration.

At the center of the RIC will be a new building that is designed and built in accordance with the **Living Building Challenge** criteria. Ultimately, the building will house the RISN Operations Center and the RISN Incubator, a niche business

Fig. 2 The Phoenix 27th Avenue Waste Transfer Station, home to the Resource Innovation Campus and RISN (Photo courtesy of City of Phoenix Public Works Department)

accelerator for entrepreneurs in the early stages of waste-to-product innovation with the goal of moving a circular economy in the Phoenix area forward further and faster.

To assist those entrepreneurs, the RISN Incubator will provide dedicated expert mentors to guide advancement, strategic advisement, introductions to industry stakeholders, and access to technical experts in the field of material reuse, technology, and the circular economy. In addition, entrepreneurs will receive business training on topics related to cost and revenue modeling, operations, and more, as well as access to waste from the Phoenix waste stream for use in the development, testing, and activation of their product or service. The RISN Incubator is already accepting applications from interested parties, made possible through a $500,000 grant provided by the US Economic Development Administration's i6 Challenge.

ASU and City of Phoenix Projects

At the core of the RISN effort in the Valley of the Sun, ASU and Phoenix work together on individual waste system related projects that may involve creating new markets and economic opportunities for waste system diversion, research and development of technologies, education and evaluation of potential solutions.

Fig. 3 Phoenix Public Works workers sorts through paper waste at the 27th Avenue Waste Transfer Station for bundling into recyclables and reusables. Feedstocks like this will be available to companies based at the Resource Innovation Campus and RISN Incubator (Photo courtesy of City of Phoenix Public Works Department)

ASU provides the direct support for the projects and pilot projects that serve Phoenix's needs, including project management by ASU staff. Much of the project task execution and research are completed by faculty experts and graduate student workers, identified on a project-by-project basis by Global Sustainability Solutions Service project managers. Several projects are summarized below to demonstrate the collaborative, holistic, and systematic approaches used to evaluate and develop sustainable circular economy solutions for the Phoenix metropolitan area.

Economic Impact Analysis

A 2014 waste characterization study estimated that approximately 14% of materials were recycled through the city's residential curbside program, but more than 55% of residential disposed garbage could be diverted through standard recycling and composting programs citywide.

The purpose of this two-phase project was to determine the potential regional economic impact of implementing a circular economy in the Phoenix metropolitan area (PMA).

The first phase estimated the economic impact of existing circular economy activities (limited to recycling activities, repair and maintenance activities, and reuse activities) in the PMA, by reviewing 43 sectors and subsectors of the local economy. In 2014, the maximum gross economic impact of circular economy firms and activities was estimated at $1.9 billion Gross State Product (GSP); 35,454 jobs paying over $1.2 billion in labor income; and $158.5 million in state and local tax revenues. Circular economy activities in the PMA as a whole in 2014 were estimated to contribute a maximum 0.9% of the annual statewide GSP.

The second phase identified and quantified the economic impact of waste diversion options for the currently recycled and additionally recoverable tons of plastic, glass, metals, and paper in the city of Phoenix municipal waste stream, using the 2014 waste characterization study data.

The 2014 waste characterization study estimated that the city of Phoenix already recycles 4,860 tons of post-consumer PET (plastic) and could potentially divert a further 4,245 tons from its municipal waste stream. This is sufficient volume to supply a post-consumer PET processing facility similar to one in Oregon. If a plant similar in size and output to the Oregon facility is established in the city of Phoenix, the construction of this facility and five consecutive operating years could cumulatively increase GSP by $113.5 million, increase real disposable personal income by $57.2 million, and directly employ approximately 50 people during each year of operation.

The 2014 waste characterization study estimated that the city of Phoenix already recycles 9,527 tons of glass and could potentially divert a further 4,591 tons from its municipal waste stream. The city of Phoenix is already home to a recycled glass processor employing 15 people and handling 50,000 tons of recycled glass each year. Seventy-three percent of this glass is currently sourced from within the county. To handle the additionally recoverable glass available in the municipal waste stream, this existing processor estimates that it will need to employ an additional five people in Phoenix and invest $1.5 million in new equipment. The economic impact of processing the additional 4,591 tons of recoverable glass alone during the 5-year study time is an estimated $11.4 million in GSP and $5.7 in million real disposable personal income.

The 2014 waste characterization study estimated that the city of Phoenix already recycles 3,975 tons of metal and could potentially divert a further 6,799 tons from its municipal waste stream. This consists of 1,026 tons of aluminum, 2,328 tons of tin and steel food cans, and 3,444 tons of other recyclable metal. The additionally recoverable metal can be easily handled by the city's current scrap metal industry, without any additional investment in jobs and equipment. At present, the city's supply of recycled metal is also greater than the demand from its aluminum can manufacturer and steel rebar mill. To generate additional local economic benefits, the city needs to either encourage existing manufacturers to expand their local operations or attract new recycled metal manufacturers to the PMA.

The 2014 waste characterization study estimated that Phoenix already recycles 53,447 tons of paper (all types) and could potentially divert a further 26,116 tons from its municipal waste stream. However, a recycled pulp plant requires a water-intensive deinking system; and there is insufficient current and additionally recoverable supply within the city's waste stream to meet the fiber inputs of a recycled corrugated box plant. If 156,000 tons of corrugated box fiber could be sourced throughout the

Southwest as part of a multi-state solution, the construction of a Phoenix facility and five consecutive operating years could cumulatively generate $437.4 million in GSP, $219.2 million in real disposable personal income, and directly employ approximately 140 people at the facility during each year of operation.

The total economic impact of a new PET processor, along with additional glass and metal feedstock recycling in the city of Phoenix over 5 years, is estimated to cumulatively generate $123 million in GSP, $61.3 million in real disposable personal income, and directly employ approximately 197–207 people during each year of operation, and increasing the recycling rate to approximately 19.11%, based on the 2014 waste characterization study.

Regional Green Organics System Design

Organic waste represents 30–60% of the residential waste stream in the PMA. In many communities, the majority of this material ends up in the landfill where organics represent both an opportunity cost as a potential market revenue to the municipalities and as a major contributor to landfill-generated methane emissions.

Nationally, a variety of technologies including anaerobic digestion, composting, and gasification have been used to convert the organic feedstock into a valued product and/or to extract energy. While these technologies have shown promise, they still face challenges to becoming economically and environmentally sound. In addition, the economic and political environment in the State of Arizona has made it challenging for municipalities to introduce new waste diversion initiatives. However, municipal leaders understand that in addition to the rising long-term economic costs of landfilling organics, environmental and social costs are also on the rise. Under this changing cost structure, the discussion is shifting from "if" it is viable to invest in an organics diversion initiatives to "when and how" it will become viable.

This project was collaboratively funded by six municipalities, two counties, and a Native American tribe and was initiated to assess the feasibility of a regional approach to building an organics management system (Buch et al. 2017). The purpose of the study was to identify plausible pathways to achieving an envisioned 2050 scenario of a regional, multi-site green organics processing system that diverts green organics in the PMA from landfills (Buch et al. 2017).

In this study, RISN developed the GIS-based Regional Circular Organic Resource System design model (RCORS) to simulate and visualize this complex problem. The model provides the functionality to analyze viable collections alternatives and the financial requirements for the organics processing technology that could support cost-effective collections for the municipalities and financial viability for the facilities (Buch et al. 2017).

Plausible pathways to achieving an envisioned 2050 scenario of a regional, multi-site organics processing system that diverts organic waste in the PMA from landfilling were identified. The scenario presented in Fig. 4 is one of many possible pathways to a PMA-wide organics management system. In the three-phase scenario, 11 organic waste processing facility sites could be implemented by 2050, with a total

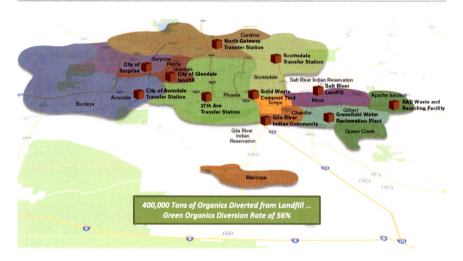

Fig. 4 Phoenix metropolitan area scenario allocation

annual organic diversion across the three phases of 400,000 tons resulting in the diversion of 56% of green organics for the participating municipalities. Eight of the sites in this scenario are municipality-owned, two are tribal-owned, and one facility is privately-owned (Buch et al. 2017).

The potential for collaborative agreements among the municipalities in the PMA would provide the opportunity to increase efficiency, reduce transportation costs, and achieve economies-of-scale on a regional basis. The RCORS model provides the functionality to help the municipalities evaluate the requirements for the siting of a financially viable organics processing facility and to explore opportunities for municipalities to collaborate in providing feedstock for each facility.

There are some external trends that are key to the success of an organics recovery program. The price and risk volatility of the commodity produced by the processing facility (compost, electricity, or fuel) are driven by the market and can affect financial performance.

Several key conclusions were reached from this study. The first and most important finding was that the regional approach is viable and that collaborations can result in processing facilities that are financially more resilient (Buch et al. 2017). A related finding was that collaborations will be based on location-specific requirements. Finally, the public sector has a longer view of capital investment and resource stewardship and is more likely to implement such organics recovery programs (Buch et al. 2017).

Sustainability Solutions Festival

The **Sustainability Solutions Festival** convenes the planet's top sustainability events and organizations to discover and explore how we can individually and collectively reimagine our lives and our planet to become sustainable. This unprecedented convening with global impact is fostered by partnerships between ASU, Phoenix, and GreenBiz Group, the leading media, research, and events organization that advances opportunities at the intersection of business, technology, and sustainability. With the additional support of partners from area public, private, and nonprofit organizations, the Festival makes Phoenix the epicenter of the "getting it done" conversation each year.

World-renowned experts in their fields leverage opportunities to collaborate with peers both within their profession and across sectors thanks to an alignment of business, academic, NGO, and local community participants – all committed to solving global challenges related to climate, waste, resources, environmental protection, employment, innovation, and equity.

Throughout the month of February, the 2017 Sustainability Solutions Festival challenged participants to (re)imagine the planet as our home through 17 separate events, focused on diverse audiences. These included GreenBiz 17, Second Nature Presidents Climate Commitment, Global Reporting Initiative, and the World Business Council for Sustainable Development conference. Public events included ASU's Night of the Open Door, City Lights Movie Nights Film Screening, and Family Day at Arizona Science Center. Results by the numbers included:

- 37,826 People were engaged in 17 events in and around Phoenix
- 2,500 Sustainability kits were distributed at public events
- Estimated 3.5 million impressions across traditional and digital media
- 5,000 Site visits by 4,280 unique users on the Festival web page and
- 1.1 Million impressions across 24,528 social media interactions on Facebook, Instagram, Snapchat, and Twitter

The event helps to raise the profile of Phoenix as a sustainability leader among diverse audiences: the global business community, government entities, thought leaders, and the public.

Parks Turf Compost Study

The premise of the multi-year project was to explore the efficacy of turning food scraps and yard waste that are currently landfilled into compost at a Phoenix-owned facility and then using that compost to improve the turf quality at Phoenix city parks, creating an internal circular economy.

Potential benefits to Phoenix and its citizens are many:

- Reducing and diverting waste from the landfill and associated methane emissions
- Exploring opportunities for cost savings
- Increasing revenue through growing the market for city-made compost
- Fostering collaboration between multiple City of Phoenix divisions and departments

- Improving quality of life by beautifying city parks
- Confirming City of Phoenix sustainability leadership nationwide

Collaboration has been key to the success of the project. The team consists of the City of Phoenix Public Works Department, which is responsible for compost production, delivery, logistics, and funding; the City Parks and Recreation Department, which is responsible for coordination, best turf practices, and application; and ASU researchers from the Julie Ann Wrigley Global Institute of Sustainability and the Swette Center for Environmental Biotechnology, who are responsible for research, sampling, testing, and third party verification.

During the first year of the study, the City's Parks Department staff applied 449 cubic yards of compost across 8.21 acres of turf at the nine City parks involved in the study. All of the compost used in the study was produced at the Public Works Department's Pilot Compost Facility at the 27th Avenue Waste Transfer Station and consisted of residential yard waste clippings, parks and landscaping clippings, large animal manure and food scraps from produce/grocery businesses, and special events. All of the compost generated at the pilot compost facility has maintained US Composting Council's Seal of Testing Assurance (STA) certification standards. During the first year of the study, Parks and Public Works Department staff received zero concerns or complaints from residents about the program.

Results of the soil and turf data analysis showed that compost does not have any negative impacts on the current turf. In fact, previous compost studies and the nature of compost lead the team to believe that compost application will, over the longer term, be beneficial to multi-use park turf. It is important to keep in mind that the study was modeled to last at least 3 years, as it often takes time for there to be measurable changes in soils. Continuation of the study over the next 3–5 years may yield significant results and enable the City of Phoenix to identify the financial and environmental benefits of compost turf application.

RISN and Regional Partners

Waste systems, material flows, and direct and indirect economic and sustainability impacts are not confined by municipal boundaries, but rather have significant regional consequences. RISN recognizes that all projects done in any of the municipalities within the region will have impacts that spill over into other municipal areas, and this can be leveraged to create more positive outcomes when the cities and towns work together.

To foster regional collaboration and stakeholder buy-in, RISN hosts biannual regional collaborative workshop meetings (Fig. 5).

Solid waste and public works officials from all municipalities in the Phoenix metropolitan area are invited to these meetings, which cover various topics that uncover barriers and challenges to create a region-wide circular economy, prioritize areas of concern, and identify opportunities for collaborative projects. The RISN regional collaborative meetings have only involved municipal stakeholders thus far,

Fig. 5 ASU staff conducted multiple workshops with regional stakeholders to develop strategic plans to implement circular economy practices (Photo courtesy of ASU Walton Sustainability Solutions Initiatives)

though there is a plan to include private sector stakeholders, such as private waste haulers, in the near future. An example of a regional RISN project is the **Green Organics System Design** project, outlined previously.

RISN and Global Partners

The highest vision and mission for RISN is to expand the sustainable circular economy model through global hubs where best practices can be shared at a global scale (Fig. 6). RISN is especially interested in delivering circular economy solutions to developing nations, as they have the highest at-risk population and have the greatest need for sustainable systems, solutions and practices, thus offering the opportunity for the greatest impact.

It is to this effort that RISN is establishing hubs in several regions around the globe. To establish a RISN hub, RISN works with a local partner who has understanding, relationships, and expertise within that specific region. The local partner facilitates and executes the development of a sustainable circular economy in their own region, with the support and expertise of RISN. RISN offers operational and informational education and guidance through the process, both formally and informally. The first RISN hub was established in Lagos, Nigeria, in 2015 and exploratory work continues elsewhere with relationships established in Guatemala and

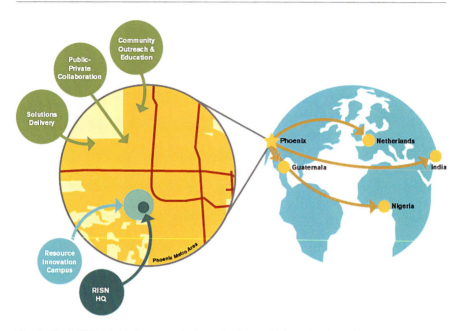

Fig. 6 The RISN global hub strategy looks to develop multiple regional collaborative partnerships at a worldwide scale to leverage and resource best practices

India. RISN staff delivered an on-site sustainable circular economy certification and workshop in Nigeria for Lagos stakeholders, to help them identify barriers and opportunities for development of the circular economy as well as strategies for advancing the initiative.

An example of RISN's global collaborative effort to advance circular economy is ASU's partnership with **Shashwat Eco Solutions Foundation** (an NGO in Pune, India) to collaborate in the development of a toolkit for assessing, planning, and implementing a local circular economy. Our partner is working with local communities at various scales (villages, towns, suburban municipalities, and cities) and collaborating with the ASU RISN team to develop a toolkit for application in other developing country municipalities.

Public–Private–Academic Collaboration

Bringing together public and private entities enables ideas and solutions that develop synergistically when organizations gain access to university solutions, research resources, and expertise. RISN facilitates public-private collaboration through focused circular economy workshops about various waste systems (such as packaging and local food systems), collaborative implementable solutions projects, and entrepreneurial services at the RIC.

RISN regularly holds workshops on various topics that can potentially be solved with circular economy strategies. Bringing together private, public, and nonprofit stakeholders allows for all entities to figure out how they may work together and leverage each other's strengths to find solutions. An example of this kind of workshop was a complex adaptive food systems workshop held by RISN and including stakeholders from grocery stores, commercial composters, nonprofit local food groups, state agencies, and municipal leaders.

RISN also serves to educate other types of organizations about the potential of the circular economy within their own business model. In the past year, RISN hosted a group of informal science education institution directors to educate them on the incorporation of sustainability and circular economy education in science museums and centers, and also held a circular economy workshop specific to the packaging industry. When packaging designers and manufacturers are educated about the circular economy, they understand how they can improve their package design to make the supply chain more sustainable. For this workshop, large packaging producers and designers were able to interact directly with the director of materials recovery facilities in Phoenix to understand exactly how package design decisions affect stakeholders and sustainability impacts way down the supply chain.

In addition to convening public, private, and nonprofit waste system stakeholders, RISN serves to bridge the public and private sector divide by providing entrepreneurial support to attract private technology companies to the RIC. Through the RISN Incubator program and other financial support provided by Phoenix, RISN seeks to closely align the power of private sector innovation with the goals of Reimagine Phoenix and RISN. The City of Phoenix and the greater metropolitan region stand to gain immensely if private companies with proprietary sustainable waste processing technologies locate to and are supported by the city.

A final component of this collaboration is outreach and engagement through a diverse range of conferences, including those focused on the solid waste industry, municipal public works, and innovation for sustainability. Examples of conferences where staff presented or attended include the American Public Works Association conference, Chamber of Commerce National Conference in Washington, DC, and the Ellen MacArthur Foundation Disruptive Innovation Festival held annually via the Internet.

Engagement Is Key to a Sustainable Circular Economy

Today, RISN envisions a global network of collaborating partners who create a world of circular resource loops that improve human well-being, generate economic value, and restore the environment for all generations of all species. The mission of RISN is to accelerate the global transition to a sustainable circular economy through a global network of public, private, and NGO partners using collaboration, research, innovation, education, and application of technologies to create economic value through sustainable resource management.

RISN at ASU aims to simultaneously serve the Phoenix metropolitan area to accommodate a rapidly increasing population with a high quality of life and sustainable communities, while being a catalyst for sustainable circular models at regional scales throughout the world.

By creating a network of academic, government, private business, and NGO partners and being inclusive of all stakeholders, RISN can create momentum for a more sustainable future by facilitating the reimagining of materials, energy, and resource systems, and balancing the needs of the community. The unique capabilities of this academic and government partnership can be a model for realizing the sustainable circular economy worldwide and solving the uniquely challenging and "wicked" problems of sustainability.

RISN has developed the following values:

- Regional Integrated Resource Management
- Continuous Innovation
- Collaborative Symbiosis
- Social and Inter-Generational Value and Justice
- Economic Development
- Education
- Inspiration

RISN is guided by the following principles:

- **Success is defined as fully embracing and implementing** the Hannover Principles and **the principles of a circular economy**, as defined by the Ellen MacArthur Foundation.
- Decision-making is to **favor long-term goals** and thinking over short-term goals or opportunities. Continuous innovation includes making choices within existing constraints today that do not inhibit or delay better solutions when they become available.
- Decision-making is to include **thoughtful consideration of all people** and **species** of this, past, **and** future **generations**, especially when their interests are not directly represented.
- **Everything is interconnected**. Any changes to activities of the network are to be evaluated for their anticipated impacts and potential unintended impacts (beneficial, benign, and adverse) on the rest of the network and the broader community.
- True **integrated resource management** and innovation **expands beyond waste** resources to include the quality, availability, responsible use, and appropriate valuation of all resources (water, air, energy, etc.).
- The network is to promote **the principles of the Living Building Challenge** and **the Living Community Challenge** for their consistency with circular economy and **zero waste, water and energy goals**.
- Inspiration is key. As a vanguard to promote behavioral transformation globally, network activities are to **include opportunities for human inspiration and education**, celebrating their roots in reimagining waste.

- A primary goal of the network is to **create circular economic value and jobs**, locally and regionally. Decisions should strengthen the market-based environment and symbiotic relationships, and **emphasize regional solutions**.
- Perfect solutions and an imperfect world often collide. **Transparency about** those collisions and the results and **lessons learned** provide enormous educational opportunities to devise better solutions for others and for the future. The network is to promote open communication and transparent knowledge sharing.

Cross-References

▶ Agent-Based Change in Facilitating Sustainability Transitions
▶ Bio-economy at the Crossroads of Sustainable Development
▶ Business Youth for Engaged Sustainability to Achieve the United Nations 17 Sustainable Development Goals (SDGs)
▶ Education in Human Values
▶ From Environmental Awareness to Sustainable Practices
▶ Moving Forward with Social Responsibility
▶ People, Planet, and Profit
▶ Relational Teams Turning the Cost of Waste into Sustainable Benefits
▶ Smart Cities
▶ Social Entrepreneurship
▶ Sustainable Living in the City
▶ Teaching Circular Economy
▶ Utilizing Gamification to Promote Sustainable Practices

References

Allenby, B. (2006). The ontologies of industrial ecology? *Progress in Industrial Ecology, 3*(1–2), 28–40.
Allenby, B. R., & Graedel, T. E. (1995). *Industrial ecology* (Vol. 16, p. 239). Englewood Cliffs: Prentice Hall.
Benyus, J. M. (1997). *Biomimicry*. New York: William Morrow.
Buch, R., Campbell, W., Osgood, K., George-Sills, D., Melkonoff, N., & Paralkar, S. (2017). Regional circular organic resource system. https://sustainability.asu.edu/wp-content/uploads/sites/18/2017/01/RegionalGreenOrganics-ProjectOverview-FINAL.pdf.
Butlin, J. (1989). Our common future. By World commission on environment and development (London, Oxford University Press, 1987, pp. 383). *Journal of International Development, 1*, 284–287.
Crane, A., & Matten, D. (2016). *Business ethics: Managing corporate citizenship and sustainability in the age of globalization*. Oxford/New York: Oxford University Press.
Cugurullo, F. (2013). How to build a sandcastle: An analysis of the genesis and development of Masdar City. *Journal of Urban Technology, 20*(1), 23–37.
Harel, T. (2013). Interface: The journey of a lifetime. The Natural Step/ The Flexible Platform, Eindhoven: The Netherlands.
Hodson, M., & Simon, M. (2010). Can cities shape socio-technical transitions and how would we know if they were? *Research Policy, 39*(4), 477–485.

Interface USA (2015). www.interface.com. Accessed 15 Dec 2015.

Lobo, R. (2014). Could Songdo be the world's smartest city? *World Finance*, 21 Jan. 2014. Web. 21 Jan. 2016.

Lieder, M., & Rashid, A. (2016). Towards circular economy implementation: A comprehensive review in context of manufacturing industry. *Journal of Cleaner Production, 115*, 36–51.

MacArthur, E. (2012). *Towards the circular economy: economic and business rationale for an accelerated transition*. Ellen MacArthur Foundation, Isle of Wight: UK, Volume 1.

McDonough, W., & Braungart, M. (1992). *The Hannover principles*. Hannover: William McDonough Architects.

McDonough, W., & Braungart, M. (2010). *Cradle to cradle: Remaking the way we make things*. London: Macmillan.

Nattrass, B., & Altomare, M. (2013). *The natural step for business: Wealth, ecology & the evolutionary corporation*. Gabriola Island: New Society Publishers.

Robèrt, K. H., & Anderson, R. (2002). *The natural step story: Seeding a quiet revolution*. Gabriola Island: New Society Publishers.

Senge, P. M., Carstedt, G., & Porter, L. P. (2001). Innovating our way to the next industrial revolution. *MIT Sloan Management Review, 42*(2), 24.

Shwayri, S. T. (2013). A model Korean ubiquitous eco-city? The politics of making Songdo. *Journal of Urban Technology, 20*(1), 39–55.

Stahel, W. R. (1982). The product life factor. An inquiry into the nature of sustainable societies: The role of the private sector series: 1982 Mitchell Prize Papers, NARC.

United Nations Statistical Commission. (2017). Report of the Inter-Agency and Expert Group on Sustainable Development Goal Indicators. New York.

U. N. World Commission on Environment and Development. (1987). Our common future, Oxford University Press: Oxford, U. K.

From Environmental Awareness to Sustainable Practices

A Case of Packaging-Free Shopping

Ragna Zeiss

Contents

Introduction	730
Approaches for Sustainable Development: From Governance to Practice	732
Packaging-Free Supermarket	734
Comparing Shopping in Regular, Organic, and Packaging-Free Supermarkets	736
Challenging Conventions?	742
Packaging-Free Shopping Without Packaging-Free Shop	744
Setup of Exercise	744
Experiencing Packaging-Free Shopping Without Packaging-Free Shop	745
Sustaining Packaging-Free Shopping? A Practice to Be Sustained?	750
Wrapping Up: Reflections and Conclusions	751
Cross-References	753
References	753

Abstract

The enormous problem of packaging materials, especially plastic, is widely recognized. Despite initiatives to reduce packaging, this recent problem is difficult to tackle. This chapter asks why we do not "simply" stop buying packaged groceries. First, it compares factors of importance and shopping practices in grocery shopping between an ordinary supermarket, an organic supermarket, and a packaging-free supermarket. Second, it reports on an exercise in packaging-free shopping in a place without a packaging-free supermarket. Which changes in everyday practices and routines were observed? In line with Garfinkel's breaching experiments, it explores what we see as normal with regard to shopping and how that normality may need to be breached in order to shop

R. Zeiss (✉)
Faculty of Arts and Social Sciences, Department of Technology and Society Studies, Maastricht University, Maastricht, The Netherlands
e-mail: r.zeiss@maastrichtuniversity.nl

packaging-free. In line with research by Elizabeth Shove, the chapter argues that changes in sustainability practices do not "simply" follow from increasing environmental awareness. It shows how efforts at sustainable living are interconnected with everyday routines and practices which are difficult to change. Investigating these practices and how they change is crucial to live a life of engaged sustainability. The chapter further reflects on the value of exercises like packaging-free shopping for understanding how practices become "normal" as well as for experiencing lived sustainability.

Keywords
Packaging-free supermarket · Grocery shopping · Sustainability exercise · Practices · Routines · Everyday life · Waste · Environmental awareness

Introduction

> Packaging – much of it single-use food wrapping – has created a rubbish problem that now pollutes every corner of the world. (Hall 2017)

Only about three generations ago, shops around the world sold mostly local products, and transportation distances were short. Shopping was done with very little, if not zero waste. Technological changes in transportation and packaging material such as the production of plastic after the Second World War have allowed preservation of food for a much longer time and import and export of food between different parts of the world.

This had led to large changes in terms of practices of shopping and standards and ideas of cleanliness, freshness, and convenience. Consumers now have access to a wide range of products all year-around. They are able to store products at home for longer times, also due to inventions such as the fridge. They can buy ready-made sauces or even entire meals. They can buy all products from the same store, a relatively new phenomenon. Self-serving shopping in the form of supermarkets only started after the Second World War and particularly in the 1970s (Tomka 2013: 238). Things are won and things are lost with all changes; these changes and changes in, for example, working life and who participates in it altered practices of shopping, eating, and cooking.

Many products became packaged to be able to transport them; to keep them clean, fresh, and safe; to convey information such as the ingredients and expiry/best before date; and to avoid tampering with the products and for marketing purposes. Whereas the benefits of packaging were clear, the enormous problem of packaging materials, especially plastic, has been widely recognized more recently. According to a Guardian investigation, consumers around the world "buy a million plastic bottles a minute and plastic production is set to double in the next 20 years and quadruple by 2050." (Laville 2017). The chapter further states that research by the Ellen MacArthur Foundation points out that the amount of plastic produced in a year is roughly the same as the entire weight of humanity and that by 2050 the ocean will contain more plastic by weight than fish. However, it turns out to be very difficult to tackle this problem that was created in only a few decades. There have been efforts to increase environmental sustainability, such as investments in recycling. However, the global production of plastic has continued to rise (Gourmelon 2015). Why then do supermarkets not simply abandon their packaging? And why do (concerned) citizens and customers not simply purchase products packaging-free? This chapter focuses on the latter question.

Can environmentally aware citizens not simply set an example? This chapter shows that environmental awareness can be important in attempting to make changes but that this is not "simply" done. Environmental awareness is one aspect of our daily lives which are largely structured by routines and practices. We cannot "simply" change the ways our lives are structured; we have a stake in maintaining these practices and routines.

An anecdote of Dave Hall illustrates the importance of practices and routines – here called habits – which are a strong obstacle to pro-environmental behavior change (Kollmus and Agyeman 2002):

> In 2003, I was told by a restaurant owner on a Thai island that local fishermen used to wrap their lunch in banana leaves, which they would then casually toss overboard when done. That was OK, because the leaves decayed and the fish ate the scraps. But in the past decade, he said, while plastic wrap had rapidly replaced banana leaves, old habits had died hard – and that was why the beach was fringed with a crust of plastic. (Hall 2017)

This chapter reports on initiatives and exercises in engaged sustainability and particularly on practices and exercises around (packaging-free) shopping. The reader is invited to reflect on his or her own shopping routines and practices. The chapter focuses in the first instance on individual practices and routines but argues that these are part larger and interconnected practices and, in line with the work of Elizabeth Shove, "normal ways of life." The chapter is in line with the idea that practices are linked to and "reproduce what people take to be normal and, for them, ordinary ways of life" (Shove 2003: 395). These "ordinary ways of life," these practices (which can be called habits at individual level), are not limited to individuals but are part of wider societal structures.

The chapter starts with a small exploration of measures and approaches that aim to reduce packaging and to develop a more sustainable society. In section "Packaging-Free Supermarket," the chapter then engages with a specific initiative as an example of what citizens and entrepreneurs can do: the packaging-free supermarket. It reports on the results of an investigation of shopping practices in a regular supermarket, an organic supermarket, and a packaging-free supermarket in Maastricht, the Netherlands, and Leuven, Belgium. It asks the question: What would it take for customers of a regular supermarket to transfer their grocery shopping to a packaging-free supermarket? However, what to do as a citizen when no packaging-free supermarket is available? The section "Packaging-Free Shopping Without Packaging-Free Shop" reflects on an exercise of engaging in packaging-free grocery shopping in a place without a packaging-free shop. The chapter ends with a reflection on the role of practices and routines as well as sustainability exercises in everyday lives in lived or engaged sustainability.

The format of this chapter is different than that of the other chapters in this book. The book cover notes that "although there is much rhetoric about sustainability, very little has really been accomplished in addressing the issue at the practical level." Yet, there have been some initiatives to reduce or abandon packaging materials, such as setting up packaging-free supermarkets. Such initiatives are rather small scale and hardly or not reported on in existing literature. To be able to address such issues as well, this chapter draws on research by and with a group of undergraduate students who investigated and engaged in lived sustainability themselves. In the course of a 5-month research project of the excellence program of Maastricht University in the Netherlands, they conducted observations, interviews, and participated in two (self-designed) exercises in the (academically largely unrecognized) field of packaging-free shopping (April 2016). This chapter draws and reflects on the information in the final report of this project: "The vision of packaging-free shopping" (Manovella et al. 2016).

Approaches for Sustainable Development: From Governance to Practice

There are many different measures and approaches aiming to develop a more sustainable society. Packaging companies have, for example, invested billions to recycle plastic. And indeed, it is tempting to "review the design and development of more efficient, less resource intensive products and technologies" (Shove 2003:395) when thinking about the environmental aspects of consumption. An example of the attempt to use less plastic for packaging was to use thinner plastic milk bottles which contained more recycled material (Laville 2017). These bottles did not prove sturdy enough and burst, creating more waste rather than less. As plastic can often not be recycled more than twice, it will end up polluting the world in the end. Despite efforts to recycle plastic, pollution is increasing rather than reduced.

There are also "simple" policy measures, regulations, and legislations, such as the EU Plastic Bags Directive. The Directive states that EU member states have to introduce national reduction targets and/or a price on plastic bags from April 2016 onward (European Parliament/Council 2015). This regulation is seen as necessary, because "in 2010, each EU citizen used an average of 198 plastic bags" and the trend is ascending (Barbiére 2015). This means "radical changes" (Barbiére 2015) for certain countries on the left side of Fig. 1, which illustrates how many plastic bags EU member states use.

According to the EU Commissioner for the Environment, Maritime Affairs and Fisheries Karmenu Vella, such simple measures can have great effects:

> In the EU we currently consume up to 200 bags per person, every year. Only about 7 % are recycled. Billions end up as litter across Europe, especially on our beaches and in the sea. This has serious environmental and economic effects. We need to tackle marine pollution, in particular microplastics. We need to save resources and move to a circular economy. Now it's

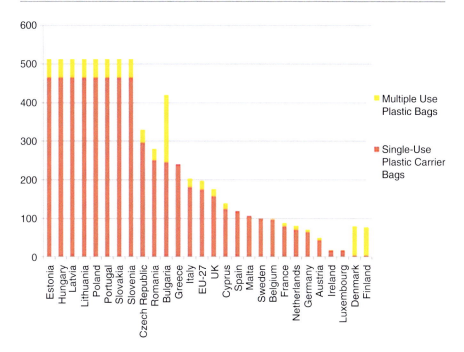

Fig. 1 The use of plastic bags in the EU. (Retrieved from Barbiére 2015)

up to the Member States to do their part. Some have already shown that simple measures can lead to big changes. (European Commission 2016)

And indeed, this Directive may help the reduction of plastic bags. Those who were interviewed and observed during the study this chapter reports on report that they already used to bring their own shopping bags to the supermarket even before the charges were introduced. Many declined a plastic bag for environmental reasons and were also unwilling to pay for a plastic bag. Nevertheless, the observations showed that the free thin plastic bags in supermarkets to collect fruit and vegetables were still used. In any case, plastic bags are only a small part of the plastic used in shopping and packaging.

Studies in science, technology, innovation, and society have developed and are developing other (academically based) approaches. These are less simple in the sense that they acknowledge the complexity of increasing sustainable development. Many groups of actors are involved at different levels, innovation trajectories are not linear, and not all consequences can be foreseen.

Responsible Research and Innovation (RRI) and transition management are two approaches that have gained increasing importance in the study of science, technology, innovation, and society as well as in policies. RRI can be discussed as a governance tool for science and innovation. It came out of a number of different discussions, approaches, and concerns such as the managing of science and innovations which are seen as ethically problematic such as genetically modified organisms and "an increasing awareness of the sometimes profound, global (and intergenerational) impacts of innovations in contemporary society" (Owen et al. 2012: 752). Often three features are distinguished. First is the idea that the purposes of research and innovation and their orientation toward the "right impacts" need to be subject to democratic governance. This means that the purposes of science, including developing a (more) sustainable society, cannot be decided by scientists alone. Rather, deciding on the direction and purposes of research should be a wide societal endeavor. Second, the process should be responsive. Approaches of anticipation, reflection, and deliberation of research and innovation and its directions should be integrated and institutionalized. The third feature revolves around responsibility as a collective activity with uncertain and unpredictable consequences (Owen et al. 2012). These consequences and the responsibility need to be continuously assessed.

Transition management was developed as a governance tool explicitly for sustainable development (Loorbach and Rotmans 2010). It outlines four very general principles, called the transition management cycle. These principles are:

1. Problem structuring, establishment of the transition arena, and envisioning
2. Developing images, coalitions, and transition agendas
3. Mobilizing actions and executing projects and experiments
4. Monitoring, evaluation, and learning

The approach has led to the development of transition policies in fields such as water management, building, and energy in the Netherlands, the UK, and Belgium.

Such governance approaches have also been criticized. The critique that is relevant and interesting for this chapter and book is the lack of attention of these approaches for practices and everyday lives of consumers. And yet "significant movement towards sustainability is likely to involve new expectations and understandings of everyday life and different forms of consumption and practice (Redclift 1996; Wilk 2002)" (Shove and Walker 2010: 471).

This chapter asks how consumers or citizens can engage in lived sustainability. This is important as much of domestic consumption takes place in practices in ordinary lives, which are often invisible (Shove 2003). Yet, these practices have consequences for sustainability. They matter. The above governance approaches tend to focus on the producers and those who govern, while consumers and practitioners are as important to change as producers (Shove 2003). Attention for practices dissolves the distinction between those who govern and those who are governed as all are part of practices.

Packaging-Free Supermarket

The World Commission on Environment and Development defines sustainable development as "a process of change" whose implementation in all areas "must rest on political will" (World Commission on Environment and Development 1987, § 30).

Political will is exercised through the adoption of top-down approaches, and the EU regulation to reduce the usage of plastic bags is an example of it (Barbiére 2015). Minami et al. (2010) recognized the existence of alternative approaches to pursue sustainable development. Packaging-free shops demonstrate how a community can initiate a "process of change" through a bottom-up approach. Several actors such the shop owners, suppliers, and customers have been mobilized to address the pressing issue of waste within the context of shopping. They identified the problem of waste and assessed the ways to reduce it. The reduction of the production of waste in our everyday practices, such as grocery shopping, as Minami et al. (2010: 3–8) suggest, is one way.

In 2007, the first packaging-free shop was opened in London. Later in Leuven, Belgium, three persons became increasingly frustrated with the amount of packaging they encountered on a regular shopping trip:

> We were very frustrated when we went to the standard supermarkets and we got home and we unpackaged what we bought and our garbage bag would be filled already with plastic. (Shop owner 12-04-16)

They were starting to look for an alternative and discovered online that packaging-free supermarkets existed, in London and in Germany. The idea of developing a shop in Leuven without packaging was born. They managed to find a number of shareholders, people with similar environmental concerns and frustrations. Facebook was essential for this. They also managed to interest suppliers, such as a local farm and jam maker, who were willing to manage (or already managed) their business in line with environmental and social responsibility. Sometimes convincing suppliers to remove packaging was hard such as in the case of convincing the shampoo supplier to refill the bottles. The location was important as customers could park their car and would not have to walk far with their jars. It was a student area, and the shop owners were aware that students were supportive of this initiative. In December 2014, the shop was opened. The illustrations show what the packaging-free shop in Leuven looks like (Figs. 2, 3, 4, 5, 6, 7, 8, 9, and 10).

With the existence of such a shop, it becomes important to understand people's considerations for where to buy their groceries. Why do people shop in a packaging-free supermarket, an organic supermarket, or a regular supermarket? What makes a packaging-free supermarket different from a regular supermarket or an organic supermarket, and what does this mean for why people do their grocery shopping in such a shop? And, importantly, what would people need to change to start buying their groceries in a packaging-free supermarket rather than in a regular supermarket? This section reports on the interviews and observations that were conducted during the 5-month research project mentioned in the introduction.

Questions to the reader
- Have you ever reflected on where you do your shopping and why, specifically regarding issues of sustainability?
- Where do you do your shopping? And why?
- What would be reasons for you to go to a packaging-free supermarket and what would be your objections?

Fig. 2 Interior

Fig. 3 Bathing products in refillable containers

Comparing Shopping in Regular, Organic, and Packaging-Free Supermarkets

The 15 interviews as well as the observations that were conducted indicate that customers in the different shops have different priorities. While location is important for all customers (most tend to shop close to where they live, work, or study), customers who shop in an organic supermarket and a packaging-free supermarket are prepared to travel longer distances, especially for the packaging-free supermarket. They support the concept and like the shop. At the same time, they also shop in supermarkets closer to home. The location also matters for how frequently they shop. Most interviewees shop once or twice a week, but those living further away from the packaging-free supermarket often shop only once a month and then buy larger quantities.

Fig. 4 Containers with seeds and nuts

Fig. 5 Containers with seeds and nuts covered in chocolate

Price is important for all customers and a reason for people to shop in a regular supermarket. The organic store and the packaging-free supermarket are slightly more expensive than the standard supermarket. A student in the packaging-free supermarket said:

> I complete my general shopping in [name of regular supermarket], a Belgian supermarket, because it is cheaper and there are more options; if I only shopped in a packaging-free shop, I would have to change my lifestyle, which would be rather difficult as I am a student. I also cannot buy fresh cheese in the packaging-free shop. (Woman, 20s, PFS (packaging-free shop))

One customer remarks how she negotiates the price: "I usually buy in smaller quantities than I would buy in a normal supermarket as I try to have less expired

Fig. 6 Different types of oil in refillable containers

Fig. 7 Glass container containing liqueur

food" (Woman, 40s, PFS). To attract customers, especially those on lower budgets such as students, they also provide discounts on older vegetables or nuts.

Another reason to go to an organic and packaging-free supermarket despite higher prices is that these customers value the quality of the food and the healthy lifestyle that it implicates. Very few of the interviewees in the standard supermarket mention that they specifically buy local or seasonal products or products that are organic, bio, or fair-trade. However, the interviews indicate that customers in the organic shop place great importance on the product to be organic, while those

Fig. 8 Vegetables

Fig. 9 Vegetables for free

interviewed in the packaging-free supermarket appreciate that most of their products are seasonal, fair-trade, and locally produced.

When the packaging-free supermarket first opened, only local and seasonal products were sold. While this offer attracts customers, many customers were used to a wider variety of products. They would go to other supermarkets to buy, for example, fruit such as bananas that simply cannot be grown in Belgium. The shop owners made a compromise and included some products that were not locally grown, ensuring that they were fair-trade and marking up the prices slightly to favor the local products and keep their philosophy of packaging-free and mostly local products. This is important as the interviews show that the variety of products,

Fig. 10 Glass container with pasta

as well as familiarity, is also a reason to shop in a particular supermarket. Supermarkets that have a wider variety of products enable customers to shop in one supermarket if they want, instead of having to shop in several different locations.

Checking labels for ingredients is an important part of the shopping practice for those who want to eat healthy and for those with allergies and intolerances. Knowing the supermarket and which products one can buy without always having to check these becomes even more important. How is this dealt with in a packaging-free supermarket? There are labels around the shop, which explicitly list the ingredients of the product. This can be seen in Figs. 4, 5, and 7, which show containers with seeds and nuts as well as a glass container with liqueur. Figure 6 shows containers containing different kinds of oil, the names of which are written on the containers. Furthermore, a sign was placed that invited the customers to ask the shop assistants any questions concerning ingredients, allergies, or expiry dates (Fig. 11).

When customers were specifically asked whether they checked the expiry date of products, one interviewee of the packaging-free supermarket replied "I've never had problems with the freshness in PFS" (Man, 30s, PFS). This illustrates that customers tend to trust the shop owners about the freshness of the products they sell. Furthermore, they rely more on their senses to assess the freshness of a product: "I can tell just by looking at the products to say whether I can still eat it or not" (Woman, 30s, PFS).

While location, prices, familiarity, quality, and variety are all important reasons to shop in a particular supermarket, there are some specific reasons for customers to go to a packaging-free shop. Many customers support the idea behind the shop and the initiative and consider the environmental impact of products. Both in regular supermarkets and in the organic store, many products, including fruit and vegetables, are wrapped in plastic. A reason for packaging organic products is to avoid fraud: stores selling standard products as organic.

In the regular supermarket, some interviewees mention "friendly shop assistants" (Woman, 40s, regular supermarket) and the comfortable design and organization of

Fig. 11 Ask shop assistants for ingredients, allergens, and expiry dates

the shop and "not to mention the free coffee" (Woman, 60s, regular supermarket). The observations and interviews in the packaging-free shop indicate that customers know the shop owners and assistants and they often stay after they finish their shopping to have a chat with them. One woman enters the shop and firstly drinks a little (glass and reusable) bottle of lemonade before she commences with her shopping. Another one has been "talking to the owner of [name packaging-free supermarket] who showed her the different spices and they chatted about the different products for about five to ten minutes" (Woman, 40s, PFS). Taking one's time seems to be as much a part of the experience as the packaging-free products themselves. The majority of the customers takes their time, looks at the different products carefully, checks the quality of the goods, talks to the owners, and asks them questions. Some customers have long chats with the owners and the interaction is part of the appeal of this kind of shop. In fact, when asked about the difference between standard supermarkets and the packaging-free supermarket, the most consistent answer was about how personal and friendly it is to shop in a place where you can talk to the owners. One woman states that it has "a more personal touch to them and you can chat to the owners, something you cannot do in a regular supermarket" (Woman, 40s, PFS), and another interviewee said "it is a much more personal experience in a packaging-free supermarket, you can taste the products, and you even get little things free if you buy a lot" (Woman, 20s, PFS).

Most customers in the organic and regular supermarket were not familiar with the concept of a packaging-free supermarket. Most were curious about the shop, the products, and the concept, and roughly half would be interested in shopping there for environmental reasons. However, there were also hesitations with regard to the variety of products, reasonable prices, and the location. Two interviewees state that there would need to be a "good variety of products" (Woman in her 50s and woman in her 40s, organic shop). A further two comment on the price, stating that they would only go to the packaging-free supermarket if it was not too expensive.

Another hesitation is related to considerations of (in)convenience. One student was worried about the transportation of containers and states that "the issue for me would be with liquids, as it would be quite challenging to bring my own containers with me when I shop" (Woman, 20s, regular supermarket). Most students in Maastricht use a bicycle as their main mode of transportation; a car is often not available. Therefore, the transportation aspect could arguably be a large factor. Another interviewee states that she could "imagine that it is a hassle to bring that many containers, therefore I would not go there if I knew that I had a lot to buy" (Woman, 20s, regular supermarket).

Further comments included "it sounds difficult" (Woman, 40s, organic shop), "too complicated" (Woman, 60s, organic shop), and "not comfortable with idea" (Woman, 50s, organic shop). They anticipated spending more time on the process of shopping. A young student in her 20s turned this around. She remarked that she has to recycle "two full bags of plastic containers" every week (Woman, 20s, organic shop) and if she were to shop packaging-free, this would no longer be necessary. So although packaging-free shopping may be more time-consuming, one may also be able to save time elsewhere. However, when the focus is on shopping practices, other related practices are often not considered.

Customers further said to only buy/purchase certain products there such as pasta and rice, yet not "shampoo and things like that" (ibid.), and that hygiene would be compromised. One interviewee states, for example, that she would not buy meat from a packaging-free store, as they think it is "more hygienic" (Woman, 40s, organic shop) to have it prewrapped. And while a male interviewee agreed with the idea that the use of plastic is excessive, he considers packaging-free products as "over the top": "why do avocados need to be wrapped in plastic? I think it's pointless. But buying everything packaging-free is over the top, especially for hygienic reasons" (Man, 20s, regular supermarket). Another interviewee says that they "think it is wasteful that they are packaged in plastic but for some products I think I would prefer to buy them [in a standard supermarket]" (Woman, 70s, regular supermarket).

Questions to the reader
After reading this, would you be more or less inclined to shop in a packaging-free supermarket and what would be your considerations?

Challenging Conventions?

What does this all teach us? Environmental awareness is important in individual considerations for choosing where to buy groceries and sometimes for the willingness to make an extra effort such as traveling further. However, although small changes in practices and routines are acceptable, issues of convenience and maintaining existing practices are demonstrated too.

The packaging-free supermarket had to broaden its assortment with products that were not local or seasonal and therewith compromise the initial idea and philosophy.

It had to accommodate existing expectations and demands of specific varieties of products to continue existing. It is apparently not easy or simple to change such expectations and practices, not for supermarkets and not for (the majority of) customers. In addition, even those who are willing to travel further to support a shop which operates in line with their environmental concerns needed to go to regular supermarkets due to, for example, demands for time in their everyday life.

An interesting question is: What would it take for customers of a regular supermarket to transfer their grocery shopping to a packaging-free shop? A packaging-free supermarket tries to go beyond developing "more efficient, less resource intensive products" (Shove 2003:395). It requires packaging-free transport, customers to bring their own containers, and suppliers willing to supply packaging-free or reuse the packaging. Yet, the supermarket and the customers are part of and subject to other structures, practices, and expectations. It seems as if a packaging-free shop can be successful as long as it neatly fits into peoples' and society's existing everyday lives and practices.

Convenience matters. And something is regarded as convenient if it fits in with existing practices and routines. Changes in time required for shopping, in mode of transportation, the organization that shopping requires in terms of bags and transportation, the price of the products, and norms of hygiene are mentioned as obstacles to consider shopping in a packaging-free supermarket. While the concept of a packaging-free supermarket is broadly supported, an anticipated change of practices and with that reorganization of other aspects of one's everyday life is not. This attachment to routines and convenience is also present among the customers of the packaging-free supermarket, even though they may have incorporated a once-a-month trip to the supermarket rather than shopping twice a week. Customers of the packaging-free supermarket stated that they might have difficulties to commit to packaging-free shopping without a packaging-free supermarket, thus showing the convenience of having a store where a variety of unpackaged products are offered.

This is in line with the idea that "domestic consumption and practice and intimately links in reproducing what people take to be normal and, for them, ordinary ways of life" (Shove 2003: 395). The packaging-free supermarket is a very valuable initiative with regard to sustainability that shows that there is more than rhetoric to sustainability and that deserves more attention. It is important to realize which initiatives exist, what they do, and how they relate to engaging with sustainability in everyday lives for ordinary citizens. In this case, the packaging-free supermarket seems to "work" as long as it is (largely) in line with existing conventions and practices and is an example of 'seeking more environmentally friendly ways of meeting given levels of service and of "eco-modernizing" society (Spaargaren, 1977)' (Shove 2003: 396). It goes beyond a focus on more efficient and less resource intensive projects (Shove 2003) and does not take "future demand and consumption as foregone conclusions" (Hand et al. 2005: 1). If packaging-free supermarkets would be nearby and budgets and modes of transportation would allow it, we can rather "simply" stop purchasing packaged groceries. However, it does not challenge the conventions or change habits significantly, while one could argue that "changing conventions and expectations have far reaching implications for the resources required to sustain and maintain them" (Shove 2003: 395–396).

Packaging-Free Shopping Without Packaging-Free Shop

Although the problems related to the waste of packaged products are clear, packaging-free shops are not mainstream. What would happen if we decide to embark on a packaging-free lifestyle and to "simply" stop buying packaged groceries? How simple would this be?

To engage in packaging-free shopping, in a world in which packaging is rather abundant, means challenging and changing one's own routines and practices as well as societal norms. This section engages with an exercise in packaging-free shopping in a city or region without a packaging-free supermarket. This exercise was inspired by the Clean Bin Project, in which three participants decide to live without any waste, including packaging, for a year. Through this exercise, the practices and challenges of packaging-free shopping are explored as well as the societal norms involved with shopping and packaging.

First, the setup of the exercise is described. Then reflections of the participants are discussed in relation to the two parts of the exercise: (1) shopping packaging-free which often resulted in avoiding packaged materials and (2) actively challenging conventions of packaging as a breaching experiment. And lastly this section discusses whether packaging-free shopping is regarded to be sustainable.

Setup of Exercise

The task of the exercise was twofold. The first task was to shop or even live packaging-free for 2 weeks, with a main focus on grocery shopping. The second task was to keep a diary of the experience of packaging-free shopping including thoughts, choices, trade-offs, feelings, and challenges and describe these in detail. In this second task, the participant becomes a researcher, both of own routines and practices and of broader societal norms. The following set of questions helped to give direction to the diary:

What food (and nonfood) did you set out to buy?
Where did you go (city, shop)? Is this/are these the places where you normally go? What did you realize with regard to these places looking through the lens of packaging-free shopping? Did you have a choice between packaged and unpackaged? Were you able to buy the product(s) that you wanted unpackaged?
Did you buy different products than you set out to buy because they were unpackaged, and which ones? How did you have to adjust your plans/dinner/daily routines to use the new product(s)? What would you normally have done and what did you have to change?
If you went to a shop/supplier you don't normally go, how did you decide and inform yourself on where to go?
How did your shopping experience in the new place differ from where you normally go? How did you go about finding your product(s) in the new place? Were they easy to locate? What did you have to do differently to bring them home now that they were not packaged? Were there things you (dis)liked more in buying and using packaging-free products?
Did you decide to buy a packaged product after all? If so, why?

(continued)

Can you describe your thoughts and feelings during your shopping experience today? (e.g., annoyed because I could not find the product, had to switch plans for dinner, was in a hurry; excited and nice to go to a new place; not knowing where to find a product in a new place; happy with a nice response; missing some ingredients while cooking at home)
Why did you decide to buy your groceries at this time of the day? How does that relate to timing in your work and/or personal life? (e.g., after work, just before collecting children from school, a calm time at the supermarket, lunch break, etc.) Is this when you normally shop? Why (not)? Did the time of the day matter for how you felt during your shopping experience?
What were the responses you encountered? (e.g., when asking for packaging-free products)
Note to the reader
You are invited and encouraged to join into this practice exercise to experience packaging-free shopping yourself. This helps you to become aware of your own routines and practices as well as societal norms in your environment. Routines, practices, and societal norms may differ between different cities, regions, and parts of the world as well as other factors such as family situations, common modes of transportation, the location of shops, frequencies of food shopping, etc. specific issues encountered below as shown in the reflections are context-dependent such as biking as a common (sustainable) means of transportation (and shopping) in the Netherlands.
• How would you go about this? • Which changes in your normal practices do you anticipate?

Experiencing Packaging-Free Shopping Without Packaging-Free Shop

The exercise clearly showed how changing one aspect of one's daily life had consequences for other parts of daily life. To shop packaging-free, participants had to change a number of their practices and routines. Which changes in routines and habits were observed? And which factors were important for a(n) (un)successful exercise? The ultimate question for the participants in the end was: Can this lifestyle be sustained?

A first important factor the participants noted was the importance of time and in particular an increase in shopping time. Every practice requires a certain amount of time. The introduction of a new practice may result in a shift of time prioritization in relation to other activities/practices. Packaging-free shopping in a place without packaging-free shop is time-consuming as there is a need to alternate between stores for products that are available unpacked. The prospects of needing more time had been a reason for potential participants not to engage in the exercise "due to lack of time" (Participant 4). One participant reported that she had to go to the city center to purchase cheese unpackaged but that she "does not have time to go there every time [she] wants to buy cheese" (Participant 8).

In addition, participants spent much time searching for unpackaged products and comparing between products. This mostly occurred at the beginning: Once the "shopping was more routinized" (Participant 4), this increase of time was reduced. For example, all participants integrated shopping at the market, as it is a place where loose fruit and vegetables can be purchased. This resulted in participants waiting for the market despite having "need to do the big grocery shopping for the last 2–3 days" and thus changing their shopping timing and schedule to integrate this new practice (Participant 5). They added new locations but also changed how and when they did their shopping.

In order to avoid packaging successfully, all participants of the experiment mentioned the necessity of planning, making sure to always bring cotton bags, reusable plastic bags, as well as containers. A lack of preparation also leads to failures in packaging-free lifestyles, such as forgetting to bring a cup to the library for the coffee machine and then having no other option except to take an un-reusable paper cup (Participant 2). Another participant reported how she had to reorganize when she forgot one's "coffee mug" and had to drink the "coffee in the café" instead of taking it away (Participant 1).

Although participants felt generally "satisfied" when the goal of unpackaged shopping was achieved and their preparation had worked out, this was not always the case. An example is the report of a participant who took a sandwich from home and later had to "[carry] an empty box" which was conceived as "annoying," since under normal circumstances the "packages of the sandwich would have been disposed" (Participant 1).

The habit of cooking and the time needed for cooking were in some cases directly impacted by the exercise. For example, the alternative to purchasing a ready-made sauce pasta sauce in a jar (which would disrupt the packaging-free experiment) is to make one's own sauce. The participants had to purchase the raw ingredients individually (such as tomatoes, garlic, and onion) and thus create less waste through packaging. However, this alternative does require more time and thus affects the shopping and cooking process and schedule of the individual (Participant 3). It also stresses the different skills and (re)learning involved in the practice of making tomato sauce.

A packaging-free lifestyle further resulted in changing their regular eating habits. When one could not buy the product without packaging, some participants substituted the products or sometimes they did not buy it at all. For instance, for the duration of the experiment, one participant did not eat pizza from the freezer as usual on Fridays because it is packaged (Participant 7). Another participant became vegan for 2 weeks, as milk could not be bought unpackaged (Participant 2).

In addition, participants were required to adjust to and compromise on the quantity and price of the products. The participants experienced costly shopping when it comes to handmade hygiene products, food such as "chocolate" (Participant 4), "meat" (Participant 8), and "environmental friendly" products (Participant 7). One participant notes on the higher price of environmental friendly washing-up liquid: "(...) it is quite pricy compared to the one I would choose if I were not making a special effort" (Participant 7). High prices were at times a reason for not buying something (Participant 1).

In cases in which there was no packaging-free option, all participants choose the larger quantity of a product in order to buy ahead, to save time throughout the week and produce less packaging-waste overall (Participant 4; Participant 7). One participant decided to buy a larger packet of "rice to reduce waste," for example, instead of "two smaller packages of rice" for these 2 weeks (Participant 4). Some participants also bought more products than usual as the shopping process is time-consuming

(Participant 2; Participant 7) or because the shop that offered a packaging-free product required them to do so, for instance, when "buying oats" (Participant 3). Moreover, as the market opens only once a week, it was necessary to buy more products at once (Participant 8). However, sometimes the participants involuntarily had to take a smaller amount as the packaging-free product would "not have fit in the container" as in the case of buying cheese at the market (Participant 1).

It became clear that trade-offs had to be made between packaging-free shopping and other aspects of sustainability. There are various examples of this. First, as participants viewed it as essential to buy more or different products in one shopping trip or to go to shops further away, a different mode of transportation would be needed or certainly be more convenient. For example, to avoid plastic, a participant purchased glass bottles of water as an alternative. She highlighted that due to the increased weight, this could prove to be an obstacle "for those, who do not have a car or cannot drive or cycle" (Participant 4). And if a store was "too far located," a lack of time prevented it from being a potential shopping choice (Participant 4). Although normally, a person would be most likely to go shopping in a store that was most close to their home, in order to avoid packaging, the participants integrated several new stores into their shopping trips. This would mean that participants who now walked or cycled to the shops would have to buy or at least drive a car. Yet, driving a car is much less sustainable than walking or cycling. Participants had to consider questions about what was worse: buying packaged products or driving cars. Second, arguably buying larger amounts of products could result on more food going to waste. Third, trade-offs were seen between, e.g., biological products and unpackaged products. One participant wanted to buy kiwis, but as "they were in a paper box and wrapped in plastic," the decision made was to buy bananas instead. However, "bio-bananas" were packaged so the participant "bought bananas of the brand [...]" (Participant 1). The brand name is taken away in the quote, but different banana companies are known to deal differently with issues such as workers' rights, environmental protection, collaborating with, e.g., paramilitary in certain countries, etc. This shows the potential consequences and potential controversy of the aim of packaging-free shopping in order to reduce waste in one's own household. One might end up supporting a company which offers its products packaging-free but operates under nontransparent business policies.

Breaching Experiment

To take the exercise one step further, participants engaged in a (further) breaching experiment. The purpose of a breaching experiment is to challenge socially constructed norms present in everyday life. The theoretical basis of breaching experiments is ethnomethodology, a study of methods which looks at the norms that are taken for granted within society (Boes 2009). It focuses on the norms which are entrenched in people's doings by creating the conditions for an unexpected event which will break these same social norms and highlight peoples' reactions. Engaging in packaging-free shopping as such is – to some extent – a breaching experiment. This second part of the exercise entailed forgoing packaging within a specific situation, one where demanding packaging-free food is considered to act directly against existing social norms. Peoples' lives are organized by social norms and have an interest in maintaining these. However, they are also able to adapt to experiences that do not align with what they normally experience (Garfinkel 1962), which is what breaching experiments explore.

The aim of this experiment was to see how understanding and accommodating our society would be toward someone trying to live a packaging-free lifestyle as well as identifying the social norms we often do not even notice. The participants went to shops and take-away venues where the respective products are routinely packaged. The breaching experiments were conducted within the same time frame as the packaging-free experiments. The participants were asked to write down their experiences, including what was requested and the reaction to the request. This section focuses on the experiences of the participants and particularly the responses they encountered.

Questions to the reader
What reactions would you expect if you would go to a take-away and request for the food and drinks to be put in a brought container?
What feelings would you expect to have when participating in such an experiment?
What happened when you did this?

To the surprise of the participants, who had been a little anxious about this experiment, not knowing what responses they would encounter, many reactions were positive. In a small take-away, the reaction was one of interest in the project and approval as the man working there felt that too much plastic was being used (Participant 3). In a chips shop, the owners approved the request of the participant in using her own container immediately, implying that they are used to this kind of demand (Participant 4). The owner of a Moroccan deli was especially accommodating when asked to store the food in the container the participant brought herself. This container was heavier than the plastic container handed out by the deli and the owner weighed the container beforehand to avoid having the participant overpay for her food due to a heavier container (Participant 2). Also at the market, a participant was told they were doing a "good job" for bringing their own bag to the market. This encouragement occurred again when the man at the market stand even said "good! We don't want to use so many bags" (Participant 3). This demonstrates that the effort to live a packaging-free lifestyle is appreciated by many.

Other responses were "neutral," even if the cashier experienced certain difficulties – such as picking up the fruit separately to weigh them at the till as they were not in a plastic bag – "she did not complain and went on weighting all the things I bought" (Participant 4; Participant 3). There were also requests that led to confusion, such as the request to not wrap a sandwich and simply hand it to the participant directly "was not a problem, although they did seem a little confused" (Participant 1).

However, the request of participants was sometimes also refused or negotiated. In a fast-food chain, after consulting with the manager, the woman at the till explained that it was not possible to bring one's own container into the kitchen as "there are hygiene regulations which do not allow [name chain] workers to bring into the kitchens tools which are not from the kitchen itself" and that it was also not possible to buy a drink with one's own cup (Participant 4). A compromise was reached, with the fries being deposited straight onto the tray instead of in a package, but the drink was not poured into the cup brought by the participant. The woman working in the fast-food chain seemed slightly shocked, while the manager was extremely surprised. Hygienic reasons are mentioned for rejecting a request to hand out food unpackaged. This also occurred when one participant "asked the shop assistant if [she] could get the cheese in [her] own container" and when the "the shop assistant at the counter in the supermarket refused to hand out meat without packaging" (Participant 8).

Participants sometimes experienced an "uncomfortable feeling." For example, while buying fish, one participant tried to explain to the seller that packaging was not needed because the participant brought the container. The seller agreed, but in the end the fish was packaged in the container. This participant reported that "this wasn't what I had intended to happen, but I didn't want to confuse her more, so I said thank you and left when she handed the box back to me" (Participant 1). Sometimes the expectation of an uncomfortable feeling led to an early end to the experiment. For example, one participant planned to reuse a pizza box (Participant 6) but in the end decided not to as she "did not want to hear their [the people working at the pizzeria] jokes" (Participant 6). Discomfort was also experienced when "other people had to wait patiently" as the cashier had to go to the backroom in order to fill a jar with oats, as this was not a standard request and "thus the employees were not as fast and efficient as they would be with other requests" (Participant 3).

What does this experiment teach us? First, that a concern about packaging seems quite widely shared – at least in the place of the experiment – and many are happy to see and encourage people's initiatives. Second, that even if people are willing to accommodate requests for packaging-free products, it may be difficult to do so. People have habits of packaging as the example of the fish seller showed, and habits are sometimes difficult to change as the example in the introduction of this chapter also illustrates. Changing habits may also raise uncomfortable feelings. And there are regulations and norms of, e.g., hygiene which oppose the ideal of packaging-free shopping. Third, that asking for something outside of what is regarded as "normal" is difficult and can raise feelings of anxiety and discomfort. In a society in which packaging is the norm, this social norm and the human desire to belong may make it difficult to challenge and change this. Requesting non-packaged products can be considered as going against other societal norms such as efficiency and not wanting to be a burden. At the same time, there are also rather widely shared environmental concerns and norms of politeness, such as treating customers respectfully even if each apple has to be weighed individually, which may facilitate packaging-free shopping. And fourth, there seemed to be difference in responses between smaller privately owned places and large chains. The latter may have stricter rules as they employ a large number of people and may have a larger need of standardizing their procedures. Initiatives that depart from the standardized expectations may be difficult to accommodate. Perhaps the smaller places tend to invest more in personal relationships with their customers as that may be a reason for why customers return, whereas chains have other ways of attracting customers – their customer know what they will get even if they come from different countries.

Sustaining Packaging-Free Shopping? A Practice to Be Sustained?

During the experiment, there were times when the participants felt "frustrated" when no alternative option seemed available or they experienced choices regarded as "contradictory" (Participant 3), such as the choice between a packaged biological product and the same product unpackaged. They also felt uncomfortable in certain situations. In addition, shopping packaging-free had proven to be time-consuming, particularly at the beginning of the experiment. Each shopping trip needed to be planned in advance, as bags or containers are necessary to purchase and carry the products. Furthermore, to avoid the packaged products from standard supermarkets, the participants had to split their purchases across several stores. The additional time that packaging-free shopping implied was perceived as a problem that, in some instances, could prevent them from looking for packaging-free options. Prices were also seen as an issue. In some cases, unpackaged products were more expensive than the packaged option "leading us to either give up on our plan to shop packaging-free or to make a financial sacrifice" (Manovella et al. 2016: 137).

However, they report: "Despite the difficulties we encountered, the experiment also revealed pleasant surprises" (Ibid.). Through the experiment, they explored places where they could buy products unpackaged. The market was, according to the majority of the participants, the most pleasant discovery. Fruits and vegetables were cheaper than in standard supermarkets and packaging was, in most cases, easily avoidable. When the goal of unpackaged shopping was achieved, the participants described that they felt "satisfied" (Participant 3): "the satisfaction that we draw from having successfully avoided packaging or from the encouraging comments we received from the vendors we approached appeared as a reward" (Manovella et al. 2016: 138). These positive aspects were shared with others in order to encourage them to act similarly, for example, "to use menstruation cups" instead of one-time female hygienic articles (Participant 5). The positive experiences and feelings encouraged creativity among the participants.

Nevertheless, these rewards do not seem to be sufficient to encourage participants to live a life of zero waste. All the participants were aware of the impact their choices have on waste production. After the experiment, the intention to continue avoiding certain packages was displayed. Some liked the adaptations they made and may maintain them, for example when it comes to the consumption of more tap water than mineral water in plastic bottles in order to save packaging-waste (Participant 8). However, a complete zero-waste lifestyle was described as "utopian" (Participant 8), "tiresome" (Participant 6), and "stressful" (Participant 4). One participant concluded that "less packaging means less "comfortableness" (Participant 2). The limited choice "can be frustrating in the long run as it gets monotonous" (Participant 3). Packaging-free shopping is not "worth the extra effort, finances, and time " and it is ultimately an "unrealistic way to live" (Participant 7). The positive aspects were not sufficient for these participants to commit to long-term packaging-free shopping although they may make different choices regarding some products and customs.

Even for these participants, who were keen on living an environmentally friendly life, this lifestyle was difficult to uphold. As in the case of the packaging-free supermarket, the exercise and experiment clearly show how the participants tried to maintain their normal lives and practices: finding substitutes for the things they normally consume and spending no more time and effort on shopping than they normally do. Spending more time on shopping and cooking would interfere with other practices which would have to be changed in turn as well. They would have less time for study or meeting friends or would perhaps need to change practices. An example of this may be to combine activities and practices that were first separated, such as making cooking time a time to spend with friends instead of first cooking and then spending time with friends. Yet, such changes in practices were not regarded as sustainable. Packaging-free shopping was considered as doable as long as it did not interfere (too much) with everyday routines and practices.

Wrapping Up: Reflections and Conclusions

This chapter started by outlining the commonly acknowledged problem of waste and in particular plastic. Much of this waste comes from food packaging. Yet, despite the wide recognition for the problem, so far the amount of plastic only grows. Measures are being taken such as investments in recycling by companies and the EU Plastic Bags Directive. Also other governance approaches focus on moving our society toward a more sustainable society, acknowledging the complexity that this brings. When it comes to citizen level, researchers often regard green consumption as "an expression of individual environmental commitment" (Shove 2003: 395). However, this overlooks a more sociological perspective on everyday life and sustainability as Elizabeth Shove outlines.

This chapter has shown that environmental awareness has made it possible for packaging-free supermarkets to exist. Environmental awareness plays a role in how customers prioritize between, for example, the price of products and their environmental friendliness. It has also demonstrated that most people are supportive of the idea to reduce waste and packaging. Yet, in line with Shove, it has also shown that a more sustainable way of life does not only depend on green beliefs. Changes in sustainability practices do not "simply" follow from increasing environmental awareness or rational choices. It is crucial to also consider the practices of everyday life. Why? It is because our habitual, customary ways of doing something are difficult to change and interconnect with other practices.

The section on the packaging-free supermarket showed how customers to a large extent maintain their everyday routine practices. This chapter further engaged with an exercise of packaging-free shopping, which did not fit neatly with the participants' everyday lives, with their ideas of convenience and comfort. Participants had to make changes to their routines and habits in order to sustain a packaging-free lifestyle. For some, a normal way of cooking included the use of ready-made packaged tomato sauce. This ties in with issues of convenience. It may also tie in with a way of life in which there is not much time for cooking. Changing to a packaging-free lifestyle implied changes in cooking and eating practices as well as diets. It implies shopping in different manners and at different times, which may have implications for other practices and activities such as work or visiting friends. Such changes were often considered as frustrating or disturbing of "normal" life. The chapter showed that it is important to understand "clusters of practice" and how these evolve as well as are held together (Shove 2003: 408).

This chapter has thus demonstrated that it is very difficult to "simply" stop buying packaged groceries and change to a packaging-free lifestyle. Even those who were willing to make changes because of environmental considerations did not see a packaging-free lifestyle as a practice they could sustain. Practices are interconnected with other practices and norms, and these are not bound to individuals. They are often institutionalized as normal activities. Also environmental policies "routinely take existing commitments and 'ways of life' for granted" (Hand et al. 2005: 1). However, as these normal activities are closely tied in to and have consequences for sustainability, Shove argues that the environmental challenge is then not one of "simply" creating environmental awareness but one of "understanding how meanings and practices of comfort, cleanliness, and convenience (or comparable services like the provision of a 'normal' diet or 'normal' forms of mobility) fall into the realm of the taken for granted, and how they change" (Shove 2003: 396). And that we need to understand the consequences of such changes for sustainability.

Understanding how practices become "normal" and analyzing their transformations (Hand et al. 2005) may lead to opportunities to intervene. Intervening would then not focus (only) on the use of packaging, water, or energy as isolated resources but on practices like those of shopping, eating, cooking, heating, and showering. It is within these practices that resources are consumed and given specific meanings. Such practices include associated technologies, materials and infrastructures (such as supermarkets, transportation, shopping bags), conventions, meanings and ideas (such as hygiene and convenience), and temporal orders in everyday life (Hand et al. 2005).

To live a life of engaged sustainability, a first step is investigating the practices we engage in and their relation to sustainability. This is not a task for researchers only. Exercises such as the one discussed in this chapter may be a useful tool to investigate and understand practices and the interconnections between them as well as often invisible social norms that help to shape and maintain such practices. Such exercises are exercises in lived sustainability at the same time and especially worthwhile if they lead to discussions and reflections among participants. This has been the case for the participants in this project, undergraduate students interested in sustainable lifestyles. And to some extent you, the reader, have accompanied us on this road as well.

Such exercises thus help us to gain new understandings of the relations between (interconnected) practices and sustainability and may lead to insights for intervention. While we are in the process of investigating and understanding, such exercises lead to experiences with lived sustainability. This in itself is worthwhile as "pro-environmental behavior change is more likely to endure in the long term if it is rooted in, and driven by, significant and meaningful experience" (Maiteny 2002: 299). Maiteny argues that "it is essential that pro-environmental behavior change initiatives work with experience and not simply continue to assume that information alone stimulates such change" (Ibid).

Changes in practices may be felt as disturbing "normal" life but also offers opportunities as recognized by the participants: pleasant discoveries, learning of new (or old) skills such as in the case of making tomato sauce from scratch, a reconsideration of what one values in life, and shifts in time spent on activities (investing more time in shopping packaging-free may reduce time needed for taking out the garbage, buying garbage bags, etc.). These practices may be sustained by participants. Such small changes in lives of (small groups of) individuals do not (immediately) change larger societal practices, structures, and norms. Some even claim that engaging in small feel-good efforts to live more sustainably leads to the problem that nobody is doing anything that matters on a larger scale (e.g., Slavoj Zizek). This chapter has another take on it. Small changes matter as well and can perhaps, in one way or another, become a diver of change for larger societal practices and norms with positive consequences for sustainability.

Cross-References

▶ Community Engagement in Energy Transition
▶ Environmental Intrapreneurship for Engaged Sustainability
▶ Supermarket and Green Wave
▶ Sustainable Living in the City

References

Barbiére, C. (2015, April 29th). Euractive: EU to halve plastic bag use by 2019. Euractive.com. http://www.euractiv.com/section/sustainable-dev/news/eu-to-halve-plastic-bag-use-by-2019. Accessed January 31, 2018.

Boes, S. (2009). *Experiencing product use in product design*. Paper Presented at the International Conference on Engineering Design, ICED '09, Stanford.

European Commission. (2016). EU countries have to drastically reduce consumption of lightweight plastic carrier bags. http://ec.europa.eu/environment/pdf/25_11_16_news_en.pdf. Accessed January 11, 2018.

European Parliament/Council of the European Union. (2015). Amending directive 94/62/EC as regards reducing the consumption of lightweight plastic carrier bags. http://eur-lex.europa.eu/eli/dir/2015/720/oj. Accessed January 31, 2018.

Garfinkel, H. (1962). A conception of, and experiments with, "Trust" as a condition of stable concerted actions. In O. J. Harvey (Ed.), *Motivation and social interaction: Cognitive determinants* (pp. 187–238). New York: The Ronald Press Company.

Gourmelon, G. (2015). Global plastic production rises, recycling lags. In *New Worldwatch Institute analysis explores trends in global plastic consumption and recycling*. Washington, DC: Worldwatch Institute. http://www.worldwatch.org/global-plastic-production-rises-recycling-lags-0. Accessed January 31, 2018.

Hall, D. (2017). Throwaway culture has spread packaging waste worldwide: Here's what to do about it. *The Guardian*. https://www.theguardian.com/environment/2017/mar/13/waste-plastic-food-packaging-recycling-throwaway-culture-dave-hall. Accessed January 8, 2018.

Hand, M., Shover, E., & Southerton, D. (2005). Explaining showering: A discussion of the material, conventional, and temporal dimensions of practice. *Sociological Research Online, 10*(2), 1–13.

Kollmus, A., & Agyeman, J. (2002). Why do people act environmentally and what are the barriers to pro-environmental behavior. *Environmental Education Research, 8*(3), 239–260.

Laville, S. (2017). Supermarkets must stop using plastic packaging, says former Asda boss. *The Guardian.* https://www.theguardian.com/environment/2017/oct/12/supermarkets-stop-using-plastic-packaging-former-asda-boss-andy-clarke. Accessed January 8, 2018.

Loorbach, D., & Rotmans, J. (2010). The practice of transition management: Examples and lessons from four distinct cases. *Futures, 42*(3), 237–246.

Maiteny, P. T. (2002). Mind in the gap: Summary of research exploring 'inner' influences on pro-sustainability learning and behaviour. *Environmental Education Research, 8*(3), 299–306.

Manovella, E., Schickendantz, D., Coenen, H., Summer, I., Di Paola, L., Freund, N., Joskin, P., (2016). *The vision of packaging-free shopping – From the idea to actuality.* MaRBLe Sustainability and Innovation, Maastricht University, Maastricht.

Minami, C., Pellegrini, D., & Itoh, M. (2010). When the best packaging is no packaging. *ICR, 9,* 58–65. https://doi.org/10.1007/s12146-010-0059-3.

Owen, R., Macnaghten, P., & Stilgoe, J. (2012). Responsible research and innovation: From science in society to science for society, with society. *Science and Public Policy, 39,* 751–760.

Redclift, M. (1996). Wasted: Counting the costs of global consumption. London: Earthscan.

Shove, E. (2003). Converging conventions of comfort, cleanliness and convenience. *Journal of Consumer Policy, 26,* 395–418.

Shove, E., & Walker, G. (2010). Governing transitions in the sustainability of everyday life. *Research Policy, 39*(4), 471–476.

Tomka, B. (2013). *A social history of twentieth-century Europe.* London: Routledge.

Wilk, R. (2002). Consumption, human needs, and global environmental change. *Global Environmental Change, 12*(1), 5–13.

World Commission on Environment and Development. (1987). Our common future. Un-documents.net: http://www.un-documents.net/our-common-future.pdf. Accessed 29 May 2016.

Community Engagement in Energy Transition

A Qualitative Case Study

Babak Zahraie and André M. Everett

Contents

Introduction	756
Theoretical Lens: Evolutionary Theory of Organizational Change	758
The Case	762
Methodology	764
Findings	766
Forming a New Entity	766
Wider Cognitive and Sociopolitical Legitimacy	769
Conclusion	772
Cross-References	773
References	774

Abstract

Energy systems and their transition to more sustainable forms of production and consumption are of interest to researchers from multiple disciplines. Community-based enterprises and grassroots innovations play a crucial role in different aspects of these transitions. They possess considerable social capital and are able to assemble a social and/or environmental vision. Some of them seek market opportunities to take action in order to construct the economic basis that will further their vision in broader societal contexts. The collective nature of these entities may add to the effectiveness of their actions. A better understanding of such entities may help foster sustainability transitions in local communities and exploration of their wider influences on national and global scales. This research

---

B. Zahraie (✉)
Research Management Office, Lincoln University, Lincoln, New Zealand
e-mail: babak.zahraie@lincoln.ac.nz; babak.zahraie@otago.ac.nz

A. M. Everett
Department of Management, University of Otago, Dunedin, New Zealand
e-mail: andre.everett@otago.ac.nz

© Springer International Publishing AG, part of Springer Nature 2018
S. Dhiman, J. Marques (eds.), *Handbook of Engaged Sustainability*,
https://doi.org/10.1007/978-3-319-71312-0_28

extends current literature on community-based entrepreneurship and grassroots innovations by investigating a New Zealand community-based enterprise, which created a network of actions and organizations that used bottom-up innovative ideas to respond to the local energy situation. Although their efforts have been partially unsuccessful to date, much can be learned from their experiences.

Keywords
Community-based entrepreneurship · Grassroots innovation · Evolutionary theory · Energy system transition

Introduction

Entrepreneurship, including social, environmental, and sustainability-driven, is considered a solution for social and environmental degradation (Dean and McMullen 2007; Pacheco et al. 2010; Rastogi and Sharma 2017; Sarkar and Pansera 2017; Zahraie et al. 2016). Entrepreneurs address social and environmental problems in their business environment through their innovative practices and therefore may become a source of variation that initiates wider changes in their business environment (Boro and Sankaran 2017; Seyfang and Longhurst 2016). Among multiple types of entrepreneurship, researchers' attention has been attracted to grassroots innovation (and the accompanying concept of community entrepreneurship), defined as "movements seek[ing] innovation processes that are socially inclusive towards local communities in terms of the knowledge, processes and outcomes involved" (Becker et al. 2017; Smith et al. 2014, p. 114). Yet, their crucial role in processes of wider change in the business environment has not been investigated adequately (Becker et al. 2017; Feola and Butt 2017; Hossain 2016; Pansera and Sarkar 2016; Seyfang and Smith 2007). Recent literature shows that the success of grassroots innovation is an emerging phenomenon that occurs as a result of dynamic interactions among three levels: individual, group, and societal (Grabs et al. 2016). Further investigation of these dynamics may result in a better understanding of these movements and help to address social and environmental degradation that threatens local and global communities (Hargreaves et al. 2013; Hossain 2016; Ornetzeder and Rohracher 2013).

Grassroots innovations are identified as appropriate spaces for experimentation, which is a necessary stage for sustainability innovations to scale up (Antikainen et al. 2017; Feola and Butt 2017; Laakso et al. 2017). They facilitate experimentation with innovations that require a social movement to diffuse and initiate a broader social change (Hossain 2016) and create a learning environment for social, cultural, and ethical values that differ from dominant norms (Monaghan 2009). This learning process creates vision and facilitates the formation of a new niche that may become stable, be adopted by salient actors (Hoppe et al. 2015), and translate to dominant trends in the societal environment at later stages of development (Martin and Upham 2016; Martin et al. 2015). This process is explained through (a) deepening, (b) broadening, and (c) scaling-up stages. Deepening presents higher-order learning among the people involved in experimenting a radical technology, structure, and/or sociocultural norms; broadening presents the imitation process where the experiment diffuses in a broader community by repetition; and finally, scaling up is when the experiment is embedded in the broader societal context and becomes mainstream (Laakso et al. 2017).

Grassroots innovation and community entrepreneurship are driven by social and environmental concerns and can provide simple solutions to address everyday life issues (Kim 2017; Sarkar and Pansera 2017). These solutions lie within the experience and skills of communities and individuals outside formal organizations and institutions (Reinsberger et al. 2015). Community-based entrepreneurship brings new dimensions relative to conventional approaches of entrepreneurship. These dimensions include, but are not restricted to, cooperation among volunteers, informal groups, and social enterprises (Hossain 2016; Martin et al. 2015). Becker et al. (2017) demonstrate that community-based entrepreneurs, in the European energy sector, usually combine renewable energy production with broader social and environmental objectives. They are collectively owned, which defines their decision-making process through engagement and democratic negotiations, requiring intense civic participation for their survival. Such participation creates their embeddedness in their surrounding social, cultural, and political systems and increases the chance of acceptance of their new practices among the wider population. Community-based entrepreneurs craft new combinations taking advantage of such dimensions and utilizing scarce resources in their communities (Sarkar and Pansera 2017). These efforts usually result in bottom-up changes that are created through nonprofit organizational forms (Blake and Garzon 2012; Ross et al. 2012; Seyfang and Smith 2007). Since the solution is provided by personally involved actors, usually not driven by financial objectives, the outcomes may be more sustainable (Pansera and Sarkar 2016).

Researchers from diverse disciplines including agriculture (Blay-Palmer et al. 2016; Rossi 2017), policy (Hargreaves et al. 2013; Smith and Stirling 2016), technology, and innovation (Sarkar and Pansera 2017) have used different theoretical lenses such as conceptual niche management (Monaghan 2009), sociotechnical transitions theory (Boyer 2014), and multi-level perspective (Ornetzeder and Rohracher 2013) to investigate grassroots innovation and community-based entrepreneurship. These studies have examined movements such as community currency (Michel and Hudon 2015; Seyfang and Longhurst 2013), the people's science movement (Kannan 1990), Honey Bee Network movement (Gupta et al. 2003), and more recently energy (Becker et al. 2017; Ornetzeder and Rohracher 2013; Reinsberger et al. 2015) and transportation (Ross et al. 2012). Research in this area has mostly focused on technical and technological aspects of these entrepreneurial actions, whereas social and cultural aspects of sustainability transitions require further attention (Becker et al. 2017; Brown et al. 2017; Ford et al. 2017; Järvensivu 2017). For example, Brown et al. (2017) in an investigation of rural energy projects in the Global South explain that success and broader influence of grassroots innovation is a result of interplay among three different but complementary types of literacy, focusing on energy systems, community projects, and politics. They placed less emphasis on technological and financial aspects of these changes. Research shows that cultural complexities related to changes associated with these entrepreneurial actions are very important (Antikainen et al. 2017). The entrepreneurial actions may have inspiring, unfortunate, and threatening aspects for the cultural dimension of societies (Järvensivu 2017). Hence, aspects such as users' involvement in sustainability initiatives are an important factor for upscaling of these entrepreneurial actions, while ability to form cooperation with regional networks and policy instruments from local and regional governments may influence the outcomes (van den Heiligenberg et al. 2017).

Although research in this area has intensified in recent years (Hossain 2016), very little is known about these entrepreneurs (Hargreaves et al. 2013), whether and how their actions may scale up (Feola and Butt 2017; Laakso et al. 2017) and what motivations and outcomes characterize them (Becker et al. 2017). This chapter aims to shed light on some aspects of this phenomenon by investigating a case study of community-based entrepreneurship in New Zealand. This investigation considers the multidimensionality of sustainability transitions and emphasizes social, cultural, and institutional aspects of these coevolutionary changes and the role grassroots innovation plays in this regard. A better understanding of community entrepreneurs, how they emerge and frame their new entities and how the dynamics between these new entities and their business environment work, is the intention of this case study. The evolutionary theory of organizational change (introduced in the following section) is utilized as the theoretical lens to show how this enterprise interacts with its business environment to pursue its communal goals.

Theoretical Lens: Evolutionary Theory of Organizational Change

Evolutionary theory is a general approach for understanding social alterations. It investigates change at different levels (individuals, corporations, and collectives) through the process of variation, selection, retention, and struggle. It is an overarching framework for several well-known organizational theories including institutional theory, resource-based theory, and organizational learning (Aldrich and Martinez 2001; Aldrich and Ruef 2006). This theory investigates the genesis of organizations and clarifies how organizations emerge through populations and communities (Aldrich and Martinez 2001; Aldrich and Ruef 2006). It explains how variations across organizations may scale up to change current populations and communities of organizations or form new ones. This theoretical lens is particularly appropriate for this case study as the main aim is to develop a better understanding of the formation dynamics of a community enterprise, what it does to remain a viable entity, and wider societal changes it may create.

Since community-based entrepreneurs develop new organizations, they can be categorized as a subgroup of "nascent entrepreneurs" in evolutionary theory (Aldrich and Martinez 2001; Davidsson 2006). A nascent entrepreneur is someone "who initiates serious activities that are intended to culminate in a viable organization" (Aldrich and Ruef 2006, p. 65). They are positioned in a continuum between reproducers and innovators (Aldrich and Kenworthy 1999; Aldrich and Martinez 2001). While reproducer entrepreneurs adopt currently accepted models of organizations, innovative entrepreneurs make alterations to those legitimate models or create entirely new combinations. The latter group can be categorized into competence-enhancing, competence-extending, and competence-destroying. Competence-enhancing and/or competence-extending improves or builds on the current trends and capabilities, while competence-destroying innovation needs to create knowledge and routines around new practices (Aldrich and Martinez 2010) and fundamentally alters the competencies for an organization (Aldrich and Ruef 2006; Kim 2017). These innovative entrepreneurs are one of the main sources of variation across organizations (Aldrich and Fiol 1994; Aldrich and Martinez 2010, 2015; Aldrich and Ruef 2006; Katz and Gartner 1988; Markard et al. 2012).

This case study, by investigating community-based entrepreneurs and grassroots innovations, is focused on innovative entrepreneurs, rather than reproducers. Moreover, since community-based entrepreneurs need to make fundamental departures from current trends and routines to address sustainability issues in their communities, often they have to utilize competence-destroying activities that may act as a spark for the formation of new organizational forms (Aldrich and Kenworthy 1999; Aldrich and Martinez 2010; Johnson et al. 2006; Tracey et al. 2011; Zeiss 2017; Zhang and White 2016). An organizational form is "a set of rules that patterns social interaction between members, facilitates the appropriation of resources, and provides an internally and externally recognized identity for an organization" (Aldrich and Ruef 2006, p. 114). It "represent classes of organizations that audiences understand to be similar in their core features and distinctive from other classes of organizations" (Fiol and Romanelli 2012, p. 597).

Considering the newness of these organizations, to be successful, nascent entrepreneurs need to create definitions of their new organizational forms and delineate their boundaries to differentiate themselves from other dominant trends (Aldrich and Yang 2014; Khaire 2014; Suchman 1995). Establishing a formal identity is one of the characteristics of emerging organizations and involves determining four properties: intentionality, resources, boundary, and exchange (Aldrich and Martinez 2001; Brush et al. 2008; Katz and Gartner 1988). In this regard Aldrich and Fiol (1994) introduce lack of legitimacy as the main obstacle for formation of new identities. "Legitimacy is a generalized perception or assumption that the actions of an entity are desirable, proper, or appropriate within some socially constructed system of norms, values, beliefs, and definitions" (Suchman 1995, p. 574). Two forms of legitimacy are recognized: (1) cognitive, defined as "how taken for granted a new form is," and (2) sociopolitical, defined as "the extent to which a new form conforms to recognized principles or accepted rules and standards" (Aldrich and Fiol 1994, pp. 645–646).

Cognitive legitimacy is about creating and spreading knowledge of new practices. It is about changing the perceptions among people in a sector and what they consider as "taken for granted" (Aldrich and Fiol 1994; Khaire 2014; Markard et al. 2016). Establishing cognitive legitimacy usually occurs during the early stages of development when new practices become accepted as legitimate substitutes to incumbents

(Bergek et al. 2008). The level of cognitive legitimacy around a method can be assessed by the level of public knowledge available on that specific activity. The highest level of cognitive legitimacy would be achieved if an approach or a new practice were to become "taken for granted" (Aldrich and Fiol 1994; Johnson et al. 2006; Suddaby and Greenwood 2005). From the producers' point of view, cognitive legitimation means new entrants may copy those trends, while from the consumers' perspective, cognitive legitimacy means they are knowledgeable about the products and services on offer (Aldrich and Fiol 1994; Khaire 2014). Finding cognitive legitimacy is the most difficult aspect of creating new organizations and organizational populations for innovator entrepreneurs (Aldrich and Martinez 2010).

On the other hand, sociopolitical legitimacy indicates that "key stakeholders, general public, key opinion leaders, or governmental officials accept a venture as appropriate and right, given existing norms and laws" (Aldrich and Fiol 1994, p. 648). New activities may not be able to rely on existing institutions for external legitimacy (Aldrich and Fiol 1994; Gustafsson et al. 2015; Markard et al. 2016), inducing entrepreneurs to either modify those institutions or create new ones better aligned with their objectives. Social context may also create windows of opportunity, eventually resulting in a change in knowledge, rules, and institutions through the process of social construction (Aldrich and Fiol 1994; Gustafsson et al. 2015; Hargreaves et al. 2013). Considering actors as individuals with "bounded rationality," who make decisions under uncertainty, emotional influence, and local information (Breslin 2008; Foster and Potts 2006; Geels 2004), highlights the crucial roles of local cognitive and social norms (Bergek et al. 2008; Powell and Sandholtz 2012) in both the strategic choices made by entrepreneurs and their efforts to gain legitimacy in their social settings. A significant transition in terms of building legitimacy – and a foundation of the decisions that create an entrepreneur – is the shift from recognizing social responsibility to actually acting on those beliefs (Marques 2017, chapter ▶ "Moving Forward with Social Responsibility").

Entrepreneurs learn and develop knowledge about their new practices by doing (Aldrich and Martinez 2010; Aldrich and Yang 2014). They need to develop knowledge of how, what, and who for different processes in their businesses, which eventually form the organization's procedural, declarative, and transactive memory (Aldrich and Yang 2014). This results in internal legitimacy, which can be defined as "the acceptance or normative validation of an organizational strategy through the consensus of its participants, which acts as a tool that reinforces organizational practices and mobilizes organizational members around a common ethical, strategic or ideological vision" (Drori and Honig 2013, p. 347). Nascent entrepreneurs have a crucial role in creating trust about their new practices among other stakeholders in their business environment (Drori and Honig 2013). The process of trust-building occurs through a self-reinforcing loop by creating a sense of self-satisfaction for founders (Gambetta 2000), which helps them to overcome social barriers to their innovative actions (Aldrich and Fiol 1994, p. 663). After this gestation period, entrepreneurs have to persuade and convince other actors in the business environment in order to find cognitive and sociopolitical legitimacy and gain access to more resources (Aldrich and Fiol 1994; Suchman 1995). They may use symbolic tools to affiliate with legitimate established institutions in their business environment so as to legitimize their new practices (Suddaby and Greenwood 2005).

Entrepreneurs, correspondingly, play an important role in shaping their desired populations by their strategic choices. They believe that the collective actions of powerful actors may allow them to take the lead regarding access to resources required to achieve their goal of system change. Collective actions may not be conducted intentionally; the cumulative effects of independent actions by self-aware individuals acting in parallel can be substantial enough to bring about systemic changes (Aldrich and Ruef 2006). Initial collective actions and networking happen, in an informal way, among the network of entrepreneurs and likeminded people and later may formalize in the guise of strategic alliances such as trade associations (Aldrich and Fiol 1994). Forming collective actions strongly influences the process of gaining sociopolitical legitimacy (McKendrick and Carroll 2001). Industry champions who step in as volunteers to form these collective actions may act as catalysts (Aldrich and Fiol 1994; Fiol and Romanelli 2012), becoming involved in institutional entrepreneurship to change the rules and regulations (Bergek et al. 2008; DiMaggio and Powell 1983). Several conditions may hamper the effort to form collective actions for new practices, including (1) divergence in design and knowledge of new practices (which may result in different competitive groups) or (2) conflicts among subgroups, which may cause confusion and uncertainty. These conditions would reduce the chance for champions to form a coalition (Aldrich and Fiol 1994; McKendrick and Carroll 2001).

New methods may expand across other populations and form organizational communities, which are defined as *"a set of coevolving organizational populations joined by ties of commensalism and symbiosis through their orientation to a common technology, normative order, or legal-regulatory regime" [all italic in the source]* (Aldrich and Martinez 2010, p. 408). The feasibility of developing communities depends on their cognitive and sociopolitical legitimacy. Perceived value arising from the core products and services of a community also influences its viability. Government agencies may evaluate the perceived legitimacy and value of new communities in their roles as potential supporters or as overseers. Dependency among different actors and organizations across communities enhances legitimacy and fosters learning processes. Mutual dependency of actors would give these activities a collective spirit, which makes them more influential on standards and regulations than isolated efforts of individual actors. Collective actions of entrepreneurs facilitate the learning process at a community level, enabling sustainability transitions (Koistinen et al. 2017, chapter ▶ "Agent-Based Change in Facilitating Sustainability Transitions"). While individual entrepreneurs may find legitimacy based on their own actions, legitimacy at population and community levels is highly dependent on the collective actions of actors. Hence, entering into a fully competitive relationship may cause problems regarding population and community level legitimacy. Governmental support plays an important role in the formation of new communities in (1) support for research and (2) enforcement of new laws (Aldrich and Ruef 2006).

The literature shows that development of sociopolitical and cognitive legitimacy occurs at different levels. Legitimate practices create positive feedback and foster double-loop learning among actors; if contextual factors nurture the adoption of these legitimate practices, they may scale up to change the dominant norms and form new populations and communities of organizations. The question of how sociopolitical and cognitive legitimacy can be accurately measured constitutes a fascinating question beyond the scope of this case study and is left for future research. The following section shows how the community-based enterprise in this case study was investigated and how the findings clarify some aspects of the aforementioned dynamics.

The Case

The community-based enterprise investigated in this research was founded in 2006 in the Blueskin Bay area of the Waitati region of New Zealand's South Island. Located approximately 20 km north of the city of Dunedin, as shown in Fig. 1, the area has a number of small settlements including Waitati, Doctors Point, Evansdale, Warrington, and Seacliff that together include around 1000 homes (Willis et al. 2012). Politically, Blueskin is part of the Waikouaiti Coast/Chalmers Wards and is within the Dunedin City boundary (Millar et al. 2015).

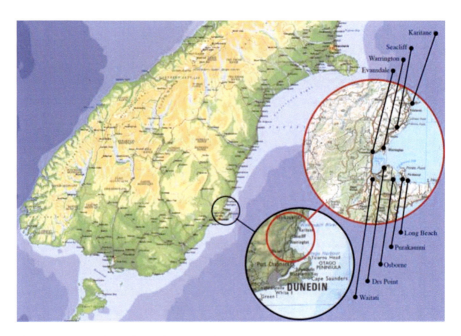

Fig. 1 Geographical location of the case study (BRCT 2017)

The community is very engaged in volunteer activities and has a rich background in social and environmental activism. This was emphasized by all the interviewees in this case study.

The community movement was started by a small number of passionate volunteers and grew over time. Severe flooding of the area in September 2006 restricted access to and from the community, resulting in strong informal support networks being formed. These informal networks continued after that event, forming the initiative for the idea of community resilience. Several community visioning exercises and forums resulted in the Waitati Edible Gardeners group (the WEGgies) and, in 2007, the Blueskin Energy Project (initially named "Waitati Energy Project"). In October 2008, these developments were formalized as the Blueskin Resilient Communities Trust (BRCT 2017; Millar et al. 2015). This official entity governs collective activities and enables the community to sign agreements/contracts and apply for funding. Moreover, it facilitates collaboration between the community and industry, local government, NGOs, universities, businesses, and landowners. In general, the enterprise focuses on enhancing the resilience of the community in response to challenges of climate change and food and energy insecurity. One of their objectives is to improve energy efficiency across different usages in their community and generate their energy locally, aligning with the growing international trend to develop localized low-carbon economies (Gingerich 2017, chapter ▶ "Low-Carbon Economies (LCEs)"). The BRCT's vision, mission, and objectives are shown in Table 1.

By September 2017, the main activities of the enterprise were defined as (BRCT 2017):

Table 1 Vision, mission, and objectives of the community-based enterprise (BRCT 2017)

Vision
We will facilitate a positive, healthy, secure and resilient future for Blueskin Bay and linked communities and promote sustainable resource use
Mission
The Trust will act to strengthen our communities in the immediate, mid and long-term future, with emphasis on energy, food, water and community resilience
Objectives
1. To develop and administer projects that provide education, support and resources to maximise locally based sustainable provision of energy, food, and water
2. To develop and administer projects that provide education, support and resources to minimise energy use, encourage healthy homes and encourage sustainable households
3. To secure and manage funding to achieve the stated goals of the Trust, and to stimulate local sustainable economic activity
4. To develop and maintain relationships to achieve the stated goals of the Trust
5. To ensure community partnership in any enterprises initiated by the Trust and to aim for the most equitable use of resources
6. To foster linkages between organisations with objectives similar to, or complementary to, the Trust's own Vision and Objectives
7. The Trust's goals and activity will always remain charitable

1. Blueskin Turbine: working toward developing a wind farm embedded in the local grid
2. Cosy Energy Advice Line: offering free independent advice regarding energy efficiency and household energy uses such as heating, lighting, and insulation
3. Firewood program: offering bulk sales of firewood for the community at a lower price
4. Affordable insulation: offering discounted insulation for residents willing and able to install it themselves
5. Cosy Home Assessments: offering independent assessments and reports on home performance for Dunedin residents and working with the Cosy Homes Trust to improve the energy efficiency of homes in Dunedin
6. Healthy Rental Certification: providing information to landlords ensuring their rental properties comply with tenancies rules
7. Community Office: a drop-in center for advice that supports community actions and remains a local hub for connecting individuals and diffusing information

The most important business dimension of the trust is intended to become electricity production via the operation of wind turbines. This project would be the first of its kind in New Zealand, if they can achieve the objectives. An exploratory case study research method was used to investigate the situation, as discussed in the following section.

Methodology

This case study is exploratory in nature as it seeks to clarify some aspects of community-based entrepreneurship. Usage of qualitative data, collected by interviewing, added to the depth of findings and enabled the researchers to narrate the story by the voice of the actors involved in the process of entrepreneurship. The objective of the interviews was to ask questions that "are sufficiently general to cover a wide range of experiences and narrow enough to elicit and elaborate the participant's specific experience" (Charmaz 2006, p. 29). The selection of interviewees commenced with purposeful sampling among the individuals who were directly involved in the core of the enterprise and continued with theoretical sampling among the actors who were identified as relevant in the previous interviews (Coyne 1997). The emerging results from the initial interviews indicated which sources to pursue next. The selection was based on the relevance of actors to the emerging themes and whether they could add details or new information. While interviews with sustainability-driven entrepreneurs and other actors in their business environment were the main source of data, other published information such as related academic literature; websites of the organizations, NGOs, or related institutes; reports; and media reports were used as secondary sources of data.

This information was used to find a deeper insight about situations under study, connect information from other sources, triangulate the previous findings, and gain detailed information about various dimensions of emerging themes. Using

Table 2 Information about the interviewees in this case study

Pseudonyms	Group	Interview time (Min)
M-I-1 Leanne	First Group	45
M-I-2 Michelle		69
M-I-3 Michael		69
M-I-4 Bruce		60
M-I-5 Sue		74
M-I-6 Kevin		70
R-I-1 Liz	Second Group	45
R-I-2 George		40
R-I-3 Christopher		50

different sources of data is aligned with theoretical sampling employed in this research (Charmaz 2006), and similar logic was used to find appropriate sources of information. This process continued until reasonable details of the emerging themes and categories were obtained (O'Reilly and Parker 2012). This was evaluated by asking questions such as when, how, and why to clarify aspects such as who was involved and what were the results/consequences. In total nine individuals were interviewed. The details of these participants are shown in Table 2 where they are categorized in two groups: main and secondary actors. In order to maintain the anonymity of the interviewees, detailed information of the participants is not provided.

Grounded theory was used to analyze the collected data. An important aspect of grounded theory coding is the bottom-up discovery of categories, themes, concepts, properties, and dimensions of the phenomenon under study that emerge from the interview data (Charmaz 2006; Corbin and Strauss 2008). This process started by initial coding. Initial coding can be conducted through word-by-word, line-by-line, or incident-to-incident analysis. With all of these methods, data will be compared with data, and codes with data and other codes, to find similarities and differences (Charmaz 2006; Goulding 2002). This research employed incident-by-incident coding, at the initial stage. This approach was appropriate for this study as the main purpose of the research is to generate an in-depth understanding of entrepreneurial actions. As such, using incident-by-incident coding retained the integrity of information about particular actions taken by actors while allowing the researcher to classify them into categories.

During the initial coding, the main goal was to stay open to emergent ideas and directions led by inductive reasoning. This was followed by focus coding and finding connections between emerging patterns, which resulted in themes that were categories of interconnected codes (Charmaz 2006; Goulding 2002). Charmaz (2006, p. 59) defines focus coding as "using the most significant and/or frequent earlier codes to sift through large amounts of data." Comparing emerging themes with new data, and themes with other themes, refined the findings and evaluated the construct of the research (Eisenhardt 1989). Finally theoretical coding was used to gain new theoretical insights. At this stage, evolutionary theory of organizational change is used to explain the findings and connect the emerging themes to relevant literature. The following sections presents the findings in this case study.

Findings

Two distinct themes emerged from the collected data. These themes are (1) forming a new entity and (2) wider cognitive and sociopolitical legitimacy.

Forming a New Entity

The participants in the case study presented their recollection of the historical events that resulted in the formation of the new community enterprise. Almost all of the participants emphasized the flooding in 2006 as a turning point in this regard. This event raised concern among individuals in the community regarding their readiness to deal with similar situations in future. This was enhanced by a rising awareness within the community regarding issues related to climate change and its consequences such as sea level rise and severe weather conditions. They discussed their isolated geographical location and how they did not receive an adequate response from governmental bodies at the time of the event. This resulted in a shared vision within the community, highlighting a gap in their surrounding environmental and societal system. Two of the participants explained this as follows:

> In 2006 there was a large flood that went through the community, and there was an understanding that these sorts of flood events will happen much more frequently in a world where climate change is accelerating. And so even though there's a risk of sea-level rise, the much more immediate threat is from the one in 50-year flood event becoming a one in 20-year flood event, becoming a one in five-year flood event, and the effect it would have on the community. And so that was a kind of a touchstone, crystallizing point for all of these other conversations to come together, and as a result of that the Blueskin Communities Trust was formed, and that has then been able to take those conversations and formalize them and run that agenda forward in a much more structured way, and probably as a result, there's been more traction with that than there has been in other places. [M-I-3 Michael]
>
> Blueskin Resilient Communities Trust is a community trust that was set up. Back in 2008, it became a charitable trust. The idea came after the Waitati floods occurred and the residents of Waitati pulled together in order to work together and try stop their homes from being completely devastated by floodwater. They had spent the early hours of the morning and into the lunchtime period working hard to try and rescue animals that were trapped behind fences, that were getting – were drowning. Working hard to stop water coming into homes, by around late afternoon or afternoon I think it was – maybe early afternoon – [The governmental group] turned up with sand bags to try and help, but by then the floodwaters were starting to go down. And what the community realized was they're separate to Dunedin. There's a big enough distance that means, when things happen they're going to have to survive on their own, ... So they came together and looked at different things, from energy, food, transport, they were some of the key issues that the group wanted to look and try and create resilience around for the community, and the Blueskin Resilient Communities Trust was formed out of that. [M-I-2 Michelle]

Finding a gap alongside the informal networks that had formed during the rescue process in the flooding event formed the initial core of the enterprise. Most of the participants then discussed the influence of passionate individuals in this process. They identified these passionate individuals as the main drivers for the formation of the formal entity, noting they allocated time and networked with other members in their community to bring them together in order to gain access to resources in their communities. This aligns with research showing that community cohesion for environmental stewardship can be achieved through passionate commitment to an agreed purpose (Sachdeva 2017, chapter ▶ "Environmental Stewardship"). The participants reported that:

> the BRCT was formed with [passionate individuals], I think perhaps as one of its drivers, but [those individuals] first amongst equals of driving this trust with the aim of creating a resilient community, a community in transition from reliance on fossil fuel and getting all its supplies from outside of the area, they were looking to design a community that was dependent much more on itself. [R-I-1 Liz]
>
> It's the right organization and the right personnel. Just being community-based hasn't done it, because you've got [a community charitable trust] at the table which is a big community-based organization, with them that we haven't been able to get that happening. [R-I-2 George]

The results show that continuous effort by these passionate individuals accompanied by support from the community resulted in the formal organization of a trust in 2008. This milestone adds to the formality of interactions between the community group and other institutions and enables them to have a more legitimate voice in their institutional environment. They could secure funding from some third parties and gain access to scarce resources in their business environment. Access to these resources enabled them to employ some of those passionate individuals, who were involved in the process since the beginning, to pursue their communal goals. This changed the nature of their organizational interactions from merely voluntary to more formal:

> Typically some of the groups that are associated with us could be regarded as very fringe and in the nicest possible way, lunatics, but they all have their place in adding to the color of a community, and the reality is that on their own they can't really do much, so by associating and being part of that BRCT umbrella, they get some grunt going in dealing with other organizations. [M-I-1 Leanne]
>
> "You can, let's do it in our spare time" [achieve community goals]. It's not that, we don't have spare time. Why not pay someone to do this work, because it's important, you know, it's make a community resilient, it's not going to happen overnight. It needs research, it needs understanding, it needs community engagement, it needs somebody to do the grunt work, and that's what I think the Community Trust is for. [M-I-2 Michelle]

Since the new enterprise differed structurally from other organizations in its institutional environment, it had to define a legitimate model to find access to resources and distinguish itself. The findings in this case study highlight different methods of resource mobilization compared to conventional models:

> As an entity, their principal problem, I think there's two problems that they face, well there's two. One is funding, so how to sustain their activities and the trust has historically been, well initially it was basically voluntary. So there wasn't a lot of costs associated with running the trust. As the trust has moved to employing people ... obviously that brings a substantial labour cost in the Trust. .... So finding the funding to sustain, I guess the salaries bill for the employees of the Trust and the overheads, in terms of accommodation and operating expenses is probably, at top level is one of the major challenges that the Trust faces and it's been very reliant to date on grants from Lotteries Grants Board, Hikurangi Foundation, various other supportive organizations. There's been a focus on the trustees to try and reduce our reliance on that to the extent that we can and the energy project is part of that strategy if you like in terms of creating some sustainable revenue source that will support the other work of the Trust. So that is one of the issues. The other issue which is more a philosophical issue I suppose, is defining the Trust's mandate within the community because the Trustees such as myself we are not elected by anybody, we're just appointed by the Trust, within the Trust so to speak, so the other Trustees effectively appoint their fellow Trustees. [M-I-4 Bruce]

As indicated in the preceding quote, in addition to funding, the participants highlighted community engagement as the main source for finding access to resources and gaining internal legitimacy. Since the initial intention of the entity was to pursue communal needs, all the decisions and strategies had to be consulted with various community groups. While this process added to the complexity facing this enterprise, it also legitimized the actions of the enterprise in a broader societal context as their requests were considered to be community demands. Hence, finding new organizational procedures to facilitate these collaborations were among the proprieties for the actors involved. Two of the participants described this:

> The first one is you have got to get the community engagement and we've adopted what is apparently more of a European model than a New Zealand model in that we frontend-loaded the community involvement right at the very start. So by that, what I mean is we've had newsletters, public meetings saying, this is what we want to do and we've invited public submissions on everything from, you know around the project, so that's to do with the ownership. So we've put up the various models of ownership that there could be. We've put up the sites. We've done, just trying to engage the community to say "what are your concerns with that?" and then constantly collected and refined those. [M-I-1 Leanne]
>
> In the commercial world that selection criteria would be done by the company, the developer. In the world of community winds, thats been a project that has been done really more as a community conversation. So there"s been a much deeper level of consultation, engagement with the community from a much earlier stage. So, you know, that participation has gone to comparing the different sites and weighing up the different merits and saying, "All right, this is the one we want to use."... It is making the process a little more complicated in one sense but the rationale is ... by the time you get to lodging the consent, the vast majority of people are comfortable with the proposition. [M-I-3 Michael]

The results in this section demonstrated how a shared experience among the community members resulted in a united vision and demonstrated a need for further actions. It showed how the shared vision resulted in formation of a trust to pursue communal values in a more formal way. The influence of passionate individuals as a driver for this process was emphasized, highlighting how these dedicated people allocated their resources to this new organization to move forward and make things happen. The results show that community engagement is an essential element in gaining legitimacy for this organization. Correspondingly, this example highlights that finding organizational procedures to facilitate collaboration among community actors that enhance inclusiveness is of great importance for securing internal legitimacy. The next section explains how the new entity mobilized community resources to find wider influence in the surrounding institutional environment.

Exercise in Practice One: Community Engagement
As one of three coleaders of the Residents Association in your near-city-center apartment complex, you have been approached by a rather shy long-term resident with an innovative solution to several of the perpetual issues facing occupants of the relatively old building: "Living Walls." The suggestion is that a combination of composting select food waste and rooftop water collection, with solar panels on the side, would enable the residents to cooperatively maintain a series of living walls that would serve to simultaneously better insulate the building, provide fresher air inside, save energy, and provide supplemental herbs (and some food) while decreasing dependence on potentially vulnerable community and commercial systems outside the building. You love the idea, but potential objections immediately occur to you: costs, odors, volunteer fatigue, leaks, maintenance, visual issues, allergies, and more. Assume that you would like to proceed with this suggestion and potentially serve as its "champion" to the leadership board and subsequently the residents as a collective. How would you go about maximizing the goodwill and cooperation of all of the affected parties, to enhance the chances that this proposal will succeed? Which tools would you utilize, and what approaches or communication techniques would you select?

Wider Cognitive and Sociopolitical Legitimacy

This section explains how different strategies were used by the enterprise to enhance its legitimacy and gain access to the resources required to achieve its goals. The enterprise had to use different methods to maintain its engagement with community groups to collect ideas and gain support. Being engaged with the community enabled obtaining some resources that otherwise were not accessible. The enterprise could define joint objectives with more legitimate organizations such as the University of Otago and the Dunedin City Council through individuals who were engaged in those institutions. These involvements added to the legitimacy of the new organization and provided leverage for their claims in negotiations with other third parties. Nevertheless, the support of the community behind those demands still played a major role backing up the enterprise's objectives:

> The Trust has worked closely with the university on a number of research projects. One of the pieces of work which led to the Cosy Homes Project was some research for energy cultures that was a community led initiative on how people react to advice about energy efficiency, and are there more effective ways of providing that information to people via the use of social networks or whatever.... The other thing that gives confidence is by having

reputed counter parties. So for example, we're in negotiations with the Dunedin City Council to sell them the energy the project produces. ... So when you talk to various other suppliers and say, "Well, we're selling the energy to an A-rated counterparty that's a territorial authority," you know, City Council, that gives them a lot of confidence that those invoices will be paid. And so the, you know, cash in the business will continue to flow. [M-I-3 Michael]

Well we're profiled in the community and we are profiled nationally high and they have built that up over time so they're doing really well at increasing that profile and it is getting to the stage as I say that they can start putting political lobbying with credibility whereas in the past they'd be just a group of greenies out there and that's what it was but that's changed over the last probably two years, three years. All of a sudden it's a political voice, the energy, when the energy plan for the city was being formulated, who are the groups that we want to be involved? [M-I-5 Sue]

Being involved in community engagement activities, they could organize some collective actions to gain social and political legitimacy. The participants discussed the pivotal role that the community enterprise had played in forming a sector group to address issues related to housing with inadequate insulation. This resulted in a sector group that was funded by government to insulate some houses in the area that needed urgent attention. This was reported:

With the Cosy Homes, they [Blueskin Trust] were contracted to run the initial session and they did all that, they did the running around. They did the invitations, they sorted that out, you know we paid with the [another Trust] for some of that stuff. And then since then [One of the people from Blueskin Trust] continued to be contracted to do that running around and has got the right people in the room. So they're doing that connecting stuff on the Cosy Homes. [R-I-2 George]

For instance, I have just joined the group with them and they are on the Chamber of Commerce so [one of the people from Blueskin Trust] is the chairman of the energy group for the Chamber of Commerce so that effectively Waitati is sneaking into the business sector and providing concepts and ideas to go to the council for decision making in regard to the energy plan for the city. [M-I-5 Sue]

Despite the progress of the new entity and finding broader influence in their institutional environment, some of the participants discussed the effect of strong minded individuals, who were involved in the enterprise, in forming and giving direction to the activities. They argued that some of the ideas pursued by these individuals were not supported by the broader community. This created a less united front and vison for the trust and left some members of the community out of the decision-making process. This issue was seen by some of the participants as a key concern that may change perceptions of the community trust, delegitimize its decisions, and split the community into separate smaller groups. One participant reported:

My understanding is that they started off okay, they started off on a series of work programs and with the group together but then quite quickly people who had come in as trustees began to fall away and people who wanted to work with the BRCT and had been hugely inspired began to drop out because there was something not working right, working within there. Quite quickly, quite a number of people who were inspired, motivated, and invigorated just began to leave and withdraw from participating. So quite quickly the community turned away from BRCT. [R-I-1 Liz]

> By and large there's always going to be, you won't please everybody, and we have some who are vehemently against it for various reasons, but, pretty much, I would say we have community support. [M-I-1 Leanne]

This tension was critical in the project to develop a wind farm and create local energy using wind turbines. This project was unique in New Zealand, and the enterprise was pioneering this approach in the New Zealand context. One of the participants reported:

> So the wind is the flagship one, in that we are working towards, well, what we have established is a community-owned wind farm that will provide the big dollop of money needed to support all this sort of stuff. So what we're looking at in the future, that's what I said, a big fund of money that comes off the sale of electricity. [M-I-1 Leanne]

The community was not united toward the final objectives of this project. While a considerable amount of resources and effort were put into this project, different institutional and legal problems question the adequacy of the evidence for stakeholders involved in it. The people passionate about this project used different strategies to solve these issues. One of the main actions was to be as transparent as possible by publishing technical data and informing various community groups on their progress. This strategy was adopted by trustees to leverage their claims against conflicting opinions in their community and the broader institutional system. As reported in a leading regional newspaper:

> The next significant matter in developing a proposed wind farm project for Blueskin Bay was "raised" yesterday. The 30m wind testing tower was raised on Porteous Hill yesterday on a near windless day. Blueskin Resilient Communities Trust manager [Name] said the tower was loaned to the trust by [Company Name] and would allow the trust to carry out more precise testing at its proposed wind turbine site, which would be important for sourcing funding for the project. ... The data collected from the tower would provide more certainty to potential investors and information that would aid the resource consent process, which was the next step to be taken, [The manager] said. The trust plans to erect up to four turbines on Porteous Hill, capable of generating 5.2GWh of energy each year, at a cost of about $5 million. (Porteous 2013)

These decisions were made to legitimize the actions based on evidence. Despite all these efforts, the project was not completely accepted by the whole community. It seems that losing community support for this project delegitimized the trust's actions so that despite initial consultations to choose the most appropriate site, the necessary resource consent was refused by the Dunedin City Council. The decision was made based on the argument that an industrial-scale turbine at the chosen site may have a negative effect on the rural enjoyment of some community landowners in the area. This was reported in the media as:

> A 110m wind turbine proposed for Porteous Hill above Blueskin Bay has been described by local residents as "the wrong project in definitely the wrong place". (Sinclair, 1 July 2017)

The results in this case study show that the new entity uses diverse strategies such as creating bonds with more legitimate institutions and organizing collective actions to find sociopolitical legitimacy. The results highlighted that community engagement and having united vision across the board are the main leverage for the new enterprise supporting their strategies. However, the outcomes highlighted that the idealism of proactive individuals and the complexities associated with community engagement have resulted in an undesirable situation. This outcome raises a question that warrants further discussion: What mechanisms should be adopted in this form of organization to balance the shared objectives of communities with the goals of the passionate individuals, involved in these organizations, who make things happen?

Exercise in Practice Two: Balancing Divergent Objectives
Pioneers have always faced difficulties and typically approach the challenges with relish. However, in practice, idealism collides with realism in an often unpredictable manner. In this case, legitimation intertwined with delegitimation as the community – united in its goals and strategies – fractured regarding one particular significant project once implementation plans were developed and their consequences better understood. The collective unison gave way to opposed individuals and factions, with both sides claiming the moral high ground and common sense. Assume that you have been appointed as an outside, neutral arbitrator to reach agreement among the various groups so that the community can heal and resume progress toward its mutual goals. What advantages and disadvantages would being an outsider, a formal neutral third party, bring to your role? How would you proceed? What time frame would you envision? What outcomes would you seek? How will you measure progress? Given hindsight, what would you advise the participants to do differently next time?

Conclusion

This case study related an example of community-based enterprise and grassroots innovation that aims to address problems related to global warming (such as severe weather conditions and lack of food security). The findings showed that realizing a gap in the surrounding institutional environment and having a shared vision among the community groups toward that gap were the main driver for the formation of the new community enterprise. Community engagement was the main source of internal legitimacy for the focal organization, with passionate individuals playing a crucial role in advancing the communal objectives in this process. The case study demonstrated that having strong roots in the community legitimizes the actions of the enterprise in its surrounding institutional environment and facilitates access to resources required for further actions. Legitimation leverages claims in negotiations with third parties and connects the enterprise with more legitimate entities that could also justify their actions. The study also raised an important concern regarding the communal interest of various community groups and passionate individuals who become a driver for the new enterprise. It showed that this paradox caused serious problems affecting one of the projects of the community enterprise in this case study. The case study encourages further discussion toward finding effective mechanisms to address this paradox.

Key Lesson for Engaged Sustainability Lesson
The case study provides an illustration of a common issue related to sustainability initiatives that typically manifest as wicked problems. Different interpretations, expectations, and worldviews among stakeholders involved in a problematic situation may result in inertia leading to nonsolution of the problems. Finding a common understanding about problems and reaching consensus regarding implementable solutions constitute the most difficult part of the transition.

Reflection Questions:

What roles can social networking play in promoting consensus toward both goals and actions in community sustainability initiatives?

Which roles in grassroots innovation efforts could be better handled by outsiders than by insiders?

Given that passionate individuals often see themselves as the expert on a given idea or project, how can other group members engage an outside expert to reduce potential acceptance problems?

How can community-based sustainability initiatives obtain the backing of sufficient capital to ensure their success? (Consider social, intellectual, and cultural, as well as financial capital.)

What are the advantages and disadvantages of linking a local community enterprise initiative to larger-scale regional or national projects, agencies, or institutions?

How would you measure such non-quantifiable aspects as community support, alignment of interest groups, likelihood of success, and degree of consensus regarding a specific grassroots innovation proposal?

What methods for determining mutually acceptable compromise strategies can you identify? Consider this within the context of an organization with multiple overlapping constituencies that share overall goals but differ in perceptions of the nature and relative importance of various types of costs and benefits. Describe potential thresholds and hurdles and how these are affected by the relative levels of incommensurable priorities.

Should a split community call in an arbitrator? When? Who? Who does the requesting? What goal parameters should be set for the arbitrator?

How does perceived conflict of interest affect the legitimation of a "passionate individual" in the collective's eyes?

Cross-References

▶ Agent-Based Change in Facilitating Sustainability Transitions
▶ Ecopreneurship for Sustainable Development
▶ Empathy Driving Engaged Sustainability in Enterprises

▶ Environmental Stewardship
▶ From Environmental Awareness to Sustainable Practices
▶ Low-Carbon Economies (LCEs)
▶ Moving Forward with Social Responsibility

Acknowledgments We thank Dr Liz Martyn for her suggestions that greatly improved the quality of the manuscript.

References

Aldrich, H. E., & Fiol, C. M. (1994). Fools rush in? The institutional context of industry creation. *Academy of Management Review, 19*(4), 645–670.
Aldrich, H. E., & Kenworthy, A. (1999). The accidental entrepreneur: Campbellian antinomies and organizational foundings. In J. A. C. Baum & B. McKelvey (Eds.), *Variations in organization science: In honor of Donald Campbell* (pp. 19–33). Thousand Oaks: Sage.
Aldrich, H. E., & Martinez, M. A. (2001). Many are called, but few are chosen: An evolutionary perspective for the study of entrepreneurship. *Entrepreneurship Theory and Practice, 25*(4), 41–56.
Aldrich, H. E., & Martinez, M. A. (2010). Entrepreneurship as social construction: A multilevel evolutionary approach. In Z. J. Acs & D. B. Audretsch (Eds.), *Handbook of entrepreneurship research: An interdisciplinary survey and introduction* (Vol. 5, pp. 387–427). New York: Springer.
Aldrich, H. E., & Martinez, M. A. (2015). Why aren't entrepreneurs more creative? Conditions affecting creativity and innovation in entrepreneurial activity. In J. Zhou (Ed.), *The Oxford handbook of creativity, innovation, and entrepreneurship* (pp. 445–456). New York: Oxford University Press.
Aldrich, H. E., & Ruef, M. (2006). *Organizations evolving* (2nd ed.). London: Sage.
Aldrich, H. E., & Yang, T. (2014). How do entrepreneurs know what to do? Learning and organizing in new ventures. *Journal of Evolutionary Economics, 24*(1), 59–82.
Antikainen, R., Alhola, K., & Jääskeläinen, T. (2017). Experiments as a means towards sustainable societies–lessons learnt and future outlooks from a finnish perspective. *Journal of Cleaner Production, 169*, 216–224.
Becker, S., Kunze, C., & Vancea, M. (2017). Community energy and social entrepreneurship: Addressing purpose, organisation and embeddedness of renewable energy projects. *Journal of Cleaner Production, 147*, 25–36.
Bergek, A., Jacobsson, S., & Sandén, B. A. (2008). 'Legitimation' and 'development of positive externalities': Two key processes in the formation phase of technological innovation systems. *Technology Analysis & Strategic Management, 20*(5), 575–592.
Blake, A., & Garzon, M. Q. (2012). Boundary objects to guide sustainable technology-supported participatory development for poverty alleviation in the context of digital divides. *The Electronic Journal of Information Systems in Developing Countries, 51*, 1.
Blay-Palmer, A., Sonnino, R., & Custot, J. (2016). A food politics of the possible? Growing sustainable food systems through networks of knowledge. *Agriculture and Human Values, 33*(1), 27–43.
Boro, R., & Sankaran, K. (2017). Empathy driving engaged sustainability in enterprises: Rooting human actions in systems thinking. In S. Dhiman & J. Marques (Eds.), *Handbook of engaged sustainability*. Cham: Springer.
Boyer, R. (2014). Sociotechnical transitions and urban planning. *Journal of Planning Education and Research, 34*(4), 451–464.

BRCT. (2017). Our settlements. *Blueskin Resilient Communities Trust*. Retrieved 6 Sep 2017, from http://www.brct.org.nz/about-us/our-settlements/

Breslin, D. (2008). A review of the evolutionary approach to the study of entrepreneurship. *International Journal of Management Reviews, 10*(4), 399–423.

Brown, E. D., Cloke, J., & Mohr, A. (2017). Imagining renewable energy: Towards a social energy systems approach to community renewable energy projects in the Global South. *Energy Research & Social Science, 31*, 263–272.

Brush, C. G., Manolova, T. S., & Edelman, L. F. (2008). Properties of emerging organizations: An empirical test. *Journal of Business Venturing, 23*(5), 547–566.

Charmaz, K. (2006). *Constructing grounded theory: A practical guide through qualitative analysis*. London: Sage.

Corbin, J. M., & Strauss, A. (2008). Basics of qualitative research (3rd ed.). Thousand Oaks, CA: Sage.

Coyne, I. T. (1997). Sampling in qualitative research, purposeful and theoretical sampling; Merging or clear boundaries? *Journal of Advanced Nursing, 26*(3), 623–630.

Davidsson, P. (2006). *Nascent entrepreneurship: Empirical studies and developments*. Boston: Now Publishers Inc.

Dean, T. J., & McMullen, J. S. (2007). Toward a theory of sustainable entrepreneurship: reducing environmental degradation through entrepreneurial action. *Journal of Business Venturing, 22*(1), 50–76.

DiMaggio, P., & Powell, W. W. (1983). The iron cage revisited: Collective rationality and institutional isomorphism in organizational fields. *American Sociological Review, 48*(2), 147–160.

Drori, I., & Honig, B. (2013). A process model of internal and external legitimacy. *Organization Studies, 34*(3), 345–376.

Eisenhardt, K. M. (1989). Building theories from case study research. *Academy of Management Review, 14*(4), 532–550.

Feola, G., & Butt, A. (2017). The diffusion of grassroots innovations for sustainability in Italy and Great Britain: An exploratory spatial data analysis. *The Geographical Journal, 183*(1), 16–33.

Fiol, C. M., & Romanelli, E. (2012). Before identity: The emergence of new organizational forms. *Organization Science, 23*(3), 597–611.

Ford, R., Walton, S., Stephenson, J., Rees, D., Scott, M., King, G., ... Wooliscroft, B. (2017). Emerging energy transitions: PV uptake beyond subsidies. *Technological Forecasting and Social Change, 117*, 138–150.

Foster, J., & Potts, J. (2006). Complexity, evolution, and the structure of demand. In M. McKelvey & M. Holmen (Eds.), *Flexibility and stability in the innovating economy* (pp. 99–118). Oxford: Oxford University Press.

Gambetta, D. (2000). Can we trust trust. In D. Gambetta (Ed.), *Trust: making and breaking cooperative relations* (Vol. 13, pp. 213–237). Oxford: Department of Sociology, University of Oxford.

Geels, F. W. (2004). Understanding system innovations: A critical literature review and a conceptual synthesis. In B. Elzen, F. W. Geels, & K. Green (Eds.), *System innovation and the transition to sustainability: Theory, evidence and policy* (pp. 19–47). Cheltenham: Edward Elgar.

Gingerich, E. (2017). Low-carbon economies (LCEs): International applications and future trends. In S. Dhiman & J. Marques (Eds.), *Handbook of engaged sustainability*. Cham: Springer.

Goulding, C. (2002). *Grounded theory: A practical guide for management, business and market researchers*. London: Sage.

Grabs, J., Langen, N., Maschkowski, G., & Schäpke, N. (2016). Understanding role models for change: A multilevel analysis of success factors of grassroots initiatives for sustainable consumption. *Journal of Cleaner Production, 134*, 98–111.

Gupta, A. K., Sinha, R., Koradia, D., Patel, R., Parmar, M., Rohit, P., ... Vivekanandan, P. (2003). Mobilizing grassroots' technological innovations and traditional knowledge, values and institutions: Articulating social and ethical capital. *Futures, 35*(9), 975–987.

Gustafsson, R., Jääskeläinen, M., Maula, M., & Uotila, J. (2015). Emergence of industries: A review and future directions. *International Journal of Management Reviews, 18*(1), 28–50.

Hargreaves, T., Hielscher, S., Seyfang, G., & Smith, A. (2013). Grassroots innovations in community energy: The role of intermediaries in niche development. *Global Environmental Change, 23*(5), 868–880.

Hoppe, T., Graf, A., Warbroek, B., Lammers, I., & Lepping, I. (2015). Local governments supporting local energy initiatives: Lessons from the best practices of Saerbeck (Germany) and Lochem (The Netherlands). *Sustainability, 7*(2), 1900–1931.

Hossain, M. (2016). Grassroots innovation: A systematic review of two decades of research. *Journal of Cleaner Production, 137*, 973–981.

Järvensivu, P. (2017). A post-fossil fuel transition experiment: Exploring cultural dimensions from a practice-theoretical perspective. *Journal of Cleaner Production, 169*, 143–151.

Johnson, C., Dowd, T. J., & Ridgeway, C. L. (2006). Legitimacy as a social process. *Annual Review of Sociology, 32*(August), 53–78.

Kannan, K. P. (1990). Secularism and people's science movement in India. *Economic and Political Weekly, 25*(6), 311–313.

Katz, J., & Gartner, W. B. (1988). Properties of emerging organizations. *Academy of Management Review, 13*(3), 429–441.

Khaire, M. (2014). Fashioning an industry: Socio-cognitive processes in the construction of worth of a new industry. *Organization Studies, 35*(1), 41–74.

Kim, Y. (2017). Mushroom packages: An ecovative approach in packaging industry. In S. Dhiman & J. Marques (Eds.), *Handbook of engaged sustainability*. Cham: Springer.

Koistinen, K., Teerikangas, S., Mikkilä, M., & Linnanen, L. (2017). Agent-based change in facilitating sustainability transitions: A literature review and a call for action. In S. Dhiman & J. Marques (Eds.), *Handbook of engaged sustainability*. Cham: Springer.

Laakso, S., Berg, A., & Annala, M. (2017). Dynamics of experimental governance: A meta-study of functions and uses of climate governance experiments. *Journal of Cleaner Production, 169*, 8–16.

Markard, J., Raven, R., & Truffer, B. (2012). Sustainability transitions: An emerging field of research and its prospects. *Research Policy, 41*(6), 955–967.

Markard, J., Wirth, S., & Truffer, B. (2016). Institutional dynamics and technology legitimacy – A framework and a case study on biogas technology. *Research Policy, 45*(1), 330–344.

Marques, J. (2017). Moving forward with social responsibility: Shifting gears from why to how. In S. Dhiman & J. Marques (Eds.), *Handbook of engaged sustainability*. Cham: Springer.

Martin, C. J., & Upham, P. (2016). Grassroots social innovation and the mobilisation of values in collaborative consumption: A conceptual model. *Journal of Cleaner Production, 134*, 204–213.

Martin, C. J., Upham, P., & Budd, L. (2015). Commercial orientation in grassroots social innovation: Insights from the sharing economy. *Ecological Economics, 118*, 240–251.

McKendrick, D. G., & Carroll, G. R. (2001). On the genesis of organizational forms: Evidence from the market for disk arrays. *Organization Science, 12*(6), 661–682.

Michel, A., & Hudon, M. (2015). Community currencies and sustainable development: A systematic review. *Ecological Economics, 116*, 160–171.

Millar, R., Bould, N., Willis, S., Lawton, E., & Singh, A. (2015). *The Blueskin and Karitane Food System Report*. New Zealand: Dunedin.

Monaghan, A. (2009). Conceptual niche management of grassroots innovation for sustainability: The case of body disposal practices in the UK. *Technological Forecasting and Social Change, 76*(8), 1026–1043.

O'Reilly, M., & Parker, N. (2012). 'Unsatisfactory saturation': A critical exploration of the notion of saturated sample sizes in qualitative research. *Qualitative Research, 13*(2), 190–197.

Ornetzeder, M., & Rohracher, H. (2013). Of solar collectors, wind power, and car sharing: Comparing and understanding successful cases of grassroots innovations. *Global Environmental Change, 23*(5), 856–867.

Pacheco, D. F., Dean, T. J., & Payne, D. S. (2010). Escaping the green prison: Entrepreneurship and the creation of opportunities for sustainable development. *Journal of Business Venturing, 25*(5), 464–480.

Pansera, M., & Sarkar, S. (2016). Crafting sustainable development solutions: Frugal innovations of grassroots entrepreneurs. *Sustainability, 8*(1), 51.

Porteous, D. (2013, March 1). Wind farm project tower goes up, *Otago Daily Times*. Retrieved from https://www.odt.co.nz/news/dunedin/wind-farm-project-tower-goes

Powell, W. W., & Sandholtz, K. W. (2012). Amphibious entrepreneurs and the emergence of organizational forms. *Strategic Entrepreneurship Journal, 6*(2), 94–115.

Rastogi, P., & Sharma, R. (2017). Ecopreneurship for sustainable development: The bricolage solution. In S. Dhiman & J. Marques (Eds.), *Handbook of engaged sustainability*. Cham: Springer.

Reinsberger, K., Brudermann, T., Hatzl, S., Fleiß, E., & Posch, A. (2015). Photovoltaic diffusion from the bottom-up: Analytical investigation of critical factors. *Applied Energy, 159*, 178–187.

Ross, T., Mitchell, V. A., & May, A. J. (2012). Bottom-up grassroots innovation in transport: Motivations, barriers and enablers. *Transportation Planning and Technology, 35*(4), 469–489.

Rossi, A. (2017). Beyond food provisioning: The transformative potential of grassroots innovation around food. *Agriculture, 7*(1), 6.

Sachdeva, S. (2017). Environmental stewardship: Achieving community cohesion through purpose and passion. In S. Dhiman & J. Marques (Eds.), *Handbook of engaged sustainability*. Cham: Springer.

Sarkar, S., & Pansera, M. (2017). Sustainability-driven innovation at the bottom: Insights from grassroots ecopreneurs. *Technological Forecasting and Social Change, 114*, 327–338.

Seyfang, G., & Longhurst, N. (2013). Growing green money? Mapping community currencies for sustainable development. *Ecological Economics, 86*, 65–77.

Seyfang, G., & Longhurst, N. (2016). What influences the diffusion of grassroots innovations for sustainability? Investigating community currency niches. *Technology Analysis & Strategic Management, 28*(1), 1–23.

Seyfang, G., & Smith, A. (2007). Grassroots innovations for sustainable development: Towards a new research and policy agenda. *Environmental Politics, 16*(4), 584–603.

Sinclair, K. (2017, July 1). Residents voice angst over wind turbine site, *Otago Daily Times*. Retrieved from https://www.odt.co.nz/news/dunedin/residents-voice-angst-over-wind-turbine-site

Smith, A., & Stirling, A. (2016). *Grassroots innovation and innovation democracy*. Brighton: University of Sussex, STEPS Centre.

Smith, A., Fressoli, M., & Thomas, H. (2014). Grassroots innovation movements: Challenges and contributions. *Journal of Cleaner Production, 63*, 114–124.

Suchman, M. C. (1995). Managing legitimacy: Strategic and institutional approaches. *Academy of Management Review, 20*(3), 571–610.

Suddaby, R., & Greenwood, R. (2005). Rhetorical strategies of legitimacy. *Administrative Science Quarterly, 50*(1), 35–67.

Tracey, P., Phillips, N., & Jarvis, O. (2011). Bridging institutional entrepreneurship and the creation of new organizational forms: A multilevel model. *Organization Science, 22*(1), 60–80.

van den Heiligenberg, H. A., Heimeriks, G. J., Hekkert, M. P., & van Oort, F. G. (2017). A habitat for sustainability experiments: Success factors for innovations in their local and regional contexts. *Journal of Cleaner Production, 169*, 204–215.

Willis, S., Stephenson, J., & Day, R. (2012). *Blueskin people power A toolkit for community engagement*. Waitati/Otago: Blueskin Resilient Communities Trust.

Zahraie, B., Everett, A. M., Walton, S., & Kirkwood, J. (2016). Environmental entrepreneurs facilitating change toward sustainability: A case study of the wine industry in New Zealand. *Small Enterprise Research, 23*(1), 1–19.

Zeiss, R. (2017). The importance of routines for sustainable practices: A case of packaging free shopping. In S. Dhiman & J. Marques (Eds.), *Handbook of engaged sustainability*. Cham: Springer.

Zhang, W., & White, S. (2016). Overcoming the liability of newness: Entrepreneurial action and the emergence of China's private solar photovoltaic firms. *Research Policy, 45*(3), 604–617.

Relational Teams Turning the Cost of Waste into Sustainable Benefits

Branka V. Olson, Edward R. Straub, William Paolillo, and Paul A. Becks

Contents

Introduction	780
Relational Team Approach	783
Relational Contracting	784
Relational Climate	787
Relational Context	792
Engaged Sustainability Through Flexible Cohesion	796
Conclusion	798
Reflection Questions	799
Akron Children's Hospital: A Case Study	799
Introduction	799
Background	799
Discussion	802
Conclusion	804
Cross-References	805
References	805

B. V. Olson (✉)
School of Architecture, Woodbury University, Burbank, CA, USA

Sindik Olson Associates, Los Angeles, CA, USA
e-mail: branka.olson@sindikolson.com

E. R. Straub
Velouria Systems LLC, Brighton, MI, USA
e-mail: edward.straub@velouria-systems.com

W. Paolillo · P. A. Becks
Welty Building Company, Akron, OH, USA
e-mail: bpaolillo@thinkwelty.com; pbecks@thinkwelty.com

© Springer International Publishing AG, part of Springer Nature 2018
S. Dhiman, J. Marques (eds.), *Handbook of Engaged Sustainability*,
https://doi.org/10.1007/978-3-319-71312-0_8

Abstract
Conventional building processes, which still dominate the industry, take too long, cost too much, and often disregard their social and environmental impacts. Over the past 50 years, while most industries have doubled or tripled productivity, the commercial building sector has experienced negative productivity growth. In order to mitigate these practice outcomes, the industry must transform from a transactional to a relational team structure that is people-centered. An integrated project model that is grounded in relational contracting, climate, and context can significantly improve building project outcomes that benefit the three Ps of engaged sustainability – people, planet, and profit. The relational project environment is built on a foundation of shared vision, values, and the basis for a common vernacular that guides human activity and generates human bonds. This chapter defines the three relational states and translates them into actionable steps with which to positively affect a building project team structure. The integrated project delivery framework lays the groundwork for a building project success model. However, the key to its realization is the level of flexible cohesion exhibited by the project team members. The collective levels of flexible cohesion determine the team's ability to create and maintain a relational project environment without which project success would be greatly diminished. The individuals' flexible cohesion potential is driven by their attitudinal and behavioral characteristics. In other words, processes alone do not deliver projects; it requires a team of engaged people.

Keywords
Project teams · Relational contracting · Relational climate · Relational context · Flexible cohesion

Introduction

Traditional management theories, many of which are still taught in business schools around the world, were developed in the late nineteenth and early twentieth centuries during the second industrial revolution. Organizations needed consistent, repeatable management methods during this explosion in mass production. In many industries, these concepts have endured with few changes. Although these methods were originally designed for industrial manufacturing, they have permeated many fields using scientific management as a basis to systematize entire segments of the economy. This approach required minimizing the impact of the human variable from the process. However, this dehumanizing effect was detrimental on performance levels as the nature of work began to change. The accelerated rate of change and the increased volume of information, that had to be processed added complexity to everyday decision-making. Work required information and knowledge workers supported by increasingly faster computing power and the Internet. Known as the third or digital revolution, it replaced mechanical tools with smart devices enabling both mobility and autonomy of workers. As a result, management researchers included variables such as worker engagement, motivation, and innovation into their study models (Crawford et al. 2010; Janssen 2005; Ramlall 2004). Academic and practitioner literature began to indicate that industries that fail to adapt to modern management and project delivery practices are at a competitive disadvantage. This led to the realization that industries that are able to capitalize on the flexibility, creativity, and passion of their workers are better positioned to succeed. In the advent of the fourth industrial revolution, as task-based work is being transitioned to automation and robotics, the role of the human worker as a relational being is ever more critical (Schwab 2017). The realization that we are currently living in the Anthropocene or Human Age, where humans control all life-sustaining systems on earth, is a daunting responsibility (Steffen et al. 2007). To advance our existence, we must employ an integrated and flexible structure to decision-making that is inclusive of all stakeholders and relational in its approach.

This phenomenon is even more relevant in the building industry. In many ways, the framework by which buildings are designed and constructed has not changed for centuries. It is historically a linear process with many points of handoff from one discipline and trade to the next. This often leads to gaps in coordination, redundancy in effort, rework, and costly overruns. The result is a confrontational and often litigious work environment. In order to realize sustained value, a different approach that results in an engaged, considered participation of all stakeholders throughout the design and delivery process of the building project is necessary. It is important to understand that processes do not deliver projects; people, working together in an interactive, cooperative context, deliver projects.

For instance, traditional construction methods typically only involve building users at two points in the process. They are involved during requirement development at the very beginning of the project, and at occupancy, near the very end of the project (Paolillo et al. 2016). The process is linear, product-focused, and transaction-oriented. Contracts are used to ensure the delivery of a building that meets specifications agreed to before the first shovel is ever turned. Value during the process is defined from the short-term perspectives of project schedule and budget. To the extent that recyclable materials, environmentally sensitive features, and efficient systems were specified during the design process, the project may achieve a sustainable building rating. Although a commendable achievement, this constitutes sustainability with a small "s." Small "s" sustainability is a focus on environmental factors that result in short-term benefits with the objective of "doing less harm." This type of sustainability is typically perceived as an added initial cost to the client intended to conserve natural resources and reduce waste. It may also include objectives to renew and recycle materials or to preserve endangered species of materials.

The results of utilizing traditional delivery methods for the commercial construction industry over the last half century have included adversarial relationships, declining productivity (-10%), and over half of all projects over budget or behind schedule (Construction 2013). This is in addition to the poor environmental track record of the industry and commercial buildings currently in use. Traditional methods are not sustainable because they fail to acknowledge the human social systems that deliver projects and the complex human systems in which final built spaces exist and are utilized.

The primary reason for this apparent failure of the building industry is the narrow and short-term perspective of how projects are defined and implemented. The involvement of the project team is perceived as temporal and finite. The metrics and values assigned to the project are typically limited to project delivery. The project has a beginning, middle, and end, with a 2- to 5-year duration. In that context, sustainability is factored in terms of building features, systems, and materials and translated into initial cost impacts. A large-scale commercial office project is often assessed upward of 25% to achieve a high LEED (Leadership in Energy and Environmental Design) rating. This curtailed view of the building project limits the industry's ability to consider the effects on the life-cycle impacts of the human, social, ecological, and economic sustainability.

Buildings have a significant impact on their owners, occupants, users, and the surrounding community. The design and delivery process implemented by stakeholders who have diverse experience, socioeconomic status, education, and desired outcomes is complex. Human social systems in general are complex systems (Ball 2012) characterized by their emergent outcomes, the self-organizing nature of their participants, and the collective intelligence greater than the sum of the individuals that make it up (Mitchell 2009). This requires an interactive project delivery process that "integrates people, systems, business structures and practices, and processes that collaboratively harnesses the talents and insights of all participants to optimize project results" (URL: American Institute of Architects, Integrated Project Delivery: A Guide 2007). Acknowledging the objective of this integrated delivery process and taking steps to enable greater cooperation for improved outcomes are the key not only to an improved efficiency in delivering construction projects but also to delivering sustainable buildings integrated with the surrounding community consistent with the long-term objectives of all stakeholders. This approach can deliver life-cycle sustainability with a capital "S" that balances the three bottom line principles of sustainability – people, planet, and profit. We call this a relational team approach based in mindful, participatory, and inclusive behavior. This approach is motivated by shared vision and values across multilevel project structures and is enabled by a common vernacular across disciplines (Olson 2016).

This broader understanding of sustainability in the context of building projects results in a complex system. Managing this complexity becomes a key element of the relational project dynamics, which is defined as consisting of many varied interrelated parts that are activated through differentiation and interdependency. Differentiation can be vertical, such as a hierarchy, and horizontal, reflected in the number of components and number of tasks. Interdependency is represented by the degree of reciprocal interaction between the components. Managing project complexity requires integration through coordination, communication, and control (Baccarini 1996).

Holling (2001, p. 391) states that "complexity of living systems of people and nature emerges not from a random association of a large number of interacting factors but rather from a small number of controlling processes." Using a small set of processes, complex systems can self-organize into complex adaptive systems. This continuous evolving process defines the meaning of "sustainable development" (Gunderson 2001). In this context, sustainability is defined as the ability to create, test, and maintain adaptive capability, while development is the method of creating, testing, and maintaining opportunity (Holling 2001). Together, these two concepts suggest the intention of inducing flexible adaption and promoting integrated benefits for the project delivery team, as well as the social, ecological, and economic factors surrounding the building industry.

Relational Team Approach

> *Inattention to social systems in organizations has led researchers to underestimate the importance of culture-shared norms, values, and assumptions in how organizations function.* (Edgar Schein 1996, p. 229)

To understand how people responsible for delivering construction projects perceive their world, we must understand the internal context in which they work. Construction project delivery practitioners have relied on the same unquestioned assumptions and metaphors to conceptualize business models and best practices with little improvement for decades. Traditionally, contracts are established to strictly enforce agreements and navigate the principal-agent problem. They are very transactional. Relational teams do not forsake contracts, but the methods for developing them and the demands of each party throughout the duration of the contract are different. "[T]eam cohesion" developed during the project definition phase "creates a relational climate of trust that is codified in the contractual requirements" (Paolillo et al. 2016). Contracts that underpin relational teams consider the long-term implications of the project and its outcomes. This type of contracting relationship means the needs of a broad group of stakeholders must be considered.

Work teams are generally composed of members with complementary specialties brought together to perform a task (Salas and Fiore 2004); however, in addition to the architects, engineers, and construction contractors, stakeholders such as space-users and community members should be engaged as part of the work team to deliver sustainable projects. Typically, the community is engaged through the permitting process or a mandated outreach effort which is engagement through compliance. Building users are involved through random surveys or represented by subject matter experts. This compartmentalized approach to building project development often results in predictable solutions based on tried-and-true ideas and archetypal categorizations. Relational project teams recognize that individual team members may not have the holistic understanding of all the ways in which a building project can impact its occupants and surroundings over time. In cohesive project teams, flexible individuals are enabled and empowered to bring integrated, new perspectives to all phases of the project (Weick et al. 2008). It is important to note that team norms and relationships develop with influence from their environment; the environment, in turn, changes as a result of the team norms and the relationships that exist within it (Bandura 1986; Giddens 1984). In other words, relational teams come together, and individuals work in an environment of both top-down constraints and bottom-up individual preferences and norms (Kozlowski and Bell 2012). Inclusion and positive outcomes require engagement from team members. When individuals are engaged, they are typically vigorous, dedicated, and absorbed in their work (Schaufeli et al. 2002, p. 74). A number of studies (e.g., Judge et al. 2001; Ryan and Deci 2000; Sonnentag 2003) have shown that engaged team members can create a virtuous circle in which motivation, dedication, and satisfaction positively feed job performance. Understanding that this dynamic exists and identifying people who can work in this context are a foundational element to the relational approach to project delivery.

An example of such a relational approach to project delivery is a recently completed children's hospital project in Akron, Ohio. The Akron Children's Hospital's (ACH) mission was comprised of three promises: (1) Treat every child as we would our own, (2) Treat everyone the way we would want to be treated, and (3) Never turn a child away (ACH Tower Team 2014: p. 21). The fact that this project was for a children's hospital facilitated the team's ability to institutionalize flexible cohesion characteristics; however, the human-centered nature of the approach was at the heart of the project's ability to deliver successful outcomes. The ACH project team implemented integrated project delivery (IPD) methodology and lean tactics of operation, design, and construction to achieve remarkable results of early delivery, under budget from conventional estimates. These methods have been shown to improve cost, schedule, and performance metrics relative to traditional design-bid-build contracting. In fact, integrated teams and lean intensity projects are three times more likely to complete ahead of schedule and two times more likely to complete under budget (URL: Dodge Data & Analytics, Owner Satisfaction and Project Performance Survey). We posit that the exceptional success demonstrated at ACH was driven by the relational nature enabled by shared vision and values and a common vernacular that manifested in the work team's flexible cohesion and produced sustainable results (with a big "S") for people, planet, and profit.

Relational Contracting

A contract enforces agreements between parties in cases of dispute and forms the basis for appeal when the dispute cannot be resolved. In traditional building projects, the owner enters separate contracts with the architect for the design phase and with the contractor for the construction phase. Each of these entities holds subcontracts with a myriad of sub-consultants, vendors, and construction specialists. Figure 1 represents the traditional relationship in construction contracting. These individual contracts all include specific scopes of work, terms and conditions, overhead and profit margins, and contingency allotments. If additional work is needed due to discrepancies in the drawings, unforeseen circumstances, or design modifications, the contracted parties must resolve it through a change order process. These can

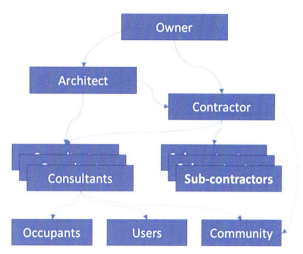

Fig. 1 Transactional project team relationship

often be costly and confrontational; furthermore, because there is no contractual relationship between the architect and contractor, the issue becomes the owner's responsibility to ultimately resolve and pay for.

In a relational contract, the parties agree upon general principles and terms of resolution when conflicts arise (Milgrom and Roberts 1992). Projects using an integrated project delivery methodology are executed based upon shared risk and shared reward under a multiparty contract. The usual actors are owner, architect, and contractor, with key sub-consultants and/or subcontractors sometimes included. The contract is executed on a representational basis where the authorized, senior-level person from each entity is the signer. This approach sets up a collaborative rather than adversarial relationship and shifts the emphasis to working in the best interest of the building project rather than individual project entities. Figure 2 represents the collaborative relationship in this type of dynamic. The order of the team members from top to bottom represents the type of role (leadership vs. supporting; action-oriented vs. consultative) that member will generally have in each phase. The involvement of stakeholders such as contractors, community members, and occupants have in the early phases of the project is worth noting. The requirements and design perspective input from these groups will inform decisions that impact areas such as constructability, aesthetics, and physical dimensions and layouts and are important for providing a long-term perspective. Later in the project, continued involvement of the contractor, architect, and consultants not only demonstrates commitment to the community and project but also facilitates ongoing changes or improvements.

While this provides a foundation for the relationship to minimize the potential for self-interested, opportunistic behavior at the top levels of the signatory organizations, it does not ensure the cooperation and trust among other members of the project team. Thus, a sub-agreement is needed that binds all other team members to the terms and conditions of the relational contract. The key condition that governs all

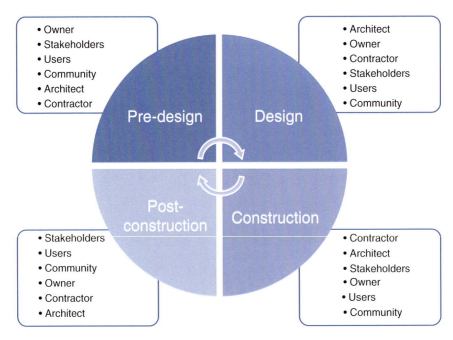

Fig. 2 Relational project team relationship

the agreements is commitment to work in the best interests of the project outcome at all times. While the legal contractual ramifications technically only apply to signers of the contract, eliciting verbal agreements from all team members that "we are in this together" can have a powerful motivation and team-building effect. Active participation like this defines the nature of relational contracting. In the case of ACH, the mantra was E^3 – Engagement, Everywhere, Everyone (ACH Tower Team 2014).

Responsibility and accountability do not only reside at the top levels of the project team utilizing relational contracting. Shared values must be communicated with and shared across all levels of the project team to be successful. Team members on the ACH project created a relational climate to support the relational contract by establishing a multilevel structure to define and drive project goals concurrently from the top down and the bottom up. The three-level team structure they developed included a senior executive team, a project leadership team, and discipline-based innovation teams. Innovation teams were comprised of subject area experts in their field and dealt with daily problems focused on (1) increasing productivity, (2) creating value, and (3) delivering the project. The project leadership team was empowered to make decisions on all matters in the best interest of the project outcome. Only issues that could not be resolved at the lower levels would be escalated to the senior executive team, which was obligated to resolve all issues prior to moving to a new project phase or subsequent task. In addition to the philosophical motivation prevalent in the project, this arrangement created a practical incentive for senior executives to resolve issues. Likewise, because teams continued interacting over a long period of time, they were incentivized to resolve issues amicably.

The risk and reward system in a relational team structure is grounded in the *all for one and one for all* mentality. While the direct costs of the project are still the responsibility of the owner, the contingency and profit allotments are pooled and distributed based upon the success level of the project delivery. The aggregated contingency is used by any team member to cover the extraordinary expenses. Any unused amount is added to the profit sum. Conversely, if the additional costs exceed the contingency allocation, the monies to cover the amount come out of the profit sum, not the owner's pocket. This incentivizes the team to minimize waste, rework, and inefficiency in project implementation, which ultimately benefits all team members.

Relational Climate

Historically, participation for most team members of a building project is typically transactional. Their involvement is for a limited period of time to complete a prescribed task(s) at a set fee. Only a very small number of individuals have bigger picture knowledge of the overall project scope and interaction with the stakeholders. The result is a disconnected and depersonalized level of project member interaction. More recently, a project delivery paradigm shift is occurring in the building industry toward an integrated implementation team approach. This approach points toward shared vision and values, as well as a common vernacular that is defined early and regularly exercised in team members' daily routines and actions in order to establish and implement common project goals and objectives (Olson 2012). The realization of this integrated approach required a relational team climate.

Social science research indicates that the way in which team members treat and motivate each other is a reflection of their understanding and acceptance of the relational climate in which they find themselves (Mossholder and Richardson 2011). The basis for the manifestation of a relational climate in a group setting has been identified as helping behavior, which has been studied as a predictor of group and organizational performance (Podsakoff et al. 2000). Helping behavior is cooperative and positively affects individuals in groups (Podsakoff et al. 2006; Settoon and Mossholder 2002). Individuals judge the value of their helping behavior based upon the problem and solution effectiveness manifested within the group context.

In order to create a relational climate that fosters this type of behavior, teams must choose their members mindfully and organize the team structure carefully. Fiske's (1992) relational models theory (Mossholder and Richardson 2011) conceptualized the relational climate types, which Mossholder (2011) equated to human organizational structures as (a) compliance systems, (b) collaboration systems, and (c) commitment systems. To apply these concepts to the project team structures, we characterized the climate types as (a) opportunistic, (b) collaborative, and (c) communal, based upon the descriptives associated with the predictive dimensions revealed in the Mossholder et al. study (2011). The opportunistic category implies a transactional approach that is self-serving. The relationships are temporal based in economic benefits for prescribed contributions (Rousseau 1995). A collaborative climate demands participation and interaction among the team members that is rooted in a mutual exchange of transactional and relational benefits. Collaborative climates typically involved knowledge sharing grounded in the achievement of shared objectives. A high relational team climate is communal. It requires collective commitment and focus toward a higher goal in a long-term context. The goal can be externally motivated, but it also creates bonds among team members with whom the goal is accomplished (Mossholder and Richardson 2011).

The predictive dimensions that influence each relational climate type are (1) motivation for exchanges among the team members, (2) justice norms that determine how exchanges between the team members are assessed, (3) perceived risks that affect the relationships, and (4) a basis for trust between individuals or organizations involved in the project. In a low relational climate or opportunistic team environment, participants will exhibit a low level of interpersonal commitment resulting in limited relational depth. The motivation revolves around the participant's perception of balance between efforts exerted and benefits received. In this situation, equity is evaluated based upon the relative ration between input and output of each person on the team rather than the absolute amount of contribution to the project. The primary interest is to derive a sufficient return on helping behavior investment. Trust is based upon the calculated risk that the benefits will be equal to the effort invested (Mossholder and Richardson 2011). A medium or collaborative relational climate required shared common interests. The team relationships are based upon common values that are both transactional and relational (Rousseau 2004). The motivations are understood to be temporary during which time the team members are working toward the same goals. The motivation to achieve mutually beneficial shared outcomes explains the need for knowledge sharing. Thus, helping behaviors support an equality-based team environment, which is manifested in common social expectations and reciprocal exchanges (Cropanzano and Mitchell 2005). Team trust is built upon willingness to engage in frequent and ongoing interactions as a relational entity over time, rather than assessing each individual event. A communal relational climate is founded in mutual high regard for team members involved in collective commitment. The team environment fostered lasting relationships between team members motivated by an attainment of mutual goals and appreciation for the team with whom the goals were achieved (Mossholder and Richardson 2011). Helping behavior is a way to exhibit mindfulness toward team members and stakeholders. This vested interest in the well-being of others reinforces the need for fairness and caring in team interactions. This ability to relate and empathize based upon common experience and familiarity leads to identification-based trust (Lewicki et al. 2006). Once identity-based trust is established, helping behavior becomes need-based, collective-directed (Perlow et al. 2004), and difficult to erode (Mossholder and Richardson 2011).

Table 1 illustrates the assessment matrix that can be used to evaluate the potential for partnership building on the part of team members. It provides some typical

Table 1 Team assessment matrix

	Relational climate types		
	Low	Medium	High
Dimensions	Opportunistic	Collaborative	Communal
Motivation	Self-interests	Knowledge sharing	Shared social values
Justice norm	Equity	Equality	Need-based
Perceived risks	Return on investment	Balanced reciprocation	Empathic response
Trust	Calculus-based	Knowledge-based	Identity-based

Source: Adapted from Motivation and Sustenance of Helping Behavior (Mossholder and Richardson 2011)

behaviors associated with these dimensions that can be used to assess the type of climate supporting a project team. A low assessment would be an indicator that the individual or organization is not a good relational fit, the assumption being that an opportunistic disposition would not support the objectives of a relational team approach. Alternatively, simply clarifying differing perspectives of stakeholders or potential team members can serve as a focal point for a dialogue around which a common vernacular can be developed and a shared vision and values can be created. A medium or collaborative assessment of the individual or organization is an indication that team participation is possible on a relational contracting level but with a low level of engagement. If the team agrees that a relational project team is the desired approach, identifying specific areas for improvement or means to facilitate these areas could lead to buy-in and improved flexible cohesion within the team dynamic as the project evolves and roles change over time. An example might be the motivation regarding knowledge sharing: as more knowledge is shared and team members understand better why knowledge was shared at a certain time and how it can best be applied, the environment is established for the convergence of values. A high communal assessment would suggest that the individual or organization is capable of full commitment to the relational climate necessary to deliver the best possible project results. This assessment matrix can be used to forecast the potential for team members' helping behaviors in a relational team climate. A relational climate indicates shared perceptions of attitudes and behaviors affecting interpersonal relationships in a team structure (Mossholder and Richardson 2011).

Enabled by the relational contracting terms and conditions within a relational climate, the project team is motivated to engage in actions that support team success and move the project forward. The ACH stated mission – to provide medical care to infants, children, adolescents, and burn victims of all ages, regardless of ability to pay – triggers an emotional response. However, it is the involvement and commitment of all the stakeholders to continuously focus on the project goals and objectives that create a relational climate. The ACH team utilized LEAN practices in operations, design, and construction to optimize the processes, systems, services, and solutions delivering effective project results. These included continuous and proactive input and participation of all stakeholders to address and solve problems during design and construction of the project. During the design phase, the staff, doctors, and parents of child patients participated in full-scale mock-ups to develop and test treatment and recovery rooms, discovering the most effective layouts. The shared vernacular was built around rewards and benefits. Construction crews acknowledged individuals that proactively undertook tasks which lead to cost and schedule improvements. Waste reduction was celebrated and rewarded.

The ACH team was an example of a communal relationship converging around a shared goal. Team members were fully committed to the values of ACH and its purpose. The project leadership team developed a thorough training program mandatory for every individual at every phase of the project. The senior executive team supported the program, and the innovation teams implemented and improved the program as it was executed. Once trained in the values, vernacular, and vision, each team member went out of their way not only to support the success of the project but also to express heartfelt emotions toward the primary stakeholders, namely, the children. Even though it was easy to empathize with and feel sympathy for the patients of a children's hospital, it was clear that the relational culture of the project team enabled this expression, which manifested itself in many ways in a stereotypically macho industry.

Practice Exercise #1: Assessing a Team's Likelihood for Success

A partnering is a framework for structuring and monitoring multiple organizations over the course of a building project that are committed to achieving shared project goals. Partnering promotes open communication and participation among members of project team. The objective is to ensure that team members are working cooperatively throughout the timeframe of project delivery. In order to sustain this cooperative state, the team members must maintain a positive relational climate. Acknowledging that it is people, not processes, that deliver a successful project, it is imperative that the project team relational climate possesses the characteristic nature of a high-performing team. This requires that the team members, at the individual and organizational level, exhibit the qualities across the four dimensions of motivation, justice norm, perceived risks, and trust toward a communal level of a relational climate.

To determine the nature of helping behaviors the team member organizations, as represented by its members, can achieve, an exercise can be integrated into the partnering agenda that assesses the team member relational capability. This would establish the baseline for the standards, policies, and expectations for the project team climate going forward. Deviations from these sets of principles can be red-flagged and mitigated through preestablished channels.

The exercise is comprised of four scenarios that target each of the predictive dimensions of relational climate. The team members are asked to discuss and then respond to the situation in the scenario that reflects their approach and attitude toward the simulated problem that may occur on the project. The predetermined core management team assesses the appropriateness and acceptability of the team and individual responses and assigns them into the low, medium, and high category of relational climate. Team weaknesses are identified, discussed, and resolved. Individuals or organizations that do not exhibit the characteristics necessary to support a high collaborative or communal team are screened out of the project.

Table 2 What is the role of each team member to resolve the situation?

Relational climate	Dimension: motivation	
Types	Characteristics	Helping behaviors
Opportunistic	Self-interests	If it doesn't impact my work, I don't need to be concerned
Collaborative	Knowledge sharing	If all other subs participate, I would be willing to take a look at an alternative solution
Communal	Shared social values	I am sure we have some ideas on how to resolve this quickly

Table 3 How does the team communicate?

Relational climate	Dimension: motivation	
Types	Characteristics	Helping behaviors
Opportunistic	Self-interests	We can address it at the next schedule project meeting
Collaborative	Knowledge sharing	Maybe the project manager can email us some options for consideration
Communal	Shared social values	We can make ourselves available for an on-site meeting to look at the situation

Table 4 How should the subcontractor work be prioritized?

Relational climate	Dimension: justice norm	
Types	Characteristics	Helping behaviors
Opportunistic	Equity	If my materials are not where I need them, I cannot do my job
Collaborative	Equality	If everybody wants to take turns, I will get in line
Communal	Need based	I can bring up extra materials in one load so I require fewer trips on the elevator

Scenario #1: Motivation for Exchange in the Relationship

An unforeseen shortage of material has forced a redesign and changes in field construction. The schedule must still be maintained. This requires a subcontractor scope change and impacts critical path schedule (Table 2 and 3).

Scenario #2: Justice Norms by Which Exchange Fairness Is Evaluated

Urgency of schedule is forcing some overlap in time and resources on-site. Access to elevators and cranes will have to be shared (Table 4).

Scenario #3: Risk That Potentially Undermines the Relationship

Urgency of schedule is negatively impacting quality control and coordination. Additional measures must be taken to ensure that work is being done properly and issues resolved quickly (Table 5).

Table 5 What should be the team members' responsibility in addressing discrepancies and maintaining quality?

Relational climate	Dimension: risk	
Types	Characteristics	Helping behaviors
Opportunistic	Return on investment	I didn't do anything wrong so I don't think I should have to invest my time into resolving the problem
Collaborative	Balanced reciprocation	Everyone should have to share in the blame when something goes wrong
Communal	Empathic response	I will coordinate with the crew behind and ahead of me and raise any concerns as soon as I see a potential problem

Table 6 What should be the team's response?

Relational climate	Dimension: trust	
Types	Characteristics	Helping behaviors
Opportunistic	Calculus based	I don't know if this person can do the job coming into it in the middle of the project
Collaborative	Knowledge based	It will take time to learn how to work with this person
Communal	Identity based	This person comes from the same background and training and will pick up where we left off

Scenario #4: The Basis of Trust Between Parties

The project manager had to go on medical leave. Another project manager from the same firm stepped in who has a good track record but has not worked on a similar project and never worked with this team (Table 6).

Relational Context

A construction project, depending on size and scale, can have a substantial impact on its surroundings during and for decades after construction is completed. This impact can be a positive or negative experience for its users, community, environment, and planet. More often than not, a building project is constrained to its property boundaries or limited to the input of only a small number of technical disciplines during the critical early phases of decision-making. The regulatory requirements only mandate basic code and zoning compliance during the project design phase. Municipality oversight agencies focus on the safety, accessibility, and use of energy. Environmental impact assessments are often driven by political priorities. A relational project team has the ability to look beyond these minimum standards and first-cost impacts. The benefits of an integrated, relational approach can have far-reaching, long-term benefits with the lowest total cost. This broader view of project delivery occurs in a relational context, which can optimize the environmental, economic, and social value of the built environment over its life cycle (Levitt 2007).

Relational context is a state that supports the creation of perceived value and goodwill for external stakeholders with benefits to the project and the internal stakeholders. It entails the identification, establishment, and maintenance of relationships with parties both internal and external to the project, such that benefits are realized, which can be balanced against the costs to the project and the community. It is widely acknowledged in many industries that the traditional, product-focused approach to delivering value is no longer enough to satisfy consumers (Grönroos 1997). Stakeholders are invested in the sourcing, production, and distribution methods of products as well as their impact on the health, safety, and welfare on people and planet. Public and private organizations must learn to assess value beyond profit-driven motivations. The notion of social value originated with the recognition of corporate responsibility and ethical economics. Public awareness of these social enterprises has been successfully used to enhance the organization's competitive edge (Watson et al. 2016).

In the building industry, an expanded view of mutually beneficial sets of expectations can engender proactive and innovative ideas and behaviors that can grow and expand over the life of the project and beyond. The need to balance the concerns of the triple bottom line (people, planet, and profit) for sustainable development projects has attracted the attention of global entities such as the World Bank's International Finance Corporation (IFC). The IFC has clearly established guidelines for projects that require engaging all affected stakeholders early in the development process to define the impacts, feasibility, and long-term benefits of projects across the globe prior to implementation (Levitt 2007). Determining the value of building projects beyond profit-driven motivations is key to sustainable building projects. However, the long-term benefits must go further than simply return-on-investment measures of building operating costs. These metrics need to extend into the qualitative nature of stakeholder interests and experiences. The stakeholder population must encompass not only the building users but also the surrounding community, as well as environmental impacts on the planet at large. While these metrics are much harder to capture and quantify, they may be just as significant for the life-cycle impact and social value of the building project (Watson et al. 2016).

These participative activities and behaviors on the part of the community can be categorized into active, passive, and non-relational modes and weighed against goodwill/relational benefits or economic/transactional costs to the project (Grönroos 1997). Defining and quantifying these intents and outcomes establish a clear and transparent relationship between the project team and all impacted parties. It provides guidelines for private and public participation in the project delivery process and defines the level and intensity of communication among the stakeholders. The leadership of the project team can assume either a relational or transactional posture relative to community involvement and value perception. Value perception entails more than just the tangible offerings of the physical project but also the experiential and emotional factors associated with the project. The established relationship can include distinctly different short-term commitments and long-term implications. Even if the project leadership determines that it is in the best short-term interests of the project to focus on the transactional, cost constraints of the project, the latent relationships with the user and professional community stakeholders still exist (Grönroos 1997). This is true regardless if the community stakeholders choose to engage with the project in an active, passive, or non-relational mode. Thus, garnering social capital for a building project can have short-term benefits and long-term ramifications and builds sustainable value.

For ACH, community participation had manifested in a number of ways. First, there was the willingness of the city to cooperate with the project stakeholders in providing transportation modes to access the hospital site. Secondly, the motivation on the part of family members to participate in the mock-ups and test runs of various procedures ensured optimal design solutions of the hospital facilities. Thirdly, the initiative and effort of the team members at every level of the project structure to engage in activities that would maintain a high energy level of the stakeholders, from dressing up as Santa Claus and distributing gifts to sick children to team lunches, helped hold each other engaged and accountable every day on the construction site. The active participation of the various stakeholders at different points in the project delivery created social value for the team member, the hospital user population, and the community at large. The team members built a level of comradery and social capital among each other, which became transferable to future projects. The hospital staff, patients, and families became vested in the success of the project outcome based upon potential mutual benefit. The community population, whether actively, passively, or not directly involved, realized a potential long-term resource, which created goodwill toward the hospital and the project team organizations.

A community relations matrix, illustrated in Table 2, provides a framework by which to assess the cost versus benefit of a strategic approach toward stakeholder and community involvement. The assumption is that a relational strategy is possible because latent relationships with stakeholders in the community always exist. However, the project leadership must discern the development costs and benefits at project conceptualization to assess its quantitative and qualitative value to the project over time. To address the life cycle and sustainable impacts, the calculus must be done based upon short-term or initial costs and benefits, as well as the ones over the lifetime of the project development. A comparison of the impacts between the relational intent of the team to be inclusive and open to stakeholder and community involvement and a transactional intent, which limits participation, needs to be assessed at the outset of the project. It is important to note that the benefits are not necessarily tangible but can also be perceptual. Furthermore, the relational intent generally produces benefits that are long-term, in contrast to the transactional intent, which is typically immediate (Grönroos 1997).

Gaging the nature and interest level of the community and stakeholders in the development project and aligning it with the project team intent can greatly affect the success and performance levels of the project outcome. For example, an active relational mode on the part of the community met by a transactional intent of the project leadership can result in a confrontational short-term situation and long-term negative perceptions of the project. Conversely, a non-relational community mode met with a transactional intent could have no consequences in the short-term. However, if the non-relational mode is latent and not identified, the long-term impact on the project

Table 7 Community relations cost/benefit matrix

		Project Team Relational intent			Project Team Transactional intent
Community stakeholders	Active relational mode	Short term	Costs	Benefits	Short term
		Long term	Costs	Benefits	Long term
	Passive relational mode	Short term	Costs	Benefits	Short term
		Long term	Costs	Benefits	Long term
	Non-relational mode	Short term	Costs	Benefits	Short term
		Long term	Costs	Benefits	Long term

Source: Adapted from Value-driven relational marketing (Grönroos 1997)

could still be negative. Similarly, a passive relational mode from the community may imply a lack of interest in the short-term. However, if the project team does not offer a relational intent, the long-term perceptions could have negative results.

The consequences of perceived social value may be significant and obvious when considering a children's hospital. The ability to create goodwill using a relational intent in the context of a community housing a children's hospital may be obvious. However, this rationale can be generalizable to any organization that provides a service and is dependent on consumers for its survival. The network of influence that the building project exerts on its surroundings both local and global is not limited to its property lines or the length of its development cycle. The building industry must recognize its impact on sustainability with the capital "S." An approach to assessing the costs and benefits of a relational versus transactional relationship between the project team and community stakeholders is illustrated in Table 7.

Practice Exercise #2: Assessing the Relational Interests of External Stakeholders and Community

A series of open forums with both the internal project team and external stakeholder community can flush out the modes and intents, which can be documented as outcomes and evaluated against short- and long-term project success metrics.

A sample of three distinctly different types of groups is self-identified using an outreach program directed at the community and user populations that represent the three categories active, passive, and non-relational mode. Each group can be further refined by selecting the level of commitment and value perception the participant identifies with for the project in question. In a moderated roundtable discussion with the groups, the team members can discern the stated and implied opportunities associated with the project (Table 8).

Once the qualitative input is collected, the project team discusses and prioritizes the influencing factors. The list of factors is then assessed based upon initial and life-cycle costs/savings as well as short- and long-term benefits/shortcomings. The trade-offs are explored and documented in the cost/benefit matrix, which highlights the project team relational versus transactional intent, as well as the consequences to community and stakeholder involvement in the project. In a public forum, the representative team members report back to the community and stakeholders the

Table 8 An assessment work sheet for individual stakeholder and community members

Community stakeholder	Active relational mode	Passive relational mode	Non-relational mode
Level of involvement			
Current			
Future			
None			
Level of commitment			
Direct			
Latent			
None			
Value needs (total value created that a user needs to feel satisfied)			
Tangible			
Experiential			
Emotional			

outcome of the analysis and negotiate the final terms of the relational and/or transactional arrangement. This ensures a common understanding and shared narrative of the project vision.

Engaged Sustainability Through Flexible Cohesion

A relational team structure supports participatory and helping behaviors of teams by creating an integrated project environment. The driver and enabler of this team environment is team member cohesion. Team cohesion is characterized by goal commitment, timeliness and extent of communication, and interpersonal attraction (Franz et al. 2016). A fourth factor of team cohesion is the willingness to compromise or adapt (Franz et al. 2016), which is an indicator of individual flexibility. Team cohesion is defined by the extent to which the project team engages with and adapts to each other to create the relational environment. Cohesion has been considered as the most important factor in the study of working teams (Carron and Brawley 2000). As projects evolve over time and move from requirement gathering phase to design phase to build phase, engaged team members adapt to new circumstances and work with their fellow stakeholders toward agreed-upon goals. Because they established a relational climate and leverage relational contracting, they share common vision and values for the project. These might be commitments to reduce waste, recycle materials, use innovative design strategies, and engage with the stakeholders and the community. In other words, a cohesive building team grounded in a relational structure results in engaged sustainability of people, planet, and profit.

Cohesion and flexibility are key determinants of team effectiveness (Burke et al. 2006). As time and the project progresses, roles change. A team member's ability to adapt to their new authority, tasks, and other stressors, combined with their ability to continue to communicate effectively and work with their teammates in the new dynamic has an outsized impact on healthy team functioning. Cohesion is the bond between members of a group holding the group together (Beal et al. 2003; Hampson et al. 1991; Kadushin 2012; Olson 2000; Pescosolido and Saavedra 2012). Flexibility or adaptability is the "capacity (and willingness) to competently engage in a variety of behaviors in response to different situations" (Lord and Hall 1992, p. 140). In our work team project delivery context, flexibility includes an individual's ability to cope or adjust to role expectations which are perpetuated through culturally established processes (Wolin and Bennett 1984). Flexibility is a factor in individual decision-making whereby the individual balances emotional response with logical or cognitive responses (Mayer and Salovey 1995).

Flexible cohesion is an individual's ability to adjust to different named roles in a team over time as the nature of the team's task evolves. Flexible cohesion is associated with an individual's willingness to engage in team environments (Straub 2015). The idea of a named role is a shortcut for connoting a level of authority and responsibility. Titles, especially in project settings, are often named roles. These can be problematic as projects evolve, especially when titles may have different meanings in various contexts. A relational project climate creates the conditions in a multilevel team structure that enables flexible cohesion among individual stakeholders. This is supported by a multiparty agreement based on a shared risk and shared reward framework that establishes the relational contract between major stakeholders. The context for this dynamic project structure stretches over time and beyond the boundaries of the project site.

It is important to note that balance is critical to a flexibly cohesive team structure in a relational framework. Too much or too little flexibility or cohesion in a team member or group can divert team efforts resulting in detrimental effects on the overall project outcome, as illustrated in Fig. 3. If the level of team cohesion is low, the team members are generally disconnected and will not participate in collaborative problem-solving efforts. As a result, team members pursue their individual agendas, and the project team fails to reach an integrated problem solution. Conversely, if team members are too cohesive or aligned, they can exhibit groupthink, focusing only on narrowly defined issues and reaching consensus too quickly.

Flexibility also has a goldilocks zone. Individuals and teams who exhibit too much flexibility tend to lack focus. When flexibility is too low, individuals and teams can become rigid and fixed in their ways. In a situation where the tendency is toward excessive levels of flexibility and cohesion, the team is likely to chase every idea and opportunity that presents itself. Low cohesion combined with high flexibility results in pet projects and lack of commitment to the project vision. Both conditions can lead to inadequate resources applied to critical tasks and schedule slippages due to lack of focus. In situations with low flexibility and low cohesion, it would be quite difficult to establish a relational project structure; however, team leadership must beware that the organizations or individual participants may not understand the expectations of a relational structure at project initiation and make commitments that cannot be kept. In the case study example, the ACH team adapted the design to accommodate construction changes not in the original documentation. In a transactional context, this would have resulted in significant rework, cost, and schedule impacts; however, with a flexible, cohesive team operating in relational context, the

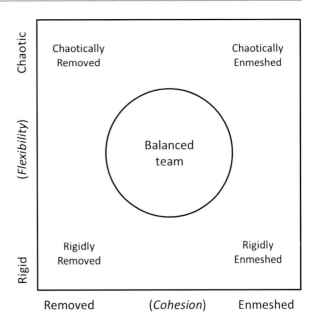

Fig. 3 Flexible cohesion matrix (Adapted from circumplex model; Olson 2000)

issue was examined, models were built, and tests were run to quantify the impact on the final product. Because the impact was not significant, the change remained, and the project stayed on schedule and under budget.

Conclusion

An essential paradigm shift in built environment delivery is the notion of a relational project team framework that is the basis for successful project outcomes. However, it is important to understand that the project delivery mechanism is only partially responsible for the ultimate project outcome. Although a project framework can significantly influence the project dynamic at both the individual and team levels, flexible cohesion among team members is the key factor in aligning the vision and values of stakeholders with project outcomes. This conclusion implies that a building project must be viewed as a long-term, relational event and team members have to be viewed as vested partners that commit to the vision, values, and a common vernacular of the integrated project delivery method.

Reflection Questions

Is engaged sustainability a worldview or a checklist of attributes?
What defines the relational environment that delivers the desired project outcomes in the built environment industry?
What effect does this relational environment have on the project team members?
What are the characteristics of the organizations or team members who are more successful at achieving the desired project outcomes?
How are these organizations or individuals identified and engaged?

Akron Children's Hospital: A Case Study

Introduction

The case study of a construction project at Akron Children's Hospital demonstrates how a project team can have transformational results when incorporating the concepts of relationship building and mindful behavior into a project delivery method. At Akron Children's Hospital (ACH), we found that using these techniques in conjunction with Lean Integrated Project Delivery methods not only delivered but also exceeded all expectations in regard to cost, time, safety, and expected operating margins (Fig. 4). The ACH construction team drove first cost 26% below market estimates ($178 million vs. the original estimate of $240 million), accelerated the schedule by 16%, and achieved a 40% better safety record than the national average and a 99.2% cash flow predictability with an increased operating margin of 33%. Furthermore, the building is 34% more efficient than any other building of its kind based on energy usage. Since the completion of the building, ACH is recognized as the fastest-growing children's hospital in the country despite the below 1% population growth in its demographic region (Fig. 5).

Background

In the last 50 years, the construction industry has seen productivity decrease by 10%, with more than 50% of projects either over budget or behind schedule. All other non-mining business productivity has increased over 250%. How is this possible? The Lean Construction Institute says one major problem is that 60% of the activity on a construction project can be classified as waste. The transactional nature of the design-bid-build contract does not support collaborative environment. In fact, it encourages a singular focus on one's own scope of work for a predetermined bid price regardless of the consequences on the total project budget. Relationships are often adversarial. As a result, construction is one of the few industries where suppliers sue customers and vice versa (Fig. 6).

Even today, the typical building company leadership believes that the transactional relationships between firms and a top-down decision-making approach to

Fig. 4 ACH New Hospital Wing

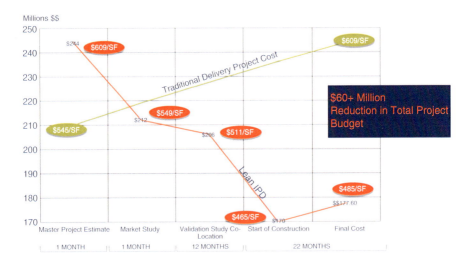

Fig. 5 ACH project data – Welty Building Co.

project delivery produce the best results. Welty Building Company took a different tack on the Akron Children's Hospital expansion project. This approach was first motivated by Bill Considine, CEO of Akron Children's Hospital, when he said that we need to design and build the ACH Critical Care Tower through the eyes of the child. Our guiding principle stated that we intend to create a place that (a) is distinctive and serves as a beacon to the community; (b) is safe and comforting to children, parents, and staff; (c) is a respectful connection to the

Fig. 6 Construction industry productivity (Smart Market Report, McGraw Hill)

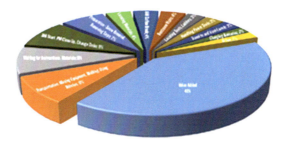

natural environment; (d) is playful and engages the imagination; (e) inspires confidence and hope; and (f) builds on a promise. If the project team was successful, the built environment created would feel like a hug to a child. A hug is a connection that fills us up, makes us feel more in control, and is a part of something bigger than ourselves. A built environment's expression of this mission enables the nurses, doctors, and staff to enhance the patient experience. What we found was this vision and values transcended the attitudes and behaviors of the project team members. Instead of a testosterone-filled environment on the jobsite, the construction team encouraged each other by awarding "bambino," small stickers with the project logo and motto. Workers proudly pasted these stickers on their hard hats in a manner akin to a football team tradition of awarding stickers for touchdowns (Fig. 7).

At the Welty Building Company, the motto is ER>=CR, employee relationships are greater than or equal to customer relationships. This people-centered attitude translated to the project team work environment. We found when a team believes it is all right to give a hug and the leadership team gives them a way to give a hug, all traditional metrics of performance improved. The integrated project delivery (IPD) method used on the ACH project maximized customer value by eliminating waste. IPD is based in a formalized relational contract between the owner, contractor, and architect to construct a building. The premise is that the three entities become stakeholders in the outcome through shared risks and shared rewards. The key to making IPD work in construction is authentic collaboration. With input and agreement from all three stakeholder groups and sub-consultants, a fully coordinated solution and project plan is devised that meets program requirements within the available schedule and budget. The difference between IPD and traditional project delivery methods is that all issues must be resolved by the project stakeholder team

Fig. 7 Bambino stickers (Welty Building Co.)

and no issues are left unresolved. Projects are integrated through collaborative leadership, coordination, communication, and interaction.

Discussion

Welty Building Company acting as the construction manager on the ACH project recognized project members as trade partners. On typical construction projects, the building trades are typically referred to as subcontractors, a second-tier participant below the construction manager. Welty embraced the integrated team. This structure enabled a 30-year tradesperson to participate in the discussion, thus, engaging all team members and the entire workforce to innovate and drive a better result. This integrated team culture was reinforced at every opportunity. On a regular basis, the entire construction crew ate lunch together. The team leadership would bring in food trucks and celebrate safety success. The project team members came in before and after work hours to play Santa Claus and directly interacted with the children in the hospital. Every time a project team member gave a hug as they handed out a present they were receiving a hug back. These connections manifested in self-reinforcing moments that enabled innovation and enhanced building project outcomes. Paul Becks, the project manager for Welty on the ACH project, spoke specifically to how they brought the team together: "I think there must be a certain amount of empathy for other people that they're human, that problems exist and hopefully they have the same respect for you ... you can't be crucified every time you make a mistake and you can't go crucifying other people every time they make a mistake. Time spent assigning blame is time not spent understanding the root cause and preventing future mistakes. Skip the blame, find a way to get it done, and find a way to make it an enjoyable experience and that happens when you stay focused on what the mission needs to be; why we are doing this. We do a stand down every Tuesday right after lunch where the entire project, all the guys, we get them all together and I let them pick the song of the week. It was the 80's song 'Safety Dance' last week. We play a song, we talk about safety. Last week there was a guy up in Cleveland that died on a jobsite and we had a moment of silence. We had somebody who knew the guy speak. It's an opportunity to bring everybody together."

What Welty has learned is in order to get the team to apply the building science of lean you need to create an environment where it is all right to give a hug. We saw this at the ACH project when the construction workers hung hearts and Christmas trees from tower cranes so the children could see them from their hospital beds (Fig. 8).

In a recently published study based on the Akron Children's Hospital project, the authors introduced the concept of upcycle construction capital (Paolillo et al. 2016). Upcycle construction capital is defined as the collateral benefits produced by a building project over time. The shift is in the emphasis from the initial construction cost to the lifetime costs created for the business, as well as the upcycle benefits for the stakeholders, the community, and the planet. In the traditional delivery processes, the creativity, passion, and expertise of the entire project team go unrealized as individuals and management are mired in transactional relationship of exchanging fees for services. The design-bid-build contractual process has focused on the initial construction costs with no reward for collaboration outside the individual project effort. An integrated project delivery method that can generate long-term benefits for

Fig. 8 Heart on the construction crane

the three Ps (people, planet, and profit) include but are not limited to safer working conditions, decreased first construction costs, reduced operating costs, lower energy costs, increased positive affect (good feelings) in the community, incremental revenues, increased operating efficiencies, and individual competencies. These constitute the integrated sustainable benefits (Fig. 9).

The Welty Building Company made a claim to the Akron Children's Hospital to be more than a builder of buildings. This promise required that the entire building team had to reach beyond expectations. They had to strive to innovate in the use of technology, be an advocate for the environment, and build community. At the outset of planning, the project was projected to cost $240 M using traditional project delivery methods. Through creative leadership and collaborative teaming, the project was brought in at a final cost of $178.3 M, 6 months ahead of the projected schedule. The lower operating costs and quality of the built environment exceeded the original expectations of the ACH.

Don Taylor, CEO of Welty, said: "I'm a believer in IPD and especially when you incorporate lean. It's got to be a better way ... I'm sure you've read this statistic that there's something like 60% waste in the construction industry, as they've defined waste, so here we are arguing about 1.75% and 2% fees and we've got 60% waste in a project. If we can get rid of some portion of the waste which benefits the client we probably can make a little more money or take the pressure of the profit side of the project and have everybody win. To me, that's where we've got to go. We can get paid what we are entitled to and the client gets the benefit of the project costing less overall. We impact not only first costs but the image of our clients in the community. The way we have succeeded as company is to focus not just on the building a building – but build the building to support the users of that building and the community effected. On this project the team did a great job working with the city to enhance public transportation access."

Conclusion

ACH's intention was to create a distinctive place in the community that is safe and comforting in the eyes of a child, parent, and staff. The culture surrounding the project delivery team was developed with these values as unifying themes. The team's sense of relatedness to the greater good of the children combined with the mindfulness of an integrated team served as a prerequisite for creating shared vision, values, and language used by all the trade partners. This created better relationships

Fig. 9 Upcycle construction capital

and improved their ability to deal with conflict, encourage open communication, and drive better project outcomes. Welty has found when a team believes it is all right to give a hug and the leadership team gives them a way to give a hug, all traditional metrics of performance improved. The Welty Building Company is currently using these same concepts in a variety of project types such as high-density mixed use, energy transmission and distribution, general office, and medical facilities. All have documented project results showing lower first cost, improved schedule, improved operating margins, and a more engaged construction team.

Cross-References

▶ Environmental Stewardship
▶ Moving Forward with Social Responsibility
▶ Responsible Investing and Environmental Economics
▶ Sustainable Decision-Making

References

ACH Tower Team. (2014). *L: Building a lean hospital facility*. Signature Book Printing. www.sbpbooks.com.

Baccarini, D. (1996). The concept of project complexity – A review. *International Journal of Project Management, 14*(4), 201–204.

Ball, P. (2012). *Why society is a complex matter: Meeting twenty-first century challenges with a new kind of science*. Berlin: Springer-Verlag.

Bandura, A. (1986). *Social foundations of thought and action: a social cognitive theory*. Englewood Cliffs: Prentice-Hall.

Beal, D. J., Cohen, R. R., Burke, M. J., & McLendon, C. L. (2003). Cohesion and performance in groups: A meta-analytic clarification of construct relations. *The Journal of Applied Psychology, 88*(6), 989–1004. https://doi.org/10.1037/0021-9010.88.6.989.

Burke, C. S., Stagl, K. C., Salas, E., Pierce, L., & Kendall, D. (2006). Understanding team adaptation: A conceptual analysis and model. *The Journal of Applied Psychology, 91*(6), 1189–1207. https://doi.org/10.1037/0021-9010.91.6.1189.

Carron, A., & Brawley, L. (2000). Cohesion: Conceptual and measurement issues. *Small Group Research, 31*, 89–106.

Construction, M. (2013). *World green building trends: Business benefits driving new and retrofit market opportunities in over 60 countries*. Bedford: McGraw Hill Construction. Smart Market Report.

Crawford, E., LePine, J., & Rich, B. (2010). Linking job demands and resources to employee engagement and burnout: A theoretical extension and meta-analytic test. *Journal of Applied Psychology, 95*(5), 834–848. https://doi.org/10.1037/a0019364.

Cropanzano, R., & Mitchell, M. (2005). Social exchange theory: An interdisciplinary review. *Journal of Management, 31*, 874–900.

Fiske, A. (1992). The four elementary forms of sociality: Framework for a unified theory of social relations. *Psychological Review, 99*(4), 689.

Franz, B., Leicht, R., Molenaar, K., & Messner, J. (2016). Impact of team integration and group cohesion on project delivery performance. *Journal of Construction Engineering and Management, 4016088*–6(6), 1–12. https://doi.org/10.1061/(ASCE)CO.1943-7862.0001219.

Giddens, A. (1984). *The constitution of society*. Berkley: University of California Press.

Grönroos, C. (1997). Value-driven relational marketing: From products to resources and competencies. *Journal of Marketing Management, 13*, 407–419.

Gunderson, L. (2001). *Panarchy: Understanding transformations in human and natural systems*. Washington, DC: Island Press.

Hampson, R. B., Hulgus, Y. F., & Beavers, W. R. (1991). Comparisons of self-report measures of the beavers systems model and Olson's circumplex model. *Journal of Family Psychology, 4*(3), 326–340. https://doi.org/10.1037/0893-3200.4.3.326.

Holling, C. (2001). Understanding the complexity of economic, ecological, and social systems. *Ecosystems, 4*, 390–405.

Janssen, O. (2005). The joint impact of perceived influence and supervisor supportiveness on employee innovative behaviour. *Journal of Occupational and Organizational, 76*, 347–364.

Judge, T. A., Thoresen, C. J., Bono, J. E., & Patton, G. K. (2001). The job satisfaction – Job performance relationship: A qualitative and quantitative review. *Psychological Bulletin, 127*(3), 376–407. https://doi.org/10.1037/0033-2909.127.3.376.

Kadushin, C. (2012). *Understanding social networks: Theories, concepts, and findings* (Kindle ed.). New York: Oxford University Press.

Kozlowski, S. W. J., & Bell, B. S. (2012). Work groups and teams in organizations. In I. B. Weiner, N. W. Schmitt, & S. Highhouse (Eds.), *Handbook of psychology, industrial and organizational psychology*, John Wiley (Vol. 2, Illustr ed., pp. 412–469), Wiley.

Levitt, R. (2007). CEM research for the next 50 years: Maximizing economic, environmental, and societal value of the built environment. *Journal of Construction Engineering and Management, 133*, 619–628.

Lewicki, R., Tomlinson, E., & Gillespie, N. (2006). Models of interpersonal trust development: Theoretical approaches, empirical evidence, and future directions. *Journal of Management, 32*, 991–1022.

Lord, R. G., & Hall, R. J. (1992). Contemporary views of leadership and individual differences. *The Leadership Quarterly, 3*(2), 137–157. https://doi.org/10.1016/1048-9843 (92)90030-J.

Mayer, J. D., & Salovey, P. (1995). Emotional intelligence and the construction and regulation of feelings. *Applied and Preventive Psychology, 4*(3), 197–208. https://doi.org/10.1016/S0962-1849(05)80058-7.

Milgrom, P., & Roberts, J. (1992). *Economics, organization and management*. Englewood Cliffs: Prentice-Hall.

Mitchell, M. (2009). *Complexity: A guided tour*. Oxford University Press. https://info.aia.org/SiteObjects/files/IPD_Guide_2007.pdf

Mossholder, K., & Richardson, H. (2011). Human resource systems and helping in organizations: A relational perspective. *Academy of Management, 36*(1), 33.

Olson, D. H. (2000). Circumplex model of marital and family systems. *Journal of Family Therapy, 22*, 144–167.

Olson, B. V. (2016). Experiential workplace design for knowledge work organizations: Worker-centered approach (Unpublished doctoral dissertation). Case Western Reserve University, Cleveland.

Paolillo, W., Olson, B., & Straub, E. (2016). People centered innovation: Enabling lean integrated project delivery and disrupting the construction industry for a more sustainable future. *Journal of Construction Engineering*.

Perlow, L., Gittell, J., & Katz, N. (2004). Contextualizing patterns of work group interaction: Toward a nested theory of structuration. *Organization Science, 15*, 520–536.

Pescosolido, A. T., & Saavedra, R. (2012). Cohesion and sports teams: A review. *Small Group Research, 43*(6), 744–758. https://doi.org/10.1177/1046496412465020.

Podsakoff, P., MacKenzie, S., & Paine, J. (2000). Organizational citizenship behaviors: A critical review of the theoretical and empirical literature and suggestions for future research. *Journal of management, 26*(3), 513–563.

Podsakoff, P., MacKenzie, S., & Organ, D. (2006). Organizational citizenship behavior: Its nature, antecedents, and consequences.

Ramlall, S. (2004). A review of employee motivation theories and their implications for employee retention within organizations. *Journal of American Academy of Business, 5*, 52–61.

Rousseau, D. (1995). *Psychological contracts in organizations: Understanding written and unwritten agreements*. Thousand Oaks: SAGE Publications.

Rousseau, D. (2004). Psychological contracts in the workplace: Understanding the ties that motivate. *The Academy of Management Executive, 18*, 120–127.

Ryan, R. M., & Deci, E. L. (2000). Intrinsic and extrinsic motivations: Classic definitions and new directions. *Contemporary Educational Psychology, 25*(1), 54–67. https://doi.org/10.1006/ceps.1999.1020.

Salas, E., & Fiore, S. M. (2004). Why team cognition? An overview. In E. Salas & S. M. Fiore (Eds.), *Team cognition: Understanding the factors that drive process and performance* (1st ed., pp. 3–10). Washington, DC: American Psychological Association (APA).

Schaufeli, W. B., Salanova, M., Gonzalez-Roma, V., & Bakker, A. B. (2002). The measurement of engagement and burnout: A two sample confirmatory factor analytic approach. *Journal of Happiness Studies, 3*, 71–92.

Schein, E. H. (1996). Culture: The missing concept in organization studies. *Administrative Science Quarterly, 41*(2), 229–240.

Schwab, K. (2017). *The fourth industrial revolution*. New York: Crown Business

Settoon, R., & Mossholder, K. (2002). Relationship quality and relationship context as antecedents of person-and task-focused interpersonal citizenship behavior. *Journal of Applied Psychology, 87*, 255–267.

Sonnentag, S. (2003). Recovery, work engagement, and proactive behavior: A newlook at the interface between nonwork and work. *Journal of Applied Psychology, 88*(3), 518–528. https://doi.org/10.1037/0021-9010.88.3.518.

Steffen, W., Crutzen, P., & McNeill, J. (2007). The Anthropocene: Are humans now overwhelming the great forces of nature. *AMBIO: A Journal of the Human, 36*, 614–621.

Straub, E. R. (2015). Flexible cohesion: A mixed methods study of engagement and satisfaction in defense acquisitions (Unpublished doctoral dissertation). Case Western Reserve University, Cleveland.

Watson, K., Evans, J., Karvonen, A., & Whitley, T. (2016). Capturing the social value of buildings: The promise of social return on investment (SROI). *Building and Environment, 103*, 289–301.

Weick, K. E., Sutcliffe, K. M., & Obstfeld, D. (2008). Organizing for high reliability: Processes of collective mindfulness. In A. Boin (Ed.), *Crisis management* (Vol. III, pp. 31–66). Los Angeles: Sage Publications.

Wolin, S. J., & Bennett, L. A. (1984). Family rituals. *Family Process, 23*(3), 401–420.

Teaching Circular Economy

Overcoming the Challenge of Green-washing

Helen Kopnina

Contents

Introduction: The Concept of Sustainability in Linear and Circular Systems	810
Transition to Circularity	811
Case Studies: Vocational Schools	815
Findings at Vocational Schools	816
Reflection on Student Projects at Vocation Schools	817
Case Studies: Leiden University College (LUC)	818
Findings at LUC	818
Reflecting on Written Assignments at LUC	827
Conclusion	829
Cross-References	830
References	830

Abstract

This chapter will introduce circular economy (CE) and cradle-to-cradle (C2C) models of sustainable production. It will reflect on the key blockages to a meaningful sustainable production and how these could be overcome, particularly in the context of business education. The case study of the course for bachelor's students within International Business and Management Studies (IBMS) and at the University College in the Netherlands will be discussed. These case studies will illustrate the opportunities as well as potential pitfalls of the closed-loop production models. The results of case studies' analysis show that there was a mismatch between

---

H. Kopnina (✉)
Faculty Social and Behavioural Sciences, Institute Cultural Anthropology and Development Sociology, Leiden University, Leiden, The Netherlands

Leiden University and The Hague University of Applied Science (HHS), Leiden, The Netherlands
e-mail: h.kopnina@hhs.nl; alenka1973@yahoo.com

© Springer International Publishing AG, part of Springer Nature 2018
S. Dhiman, J. Marques (eds.), *Handbook of Engaged Sustainability*,
https://doi.org/10.1007/978-3-319-71312-0_48

expectations of the sponsor companies and those of students on the one hand and a mismatch between theory and practice on the other hand. Helpful directions for future research and teaching practice are outlined.

Keywords
Circular economy · Cradle to cradle · Environmental education · Sustainability

Introduction: The Concept of Sustainability in Linear and Circular Systems

The mainstream sustainability models tend to focus on minimizing environmental damage, with negative effects merely delayed but not eliminated by eco-efficient technologies (McDonough and Braungart 2002). John Foster (2012) has noted that consumer capitalist thinking has largely allowed unsustainability to persist. In the words of Blühdorn (2007), symbolic politics of sustainability only "sustains the unsustainable." Despite international conferences and talks, any large environmental challenge from climate change to biodiversity loss (Crist et al. 2017) to pollution (e.g., The Economist 2016a) has not been solved. The key challenge is overcoming the cult of economic growth (O'Neill 2012) and associated anthropocentrism (Crist 2012). These ideologies perpetuate themselves in practice that subordinates environment to industrial development demands (Washington 2015) and in business education that tends to present the triple Ps of *p*eople, *p*rofit, and *p*lanet as equal "partners" (Kopnina 2012, 2015a, b). Production systems seem to be oriented toward the only bottom line, *p*rofit, and in fact subordinate *p*eople and *p*lanet to industrial development and economic growth (Rees 2010; York 2017). Critical sustainability researchers have proposed degrowth (O'Neill 2012) and the transition to the steady-state economy with fixed population and a constant sustainable throughput of resources (Washington 2015). However, as long as human population continues expanding and material needs remain unsustainably high or continue to rise, there is also an urgent need for the systems of production that radically overhaul the take-make-waste ("cradle-to-grave") manufacturing.

To further this aim, cradle-to-cradle (C2C), the concept developed by McDonough and Braungart (2002), and the circular economy (CE) (Webster 2007) models offer alternative perspectives (Kopnina 2015b, 2017). C2C framework adheres to "waste equals food," "respect diversity," and "use solar income" principles (McDonough and Braungart 2002). "Waste equals food" refers to materials that circulate either in biological (organic, biodegradable) or technological cycles, which are never wasted in the landfills or destroyed but endlessly reused. "respect diversity" implies to both natural (biodiversity) and cultural diversity or locally informed ways of knowing and making products. The "use solar income" also extends to other forms of endlessly renewable energy, including wind, kinetic, and tidal waves. The C2C framework suggests that current efforts to mitigate harm through eco-efficiency do not reach deep enough to promote fundamental change.

Current research in sustainability education often involves the conventional models of "sustainable consumption" based on the diffused notion of sustainable development

(Kopnina 2012, 2014; Bonnet 2013; Washington 2015). While sustainable consumption in education has also been addressed through the circular models (Webster 2007; Boer et al. 2011; Huckle 2012; Kopnina 2013, 2015a, 2016b), the critical evaluation of these transformative models in education has not yet been realized.

This chapter will focus on transformative models of sustainability and discuss the associated challenges at teaching about CE and C2C at vocational and liberal arts school levels. This chapter will first discuss the background of circular frameworks and then turn to the case study of experiential (practical) and theoretical education in the Netherlands. The sections below will examine production models that reach beyond conventional sustainability with its focus on closed-loop models and then turn to examples from educational practice. In this chapter, I shall explore how students can be taught to distinguish between linear and circular models and how the pitfalls of subversion can be avoided. The following sections will discuss different sustainability models including areas in which subversion is possible. The implications for teaching circular economy to bachelor's students will be discussed within the case study of an experimental online course piloted by vocational schools in Rotterdam, Utrecht, and The Hague and Leiden University College in the Netherlands. This course was targeted at increasing students' awareness of alternatives to mainstream production models.

Transition to Circularity

Waste has become one of the prominent features of industrial development. According to UN's Food and Agriculture Organization, a third of all produced food, about 1.3 billion tons, is wasted before it reaches the consumer, and about 2–5% of meat is thrown away (The Economist 2016b). According to the European Commission, about 162.9 kg of packaging waste was generated per inhabitant in the European Union in 2014, including "paper and cardboard," "glass," "plastic," "wood," and "metal" (EC). One of the measures to address this waste is circular economy.

Based on C2C framework, Ellen MacArthur Foundation has popularized the concept of circular economy in Europe and Northern America. Based on the Circular Economy Stakeholder Conference held in Brussels in March 2017, the European Commission and the EESC jointly launched the European Circular Economy Stakeholder Platform (http://ec.europa.eu/environment/circular-economy/index_en.htm). The organizers of this platform bring together circular economy-related networks and platforms; groupings and organizations of businesses, of trade unions, and from the civil society; networks of national, regional, and local public authorities/bodies; and organizations from the knowledge and research communities, think tanks, and universities.

The term circular economy underlies the role of natural diversity as a characteristic of resilient and productive systems and underscores the importance of the process of production in which "waste" from a production process is used as a resource for new products. This process is known as end-of-waste (EoW) criteria (Zorpas 2016) combining environmental protection and public health, take-back requirements, and extended producer responsibility.

Ideally, circular system not only improves/optimizes resource yields and reduces production risks by managing renewable flows but also transforms the system of production from the onset. The crucial point is not dealing with waste after it has been already produced, as most of conventional sustainability proponents do (e.g., G-Star, a denim producer, makes some of their jeans out of plastic from the oceans) but *not producing* anything that results in waste in the first place. We note that this is an aspiration, which has yet to be demonstrated by government policies, corporate leaders, and general public.

The idea of circular economy stems from the 1976 report to the European Commission, *The Potential for Substituting Manpower for Energy* written by Walter Stahel and Genevieve Reday (1976). Stahel (1997) has further developed a concept of "self-replenishing economy" through cycling materials and has developed the foundations of what came to be known as "functional economy." The functional economy emphasizes turning products into services through leasing contracts. Stahel (1997) also argued that functional economy should lead to an increase in jobs as labor is required to keep products in use through each use phase. Murray et al. (2017) define circular economy as "an economic model wherein planning, resourcing, procurement, production and reprocessing are designed and managed, as both process and output, to maximize ecosystem functioning and human well-being." The concept of circular economy is intertwined with other subfields and disciplines of industrial ecology and environmental economics where it is seen as a way to limit and ideally keep low and constant the amount of resources extracted. C2C and CE framework works with the concepts of biological (organic) and technical (synthetic) nutrients or metabolisms, all materials that operate in regenerative cycles. Understanding of these cycles requires understanding of life cycle analysis (LCA), also known as cradle-to-cradle analysis, a method to assess a product's environmental impacts taking into account all stages of the product's life.

The circular approaches are basically critical of mainstream approaches to sustainability that focus on "eco-efficiency" that merely reduce but do not eliminate damage (Brennan et al. 2015; Lieder and Rashid 2016). For example, recycling involves down-cycling, as this process requires energy and transportation and is in actuality converted from valuable products into low-value raw materials (McDonough and Braungart 2002). Another example includes "waste to energy" initiatives. While burning trash to generate electricity may appear sustainable, it is still cradle-to-grave model in which both organic and technical resources are wasted for a short burst of energy (McDonough and Braungart 2002). The same can be said about the use of biofuels (largely advertised as "green" renewable energy source) that are derived from trees; wooden pallets or algae also have a negative impact on forests that would otherwise absorb carbon (The Economist 2015a). Also, biofuels compete for productive agricultural land and threaten biodiversity (Kopnina 2016a). Instead, circular economy and C2C propose production systems and materials without the loss of value, where materials are *endlessly reused* rather than (slowly) degraded. Examples range from preindustrial production and distribution systems (e.g., milk directly from the farmer distributed in glass milk bottle that is collected and refilled by the milkman) to hyper-modern designs, including the use of "true" renewable energy sources (Kopnina and Blewitt 2014). By "true" renewables, it means an endless supply of it, namely, from wind, sun, or air pressure – which are (to be) used in inventions ranging from Hyperloop and solar airplanes.

Another feature of C2C circular economy is dematerialization and transition from manufacturing to services. In product-service system or PSS (Mont 2002), it is recognized that a consumer does not need a drill but a hole in the wall and companies can provide a drill-leasing service. This means that the product (in this case, the drill) should be durable but also easily repairable, so it can be shared by hundreds of customers and, ideally, made of cradle-to-cradle materials.

The application of these ideas is evident in Europe, where PSSs were largely promoted by environmental agencies (Tukker and Tischner 2017), but also in the rest of the world. In China in 2002, the government has adopted the circular economy as a new strategy for development (Yuan et al. 2006; Geng et al. 2016). Part of these circular economy strategies were establishment of educational centers, programs, or curriculum. Ellen MacArthur Foundation, for example, offers free teaching materials and power points to enable engineering, design, and business students and professionals to learn more about sustainable consumption and production. In Chinese *Law for the Promotion of the Circular Economy*, the government agencies stimulate research, development, promotion, and international cooperation of science relating to circular economies, as well as supporting the education (http://www.lawinfochina.com/display.aspx?id=7025&lib=law).

To further these aims, C2C uses LCA for its certification process. The cradle-to-cradle certification spans over five categories, which include material health, material reutilization, renewable energy and carbon management, water stewardship, and social fairness. Based on these categories, the product or process can receive five different levels (Basic, Bronze, Silver, Gold, Platinum), with the lowest level reached in any of these categories determining the final certification. Such assessments need to be both rigorous and continuous in order to avoid the danger of slipping back into unsustainable patters.

Potential Risk of Subversion of Transformative Frameworks

The circular economy models are not without limitations. First, transition cannot be only top-down. Stegeman (2015) notes that the decisive factor is consumers' behavior change, longer use, and lower consumption of products, consumers need to change their behavior. Despite the success of Dutch and international collaborative platforms like Marktplaats, Snappcar, Peerby and Airbnb that offer 'shared' services their popularity does not automatically lead to circular economy (Stegeman 2015).

Secondly, circular economy is restricted by traditional economic reasoning. Ellen MacArthur Foundation as well as European Union and Chinese circular economy initiatives see circular economy as a "new engine of growth," and the top "best practice" examples from its website indicate that business as usual is happening. In this framing, the circular economy contradicts degrowth discourse (O'Neill 2012) and rejects the need for critically addressing economic growth (Jackson and Senker 2011; Washington 2015). China, for example, uses its CE policies which are presently primarily targeted at reduction of waste (Ghisellini et al. 2016), as illustrated by the Chinese documentary films "Beijing Besieged by Waste" and "Under the Dome." Chinese industry at the moment shows no clear signs of overall transition to circular economy and maintains unsustainable levels of production and consumption, although some significant strides in sustainability have been made (The Economist 2015b). The unique points of circular economy framework are put in contrast to conventional economic models in the Rabobank report:

> Circularity does not refer to the macroeconomic process (which is already circular), it refers to making the circuits of materials and goods circular. It is thus not a new economic model, it is foremost about practical and useful thinking to structure an effective economy based on the efficient use of materials and reducing and ultimately eliminating waste flows.... And here we encounter an essential problem in our analysis. In the macroeconomy, the macroeconomic cycle is the focus of the analysis. Not the materials cycle, or the effect on inventories of natural resources or waste. (Stegeman 2015)

Thirdly, circular economy calls for the need to reorganize production processes and product expectations. Rejecting the built-in obsolescence principles of modern production based on continuous purchases means that direct sales of new products will decrease (Brennan et al. 2015). Also, currently, remanufacturing seems to be viable in business-to-business niche markets and not so much in consumer markets (Vogtlander et al. 2017). Indeed, to enable and accelerate circular economy transition driven by industry, it is necessary to "tap potentials of CE transition scenarios on company and inter-company level" (Lieder and Rashid 2016, p. 13).

Fourthly, the economy of scale necessary for managing "waste" of almost eight billion consumers is overwhelming. Even if human excrement is fully reused for fertilization to complete the biological cycle, the sheer volume of food produced and consumed cannot realistically be turned into "service economy." Consumption of meat, for example, results in methane emissions that contribute to climate change and violate animal welfare through CAFOs or intense feeding operations (Crist 2012). While "feeding the world" is possible, given a better distribution system and more intensification of agriculture (The Economist 2009a), food production and consumption are far from "circular." Also, on a greater scale, while consumption and fertility might be falling in some parts of the world, they are increasing in others (The Economist 2009b, c, 2013). For example, consumption of meat is supposed to increase by three quarters in a few decades as new middle classes in China, India, and many countries in Africa (The Economist 2016b). Circular economy alone cannot reduce the mere volume of food produced and consumed (Rammelt and Crisp 2014). As long as the economy keeps growing, the adoption of circular economy can postpone the time it takes to reach the boundaries of the biophysical envelope (Rammelt and Crisp 2014). Thus, circular economy in business needs to be understood as most effective in the context of degrowth (O'Neill 2012) and steady-state economy (Washington 2015). In education, Huckle (2012) has noted that Ellen MacArthur Foundation's teaching materials while they are strong on technical aspects of sustainability literacy were seriously deficient with regard to critical aspects. As the case studies below will illustrate, CE and C2C can also be coopted to justify "business-as-usual" economic growth models, wrongly assume that easy consumer behavior change will be easy, and ignore the challenge of increasing population and material demands.

Case Studies

The Netherlands has been involved in a number of educational activities already involving awareness of circular economy, some of them described by Boer et al. (2011). More recently, the initiatives have included the Dutch foundation Stichting Duurzaam Geleerd (SDG; Learning for a Sustainable Future) that specifically promotes C2C and circular economy in secondary education (http://mycircularfuture.com/en/). The courses below represent how integrating C2C and circular economy can be achieved in general sustainability courses, targeted both at practical skills and theoretical knowledge development. What makes the courses described below unique is that aside from introducing general concepts of sustainability and circularity, the curriculum was focused on development of critical thinking also in regard to issues mentioned in the introductory sections (detailed description of the content is found on Kopnina in print). In the case of IBMS departments of vocational schools, the method of instruction was a series of lectures discussing circular framework in the context of business operations. A more theoretical course at Leiden University College delved deeper into some of the literature and debates mentioned in the Introduction and "Potential Risk of Subversion of Transformative Frameworks" sections of this chapter.

Case Studies: Vocational Schools

This case study is based on the minor Circular Economy in the Cloud, an online course piloted by Universities of Applied Science (vocational schools) in Rotterdam, Utrecht, and The Hague International Business and Management Studies (IBMS) departments. The author of this chapter was in a position of tutor/assessor. The main objective of this course, given between September 2014 and February 2015, was to teach students what circular economy is and how small- and medium-sized enterprises (SMEs) could make the transition from a linear to a circular model of operation. Sixty-eight students were initially enrolled in the minor. The author of this chapter has supervised two teams of students with four students in one group and five in another group. All students had a background in international business, marketing, finance, and branding. Seventeen companies were selected on the basis of their interest to participate as they expected to benefit from the practical solutions. The intention was that students help companies to enable transition to circular production model and simultaneously offer competitive advantage over their "linear" competitors. The two teams that the author has supervised cooperated with a company that made bridges ("Bridges") and a company specialized in renting camping equipment ("Tents").

Bridges was founded in 2012 and separated into three departments: engineering, research and development/innovation, and products (bridges and pipes). The student group worked with the products division. As outlined on the company's website, the bridges of Bridges company have the following advantages: "Expected lifetime of hundred years, no maintenance required, easy to implement, and zero erosion from nature" (company website is anonymous). The bridges are manufactured from steel, plastic, resins, and fiber, which are provided from two main suppliers. In addition to preparing a report advising how the company should apply the circular economy concept to their business, market expansion was one of the central aims of the student project. Apart from the conventional wooden bridge manufacturers in the Netherlands, there was one main composite bridge manufacturer with a capacity of 100 bridges annually, leaving a capacity of 400 bridges to be targeted. In the long term, the company wanted to expand their operations to other European countries.

Tents company's main business is renting airbeds, tents, and sleeping bags, delivered to festivals. Tents claimed that it was able to reduce the carbon footprint of festivals by recycling the tents into a high-quality granulate used for future tents. According to the company's website, the tents are made from 100% recycled film or plasticulture that is also used in agriculture. Through a deposit scheme, Tents encourages the users to return the used tents to the materials' supplier, claiming to be a green company. One of the tasks that the company assigned to students was to help further improve its green credentials by providing advice as to how to make its operations more "circular." The company's greatest interest, however, was in expanding its market to the United Kingdom (UK). According to the guidelines given to students, the plan needed to contain an outline of festivals, attendees, turnover, margins, archetypes of the UK festival market, a list with possible partners, and analysis of the competition.

Findings at Vocational Schools

In the case of Bridges, in their evaluation interview with the instructor, the students reflected that they were refused information about the material use in the whole supply chain. One of the students has reflected: "I think they [the company] had no interest in circular economy in the first place. I think they were just thinking of expanding... and we [as business students] could help them. ..." Indeed, as reported by other members of the group, the most important objective of student participation, as far as their sponsor was concerned, was to make a comparison between wooden bridges and composite bridges and provide the client with an analysis that focused on sustainability. This report was to be based on the research of the business processes of two competitors, specifically the value chain. In their research, in accordance with the company's wishes, the group was set out to advise Bridges how to create a competitive advantage based on the value chain and material analysis, no matter how circular or linear the processes are. In their desk research about materials used for bridges, students distinguished synthetic fibers and the natural raw materials used to create synthetic fibers. The former include raw petroleum (carbon), silica (glass), and basalt (ceramic) and natural fibers found in plants. The latter include cellulose fibers, animal (protein) fibers, and minerals (such as asbestos, which is now banned due to its carcinogenic qualities). Students have found that while some materials have offered good examples of either biological or technical nutrients, it was the "hybrid" synthetic materials that were preferred by the company. However, changing materials to more sustainable ones would require an overhaul of the existing business model. The company supervisor has communicated to the students during the last stage of their project that the Bridges will remain true to its original business model as it "proved to be profitable." The feasibility assessment consisted of an evaluation of the company's and competitors' value chains and suggestions how Bridges can be promoted as being more sustainable than competitors.

As for the Tents, in analyzing inbound/outbound logistics of Tents, the students noted that the current products are brought from China to be transported to the targeted locations at the festivals or the homes of the clients in the Netherlands. The materials used were mostly recyclable plastics and cotton fiber mass. The students noted that the current way that the products are produced and transported results in high costs and CO2 emissions. The students have contemplated production at local sites, finding local suppliers and revising outbound logistics. They have also suggested the use of materials that could extend the longevity of the tents to make them more suitable for refurbishing, using a tent repair kit. The main renting service could be supported by additional services, such as cleaning, repairing, and, if needed, replacing inner tents. The students have noticed that domestic production, as well as repair and other services, will be costlier than the currently employed system. The company's director has admitted that the current model is far from being a complete circular economy operation but that his company would be willing to take the necessary steps if a transition was feasible within the current business model. All student suggestions, however, have led the team as well as company supervisor to a conclusion that a complete revision of the presently used business model will be necessary to achieve recommended improvements. The company seems to have followed the habitual course without notable changes.

Reflection on Student Projects at Vocation Schools

In student projects, the "business as usual" qualities were sought after – students' expertise in marketing, branding, and finance. Perhaps unintentionally, it appeared that the reports of business students could help companies to underline their comparative advantage of being green. The Tents company supervisor, in a meeting with this researcher, has admitted that changing production processes, including materials and transportation, will not be considered, as the company has no surplus capital to invest in this costly change. Confronted with the need for the radical overhaul of established practices within the entire supply chain and the question of financial viability, the company practically withdrew its ambition for transition. The students have reported that they have learned that the transition was not really feasible.

The real-life examples have confronted both students and companies with difficulties of linking one small company to global supply chains in such a way that all stakeholders – from producers of raw materials to consumers – could contribute to the overall reorientation of the manner of production. As a result, the students felt disappointed, reporting that the companies were not willing to take their advice on board and doubting in general whether transition to circular economy is possible. While the students were able to make recommendations for expansion of the existing business into international market, they could not contribute to the circular mode of operation. However, in the evaluative meeting with the students, they have reported that they did gain critical learning experience.

Case Studies: Leiden University College (LUC)

The author was also involved in instruction of students of the advanced course environment and development (E&D) at Leiden University College (LUC), for the course was offered between August 2015 and March 2017. As in the case of vocational schools, the author of this chapter was in a position of tutor/assessor. In the course's individual assignments, the students were asked to reflect upon the difference between conventional (eco-efficiency, reuse, reduce, recycle) and circular sustainability frameworks. These assignments as well as detailed course structure are described in Kopnina (in print).

Findings at LUC

The students demonstrated that they have grasped theoretical background and origin of the concepts. One student has reflected in their writing assignment:

> Instead of focusing on continuous growth and expansion, the circular economy envisions a closed loop of regeneration and reuse (Kopnina and Blewitt 2014). The concept is widely used in the fields of industrial ecology and environmental economics, where it is seen as a way to both minimise the waste-flow into the environment (Andersen 2007). Inspired by circular systems in nature, the proponents believe that the current volumes of human waste are unnecessary (Stahel 2016). Reuse of products if possible, and recycling the remnants of a product when it can no longer be refurbished are at the core of the circular economy's logic (ibid.). There is also a need to change thinking and behaviour. For the circular economy to be successful, definitions of ownership and the positive connotation of having something new need to be reconsidered. Collaborative models of consumption ('the sharing economy') can significantly reduce the human environmental impacts by decreasing levels of consumption (Ghisellini et al. 2016). The effectiveness of many products could increase spectacularly if they were to be shared instead of stored somewhere without a purpose. Next to that, because ownership gives way to stewardship, systems where the producer retains ownership over a product while the consumer only pays for its usage, give producers more incentives to make durable products that can be reused and ultimately recycled. (Stahel 2016)

The distinction between eco-efficiency and eco-effectiveness was identified. One student wrote in his essay:

> Focusing on production, two schools, broadly speaking, can be identified: eco-efficiency and eco-effectiveness (Golinska 2014). The former can be called the more traditional and

conservative, which has as its core aim to reduce harmful effects of production and consumption. Arguably, eco-efficiency supports exceedingly modest goal of making business as usual "less bad". Braungart and McDonough (2002) argue that eco-efficiency harms the cause it is supposed to serve as it fails to acknowledge the urgent necessity of radical sustainable innovation and confront issues like toxicity in material flows.

One student wrote her essay about Narayana Peesapaty's Bakeys' edible spoons and other utensils. These spoons are made of water, sorghum flour, rice flour, and wheat flour, without preservatives and pesticides (http://www.bakeys.com/). The locally produced millet can be naturally flavored and turned into cutlery which requires little water compared to rice (Sadras et al. n.d.). Bakeys' business model is based on Kickstarter, a crowd-funding organization that represents mostly American investors and consumers, and as the majority of interested donators come from the United States (Table 1).

In her report, this student has reflected that it is unknown how the factory producing spoons is run and whether it uses solar energy is unknown. Also, the "exact manufacturing process remains a mystery." The student found the Bakeys' distribution system especially worrying as mostly using lorries within India. Yet, the major demand for spoons comes from Americans that have sponsored the company via Kickstarter. While according to the Bakeys' website the spoons "should be used in rural India by people who sell food on the street" (Bakeys n.d.), flying spoons to America defeat their purpose. Also, due to American food and safety regulations, the spoons had to be delivered in plastic packages. In reflecting on this paradox, the student makes a more general comment about the risk of subversion of C2C:

> It is indeed their consumption that allows Peesapaty's edible spoons to be a viable business concept at all, but it is also this consumption that makes it problematic. It highlights the problems that I have with the Cradle to Cradle concept. I would prefer people to use metal cutlery that lasts for decades and not because current consumers use plastic cutlery instead.

Another aspect of this criticism is the sociocultural system of inequality:

> It is mostly people of lower caste who use plastic cutlery when consuming street food because, after using it, the things they touched are considered no longer suitable for those of higher social standing. The real problem here is the use of plastic cutlery – the result of caste discrimination in India where it is simply not possible to grant access to metal cutlery to untouchables – and the fact that demand for edible spoons is currently highest in the United States.

The student noted that for polypropylene spoons, polypropylene is created first by steam cracking propane and butane, steam cracking naphtha, or catalytic cracking of gas oil. This is a process using several chemicals. The propylene gas created then has to undergo a process with the Ziegler-Natta catalyst and a solvent, which is called polymerization, and then, under high temperature, molded into a spoon shape. A schematic depiction of the process from propylene to polypropylene can be seen in Fig. 1 below.

Table 1 Table listing ingredients of Bakeys and polypropylene spoons and the environmental impacts of those ingredients

Type of spoon	Ingredients	Origin	Renewable?	Environmental impacts
Polypropylene	Ethane	Petroleum oils and coal	No	Usage of these kinds of compounds leads to carbon emissions and thus contributes to climate change
	Propane	Natural gas, light crude oil, and oil-refinery gasses	No	Usage of these kinds of compounds leads to carbon emissions and thus contributes to climate change
	Butane	Natural gas and crude oil	No	Usage of these kinds of compounds leads to carbon emissions and thus contributes to climate change
	Naphtha	Distillation of hydrocarbons	No	Usage of these kinds of compounds leads to carbon emissions and thus contributes to climate change
	Petroleum	The earth's crust	No	Usage of these kinds of compounds leads to carbon emissions and thus contributes to climate change
	Ziegler-Natta catalyst	Halides of titanium, chromium, vanadium, and zirconium. Alkyl aluminum compounds	No	Toxic to the environment and humans
Bakeys	Water	Rivers and other fresh water basins	Depends on abundance (UN-water WWAP)	Misuse of water could lead to water scarcity, for both humans and ecosystems, especially in arid regions
	Sorghum flour	Agriculture	Yes	Land use change, water use
	Rice flour	Agriculture	Yes	Land use change, water use
	Wheat flour	Agriculture	Yes	Land use change, water use

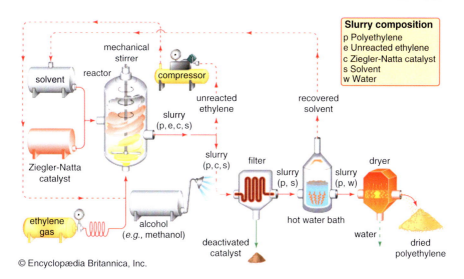

Fig. 1 Schematic depiction of the Ziegler-Natta polymerization of ethylene; for polypropylene, replace ethylene gas with propylene gas (Source: adapted by student from Encyclopaedia Britannica)

Indeed, one of the issues with C2C, as noted by the student, is that certain "good" products, such as metal spoons that last for generations, already exist. As the student notes, the use of existing products may be more sustainable than the consumption of more and more "revolutionary" C2C products. The larger question of economic growth and consumerist culture needs to be addressed:

> If one were to criticise manufactures, it should be planned obsolescence they target, not necessarily their production methods. A disposable shoe may not have as much of an environmental impact when disposed of than a pair of hard-leather shoes, but in the end, more resources – with the pollution that often accompanies it – have to be spent producing new pairs. Manufactures should make products that are built to last, and consumers should use them until they break – or give them away when they no longer like them – rather than throw them out.... The point is to decrease our ecological footprint; then why do we not implore the consumer to buy second-hand – even if such products are not strictly C2C – and give away rather than dispose products that are no longer wanted? Even if one argues that they are just working within the framework of contemporary consumerist culture, Braungart and McDonough do then not take into account economies of scale. In other words, C2C either does not properly address consumer culture as a theory, or it does not address economic reality as a practice.... Cradle to Cradle draws our attention away from the sustained use of long-lasting products to products that may sustainably be disposed of. As the authors of the C2C like to accuse other approaches to sustainability of greenwashing, it is only right to point out that they are essentially guilty of the same practice.

Adding to this, the student reflects that while changing consumer behavior is difficult, the responsibility of consumers is necessary to cut down production. Reflecting on the role of consumer and not just producer needs, this student wrote:

The problem does not necessarily lie with the manufacturers of cradle to grave products; it is instead the consumers we should blame for the current situation. Braungart and McDonough claim that it is up to industrial designers and manufacturers to produce goods with a small footprint that are easily re-used, or that we return to pre-industrial ways of manufacturing that, in my experience, nobody really desires. The first is unattainable due to economies of scale – such products will remain luxury goods until we are able to mass-produce them, after which they will no longer be desirable because they are then very commonly-used and the second is rather patronising to those who do not enjoy the same standards of living that we in the West do...

Another student investigated Ellen Mac Arthur Foundation website and found the Coca-Cola Enterprises' "base case study" commitment to "maximising the usage and value of the plastics used in bottle production" (EMF 2016a, b). The student discovered that the company's trademarked "PlantBottle" is in fact *partially* (up to a maximum 30%) based on plants (Coca-Cola 2012). While Coca-Cola claims that the PlantBottle packaging has reduced emissions (Coca-Cola 2015), the company seems to have no serious plans about recycling let alone reusing the bottle after its content has been consumed (Terry 2011). In fact, the PlantBottle represents a case of "monstrous hybrid" (Braungart and McDonogh 2002), a mix of biological and technical nutrients that can neither be recycled nor safely disposed (Fig. 2).

In analysis of Coca-Cola's efforts to adapt a more circular economic-model for its plastic bottles shows that some initiatives are not as transformative as one would expect them to be based on the CE framework. The paper concludes that for the circular economy to have impact, all actors should really start to walk-the-talk and go beyond symbolic language and small measures. Truly transformative change by both government and private sector is necessary to reduce human environmental impact and to protect ecological systems.

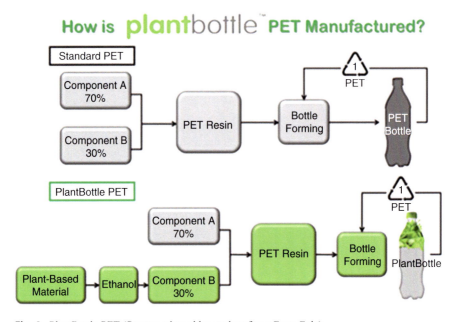

Fig. 2 PlantBottle PET (Source: adapted by student from Coca-Cola)

Another case discussed by the student is that of Satino Black hand towels produced by the Dutch company Van Houtum. The student notes that presently, in the United States alone, more than 13 billion pounds of paper towels, created of tree pulp, water, chemicals, and energy, are used annually. Globally, 254 million tons of paper towels is disposed every year. The paper towels are distributed to the consumers by first delivering it to grocery stores in trucks or ships, to be consequently used once and dumped in the mixed garbage container. In Van Houtum's product, all hazardous chemicals have been phased out and substituted by biodegradable alternatives. To reduce transportation, Van Houtum exports no further than Greece. The company is also not planning on expanding their exports but rather aims for "value increase." ("Van Houtum Aims to Better Substantial Value Growth in Washroom Solutions | Tissue World Magazine." Accessed October 25, 2016. http://www.purpleprint.eu/twmagazineclon/featured-slider/van-houtum-aims-to-better-substantial-value-growth-in-washroom-solutions.) The hand towels were certified in 2010, and its C2C scorecard is displayed in Fig. 3.

The student further discussed C2C certification impacts on the five quality categories and outlined the following:

(i) *Material health:* This quality category assigns each material used for the product a material health rating. The rating is based upon a thorough analysis of toxicity of the materials, including the hazards associated with the disposal of the product. A "green" rating means that the material is supporting C2C's objectives, a "yellow" rating means that there are moderately problematic properties of the material, and a "red" rating means that there are highly problematic properties of the material, meaning that it should be phased out in order to optimize a product. Material health in Satino Black paper towels has been rewarded the "Silver" stamp, making it the least score of all five categories. When compared to the material health of its non-C2C equivalent, it becomes clear that the Satino Black hand towels scores way higher: all "X" or red-rated chemicals are removed from the product and packaging. Figure 4 gives an overview of the material health's performances of both products.

(ii) *Material reutilization:* The Satino Black hand towels are produced almost fully from recycled paper. This was identical before and after certification. In this category, the hand towels have been rewarded the "Platinum," which means that the product is designed in such a way that it is intended for the technical and/or biological cycle, and more than 80% of the nutrients used are reutilized (Fig. 5).

(iii) *Renewable energy and carbon management:* The current hand towels score "Gold" on this category, which is achieved through energy management optimization (i.e., less demanding production and better energy sourcing). In 2008

Fig. 3 C2C scorecard for Black Satino hand towels

(before certification), the consumption of renewable energy for the production of the hand towels was at 8%. In 2012, the consumption of nonrenewable energy was at 100% (from green gas and hydroelectricity) (Fig. 6).

(iv) *Water stewardship:* The hand towels have been rewarded the level "Gold" for this category. Both the amount of abstracted water and the amount of wastewater per ton of product have remained nearly the same before and after certification. The amount of abstracted water even appeared to be a bit higher after certification, but according to Van Houtum, this is not related to certification but rather to the fact that water consumption is measured at the site itself, which allows for discrepancies (Fig. 7).

(v) *Social fairness:* Van Houtum has been rewarded the level "Gold" for this category. Van Houtum has not changed its view/policy toward social responsibility since

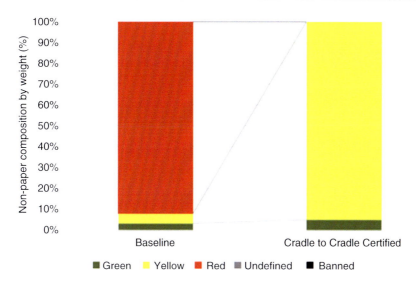

Fig. 4 Material health of hand towels before and after certification

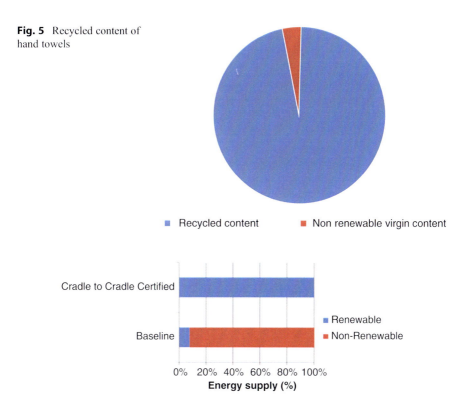

Fig. 5 Recycled content of hand towels

Fig. 6 Percentage of renewable energy used in production of hand towels before and after certification

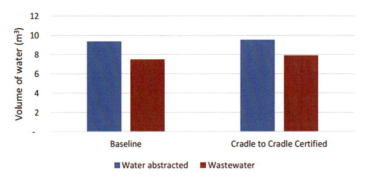

Fig. 7 Water consumption of product before and after certification

certification of the hand towels. Every year, the company produces a social responsibility report (Van Houtum. "CSR Report 2015." http://www.vanhoutum.nl/en/csr/2594/csr-facts-and-figures.html.). Furthermore, the two processing sites are monitored and certified, making sure that the sites meet the minimum safety and health requirements ("People – Van Houtum." Accessed October 25, 2016. http://www.vanhoutum.nl/en/csr/vision-mission-policy/2586/people.html.).

The student has noticed that since Satino Black hand towels' certification in 2010, sales have increased significantly. According to Van Houtum, the C2C certification has been a competitive advantage, and suppliers down the chain became more involved by offering ideas, thoughts, and innovations on further improvements ("Succes Met Cradle-to-Cradle Toiletpapier | Supply Chain Magazine." Accessed October 25, 2016. http://www.supplychainmagazine.nl/succes-met-cradle-to-cradle-toiletpapier/.). Figure 8 shows that between 2010 and 2012, sales went up significantly.

In conclusion, the student found that C2C certification for the Sativa Black hand towels had a positive influence on all five of the quality categories as well as resulted in a significant increase in sales of the product. The student further reflected that although the C2C certification is not meant to be an innovation driver, it is a way for companies to show their progress toward a more sustainable product. The student noted that:

> a heavy emphasis often lays on the business benefits of C2C certifications, i.e., promoting the financial benefits to encourage environmental awareness. This could be explained by looking at the notion of anthropocentrism: due to our nature, humans are more likely to respond to self-interest motives.

Some government agencies also use the Ellen MacArthur Foundation's platform, even though their policies might not be strictly "circular."

One of the policy-oriented case-studies is the Dutch 'Green Deal' approach (EMF 2016a). The Green Deal is an initiative by a couple of Dutch governmental ministries to facilitate sustainable initiatives... Although the Foundation's website (EMF 2016a) makes it look like the Green Deal is meant only for projects that fit the circular economy model, the Green Deal website (2016) explains that the project is meant for a broader range of sustainable initiatives. In fact, most of the Green Deals at this point are focused

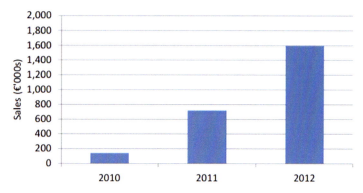

Fig. 8 Sales of product since C2C certification

on emission reduction, mainly addressing symptoms rather than their underlying causes of unsustainability. For example, one of the recently established green deals is focused on reducing emissions of mobile machinery by 10% in various construction sectors. It aims to do so mainly by teaching the (engine) drivers how to operate the machines in ways that produce fewer emissions. However, to get closer to the core of the problem one would have to think about how these machines could not be fuelled by fossil fuels. To be truly transformative one should consider whether many of the construction projects should be initiated in the first place, when there are also many existing structures that could be refurbished. Decreasing the number of demolition- and construction projects would reduce emissions, resource extraction and waste.

Pondering applicability of C2C and CE, one student reflected:

> There are certain limitations to the implementations of these models and their measures. Cradle-to-Cradle and the Circular Economy certainly actualize theoretical conceptions and enhance one's ability to partake in private sphere environmentalism (Kopnina and Blewitt 2014). However, these measures alone cannot solve the paradoxes and complexities of environment and development. Cradle-to-Cradle creates a more sustainable method of production, yet Circular Economy alone cannot address rampant consumerism and overpopulation. These efforts must be implemented amidst a growing ecocentric dialogue that also accounts for macro-environmental responsibility, and only when this convergence of private and public measures to create sustainable development can this goal truly be achieved.

Reflecting on Written Assignments at LUC

Promising directions in achieving circular production and consumption were outlined by the students, such as greater sensitivity to sociocultural context, as in the case of utensil use for lower castes in India. Many students at LUC have called for more attention to consumer responsibility. As one of the students suggests the necessary change requires that consumers reconsider their "definitions of ownership and the positive connotation of having something new" and willingly engage in sharing economy. When comparing "ideal" frameworks and real-life products, the students noted that political and economic constraints need to be taken into account. Furthermore, some students have reflected that economic motive is indeed important, as noted by the student in the case of C2C certifications where "heavy emphasis often lays on the business benefits." When profit and sustainability go hand in hand, this should not be a problem.

Less optimistically, some students felt that circular economy framework does not account for material consumption that cannot be serviced, shared, or "dematerialized." Indeed, as one of the students reflects, "Circular Economy alone cannot address rampant consumerism and overpopulation." One of the most risks identified by students involves the "rebound effect" (Greening et al. 2000) – consuming more products because they appear to be green. This point was made by the student commenting on Bakeys' edible spoon and the fact that using one metal spoon for a lifetime might have been a better solution than to keep producing new materials. Indeed, as one student states in the case of construction, "one should consider whether many of the construction projects should be initiated in the first place, when there are also many existing structures that could be refurbished," or spoons, "use of existing products may be more sustainable than the consumption of more and more 'revolutionary' C2C products."

"Rebound effect" can be also seen in the way this student has critically presented the desire of C2C to produce more products and questioned whether consumption can be limited by changes in consumer behavior. One of the largest risks identified by the students in relation to company operations or government policies is greenwashing and window-dressing. In the case of Coca-Cola, the student observed that the "plant bottle" presents a perfect case of "monstrous hybrid" without any plans for reuse. In a similar way, entire government policies, such as the Dutch Green Deal discussed by another student, can be subordinated to the service of business as usual. The student evaluating Satino Black towels related her doubts about "green gas" energy.

The student assignments illustrate that many companies or government initiatives continue to produce and sell new products, as in the case of edible spoons, while metal cutlery that can last for generations already exists. But using circular economy for economic gain misses the point of critique of economic and population growth as the driving force behind unsustainability (Washington 2015). Significant in this regard is that governments channel the economic growth ideology to consumers via the media (e.g., Jackson and Senker 2011). This economic reasoning is often internalized by consumers, which makes "consumer-driven change" and individual responsibility suspect (Bansel 2007). The Dutch political initiative the Green Deal, for example, in the words of student attempts to convince the public that environmental measures currently in place are sufficient without the need for radical change.

All this has implications of how students are taught, not just about revolution in production – although this is necessary – but also about larger demographic and societal challenges. As Huckle (2012) noted, development of critical literacy is crucial in this respect, in understanding the cultural practices that may help or hinder the related forms of production and consumption. It is also crucial to engage students in discussion of addressing population growth – through the exercise of women's rights including prohibition of child marriages, free provision of contraception, and raising educational levels in countries with high level of unwanted pregnancies (Crist et al. 2017). After all, keeping a balance between different elements in the natural ecosystem is part of cradle-to-cradle philosophy. Indeed, in the courses described (see Kopnina in print), the lecturer employed a discussion of underlying causes of unsustainability and alternative ways of measuring growth that include the well-being of humans and the environment. This includes economic system in balance with people and the planet, such as a steady-state economy (York 2017). The assignments discussed in this chapter point out that students are already engaged in critical thinking and exhibit healthy skepticism. Instruction can help them to remain pragmatic but also, crucially, committed to transformative measures.

Indeed, what is meant by pragmatism here is that transition from linear to circular model requires "economically feasible value recovery activities" (Lieder and Rashid 2016, p. 13).

Conclusion

This chapter has raised a number of questions related to ideal and feasible options for, in general, sustainable consumption and production and for approaching these options in educational practice. When noting potential pitfalls in both practice and teaching of circular frameworks, it is important not to become overly committed either to the optimistic win-win scenarios that might be unrealistic or to all-down skepticism. Indeed, in order to address sustainability challenges, we may need to reengage individuals – and in this case students – in their roles as citizens, helping to effectively and fairly regulate resource use and waste production (Isenhour 2015). The role of the student-citizen is in this case related to both staying critical – especially in cases regarding greenwashing – and yet able to remain open to the possibility of implementing "ideal" practice. Such reorientation would require the willingness of schools, sponsor companies, and students radical to engage intelligently with evaluating both feasible and ideal sustainability options. The emerging studies of circular economy in education indicate that students need to be taught about both consumer-based (thus, their own) and production-focused (thus, that of larger players) behaviors in order to address transition to circular economy (Huckle 2012; Kopnina 2015a).

For example, while it might be impossible to build a bridge using 100% organic fibers or to produce enough edible spoons without using fossil fuels, this does not mean that "green" bridges or edible spoons cannot be sustainably produced on a smaller scale. A compromise between what is financially feasible and ideally desirable is the core challenge that both the companies face. Evaluating the "circularity value" of actual products and services and entire supply chain needs to be part and parcel of both corporate and teaching practice.

The "sustainability value" of circular economy courses requires the students' ability as conscious consumers or future producers and managers to come up with solutions without resorting to pure economic reasoning. However, even the most conscious consumers can hardly influence large players such as governments and corporate decision-makers, as their power is limited to small committed groups (Isenhour 2015). Such committed groups are indeed the very students who have followed sustainability courses. Education for sustainability can certainly help to "engage the unengaged" – students who just want to get a good internship and eventually a job without the burden of thinking about the trade-offs of economic development and humanity's dire predicament. Being in a classroom with like-minded individuals, being able to reflect on and discuss the barriers and bottlenecks critically, might help students to carry their insights into the real world. Hopefully, together with lecturers, students can align around the intention to transition to a circular course of production but also be able to engage meaningfully in global debates about population and consumption. The students' combined networks are potentially global; they include families, friends, neighbors, and business colleagues as well as more remote online networks. It is my hope as a lecturer that these students will be able to carry their critical ability and insights forth.

Cross-References

- ▶ Bio-economy at the Crossroads of Sustainable Development
- ▶ Business Youth for Engaged Sustainability to Achieve the United Nations 17 Sustainable Development Goals (SDGs)
- ▶ Expanding Sustainable Business Education Beyond Business Schools
- ▶ From Environmental Awareness to Sustainable Practices
- ▶ Intercultural Business
- ▶ Relational Teams Turning the Cost of Waste Into Sustainable Benefits
- ▶ Social License to Operate (SLO)
- ▶ To Be or Not to Be (Green)
- ▶ Transformative Solutions for Sustainable Well-Being

References

Andersen, M. (2007). An introductory note on the environmental economics of the circular economy. *Sustainability Science, 2*(1), 133–140.
Bakeys Foods Private Limited. (n.d.). Why edible cutlery. http://www.bakeys.com/why-edible-cutlery/. Accessed 26 Oct 2017.
Bansel, P. (2007). Subjects of choice and lifelong learning. *International Journal of Qualitative Studies in Education, 20*(3), 283–300.
Blühdorn, I. (2007). Sustaining the unsustainable: Symbolic politics and the politics of simulation. *Environmental Politics, 16*, 251–275.
Boer, P., van Heeswijk, J., Heideveld, A., den Held, D., & Maatman, D. (2011). *Inspired by cradle to cradle: C2C practice in education*. Hilversum: Hiteq.
Bonnett, M. (2013). Sustainable development, environmental education, and the significance of being in place. *Curriculum Journal, 24*(2), 250–271.

Brennan, G., Tennant, M., & Blomsma, F. (2015). Business and production solutions: Closing the loop. In H. Kopnina & E. Shoreman-Ouimet (Eds.), *Sustainability: Key issues*. New York: Routledge.

Coca-Cola Company (Coca-Cola). (2012). 2011/2012 sustainability report: Sustainable packaging. http://www.coca-colacompany.com/sustainabilityreport/world/sustainable-packaging.html#section-managing-packaging-to-manage-risk. Accessed 27 Oct 2017.

Coca-Cola Company (Coca-Cola). (2015). Introducing PlantBottle. http://www.coca-colacompany.com/videos/introducing-plant-bottle-ytaevvjxqwaz8. Accessed 26 Oct 2017.

Crist, E. (2012). Abundant earth and population. In P. Cafaro & E. Crist (Eds.), *Life on the brink: Environmentalists confront overpopulation* (pp. 141–153). Athens: University of Georgia Press.

Crist, E., Mora, C., & Engelman, R. (2017). The interaction of human population, food production, and biodiversity protection. *Science, 356*, 260–264.

EC (European Commission). Packaging waste statistics. http://ec.europa.eu/eurostat/statistics-explained/index.php/Packaging_waste_statistics

Ellen MacArthur Foundation (EMF). (2016a). History. https://www.ellenmacarthurfoundation.org/about/history. Accessed 25 Oct 2017.

Ellen MacArthur Foundation (EMF). (2016b). Case studies. https://www.ellenmacarthurfoundation.org/case-studies. Accessed 25 Oct 2017.

Foster, J. B. (2012). The planetary rift and the new human exemptionalism: A political-economic critique of ecological modernization theory. *Organization and Environment, 25*, 211–237.

Geng, Y., Sarkis, J., & Ulgiati, S. (2016). Sustainability, wellbeing, and the circular economy in China and worldwide. *Science, 6278*, 73–76.

Ghisellini, P., Cialani, C., & Ulgiati, S. (2016). A review on circular economy: The expected transition to a balanced interplay of environmental and economic systems. *Journal of Cleaner Production, 114*, 11–32.

Golinska, P. (2014). *Logistics operations, supply chain management and sustainability*. Cham: Springer.

Green Deal. (2016). Aanpak and GD 203: Het Nieuwe Draaien. http://www.greendeals.nl. Accessed 26 Oct 2017.

Greening, L. A., Greene, D. L., & Difiglio, C. (2000). Energy efficiency and consumption – The rebound effect – A survey. *Energy Policy, 28*, 389–401.

Huckle, J. (2012). Even more sense and sustainability. *Environmental Education Research, 18*(6), 845–858.

Isenhour, C. (2015). Sustainable consumption and its discontents. In H. Kopnina & E. Shoreman-Ouimet (Eds.), *Sustainability: Key issues*. New York: Routledge.

Jackson, T., & Senker, P. (2011). Prosperity without growth: Economics for a finite planet. *Energy & Environment, 22*(7), 1013–1016.

Kopnina, H. (2012). Education for sustainable development (ESD): The turn away from 'environment' in environmental education? *Environmental Education Research, 18*, 699–717.

Kopnina, H. (2013). An exploratory case study of Dutch children's attitudes towards consumption: Implications for environmental education. *Journal of Environmental Education, 44*, 128–144.

Kopnina, H. (2014). Future scenarios and environmental education. *The Journal of Environmental Education, 45*(4), 217–231.

Kopnina, H. (2015a). Sustainability in environmental education: New strategic thinking. *Environment, Development and Sustainability, 17*(5), 987–1002.

Kopnina, H. (2015b). Neoliberalism, pluralism and environmental education: The call for radical re-orientation. *Environmental Development, 15*, 120–130.

Kopnina, H. (2016a). Energy policy in European Union: Renewable energy and the risks of subversion. In J. De Zwaan, M. Lak, A. Makinwa, & P. Williams (Eds.), *Governance and security issues of the European Union: Challenges for the future* (pp. 167–184). The Hague: TMS Asser Press/Springer.

Kopnina, H. (2016b). Animal cards, supermarket stunts and world wide Fund for Nature: Exploring the educational value of a business-ENGO partnership for sustainable consumption. *Journal of Consumer Culture, 16*(3), 926–947.

Kopnina, H. (2017). Sustainability: New strategic thinking for business. *Environment, Development and Sustainability, 19*(1), 27–43.

Kopnina, H. (in print). EE and ESD programs in The Netherlands: Towards sustainable development goals? *Environmental Education Research*. http://www.tandfonline.com/doi/full/10.1080/13504622.2017.1303819

Kopnina, H., & Blewitt, J. (2014). *Sustainable business: Key issues*. New York: Routledge Earthscan.

Lieder, M., & Rashid, A. (2016). Towards circular economy implementation: A comprehensive review in context of manufacturing industry. *Journal of Cleaner Production, 115*, 36–51.

McDonough, W., & Braungart, M. (2002). *Cradle to cradle: Remaking the way we make things*. New York: North Point Press.

Mont, O. K. (2002). Clarifying the concept of product – Service system. *Journal of Cleaner Production, 10*(3), 237–245.

Murray, A., Skene, K., & Haynes, K. (2017). The circular economy: An interdisciplinary exploration of the concept and application in a global context. *Journal of Business Ethics, 140*(3), 369–380.

O'Neill, D. W. (2012). Measuring progress in the degrowth transition to a steady state economy. *Ecological Economics, 84*, 221–231.

Rammelt, C., & Crisp, P. (2014). A systems and thermodynamics perspective on technology in the circular economy. *Real-World Economics Review, 68*, 25–40.

Rees, W. (2010). What's blocking sustainability? Human nature, cognition, and denial. Sustainability: Science, Practice, & Policy 6(2):13–25

Sadras, V. O., Grossini, P., & P. Steduto. (n.d.). Status of water use: Efficiency of main crops. SOLAW Background Report – TR07.

Stahel, W. R. (1997). The functional economy: Cultural and organizational change. In *The Industrial green game: implications for environmental design and management*, National Academy of Engineering, Washington, DC, pp. 91–100.

Stahel, W. R. (2016). The circular economy. *Nature, 531*(7595), 435.

Stahel, W. R., & Reday, G. (1976). The potential for substituting manpower for energy, report to the Commission of the European Communities. Available via http://cordis.europa.eu/publication/rcn/1989119069800_en.html

Stegeman, H. (2015). From circular materials cycles to a circular macroeconomy with scenario's for the Netherlands. https://economics.rabobank.com/publications/2015/july/the-potential-of-the-circular-economy/

Terry, B. (2011). The truth about pepsi's new plant-based PET plastic bottle. http://myplasticfreelife.com/2011/04/the-truth-about-pepsis-new-plant-based-pet-plastic-bottle/. Accessed 26 Oct 2017.

The Economist. (2009a). Feeding the world: If words were food, nobody would go hungry. November 21, pp. 61–63.

The Economist. (2009b). The baby bonanza. August 29, pp. 20–22.

The Economist. (2009c). Falling fertility. October 31, pp. 13.

The Economist. (2013). Demography in Latin America: Autumn of the patriarchs. June 1, p. 47.

The Economist. (2015a). Thin harvest: Biofuels. April 18, pp. 69.

The Economist. (2015b). Saving fish and bearing teeth. April 18, pp. 52.

The Economist. (2016a). Worse than Beijing: Pollution in India. November 12, p. 48.

The Economist. (2016b). Wrap stars: Retailing and the environment. December 17, pp. 56–57.

Tukker, A., & Tischner, U. (Eds.). (2017). *New business for old Europe: Product-service development, competitiveness and sustainability*. New York: Routledge.

Vogtlander, J. G., Scheepens, A. E., Bocken, N. M., & Peck, D. (2017). Combined analyses of costs, market value and eco-costs in circular business models: Eco-efficient value creation in remanufacturing. *Journal of Remanufacturing, 7*(1), 1–17.

Washington, H. (2015). *Demystifying sustainability: Towards real solutions*. London: Routledge.

Webster, K. (2007). Hidden sources: Understanding natural systems is the key to an evolving and aspirational ESD. *Journal of Education for Sustainable Development, 1*, 37–43.
York, S. (2017). Confronting our global growth obsession. http://mahb.stanford.edu/blog/global-growth-obsession/
Yuan, Z., Bi, J., & Moriguichi, Y. (2006). The circular economy: A new development strategy in China. *Journal of Industrial Ecology, 10*(2), 4–8.
Zorpas, A. A. (2016). Sustainable waste management through end-of-waste criteria development. *Environmental Science and Pollution Research, 23*(8), 7376–7389.

Part VI

Sustainable Living and Environmental Stewardship

Smart Cities

New Urbanism and New Agrarianism as a Path to Sustainability

John E. Carroll

Contents

Introduction	838
The How and Why of the Smart City	840
The How of the Smart City	840
The Congress for the New Urbanism (CNU)	840
What Constitutes the Smart City?	842
The Charter	842
Components of the Smart City and Town: By the Numbers	843
The Canons	845
A Step Further: The Case for the Agrarian Urbanism of Andres Duany	849
The Transect	849
In the Transect	849
The Transect Zones	850
The Rationale for Agrarian Urbanism	852
Food and the Smart City	854
The Smart City in Relation to Its Periphery and Hinterland	858
Protein in the Smart City	858
A Pattern Language	860
Two Concepts: Small Is Beautiful and Limits to Growth	863
Conclusion: City- and Town-Building in a Carbon-Constrained World	865
Cross-References	867
References	867

Abstract

A smart city is a city that prepares for the future. It can be defined as a city that adopts the principles of New Urbanism and New Agrarianism in its design and organization. The charter and canons of New Urbanism are presented and

---

J. E. Carroll (✉)
Department of Natural Resources and Environment, University of New Hampshire, Durham, NH, USA
e-mail: John.Carroll@unh.edu; carroll@unh.edu

© Springer International Publishing AG, part of Springer Nature 2018
S. Dhiman, J. Marques (eds.), *Handbook of Engaged Sustainability*,
https://doi.org/10.1007/978-3-319-71312-0_13

described, as is the working tool of the transect. An important aspect of New Urbanism and the smart city is walkability, with accompanying population density and mixed use. Food is fundamental to all forms of human life and is a principal organizing principle of the smart city. Key to the success of the smart city is Urban Agrarianism: the production, processing, preparation, and distribution of food within the city, including the use of the periphery for meat and dairy, while focusing fruit and vegetables within more densely populated inner transect sectors. The smart city is the food secure and relatively diverse city that follows nature as guide, organizing itself by ecological principles which support diversity and variety and which encourage interaction of the population with one another and with their surroundings. While successfully combining new forms of urbanism and new forms of agrarianism, the smart city builds on older time-tested forms of urban and rural life. The smart city is a necessary answer to the twin challenges of energy dependency and climate change in a carbon-constrained world. Such a city ever seeks to redefine the edge of possible.

Keywords
Smart city · New Urbanism · Urban Agrarianism · Transects · Urban agriculture · Food security · Sustainability · Decentralization

Smart Having or showing intelligence, bright
Canny and shrewd in dealings with others
Fashionable, elegant
Capable of making adjustments
Showing sound judgment and rationality
Mentally alert, bright, knowledgeable, and shrewd
Very good at learning or thinking about things

Introduction

What is a smart city? Applying the definition above, a smart city is one that is sufficiently intelligent and wise that it prepares for the future and for the future realities which will govern the success or failure of the city. A city exists first and foremost to serve the needs and enrich the lives of its inhabitants. A city's underlying purpose is not to serve its state, province, nation, or the world. It can, of course, do all of those things, but not without first serving its own population and, through that population, its broader surroundings. It engages in this service with a particular eye on its hinterland, which is to say its foodshed and watershed, with which it is interdependent. I am native to a very large city, New York City. Simultaneously, I am native to a very small town, the neighborhood within New York City in which I was raised and in which my forebears lived for at least a century: Woodside, Queens, just across the East River from midtown Manhattan. Today, I see that neighborhood, and that great city, as gradually, but slowly, becoming a smart city. It appears to be moving in the right direction, but perhaps not rapidly enough, given the circumstances of food, energy, climate change, and the social needs in which it functions.

So, then, what makes a city "smart"? Sustainability, which means to sustain life, society, and culture, causing it to survive into the indefinite but far-off future, is an important part of the answer. Sustainability has been defined as economic development activity that "meets the needs of the present without compromising the ability of future generations to meet their own needs" (World Commission on Environment and Development 1987). For sustainability to occur, all necessary resources, including food, fuel, materials, energy, and the economic means to obtain them, are required. But sustainability also requires a mindset that is relational, indeed ecological. Material resources alone are not sufficient, as they are not operable without communitarian resources, and without genuine respect for our interdependent relationship with one another as well as with all that is. Thus, New York, among other places, is becoming a smart city by favoring the bicyclist and the pedestrian. It is also becoming a smart city by favoring public transit over individually owned motor vehicles and increasing community space while reducing space devoted to the private vehicle. By supporting its wide array of farmers markets and direct linkages to many farms in its foodshed, this city is additionally smart, and a growing sense of community is resulting. Strong community means a strong town, a strong city. Enough such strong places make a strong nation.

> **The New Challenge of the Coasts and Climate**
> For some decades, the US coastal regions have experienced greater levels of population growth and economic development than has the continental interior, sometimes denigrated as "flyover country." But coastal regions now face particularly strong challenges to their well-being, economic and otherwise, from sea level rise. As if this challenge, with its consequent flooding and storm damage, were not serious enough, we are now coming to witness the phenomenon of both saltwater intrusion into fresh groundwater aquifers upon which our cities and towns depend and the still lesser-known phenomenon of seawater pressure against fresh groundwater aquifers, which are causing those groundwater aquifers to rise to the land surface and create flooding challenges of their own. As a result of these phenomena, population (and development) may again move at least somewhat more inland from the exposed coasts. Inability to obtain (or afford) storm and flood insurance will undoubtedly hasten a retreat from the coasts.
>
> For some time into the future, for both coast and inland, climate change adaptation and mitigation as well as the associated need for resilience in organization and design will become a central organizing principle for our cities and towns. We likely have little choice in the matter. And in our climate-altered future, food production and access to food will always be our central need.

The How and Why of the Smart City

The word "smart" carries a positive connotation. Presumably, the smart city is more desirable than the not-so-smart city. With this bias in mind, I will define the smart city and illustrate the path to making cities smart. I will then approach the "why" question to reinforce the importance of shifting cities in the direction of smart, regardless of how difficult the shift may prove.

The How of the Smart City

Smart cities feature the following characteristics:

1. A smart city looks to nature as a teacher and guide, embracing nature's example as a fundamental tenet of ecology that we ignore only at our peril.
2. A smart city recognizes the need for vital production of food both near and within the city. For such a city, local food production carries an importance equivalent to a municipal water service and is perhaps the biggest single key to the idea of a smart city's sustainability. To the extent that a city can reduce its dependency on food from away, particularly from far away, the city is, by definition, more sustainable, not to mention more secure. This is Agrarian Urbanism, and improved human health is a byproduct.
3. A smart city features a high degree of walkability, the ability of its residents to fulfill their most basic needs on foot and to enjoy doing so. Like Urban Agrarianism, walkability can only result in improved health – physical, mental, and spiritual.
4. A smart city accommodates the need for human beings to fulfill their curiosity about other human beings (particularly those of similar age), while simultaneously maintaining a necessary standard of privacy. According to urban planner William Whyte, what attracts people most is other people.
5. In addition to providing avenues for socialization, a smart city also offers intellectual stimulation, which human beings as a species both require and thrive upon.

The Congress for the New Urbanism (CNU)

Perhaps the best example of sustainability applied to cities and human communities is the charter and canons of the new urbanist movement as put forth by the Congress for the New Urbanism (CNU) and as applied to city- and town-building as well as to suburban repair. The terms *city-* and *town-building* here do not refer to construction of new cities or towns in the countryside, but rather the rebuilding of existing towns and cities in their existing population centers. But the makeover prescribed by New Urbanism is so fundamentally different from still prevailing twentieth-century ideas that these centers acquire such adjectives as "new" and, indeed, "smart." They are smart because they adequately respond to the new realities regarding food, fuel, environment, ways of living, ways of working, ways of transit, and ways of human and spiritual fulfillment. These are happy places that are made for happy people. The New Urbanism of the CNU is an appropriate response to the question, "What is the smart city?"

Importantly, the canons, or principles, referred to create the pathway for assimilation of the food system, essentially the New Agrarianism, into the actual physical makeup of the community, providing a framework for how we must build, rebuild, and operationalize our live and work space. Leading American agrarian Wendell Berry recently wrote, "The gait most congenial to agrarian thought and sensibility is walking. It is the gait best suited to paying attention, most conservative of land and equipment, and most permissive of stopping to look or think" (Berry 121). New Urbanism and its most significant founder, Andres Duany, equally champion "walkability" as central to the organization of our society. Even Pope Francis has written of these things in his papal encyclical, *Laudato Si: On Care for Our Common Home* (see my chapter, "The Environment Is a Moral and Spiritual Issue" in *Spirituality and Sustainability*, Springer, 2016). In his encyclical, Francis argues that humanity must integrate its relationship with nature and the relationship of humans, one to another. On the topic of urban planning and "smart cities," Francis writes:

> Given the interrelationship between living space and human behavior, those who design buildings, neighborhoods, public spaces and cities ought to draw on the various disciplines which help us to understand peoples' thought processes, symbolic language, and ways of acting. It is not enough to seek the beauty of design...[I]t is important that the different parts of a city be well integrated and that those who live there have a sense of the whole, rather than being confined to one neighborhood and failing to see the larger city as space which they share with others. (Carroll 2016, 64)

In his encyclical, Pope Francis further rejected the centrality of the car, prioritizing public transit not only for environmental and energy reasons but because of his conviction that cars destroy community. It is important to oppose that which destroys community and to support that which builds community.

Significantly, each of the pope's points outlined above synchronizes remarkably with the goals of New Urbanism and therefore with the essence of the smart city. We may conclude that Pope Francis is a proponent of New Urbanism and a visionary when it comes to smart cities.

The local food and farming movement, a movement of revolutionary proportions in the northeastern United States and other parts of the country, is currently attempting to scale up in order to meet the large-scale corporate needs of the institutionalized food system, which has been in place for the past half century or more. And yet, with respect to both New Urbanism and urban agriculture, to ask the question "How can we scale up?" is to ask the wrong question. The right question is: "How can we scale down?" We should not be asking, "How can we further centralize the system?" but rather, "How can we decentralize the system?" While scaling up may be tempting, especially if we speak of food provision to larger institutions (chain supermarkets, large schools, and other institutions), such "scaling up" is to force the philosophy of local food into a countervailing philosophy, the philosophy of bigness, of corporatization, and of centralization – it just won't work. Local food, by definition, is decentralized and therefore inherently small scale. New Urbanism and true sustainability – indeed, most forms of urban agriculture – argue for local (in food as well as in all other goods and services) in every local place, a model of decentralization of which both philosopher E. F. Schumacher and esteemed agrarian philosopher Wendell Berry would be proud. The existing large-scale corporatism to which we have become accustomed runs deeply counter to New Urbanism since New Urbanism embraces a neighborhood scaling in order to honor the precept of walkability. Reorganizing our society to adapt to new energy and environmental realities starts with reorganizing the way we think. Everyone knows you can't force a square peg into a round hole. Similarly, you can't put localization, which is small scale by its very nature, into corporate thinking, which is inherently large scale. The two are in perennial opposition to one another.

What Constitutes the Smart City?

The Congress for the New Urbanism (CNU) has given us three tools with which to build the smart city: the Charter and the Canons, which constitute the philosophy and governance of New Urbanism, and the transect. Examining these short, simple, and clear documents will help us see the contrast New Urbanism represents to our nation's approach to housing design as well as town- and city-building over the past half-century.

The Charter

The entity which has designed and sponsors the Charter, the Congress for New Urbanism (CNU), is an informal national organization of new urbanists that meets annually and sponsors chapters around the country. Concerned about disinvestment in central cities and downtown districts, the spread of sprawl (which it calls *placelessness*), increasing inequality in both race and income, environmental deterioration, loss of agricultural lands and wilderness, and the erosion of society's built heritage – all of which it sees as one large interrelated challenge – the CNU has outlined the following values and practices, divided into three sections, as the cornerstone of New Urbanism. New Urbanism:

- Stands for the restoration of existing urban centers and towns, the reconfiguration of sprawling suburbs into communities of thriving neighborhoods and diverse districts

- Recognizes that physical solutions can only be applied with economic vitality, community stability, and environmental health
- Advocates for public policy that supports neighborhood diversity in both use and population, design favoring the pedestrian and transit over the car, the shaping quality of physically defined and universally acceptable public spaces and community institutions, and an architecture and landscape design that celebrate local history, tradition, and ecological reality
- Represents a broad-based constituency committed to reestablishing the relationship between the art of building and the making of community
- Dedicates itself to urban reclamation from the city and suburbs, from homes, blocks, and streets to parks, neighborhoods, districts, towns, cities, regions, and environments

To carry this out, New Urbanism asserts the following principles for urban planning and design.

Components of the Smart City and Town: By the Numbers

The region (metropolis, city, and town):

1. The finite nature of the physical environment and the multiple centers of the metropolis, each with its own identifiable center, must be recognized.
2. The metropolitan region as a single fundamental economic unit must be recognized, and policy must reflect that reality.
3. The metropolis has a necessary and fragile relationship to its agrarian hinterland and natural landscapes: farmland and nature are equally important to the metropolis as the garden is to the house.
4. Infill development must be favored over peripheral expansion. Development patterns should not blur or eradicate the edges of the metropolis, and infill development conserves environmental resources as well as economic investment and social fabric.
5. New, contiguous developments should become neighborhoods and districts, while non-contiguous development should be organized as towns and villages of their own with their own edges (and planned for a jobs-housing balance and not as bedroom suburbs).
6. All development and redevelopment should respect historical patterns.
7. Cities and towns should bring into proximity a broad spectrum of public and private uses to support a regional economy, including affordable housing, to benefit people of all incomes and avoid concentrations of poverty.
8. A variety of transportation modes and infrastructure should be available: public transit, pedestrian, and bicycle.
9. Revenues and resources should be shared cooperatively and destructive competition avoided.

For the neighborhood, the district, and the corridor:

1. The neighborhood, the district, and the corridor are the essential elements of development and redevelopment.
2. Neighborhoods need to be compact, pedestrian friendly, and mixed use.
3. A variety of daily living activities should occur within walking distance, offering residents independence.
4. Within neighborhoods, there need to be a broad range of housing types and price levels, ensuring diversity of population and strengthening both personal and civic bonds.
5. Transit corridors can help organize metropolitan structure, the opposite of highway corridors which displace investment from and damage existing centers.
6. Buildings should be within walking distance of transit stops, making public transit a viable alternative to the car.
7. Concentrations of a mixture of civic, institutional, and commercial activity should be embedded in neighborhoods and districts, never isolated in "single-use complexes." As an example, every school should be within walking and biking distance of all students.
8. Graphics of urban design codes are important so that users/residents can envision them.
9. A range of large, small, and very small parks should be well distributed within neighborhoods, with larger open space defining and connecting different neighborhoods and districts.

For the block, the street, and the building:

1. The physical definition of streets and public spaces should always signify shared use.
2. All architecture should seamlessly link to its surroundings.
3. The design of streets and buildings should reinforce safe environments, but not at the expense of accessibility and openness.
4. The accommodation of cars should respect the pedestrian and the form of public space.
5. Streets and squares should have visual interest and encourage walking, enabling neighbors to get to know each other.
6. Architecture and landscape design should be developed based on local climate, topography, history, and building practice (as Seaside, Florida, the first new urbanist community in the United States, has done).
7. Civic buildings and public spaces should reinforce community identity and the culture of democracy.
8. All buildings should rely on natural methods of heating and cooling.
9. Preservation and renewal of historic buildings, districts, and landscapes should be given high priority.

These stated values and guidelines summarize the now decades-old Charter of New Urbanism, as put forth by the CNU.

The Canons

The Canons of Sustainable Architecture and Urbanism stand as a companion to the Charter of the New Urbanism and reflect a further evolution of the Charter.

A worsening of the environmental situation globally, as evidenced notably through climate change and habitat destruction, has caused new urbanists and the Congress for the New Urbanism to add to the Charter a set of Canons designed to recognize and address the deteriorating environmental situation we face. The Canons make clear, as Pope Francis has likewise articulated, that environmental solutions must address poverty, health, and underdevelopment, while simultaneously addressing environmental concerns. The CNU's proposed solution integrates (1) smart growth, (2) green building, and (3) New Urbanism. These Canons harken to a higher level of sustainability and seek to provide a set of operating principles for, in their own words, "addressing the stewardship of all land and the full range of human settlement: water, food, shelter and energy," with a simultaneous engagement of urbanism, infrastructure, architecture, landscape design, construction practice, and resource conservation.

There are presented seven general Canons:

1. Design and financing must be refocused on long life and permanence rather than transience, enabling reuse as well as long-term use.
2. While recognizing the need to address both climate change and affordability, investors should be rewarded by greater returns over the long term.
3. Indigenous urban, architectural, and landscape patterns are necessary to any truly sustainable design.
4. Design must achieve ecological and resource conservation goals, including provision of local food, protection of local watersheds, preservation of clean air, and conservation of natural habitat and biodiversity.
5. Human settlements must be recognized as part of the earth's ecosystems.
6. The rural to urban transect, a transect drawn across the rural-urban divide which delineates zones of agrarian and urban characteristics, is the essential framework for the organization of the natural, the agricultural, and the urban realms.
7. Social interaction, economic and cultural activity, spiritual development, energy, creativity, and time are all to be maximized by buildings, neighborhoods, towns, and regions.

The Canons further lay out seven points of a general nature:

1. The fabric and infrastructure of the city must enable reuse, accommodating growth, change and long-term use.

2. Patient investors seeking long-term rather than short-term return should be encouraged, particularly in responding to the long-term impacts of climate change and increasing affordability.
3. Sustainable design must be rooted in adaptations to the local natural and cultural heritage.
4. Design must preserve the proximate relationships between urbanized and adjacent rural areas so as to ensure local food production, watershed and water supply protection, and protection of local natural resources, ecosystems, and biodiversity.
5. Human settlements are part of the earth's ecosystem, not separate.
6. The rural-to-urban transect developed by the architectural firm DPZ and the CNU provides the design framework.
7. All components will maximize social interaction, economic and cultural activity, spiritual development, energy, creativity, and time.

These Canons then provide a prescription for buildings and infrastructure.
With regard to the building and infrastructure:

1. The objective of building design is to create a culture of permanence in structures of enduring quality (longevity is promoted, as is stewardship of land and buildings).
2. Architecture and landscape design derived from local natural, historic, and cultural conditions.
3. Exterior building shells must be designed to be enduring, while interiors must be designed to be flexible and adaptable.
4. Preservation and renewal of historic buildings saves embodied energy as well as cultural continuity.
5. Buildings shall conserve and produce renewable energy wherever possible to reduce demand for fossil fuels.
6. Building design and configuration must reduce energy usage and promote walkability, both vertically and horizontally.
7. Renewable energy sources shall be used to reduce carbon and other greenhouse gases.
8. Water shall be stored and reused on-site and let it percolate into local aquifers.
9. Water uses shall be minimized and conserved through landscape strategies, including emphasis on native vegetation.
10. Building materials shall be locally sourced and recycled and contain low embodied energy – or chosen for durability and exceptional longevity, taking advantage of thermal mass to reduce energy usage.
11. Building materials shall be nontoxic and noncarcinogenic.
12. Food production of all kinds should be encouraged in and around all buildings to promote the values of decentralization, self-sufficiency, and reduced transportation.

With regard to the street, block, and network:

1. Street design shall encourage the shaping of a positive public realm, always encouraging shared pedestrian, bicycle, and vehicular use.

2. The pattern of blocks and streets shall be compact and well-connected for walkability and minimize material and utility infrastructure.
3. The shaping of the public realm shall focus on creating thermally comfortable spaces.
4. All design of streets, blocks, etc. shall be configured for reduced overall energy usage.
5. Roadway materials shall be nontoxic and provide for water reuse through percolation and retention while maintaining street connectivity.
6. The supply of parking shall be constricted in order to induce less driving and create more human-scaled public space.

With regard to the neighborhood, town, and city:

1. The balance of jobs, shopping, schools, recreation, civic uses, institutions, housing, and food production shall all be at the neighborhood scale and within easy walking distance of one another or of public transit.
2. All new development shall be on underutilized, poorly designed, or already developed land (i.e., urban infill or urban adjacent).
3. Prime and unique farmland shall be protected and conserved, and additional agriculture shall be promoted on already urbanized/underutilized land.
4. Neighborhoods, towns, and cities shall be as compact as possible, promoting lively mixed urban places.
5. Renewable energy shall be produced at the scale of neighborhood and town, as well as individual buildings, in order to maximize energy decentralization.
6. Brownfields shall be redeveloped.
7. Wetlands shall be protected and recharge of aquifers should be restored.
8. Natural places shall be within easy walking distance of everyone.
9. Within neighborhoods there need to be a broad range of housing types, sizes, and price levels in order to create diversity, self-sufficiency, and social sustainability.
10. Nearby rural agricultural settlements shall be promoted to preserve local foods and food culture, to ensure a wide range of locally raised foods within a short distance, enable food self-sufficiency, and cap the overall size of neighborhoods.
11. Light pollution and noise pollution should be minimized.
12. Neighborhood design should use natural topography and avoid the import and export of fill.

With regard to the region:

1. The finite boundaries of the region shall be governed by natural features.
2. Whole regions shall strive for self-sustainability in food, goods and services, employment, renewable energy, and water supplies.
3. The region's physical organization shall promote transit, pedestrian, and bicycle systems while reducing dependence on cars and trucks.
4. Development shall be primarily organized around transit lines and hubs, enabling a spatial balance of jobs and housing.

5. The siting of new development shall prefer already urbanized land, with rural undeveloped land development allowable only under stringent circumstances.
6. Sensitive habitats and prime farmlands shall be conserved and protected, and projects to regenerate and recreate additional agricultural areas and habitats shall be promoted.
7. Wetlands and water bodies shall be protected.
8. Development shall be avoided in locations that disrupt natural weather systems and induce heat islands, flooding, fires, or storms.

Rather than defined by technology ("smart" or otherwise), smart cities take their shape around questions of design that lead to specific ends. Walkability is central among those ends and flows into the desired goals of variety in streetscapes designed to hold visual interest and facilitate the mixing of interesting people. Also central among those ends are nearness to all necessary goods and services, thriving cultural opportunity amidst an equally thriving street culture, and proximity to plenty of good food. Such design offers opportunity to avoid investing in car ownership and sets limits on private vehicular traffic, emphasizing its alternatives: walking, bicycling, and public transit, along with the option of seriously reducing the necessity of travel itself. And, most comfortingly, such design is fundamental to security. These are the characteristics of the smart city, especially descriptive of the smart city as seen through the eyes of millennials, those adults under 40 who now constitute the largest single segment of the population. Thus, these characteristics are descriptive of the future.

> **"Rear Window Ethics" and Human Desire**
> Film director Alfred Hitchcock introduced us to the notion of "rear window ethics," raising questions about the natural voyeuristic tendencies of all humans to want to know what is going on in the life of fellow humans. New Urbanism, which emphasizes dense centers where humans live side-by-side in order to enjoy a greater variety of necessary services in a sustainable and less car-dependent manner, presents certain challenges in this regard. But New Urbanism founder Andres Duany sees this as an asset. He believes there is nothing human beings are more interested in, or more curious about, than other human beings, particularly people of the same generation – our peers. The need to satisfy this natural curiosity is well answered by the very nature of New Urbanism, from high walkability to street-facing porches and vegetable gardens, not to mention the large amount of public space interspersed with living space. Such characteristic features not only contribute significantly to a higher quality of life for those who live in new urbanist communities and neighborhoods, but they also make for an infinitely more interesting life. What are our neighbors wearing? What are they eating? What are they doing? What are they thinking? How are they coping with life, and what can we learn from them? And so, while fulfilling this curiosity toward our fellow human creatures, the new urbanist approach designs privacy options into its enhanced public space.

A Step Further: The Case for the Agrarian Urbanism of Andres Duany

Out of the unlikely subtropical urban environment of South Florida, and from the mind of an urban planner of world renown, comes the unlikely and seemingly contradictory idea of Agrarian Urbanism. As he would be quick to tell you, the idea wasn't born with Andres Duany, a contrarian of sorts with an interest in agriculture. Rather, it was born out of the creativity of an early twentieth-century planner trying to find an inspiring antidote to the tired and depressed spirit of time-worn towns in industrial England. I refer to the Englishman Ebenezer Howard and his idea of garden cities. Howard was responding to the need for a better way to live at the beginning of the twentieth century. Duany is responding to a necessary way to live in the early stage of the twenty-first century. He charismatically refers to the garden city as a model of equilibrium and the only way to think about cities. Such cities, he claims, also represent a form of adaptation to zero-carbon living and thus to resolution, or at least amelioration, of the effects of climate change. To better understand Agrarian Urbanism, one must understand the basic tool of the Charter and the Canons, the transect.

The Transect

Pioneered by Andres Duany and inspired by examples from nature, the transect is a tool for translating the Charter and the Canons into practical expressions, bringing these documents to life. Drawn from ecological reality itself, the transect is a way of envisioning, using nature as our teacher and guide. Working with nature and never Against nature, the transect merges ecological thought with city-building. The designed and built city, in this way, reflects nature – a wise goal.

The transect recognizes a progression of essential habitats that make up the transition from one natural ecosystem to another (e.g., from field to forest, land to sea). Ecologists use the term "ecotone" to describe the intermediate steps in this progression. In Duany's CNU model of city- or town-building, the transect ranges from T-1 (the natural environment of the rural countryside), through T-2, T-3, T-4, T-5, (various stages of suburbia through a progressively more densely settled urban habitat), to T-6 (the urban core, also called the center city, the central business district, or, variously, downtown – and, in larger urban agglomerations, midtown or uptown).

In the Transect

T-1 (outer periphery) has high natural diversity and is a polyculture (i.e., many cultures of diversity which is, in fact, ecological), a reflection of ecological reality.

T-6 (inner core) has high social diversity and is also a polyculture (and hence ecological, in spite of its developmental density).

T-3 (in between), the core of suburbia, essentially lacks diversity and, according to Duany, is therefore a social and natural monoculture (i.e., a uniform single culture, which is anti-ecological in that it does not reflect nature).

The six transects can be formally identified as:

T-1: Natural
T-2: Rural
T-3: Suburban
T-4: General urban
T-5: Urban center
T-6: Urban core

Such a tool as the transect can serve as a guide to the direction and development of each segment of the new city or new town (Duany).

The Transect Zones

The transect zones may be described as follows:

T-1 – Natural zone consists of lands approximating or reverting to a wilderness condition, especially where unsuitable for settlement due to topography, hydrology, or vegetation. Natural diversity. A polyculture.

T-2 – Rural zone consists of sparsely settled lands in open or cultivated conditions. These include woodland, farmland, grassland, and irrigable desert. Typical buildings are farmhouses, agricultural buildings, cabins, and villas. Roads and trails are common.

T-3 – Suburban zone consists of low-density residential areas and some retail at corners. Home occupations and outbuildings are present throughout. Planting is naturalistic, and building setbacks are relatively deep. Blocks may be large and the road pattern irregular to accommodate natural conditions. No diversity. A social and natural monoculture.

T-4 – General urban zone consists of mixed use but primarily residential areas. It includes a wide range of building types: shops, houses, row houses, and small apartment buildings. Setbacks and landscaping are variable. Streets with raised curbs and sidewalks define medium-sized blocks.

T-5 – Urban center zone consists of higher-density, mixed-use buildings, with shops, offices, row houses, and apartments. Streets have raised curbs, wide sidewalks, steady tree planting, and buildings with short setbacks.

T-6 – Urban core zone consists of the greatest density and building height, most being mixed use. This is the setting for civic buildings of regional importance. Blocks may be large to accommodate parking within. Streets have steady tree plantings and buildings set close to the wide sidewalks. Only large towns and cities have urban core zones. High social diversity. A polyculture.

According to Duany, agrarian urban theory correctly retains the New Urbanism's high natural diversity of T-1 (natural) and the high social diversity of T-6 (urban core), but it also radically improves the performance of T-3 (suburban). How so? By integrating it technically into a green regime of which agrarianism is the most thorough manifestation.

Duany thus credits his theory of Agrarian Urbanism as equalizing environmental performance along the transect while at the same time retaining lifestyle choice, a positive market benefit. In fact, he sees lifestyle choice as improving the combined diversity of all T zones. In other words, "The New Urbanist Theory performs better by valuing both the natural and social diversity at T1 and T6, while correctly but problematically de-valuing the suburban point of T3, which has the lowest indices of both. Agrarian Urbanism mitigates T3 Sub-Urban so that all transect zones are equalized" (Duany, 74). This is in sync with ecological thought (i.e., nature chooses polyculture/diversity and always abhors monoculture/lack of diversity). It also saves the suburbs by offering them a positive image, as the suburban house of T-3 can become a locus of both food and energy production, as well as of composting and recycling.

Duany differentiates Agrarian Urbanism from Agricultural Urbanism, seeing Agrarian Urbanism as a conceptually more complete version of the two since it transcends mere production of food to incorporate processing, value added, and preparation, providing a "social condenser," which food and a farm alone cannot provide.

As a business proposition, an agrarian program in a residential development functions as an amenity, and it counts as a "green development." Such amenities cost less to build and are easier to phase in for the developer. "Is there a future for this?," Duany asks. Yes, he responds: "This kind of development is all about the future. Sustainability to the point of self-sufficiency is where the market is going, especially if it becomes apparent that the campaign to mitigate climate change is being lost" (Duany, 79). He further reflects, "If this conclusion seems early, remember that urbanism operates on the time frame of decades. The present is a distortion field that should be examined skeptically" (Duany, 79).

A further incentive to Agrarian Urbanism is planning approvals that are otherwise difficult to secure. For example, in my state of New Hampshire, we now have the incentive of the new state statute "Granite State Farm to Plate" (RSA 425:2-a), which encourages but does not require planning boards and other municipal entities, as well as all state agencies and the courts, to both support and promote local food and farming in their decision-making. The statute is designed to make the path easier. Easier planning approvals would be one way to make the path smoother. And since Agrarian Urbanism is part of the solution, not part of the problem, "It brings along the moral authority of much of the environmental agenda" (Duany, 80). Regarding acceptability, Duany importantly claims that "... Agrarian Urbanism can be presented to a financial institution as conventional" for "a house with a yard, a townhouse (skip the description of the roof garden), or an apartment building next to a public park (don't necessarily call it the community garden) are not difficult to explain" (Duany, 80).

In answer to the question, "Why should I do this?" Duany offers four reasons relating to health, environment, economics, and social values:

> Regarding health, Agrarian Urbanism provides recreational and productive physical activity; fresh humanely raised food; control over food supply and quality; and it preserves open space.
> Environmentally, with Agrarian Urbanism there is less pollution of water and soil; closed cycles of food-to-waste-to-food; retention and restoration of marginal agricultural land; an intensification of crop yields; and less energy applied to, and pollution resulting from, food transportation.
> Economically, Agrarian Urbanism efficiently delivers to a variety of emerging economic realities; it positions recreational time as productive rather than consumptive; it saves cost in waste disposal, fertilizers and herbicides; it allows maintenance budgets to be reduced; and it appeals to a large and growing market.
> Socially, Agrarian Urbanism supplies social gathering places enriched by utility; provides training in a useful craft, meaningful jobs for some, and participation for all, including children and seniors; it fosters cultural traditions based on mutual reliance; and it aggregates individual effort to the level of an economic system – not just amateur performance. (Duany, 81)

One might conclude, in other words, that we should embrace Agrarian Urbanism because there are few better ideas around.

The Rationale for Agrarian Urbanism

Duany has articulated a support system for Agrarian Urbanism that includes:

– The Internet, which provides both basic knowledge and social connectivity. These components are especially valuable to the farming/gardening population in T-1, the periphery.
– The value of the system as a "social condenser."
– The transfer of funds from landscaping to agrarian pursuits.
– The property owners' association or co-op as a management framework.
– The diversity and vigor of the market for Agrarian Urbanism.

Duany has written that the Internet shores up Agrarian Urbanism today as a viable possibility. The Internet provides cover for four of the perennial disadvantages of life in more remote places, as it offers surrogates for (1) limited social networks, (2) scant entertainment, (3) threadbare shopping, and (4) the absence of "medical diagnostics" (Duany, 46). With the Internet, agrarian life becomes much more acceptable to many. Prospective agrarian residents are willing to sign off on certain lifestyle obligations necessary for successful food production that are not otherwise acceptable.

In Agrarian Urbanism, "the market square is the locus of agricultural processing as well as the social and commercial center of the community. Socializing takes place around food and its consumption rather than being based on recreational shopping" (Duany, 55). The market square is thus the "primary social condenser of Agrarian Urbanism." And the community garden can also serve as an important agrarian condenser.

As Duany further notes, today we expend much land, money, and labor on ornamental landscaping. This investment includes nothing less than (1) the voluntary work of private gardeners; (2) hired help that performs maintenance tasks; (3) money for tools, pesticides, fertilizers, etc.; (4) municipal budgets for public areas; and (5) high fees collected for semipublic landscaping of suburban developments. Agrarian Urbanism reassigns these public, semipublic, and private funds away from ornamental planting to the more demanding aspects of agriculture and food production. Resources now spent on lawns and exotic plants requiring constant attention are redirected much more efficiently to edible landscapes. Thus, the more demanding, not to mention more boring aspects of agriculture, are handled by contract workers who are compensated with redirected monies, while the more satisfying and pleasant tasks are undertaken by willing residents in their spare time.

Duany and others articulate a variety of motivations with which people come to gardening. Duany writes that "[G]ardening has become the most popular hobby among Anglo-Americans. It is more than a niche and it is growing" (Duany, 63). An indispensable tool enabling this development, Duany notes, is the property owners association or co-op. Three levels of personal commitment are involved: the decision to be a dues-paying resident in such a community, participation in a loosely organized gardening association, and the agreement to cooperate in plant-to-table activity.

Duany locates the market for Agrarian Urbanism in four categories of people:

– Ethicists, because they believe it's the right thing to do
– Trendsetters, also known as the "cool greens"
– Opportunists, the people who ask the question, "What's in it for me?"
– Survivalists, or those who "circle the wagons," believing that security comes first (Duany, 66–69).

Michele Owens, author of *Grow the Good Life: Why a Vegetable Garden Will Make You Happy, Healthy, Wealthy and Wise*, articulates a somewhat different motivation: "Thanks to my garden, I can take a small stand against everything I find witless, lazy, and ugly in our civilization" (Duany, 70). (Wendell Berry took this a step further when he called gardening a revolutionary act.)

Dominique Browning, former Editor-in-Chief of *House and Garden* and a noted author of gardening books, sums up these motivations elegantly: "The vegetable garden, it turns out, is a ripening political force: the best response to the enemy crisis, the climate crisis, the obesity crisis, the family crisis, and the financial crisis. It will be no small irony, if suburbia becomes the locavore's home of choice … and growing backyard veggies could be the answer to the crisis of disaffected suburban youth" (Duany, 82).

A Matter of Words
"Agricultural" is concerned with the technical aspects of growing food. "Agrarian" emphasizes the society involved with all aspects of food. "Agrarian Urbanism" addresses the problem of food production and the problem of sprawling communities simultaneously.

"The transformative power of food is absolute."
   Janine de la Salle and Mark Holland

Will Allen Has It Right
"The (food) revolution comes in breaking these long chains in the food system and in the minds of buyers and in building new tight webs around the centers of communities."
   Will Allen

Food and the Smart City

Author and new urbanist James Howard Kunstler has prophetically remarked, "Agriculture is going to come back to the center of the American life in a way that we couldn't imagine" (Carroll 2008). This is precisely what has been happening over the past decade in towns and cities, including the largest cities, all over America. We are witnessing the rise of intown and inner-city food production from patch gardens to sizable acreage rooftops. And we've been witnessing a healthy growth in small-scale farming at the urban edge, with farmers markets and CSAs as the current method of choice for marketing farm product to hungry city and intown stomachs. Urban agriculture's day has come, as Kunstler predicted in his statement that was once viewed as radical. And its growth shows no sign of abating. Food – local food locally produced, processed, and prepared – is the largest single component sustaining the smart city. And it is the one component that offers the city or town resident the opportunity to make a difference, to have a hand in the city's direction and destiny. Local food locally produced is a near-perfect tool for achieving the character and promise of a truly smart city. Thus, Agrarian Urbanism becomes a reliable path to a smart city.

   Agrarian Urbanism tells us that the production of food is coming home. But are we ready? And what does "home" mean? Does it mean a return of agriculture to the regions it feeds? Yes, it means that. But, importantly, it means a return of agriculture, including small-scale backyard farming and market gardening, within view of the mouths to be fed. It means rooftop gardening, vest pocket gardening (also known as small patch gardening), and vertical gardening on walls – gardening wherever sunlight shines and water is available and where the mouths to be fed are located, at home, at work, or at play. In short, agriculture's return home means Urban Agriculture and, more specifically, Agrarian Urbanism, the practice of urban planning with food production in mind. When Urban Agrarianism is achieved, it is lodged in the mindset of city dwellers who routinely apply it to their use of land, exercise, health, recreation, and pleasure. Indeed, Urban Agrarianism is a form of lean urbanism (town planning that makes do with what is naturally and cheaply available) and of lean agrarianism (which likewise makes do with the land, water, sunshine, and human will present on the scene).

Andres Duany has outlined a path to Agrarian Urbanism. Starting with "agrarian retention," that is, simply saving existing farmland, one moves to the increasingly popular production level of "urban agriculture" in its many forms but simply described as food production within the city. The progression continues to "Agricultural Urbanism," or settlements equipped with a working farm economically connected to the community's residents. Finally, one arrives at the truest form, agrarianism, wherein settlements are involved with food in all its aspects (production, processing, preparation and distribution), and where the physical pattern of the settlement supports the working of an intentional agrarian society. Ever concerned about sufficiency of urban population density, Duany writes, "As the primacy of shopping necessarily wanes, agrarian activity can provide a surrogate urban condenser" (Duany, 35). And he gives us the image: "Agrarian sociability is based on the organizing, growing, processing, exchanging, cooking, and eating of food – much of it taking place around a market square" (Duany, 35). Such squares then become indispensable as third places for social interaction, supplementing those of home and workplace.

Convinced that the smart city is a city of high social interaction, Duany comments on the discomfort of people exposing themselves in acts of leisure, but the relative comfort that people in our society feel in being seen in the act (or pretense) of working.

The requirements of local food and farming routinely lead to extended social networks, for "All ages can work together to accomplish many of the tasks required to make preserves, flash-freeze vegetables, fill boxes for CSAs, or administer a farmers market" (Duany, 36). And planting, tending and harvesting, he notes, impose a routine of great activity akin to a considerable amount of walking, resulting in a higher level of walkability in the community. Such hand-tended crops have the added benefits of being less toxic (or nontoxic), lowering the carbon footprint, and offering higher-quality flavor since they need not be travel-resistant or have a long shelf life. And food production in the community can include energy production along with recycling and composting, all of which activities are additional social condensers leading to those same extended social networks.

The Insight of Andres Duany

Andres Duany tells us that a neo-agrarian way of life should be made available to as many as possible because of its mitigating effect on climate change – in other words, for ethical reasons no less than for practical ones. But, as the following ideas from his book, *Garden Cities: Theory and Practice of Agrarian Urbanism,* testify, Duany's arguments do not lack the practical side as well:

> "Agrarian Urbanism transforms lawn-mowing, food-importing suburbanites into settlers whose hands, minds, surplus time, and discretionary entertainment budgets are available for food production and its local consumption."
>
> "By bringing most ordinary daily activities within walking distance, those who do not drive (usually the young, the old, the poor and the principled) gain independence ... The possibility of not owning an automobile provides a virtual subsidy that can be applied to housing costs. There is no more organic way to increase affordability."
>
> "The financial model of agrarian urbanism also includes the subsidy that is transferred from the debit column of landscape maintenance."
>
> "Agrarian activity provides a surrogate urban condenser to the primacy of shopping."
>
> "The traditional village was a machine to grow food."
>
> "Agricultural urbanism is a new kind of real estate. Growing food is a very good way to sell real estate."
>
> "Development structured on useful and productive activity that supports the common health and wealth is a compelling vision not least for the marketing of real estate."
>
> Agrarian urbanism should become a "normative real estate product."
>
> "Agriculture is the new golf."
>
> "Garden frontage generates social contact and well-being. Gardening is fantastically sociable – even more than the porch. There's nothing that generates conversation more than vegetables."
>
> "Agrarian sociability is based on a foundation of organizing, growing, processing, exchanging, cooking and eating of food - much of it around a market square."
>
> "A tomato plant may be a vector of conversation no less effective than a baby in a pram or a cute dog on a leash."
>
> "The village of yore was essentially an apparatus to grow, process and distribute food."

From the point of view of expense and investment, "The barn is cheaper than the clubhouse, community gardens more incremental than fairways – so the barn and gardens can be spread over time, thus sealing the case for community farms over golf courses."

Ultimately, for local food production to scale up, to become much more common, Duany believes that it must engage the same market-oriented system "that so efficiently delivered conventional suburban development during the recently concluded century" (Duany, 45).

"If you build a garden, you believe in the future."
Linda Isaacson

Urban agriculture, including urban gardening, forces one away from exclusively thinking of and interacting with fellow humans and reorients one toward nature and its systems. It moves one "out of the ruts which an exclusive association with the human animal produces in the mind of man" (Grunwald, 208).

Stapleton/Denver and Making the Local Foodshed Independent

The high plains and prairie grasslands of interior North America bespeak food and agriculture, whether grass into meat and dairy or grass into grain and mixed farming. Stapleton, Colorado, established on the land of a former international airport and positioned within the largest city on the high plains of North America, Denver, offers a model with high potential. In the shadow of the majestic Front Range of the Rocky Mountains, Stapleton is well situated to feed itself in virtually all food categories. Its proximity to Denver makes it a potentially vibrant player in the Denver foodshed, both as a producer and trader in vegetables and fruit, and even more so as a producer of meat, dairy, and grain production – all on a scale to feed the local community, creating a truly independent local foodshed. The newer northern sector of Stapleton is particularly well suited to meat, dairy, and grain production, while both portions of the former airport, including the older southerly and more urbanized portion, are suited for vegetable and fruit production. Stapleton could serve as a beacon when it comes to providing food security to a large urban population far removed from other large urban conurbations.

Food Security

Shortening the supply chain is perhaps the single most important step toward ensuring food security. Positive byproducts of a shortened food chain are many: meaningful local work, opportunities for recreation and pleasure, enhanced human health due to the availability of more highly nutritional food, and a reduced carbon footprint.

A Progression of Urban Food
The power of gardens
  The place of backyard farming

(continued)

The place of protein at the periphery

These three subheadings represent a progression along the transect from T-1 to T-6. Gardens can be created anywhere along the transect, even in the densest downtown development of transect 6. Backyard farming represents an intermediate stage of food development within the middle of the transect (T-3, T-4, and beyond). Dairy and meat production, because they involve larger animals, is logically restricted to the outer zones of the transect (T-1 and T-2). Thus, within the transect, most food needs of the city's population are accommodated.

The Smart City in Relation to Its Periphery and Hinterland

All cities and towns depend upon their hinterland for food, water, energy, and many other goods and services. The hinterland of a city may consist simply of its watershed, whether large or small, or it may constitute a sizable portion of a continent. Because of technology, a hinterland may even be thought of as the globe, a broader but much less secure area from which food and other goods are procured.

In contrast to cities with sprawling hinterlands, the smart city focuses on its immediate periphery (T-1 T-2), which has the capacity to provide food to the city's residents. This peripheral area can also supply needed space for recreation and, respite to the city's population, sometimes providing food and recreation simultaneously.

Protein in the Smart City

A particular food value of the urban periphery relates to its capacity to provide meat and dairy, basic proteins, and, to some extent, grains – all necessary food products that are somewhat more difficult to produce within the densely populated areas (T-5 and T-6). Thanks to intensive rotational grazing, also known as management intensive grazing, virtually all meat and dairy needs of the urban population can be provided for within close proximity of mouths to be fed. Additionally, meat and dairy production along the immediate periphery of the city supplies employment, recreation, and a higher quality of life for the urban residents.

Grass-based animal production for meat and dairy can be accomplished on relatively small acreages at the urban-rural interface (the periphery) in a manner that provides open access to the city's residents, facilitating a sense of connection to food sources and, to the extent desired, opportunities to participate in food production and preparation.

For ecological reasons as well as for human and animal health, all meat and dairy production should be grass-based and not, as is so often the case, grain-based. Grass-based represents sustainability for the land, for animals, and for people. This approach represents, along with the factor of scale, the epitome of sustainable agriculture. When it comes to animal production, the focus is on cows, sheep, and goats – all three for both meat and dairy. Pigs, poultry (including chickens, turkeys, ducks, and geese), and smaller animals such as rabbits can be added to the mix to enable an ecologically secure interspecies mix of dependency. As we have seen, nature requires diversity, abhorring monoculture and embracing polyculture, so following this principle of nature is a sound recommendation for both ecological health and economic production.

Intensive rotational grazing is the design mechanism best suited for providing meat and dairy (and thus protein) to town and city populations (Carroll 2008). This technique, which takes its cue from centuries-old agricultural models and ultimately from ecological principles, finds its basis on newly available but well-tested light-weight electric fencing. In addition to delivering meat and dairy product to the urban core, intensive rotational grazing serves as an important source of fertility for the production of fruit and vegetables within the city core and all adjacent transects. By mimicking nature, this method represents the epitome of sustainability and food security, while helping to meet an urban population's need for open space, access to nature, physical exercise, meaningful employment, and mental as well as spiritual health.

While land needs are minimal in this system, management of both the land and the animals is intensive. Consequently, this approach requires not only close observation of the land and the development of a deeper, more intimate relationship with the ecosystem but also more eyes to acres and more hands to acres than we now witness in food production. The community enjoying the food becomes connected to the systems that sustain them, involved in the production, processing, and preparation of its food. Involvement on the part of the recipient of food is a key to sustainability. Sustainable food systems such as this do not require the participation of all, only of those who desire involvement. In fact, a relatively small percentage of the population (perhaps around 15%) is needed, and at varying stages of intensity. (See my four books on these topics, available to the public for free See references.)

Livestock in Town
To most people, the image of a farmstead is a homestead in the countryside. But traditional German and some other Central European farmsteads paint a different picture. In Germany and other Central European countries, farmsteads are often located right in town, as can still be seen today in farm villages. These were (and, in some cases, still are) integrated animal operations featuring cattle and other livestock at ground level or lower, and farm living quarters above, benefiting from the rising heat of the animals. Although not the

(*continued*)

agrarian design mode of today, this remains an outstanding example of safely maintaining farm animals in an urban setting. Many ancient civilizations on all continents evolved the same kind of integrated living and farming systems. It is obvious today that small livestock (poultry, rabbits, goats) can be kept in close quarters and integrated with vegetable and fruit production, but few realize that large animals have historically been kept in this way as well. While I do not necessarily advocate a return to this system of agriculture, it is quite obvious that large and small livestock, whether for meat or dairy, can at least be raised at the immediate edge of settlement, close enough to provide a rounded as well as secure diet.

A Pattern Language

A forebear of Duany is Christopher Alexander, who also created an important conceptual platform for food in the city. His path-breaking book, *A Pattern Language*, has had no small influence on New Urbanism. According to Alexander, ecology teaches that

> No thing is an isolated entity. And a pattern language teaches that no pattern itself is an isolated entity. Each pattern can exist in the world only to the extent that it is supported by other patterns: the larger patterns in which it is embedded, the patterns of the same size that surround it, and the smaller patterns which are embedded in it. This is a fundamental (and fundamentally ecological) view of the world. It says that when you build a thing you cannot merely build that thing in isolation but must also repair the world around it and within it, so that the larger world at that one place becomes more coherent and more whole; and the thing which you make takes its place in the web of nature, as you make it. (Alexander, xiii)

As would any ecologist, Alexander tells us that patterns are alive and evolving and that there are as many pattern languages as there are people, creating a sense of fragmentation. "The (pattern) languages which people have today are so brutal, and so fragmented, that most people no longer have any (pattern) language to speak of at all – and what they do have is not based on human, or natural, considerations" (Alexander, xvi).

The highly ecological concept of "pattern language" is directly applicable to urban planning. According to Alexander, neighborhoods need boundaries. Agrarian Urbanism can help to set those boundaries. Alexander writes, "People need green open places to go to; when they are close, they use them. But if the greens are more than three minutes away, the distance overwhelms the need" (Alexander, 305). He advises, "Build one open public green within three minutes walk – about 750 feet – of every house and workplace. This means that the greens need to be uniformly scattered at 1,500 foot intervals, throughout the city. Make the greens at least

150 feet across, and at least 60,000 square feet in area" (Alexander 308–309). Such "greens" could easily be gardens of various sorts including pocket gardens and gardens with walking paths. And rainwater can easily be collected from the roof gutters of buildings to keep the needed gardens watered. In spite of their compact size, Alexander's "greens" can be significant food producers.

Alexander offers advice on public squares, asserting, "A town needs public squares; they are the largest most public rooms that the town has. But when they are too large, they look and feel deserted" (Alexander, 311). He advises, "Make a public square much smaller than you would at first imagine; usually no more than 45 to 60 feet across, never more than 70 feet across. This applies only to its width in the short direction. In the long direction it can certainly be longer" (Alexander, 313). Such public squares can also usefully accommodate public and community vegetable gardens as well as fruit trees. Public squares with community gardens would most certainly neither look nor feel deserted, and the presence of food production in such a prominent local place would serve to raise interest in both the food produced and the craft of gardening itself. The addition of poultry or rabbits would not only provide fertility but add significant interest.

Regarding the importance of the public square and shared land, Alexander further asserts, "Without common land no social system can survive (Alexander, 337). Therefore, he says, "Give over 25% of the land in house clusters to common land which touches, or is very, very near the homes which share it" (Alexander, 340). He further cautions, "Be wary of the automobile; on no account let it dominate this land." It is not unusual for common land to be used for gardening and food production. This should be encouraged for food security as well as many other reasons.

With respect to children and the need for play space, Alexander writes, "Lay out common land, paths, gardens and bridges so that groups of at least 64 households are connected by a swath of land that does not cross traffic. Establish this land as the connected play space for the children in their households" (Alexander, 346–347). This recommendation recognizes the connectivity between gardens and play, for supervised gardening can be considered play (and important play) for children.

With respect to the need for animals in our lives, farm animals count! Alexander writes, "Animals are as important a part of nature as the trees and grass and flowers. There is some evidence, in addition, which suggests that contact with animals may play a vital role in a child's emotional development" (Alexander, 372). Securing ready access for children to common farm animals, including poultry, pigs, sheep, cows, and goats, can well fulfill children's need for contact with animals. Because of the benefit to children, keeping farm animals beyond their obvious food and fertility value is preferable. Alexander advises, "Make legal provisions which allow people to keep any animals on their private lots or in private stables. Create a piece of fenced and protected common land, where animals are free to graze, with grass, trees and water in it. Make at least one system of movement in the neighborhood which is entirely asphalt-free – where dung can fall freely without needing to be cleaned up" (Alexander, 373–374).

Since, as Alexander reminds us, "The nuclear family is not by itself a viable social form," we need to embrace a communitarian approach that finds family in community" (Alexander, 377). Thus, community-building through gardens and other measures is always vital. Further, a true community approach nurtures opportunity for learning through apprenticeship, particularly in gardening, in food production, and in food preparation. As Alexander states, "The fundamental learning situation is one in which a person learns by helping someone who really knows what he is doing" (Alexander, 413). Alexander asks us to "Treat every piece of work as an opportunity for learning. To this end, organize work around a tradition of masters and apprentices..." (Alexander, 414).

On farmers markets, Alexander writes, "The simple social intercourse created when people rub shoulders in public is one of the most essential kinds of social 'glue' in society" (Alexander, 489).

On placing gardens, Alexander writes "If a garden is too close to the street, people won't use it because it isn't private enough. But if it is too far from the street, then it won't be used either, because it is too isolated" (Alexander, 545). Half-hidden is thus ideal, albeit fully hidden will, I believe, work, if the gardens in question are serious food production gardens, also known as market gardens. And a courtyard or plaza can be brought to life socially and psychologically, as well as become biologically organic, by placement of food-producing gardens within.

Gardens do not offer enough relief from noise unless they are well protected, so an enclosure is needed. In a very small garden, Alexander tells us, the enclosure should be formed of buildings or walls. Because an enclosure around garden space gives the garden the feeling of a room, the gardened outdoor space can become a special outdoor room, increasing its psychological value and thus its desirability as a place in which to spend time. Because orchards give the land a special, almost magical quality, so gardens should include fruit trees, as should streetscapes.

> "A garden which grows true to its own laws is not a wilderness, yet not entirely artificial either" (Alexander, 802).

A greenhouse/glass house efficiently traps heat and turns it into the production of energy in vegetation, while at the same time providing a space to live in and enjoy. The value of the garden can be much enhanced with seating – an important place to both enjoy and commune with nature. "In a healthy town every family can grow vegetables for itself. The time is past to think of this as a hobby for enthusiasts; it is a fundamental part of human life" (Alexander, 819). "Set aside one piece of land either in the private garden or on common land as a vegetable garden. About one-tenth of an acre is needed for each family of four" (Alexander, 821). "Arrange all toilets over a dry composting chamber. Lead organic garbage chutes to the same chamber, and use the combined products for fertilizer" (Alexander, 826). Alexander warns that we must not waste this good fertility source, and we should not poison bodies of water with its excess.

Two Concepts: Small Is Beautiful and Limits to Growth

There are two seminal books on the topic of growth and decentralization. In his 1973 book, *Small is Beautiful: Economics As If People Mattered*, E. F. Schumacher forges another path to New Urbanism. An important champion of small-scale agriculture and the central importance of soil in human life, Schumacher was a successful industrial economist who criticized large-scale economic systems. He was perhaps the leading thinker in our time when it comes to decentralization, and decentralization is key to the idea of the smart city. For decades now, North America and Europe have witnessed the decline and shrinkage of large-scale industrial systems (Detroit being an outstanding, if extreme, example).

The infamous 1972 study, *Limits to Growth* by Meadows et al., a team out of MIT, whose long-criticized (and even rejected) work is now being vindicated, foresaw the decline of large-scale industrial agriculture. The decline of faith in large-scale systems in both Europe and North America and, in some locales, a rise in libertarianism, may be a further consequence. And in planning (whether regional, city, or town), we see a further expression of such thinking, with both New Urbanism and Agrarian Urbanism emerging as responses to these circumstances.

I would take the concepts of decentralization and limits to growth one step further by embracing lean agrarianism, as necessitated by peak oil and other energy realities. (The term *peak oil* refers to the global peak in global production and demand and the resultant end of cheap and easy-to-obtain oil.) Also necessitating lean agrarianism are climate change and human response generated by fear of the consequences of these events and fear of a future likely more fundamentally different than has been experienced by humans in many centuries. It appears that the lives of our children and grandchildren will not simply be a continued evolution of our own lives and times, as has long been the case since the start of the Industrial Revolution three centuries ago. Our descendants will face a brave new world where survival will be a challenge and there won't be a lot of guidance from the generations preceding them. The adoption of "small is beautiful" as a controlling philosophical tenet, expressed in New Urbanism and in other ecological approaches to the future, is becoming necessary.

> **Institutionalizing Local Farmers Markets**
> Institutionalizing local farmers markets at a local scale, borrowing from models in Santa Fe, New Mexico; Woodstock, Vermont; and other places, will prove critical to the flow of local food as a key component of Agrarian Urbanism. The Santa Fe Farmers' Market Institute may be the gold standard when it comes to building and supporting a year-round indoor/outdoor farmers market in a small city or large town. This Institute had the wisdom to locate in the Santa Fe Railyard District, a location with a very attractive feature for the new reality: passenger rail transportation across the region. (The "Rail

(continued)

Runner" trains, which actually enter the farmers market, serve Albuquerque, the state's largest city, and other points south of Santa Fe.) The market's location in this district is as much a symbol for the younger generation as the market itself, making an important statement about the future, when affordable ecologically desirable transportation will be increasingly in order.

Boston Urban Agriculture and Article 89

As articulated in Article 89 of the municipal zoning code, Boston, Massachusetts, the largest city in New England, believes that all city residents who wish to grow their own food should have access to appropriate space to do so. This means access to clean soil and water within walking distance of homes. With the increase in urban agriculture likely to result, more residents will likely also want access to more farmers markets and local food in neighborhood stores. Gardening, including chickens, bees, rabbits, etc., will become openly available to everyone, either in their own space or as a community flock, herd, or garden. No formal goals or targets have yet been set for the Boston urban agriculture effort, but the primary themes are (1) inclusion of biodiversity, (2) determining what is the economic viability of urban agriculture for "for profits," (3) the needs of community gardening in Boston, and (4) youth involvement and intergenerational engagement. The Massachusetts Commercial Food Waste Ban is resulting in increased compost for city residents. The local Kendall Foundation estimates that community gardening displaces $1 million per year in retail food purchases. And since transport of food accounts for about 80% of congestion on city streets, Article 89 will likely lower this statistic. Boston's urban agriculture plan offers many benefits beyond food production and is essentially an economic as well as a social plan designed to keep capital invested at home in the city.

Two Positive Trade-Offs

The higher population density associated with the New Urbanism frees up much land on the town periphery and beyond, which can be kept natural and provide ecological services (including storm water containment, flood alleviation, water and air purification, and nature protection for human recreation, edification and enjoyment). Likewise, urban agriculture at the city core, where the people are, means less food needs to be produced on farmland beyond the cities and towns, likewise freeing up that land for the protection of ecological services. And both New Urbanism and urban agriculture equate to a reduced

(*continued*)

carbon footprint, reduced greenhouse gas emissions, and thus action against climate change.

Powerful Economic Development
Every molecule of food consumed from a local source displaces a molecule of food from away. Local food, then, is an inherently powerful form of economic development. This frees vast acreages of land now devoted to large-scale food production, whether in California, Florida, Iowa, or overseas, some of which can be returned to its natural state, providing important ecological services. Locally sourced food also means that people living closely together experience a higher quality of living, opening the possibility of releasing more land in rural and suburban zones to its natural state and freeing it to produce the many rich ecological services of which it is capable. This model helps return people and land resources to the natural equilibrium they enjoyed over hundreds of thousands of years.

"[T]he grounds for hope are in the shadows, in the people who are inventing the world while no one looks" (Solnit, 164).

Lean Urbanism Is All About Making Small Possible
Because I agree with my mentor, E. F. Schumacher, that small is beautiful, I must then believe that anything which makes small possible is also beautiful. Therefore, "lean urbanism," which helps to make small possible, must also be beautiful. It is also practical, timely, appropriate, and, in the present economic reality, necessary.

Conclusion: City- and Town-Building in a Carbon-Constrained World

Planners and politicians will be severely constrained by reality in plotting their future course. And that reality includes living in and coping with a carbon-constrained world, a world which will permit significantly less carbon emission than we have been accustomed to enjoying. Such a world requires locally produced food rather than food from away. It means significantly less dependence on the personal motor vehicle. And it will look like significantly less access to aviation, indeed, less mobility in general, whether for goods or people. For these reasons, our cities and towns will come to bear a more significant resemblance to their history than we might imagine, albeit with the combination of more sophisticated but less carbon-dependent technology. New Urbanism, with its detailed charter and its detailed canons, offers a clear path to this future and will, in its way, become an organizing principle for society, focused, as we must be, on the local. For this reason, models of New Urbanism, and especially those with an agrarian (i.e., local food) component, will be particularly valuable. The twin challenges of oil volatility in the era of peak oil and the necessities of adaptation to climate change realities (i.e., reduction in carbon, methane, and other greenhouse gas emissions) lead directly to models of life and societal organization akin to the model organized by the designers, planners, and architects of the New Urbanism.

New Urbanism forms the "smart city," and it will emerge, I believe, whether by choice or necessity. Smart cities will be among the first cities to embrace the New Urbanism, and especially so in its agrarian form.

From its inception in the early 1980s, New Urbanism has been a movement of choice for those attracted to it. But today, in the second decade of a new millennium, while it remains a choice for some, it has become for everyone a necessity. (Of course, all may not see it that way yet.) It has become a necessity because it is the only way of organizing our society, our settlements, and our towns and cities, in such a manner as to reflect the realities of two of our greatest societal challenges: climate change and energy uncertainty. The effects of climate change are becoming more evident every day, with record-breaking increases in temperatures, violent storms of many kinds, heavy precipitation, extreme drought, rising sea levels, intense forest and grassland fires, and flooding on a massive scale – all well beyond established climate records. The evidence abounds. Simultaneously, we face energy circumstances in our fossil fuel-dependent society of a kind unknown in nearly a century: the volatility and unpredictability of oil prices and perhaps soon of oil availability.

Our awareness of both peak oil and of the meaning and magnitude of climate change has emerged simultaneously, forcing us to adapt and build resilience in support of our very survival – economically, socially, and physically. Our survival necessarily requires fundamental change in our values and way of life, not only for the moral reasons about which Pope Francis regularly reminds us but also for reasons of physical necessity. To be sensitive to intergenerational justice and equity, I believe, is a serious moral and spiritual obligation. But, given the rapidity of events, both environmental and social, we may be upended by what we increasingly know we must accomplish if there is to be any livable society or livable planet to pass on to those future generations.

This is where New Urbanism enters the picture. This model is an entirely new tool in city and town planning that borrows from the past (even the distant past of ancient civilizations), but which upends if not outright repudiates the post-World War II love affair with automobile-driven sprawl and with zoning designed to separate rather than join. The New Urbanism approach embraces decentralization. And it is coming into its own just in time to pragmatically address the two realities of peak oil and climate change.

How does New Urbanism accomplish such a critical task? It does it through the force of decentralization, championing what Schumacher referred to by his phrase, "Small is beautiful." While some may prefer the benefits and costs of centralization, and others prefer the benefits and costs of decentralization, we appear to have passed the point where we have the luxury of choosing. All signs now point to smaller-scale systems and decentralization as an imperative. Increased self-sufficiency at the community level is a necessity. Our circumstances point in one direction: decentralization, for such is the path to resilience and adaptation and very possibly the path to survival itself.

Because they represent our flourishing, if not our survival itself, we need a clear understanding of the terms "New Urbanism," "smart city," and "agrarianism." Wendell Berry and other agrarians have written that agrarianism, in many respects the compatriot of urbanism, is a practice before it is a theory – that the theory of agrarianism can only be understood as a result of practice and not study. I suggest the same holds for urbanism. To conduct that practice, we need many more smart cities and smart towns. And we must redefine the edge of the possible.

Cross-References

▶ Community Engagement in Energy Transition
▶ Sustainable Living in the City
▶ The Spirit of Sustainability
▶ Transformative Solutions for Sustainable Well-Being

References

Alexander, C. (1977). *A pattern language*. New York: Oxford University Press.
Berry, W. (2015). *"Our deserted country." In our only world: Ten essays by Wendell Berry*. Berkeley: Counterpoint.
Carroll, J. E. (2005). *The wisdom of small farms and local food: Aldo Leopold's land ethic and sustainable agriculture*. Durham: University of New Hampshire.
Carroll, J. E. (2008). *Pastures of plenty: The future of food, agriculture and environmental conservation in New England*. Durham: University of New Hampshire.
Carroll, J. E. (2010). *The real dirt: Toward food sufficiency and farm sustainability in New England*. Durham: University of New Hampshire.
Carroll, J. E. (2014). *Live free and farm: Food and independence in the granite state*. Durham: University of New Hampshire.
Carroll, J. E. (2016). The environment is a moral and spiritual issue. In *Spirituality and sustainability: New horizons and exemplary approaches*. Springer Publishers.
Congress for the New Urbanism. (n.d.). *Charter and canons of the New Urbanism*. Washington, DC: Congress for the New Urbanism.
Duany, A. (2011). *Garden cities: Theory and practice of Agrarian Urbanism*. Miami/London: DPZ and the Prince's Foundation for the Built Environment.
Grunwald, M. (2006). *The swamp*. New York: Simon & Shuster.

Meadows, D., & Meadows, D. (1972). *Limits to growth*. New York: Universe Books.
Schumacher, E. F. (1973). *Small is beautiful: Economics as if people mattered*. London: Blond & Briggs.
Pope Francis (2015). *Laudato si: On care for our common home*. The Vatican: Papal Encyclical of Pope Francis.
Solnit, R. (2006). *Hope in the dark*. New York: Penguin.
World Commission on Environment and Development. (1987). *Our common future: Report of the World Commission on Environment and Development.* New York: World Commission on Environment and Development.

Sustainable Living in the City

The Case of an Urban Ecovillage

Mine Üçok Hughes

Contents

Introduction	870
What Is an Ecovillage?	871
Case Study: An Urban Ecovillage	873
Sustainable Living Practices	874
Ecological Dimensions	875
Economic Dimensions	876
Socio-Cultural Dimensions	877
The Challenges	879
Conclusion	880
Exercises in Practice	880
Engaged Sustainability Lessons	881
Chapter-End Reflection Questions	881
Cross-References	881
References	882

Abstract

On December 2015, heads of states from all over the world got together at United Nations Climate Change Conference in Paris where they agreed that climate change is not only an unprecedented challenge for humanity but that it requires urgent global agreement for action. Urbanization is particularly relevant to the implementation of the Paris Agreement as of 2007 the majority of world's population lives in urban areas. In 2016, the world's cities occupied just 3% of the Earth's land, but accounted for 60–80% of energy consumption and 75% of carbon emissions. While on one hand the rapid urbanization is exerting pressure on the living environment and public health, on the other hand high density of cities can bring efficiency gains in solving those problems.

M. Üçok Hughes (✉)
Department of Marketing, California State University, Los Angeles, Los Angeles, CA, USA
e-mail: mine.ucokhughes@calstatela.edu; mineucok@gmail.com

Ecovillages are intentional communities whose members holistically integrate ecological, economic, social, and cultural dimensions of sustainability. An urban ecovillage situated in the heart of a Western metropolis is taken as a case study to present how its members strive to lead more sustainable lives in the city by altering their consumption practices and adopting more sustainable ones. Specific examples of more sustainable solutions are given to some consumption practices that can be implemented in an urban setting.

Keywords
Ecovillages · Intentional communities · Sustainability · City

Introduction

The 2015 United Nations Climate Change Conference in Paris, France, "was the largest UN conference ever seen, with 37.878 participants, including 20.000 government/parties representatives, 8.000 IGOs/NGOs observers, and 3.000 media" (UN-Habitat 2016, x). It was deemed to be an historical event not only due to its size but also because 175 parties (174 countries and the European Union) all agreed that "climate change is a real and unprecedented challenge for humanity" that requires an urgent global agreement for action (UN-Habitat 2016, x).

Urbanization is particularly relevant to the implementation of the Paris Agreement (also known as COP 21) as "human activities in cities, are in large part responsible for the current climate change trends and dynamics" while at the same time, urban populations are vulnerable to the increasingly negative effects of climate change and air pollution mainly generated by greenhouse gas emission of transportation and heating/cooling systems (UN-Habitat 2016, 3).

In 2016, the world's cities occupied just 3% of the Earth's land, but accounted for 60–80% of energy consumption and 75% of carbon emissions (World Bank 2009). This reality poses many threats and many opportunities simultaneously. While on one hand the "rapid urbanization is exerting pressure on fresh water supplies, sewage, the living environment, and public health," on the other hand "high density of cities can bring efficiency gains and technological innovation while reducing resource and energy consumption" (UN-Habitat 2016, 1).

The United Nations Human Settlements Programme (UN–Habitat) estimates that by 2030, almost 60% of the world's population will live in urban areas and 95% of urban expansion in the next decades will take place in the developing world, adding an estimated 70 million new urban residents each year (UN-Habitat 2016). This will result in more people living in urban, rather than rural areas in all developing regions, including Asia and Africa (World Bank 2009).

Increasing urbanization has acute environmental, economic, and social implications for the world's future (Wu 2010). It has a direct impact on climate change as "it creates major producers of greenhouse gases and air pollutants; and it is the most drastic form of land transformation, devastating biodiversity and ecosystem services" (ibid). Yet, "[c]ities have lower per capita costs of providing clean water, sanitation, electricity, waste collection, and telecommunications, and offer better access to education, jobs, health care, and social services" (ibid). The high population density of the cities makes them much more cost-effective in terms of implementing more sustainable processes. Wu (2010) strongly believes that "urbanization offers a number of things that are critical to achieving sustainability" as "cities epitomize the creativity, imagination, and mighty power of humanity. Cities are the centers of socio-cultural transformations, engines of economic growth, and cradles of innovation and knowledge production."

Among the 17 Sustainable Development Goals (SDG) set by the United Nations which are part of the 2030 Agenda for Sustainable Development is "Goal 11: Sustainable Cities and Communities" (UN 2015). This goal aims to "[m]ake cities and human settlements inclusive, safe, resilient and sustainable." Among its 11 goal targets are reducing "the adverse per capita environmental impact of cities, including by paying special attention to air quality and municipal and other waste management" and providing "universal access to safe, inclusive and accessible, green and public spaces, in particular for women and children, older persons and persons with disabilities" by 2030 (UN 2015).

It is evident from these reports that as the majority of the world population is moving to urban areas and contributing to negative environmental impact, cities play an important role in the future of our planet regarding the implications of climate change. Therefore, the question arises: Is it possible to lead sustainable lifestyles in the city that are respectful of the ecological environment and human relations? According to John Wilmoth, Director of United Nations Department of Economic and Social Affairs's Population Division, "[m]anaging urban areas has become one of the most important development challenges of the 21st century. Our success or failure in building sustainable cities will be a major factor in the success of the post-2015 UN development agenda" (UN 2014).

This chapter looks at ecovillages as one way of tackling the complex phenomenon of sustainability, especially in an urban setting. The chapter begins with a definition of ecovillages and the core practices and values shared by ecovillages around the world. This section is followed by a case study of an urban ecovillage in the western United States which describes ways in which the members of this intentional community strive to achieve and maintain a lifestyle that is sustainable in ecological, economic, and socio-cultural ways in the heart of a metropolis. Specific examples of more sustainable solutions are given to some consumption practices.

What Is an Ecovillage?

Ecovillages are models of sustainable living. They are a grassroots movement of cohousing projects where its members live in intentional communities. Global Ecovillage Network (GEN), founded in 1995, defines ecovillages as "an intentional, traditional or urban community that is consciously designed through locally owned, participatory processes in all four dimensions of sustainability (social, culture, ecology and economy) to regenerate their social and natural environments" (GEN n.d.). There are over 1000 ecovillages in rural, suburban, and urban settings around the world that are part of GEN which works as a hub to share and disseminate information to ecovillages around the globe. In 1998, the first ecovillages were officially named among the United Nations' top 100 listing of Best Practices, as excellent models of sustainable living.

Karen Litfin, a political scientist who specializes in global environmental politics and has visited and studied over a dozen ecovillages around the world, characterizes ecovillages as "emerging in an astonishing diversity of culture and ecosystems, as a planetary knowledge community grounded in a holistic ontology and seeking to construct viable living systems as an alternative to the unsustainable legacy of modernity" (2009, 125). Ecovillages show diversity according to the cultural and ecological context they are part of, but what unites them is their "members' commitment to a supportive social environment and a low impact of life" (Litfin 2009, 125). What distinguishes ecovillages from mainstream neighborhoods, towns, and cities is that its members intentionally become part of a community that has shared ideals and goals with emphasis on ecology (Van Schyndel Kasper 2008).

Ecovillages vary in many ways, as they are shaped by the particular objectives, values, belief systems, and cultures of their founders and members as well as local laws and geographic contexts among other variables. Despite their differences, GEN outlines the following as three core practices shared by most:

- Using local participatory processes
- Integrating social, cultural, economic and ecological dimensions in a whole-systems approach to sustainability
- Actively regenerating natural and social environment. (GEN n.d.)

The principles of permaculture overlap with the ecovillage ideology. Initially developed in the 1970s in Australia by Bill Morrison and David Holmgren, the permaculture movement advocates creating virtuous circles rather than vicious ones within any community. Permaculture can be described as a "creative design process based on whole-systems thinking informed by ethics and design principles" including land and nature stewardship, health and spiritual wellbeing, education and culture, finance and economics, and tools and technology (Permaculture Design Principles n.d.). Many permaculture principles are adopted by ecovillage communities as they both advocate to reduce waste, store energy, use small and slow solutions, use and value diversity, and integrate rather than segregate (Litfin 2009).

There are many reasons people choose to join an ecovillage or an intentional community. Among those are the desire to move away from consumerism and materialism (Cunningham and Wearing 2013; Bang 2005), to achieve sustainability, and lead a transparent, low-key, low impact lifestyle (Hong and Vicdan 2016), search for a more spiritual life as part of a community (Cunningham and Wearing 2013), and longing for a safe environment to raise children (Van Schyndel Kasper 2008). As Litfin puts it "ecovillagers see themselves pioneering an alternative socioeconomic system to the unsustainable legacy of modernity" (2009, 132).

The extant literature on ecovillages spans a wide range of academic disciplines (See Kunze and Avelino 2017 for a rather comprehensive list of publications). These include human ecology (Van Schyndel Kasper 2008), anthropology (Chitewere 2006; Chitewere and Taylor 2010; Holleman 2011; Standen 2014), environmental psychology (Kirby 2003), public policy (Sherry 2014), and ecopolitics (Kirby 2004; Cunningham and Wearing 2013) focusing on the human relations in communal settings, issues of social justice, and role of boundaries. Others include urban planning (Whitfield 2001), environmental governance (Gesota 2008), environmental studies (Breton 2009a, b), sustainability studies (Miller and Bentley 2012; Winston 2012; Dawson 2006), and practical guides to living in sustainable communities (Bang 2005). Finally, some scholars also approached ecovillages as utopias posing the question whether they are modern utopias (Andreas 2013; Bossy 2014; Hedrén and Linnér 2009; Meijering 2012; Sargisson 2012). As can be seen from these examples, an interdisciplinary approach is necessary to understand the full scope of issues surrounding ecovillages and for developing solutions.

Case Study: An Urban Ecovillage

Los Angeles Eco Village (LAEV) is situated in one of the most densely populated downtown neighborhoods of Los Angeles, California. Founded in 1993, the members' original concept to establish an ecovillage in a rural setting was quickly abandoned in favor of creating an urban "demonstration project." The ecovillage consists of three apartment buildings surrounded by garden plots, outdoor patios, a communal room, a bike room, and garages converted for other uses, such as a tool shed and an art studio.

LAEV "intends to demonstrate processes for lower environmental impact and higher quality of living patterns in an urban environment" by sharing its "processes, strategies and techniques with others through tours, talks, workshops, conferences, public advocacy and other media" (About Los Angeles Eco-village n.d.). The LAEV website describes the intentions behind its foundation and the practices of the community:

> Approximately 35 neighbors from diverse backgrounds and income levels have moved to the neighborhood intentionally to learn, share their knowledge and to demonstrate EcoVillage processes. Many attend regular community potluck dinners, community meetings, workshops on permaculture approaches to sustainable urban living, community work parties, and provide a variety of public services to the neighborhood and the city at large on a broad range of sustainability areas [...] While 35 neighbors moved to LA Eco-Village intentionally, we share our buildings and the neighborhood with many pre-existing neighbors. (About Los Angeles Eco-village n.d.)

The data for the case study were collected as part of an 18-month long prolonged ethnographic study. The methodology consisted of long unstructured interviews with 20 LAEV members, participant observation of community events, potlucks, meetings, photographs of the ecovillage, and online archival data – video clips, news articles, blog posts, etc., written by and about the ecovillage and its members. All interviews were transcribed verbatim and analyzed by using a qualitative data analysis software program.

Sustainable Living Practices

Ecovillages strive toward increasingly sustainable consumption practices. Material goods are minimally consumed, shared, and recycled. LAEV residents make conscious decisions about their individual as well as communal consumption choices. A member explains:

> I think we're better at waste reduction. We compost. We greywater. We've got a smaller trash bin. It's both giving ourselves limits and also showing that we don't need as much. There's a strong culture of non-consumption. We have clothing swap parties. There's a strong... thrift shop and used secondhand purchasing, whether it's through Craigslist or whatever. There's a strong culture around reuse and reduction. There's a lot more food growing that's happening.

A communal space in the literal and metaphorical sense of the word allows people who share similar values to gather under one roof. One member describes his worldview:

> Well, I'm generally under the umbrella of sustainability. I think at that time [when he joined LAEV] that was big on my mind in terms of how I felt like I was spending energy. So like, choice of what products people consume or how they relate to materials over the course of the day.... or how people relate to transportation over the course of the day. How people relate to energy-consumption over a few days. When it comes to my values I believe that the world has too many humans and at the average rate of consumption we don't have enough resources to support the lifestyle that the average human has. The way that impacts humans in the unjust society we live in is that there is a tremendous number of people that are suffering in an unnecessary way. There're a bunch of different things we can do to help change that. For me, it involves everything from, you know, having a perspective on population, having a perspective on...how we get around and what we value in society. I have a lot of issues with how people prioritize money-making and how making money puts pressure on people to make bad decisions that are not based on the larger good. I'm generally very left-leaning, politically, so I believe that we should be sharing resources and that we should be supporting each other and we should be making it easier for people to participate in political processes.

Another LAEV member explains how she desired to lead a sustainable lifestyle in a city like Los Angeles which is notorious for having a car-centric culture:

> Just an interest in the concept of people doing something different, and doing something that was potentially beneficial both to the people who are part of it, but also a more sustainable way to live, in general, but also in LA, which can be very challenging because it's a car-centric city, and it doesn't really have much culture embedded in the city around ecological sustainability practices. So it was like trying to figure out how to live in this city, how to find

my place in this city. It sort of seemed like a bit of its own identity within a much bigger identity that I was worried about how well I would relate to this city in general.

It is not possible to distinctly separate the ecological, economic, and socio-cultural dimensions of sustainability one from another as they are closely intertwined and overlapping. However, to organize the various practices that ecovillagers adopt and ideologies for which LAEV advocates, these dimensions will be examined individually.

Ecological Dimensions

> I was looking for a place to live where I could continue to live my values, which was around an ecological approach to living, living lightly on the earth. And whenever I would calculate my carbon footprint, it would be very, very, very low until I added a trip to California [laughs] and then it went flying up the chart. So I decided that's not sustainable. I don't want to keep doing that. Plus it was tiring. So I went searching for intentional communities and found Los Angeles Ecovillage in the middle of Los Angeles.

This quotation is from an LAEV member who moved to Los Angeles from the East Coast United States to be closer to her son. Like many other members of the community, what attracted her to the ecovillage was the joy of finding a place in the city where she can continue to pursue her ideals living among like-minded people.

In the midst of one of the most densely populated neighborhoods of Los Angeles, as one enters through the gate of LAEV and moves through the lobby to the back yard, one finds oneself in a communal garden full of fruit trees, edible flowers and plants, and a chicken coop. The banana tree, among other plants, is irrigated with the discarded water from the washing machine located in the yard as part of a greywater system installed by one of the members, a prominent expert in the field of greywater studies. The community garden is tended by some members who have their own plots; the produce is shared communally and eaten at weekly Sunday evening potlucks where the residents share homemade food and stories. Members can choose to join the food co-op that delivers weekly fresh, organic vegetables, fruits, grains, legumes, and other food items. Food scraps do not go to waste but end up in the compost area where people experiment with different types of composting. A member describes her daily food source:

> Our food systems are better in that we're growing more food. We have a food co-op that we get food from – a relatively local organic farm – and a bulk room where we can buy grains and dry goods. So most of my food comes from one of those. Almost all of my produce comes from either the garden or from the farmer, and part of the food co-op. It's a buying co-op. We buy inventory. We buy goods, like all kinds of grains and nuts and even soaps from [name of company], which is a natural food distributor, and we get a truck delivery …whenever we need more. We store it all in the bulk room, and we have that open two times a week, and I think it's going down to one time a week. Then we all take shifts. We all volunteer working, either the produce sorting – that's the weekly farmer's delivery. The produce delivery, that's every week on Sundays, after they go to the farmer's market, they bring by a bunch of produce and we split it up in boxes. That's where we get a lot of organic produce. It's very inexpensive when you do it that way. It's $10 a box, and it usually takes me two weeks to go through it. That's a lot of organic produce for $10.

As described by this particular member, food co-ops and communal gardens can be both more economical and ecological methods of obtaining fresh, organic food in the city. Another area where these two dimensions of sustainability overlap is transportation. With twelve-lane highways weaving across the city like a webbed network, the prominence of car culture in Los Angeles is undeniable. Despite efforts to improve public transportation in the city with an expanded system of light rail and bus routes, adding High Occupancy Vehicle (HOV) freeway lanes to promote carpooling, and extending bike paths to encourage bicycling, car dependency is still high. Of the 10 most congested highway corridors in the United States, 7 of them are in Los Angeles (Texas A&M Transportation Institute 2011). Traffic is determined to be among the biggest contributors of urban air pollution in the Western world in some regions (EU Science Hub 2015). According to the American Lung Association's "State of the Air 2016" report, "Los Angeles remains the metropolitan area with the worst ozone pollution, as it has been for all but one of the 16 annual reports, despite having had its best air quality ever in the State of the Air report's history" (5). It is evident that air pollution and traffic are major points of concern for the environment in LA. It is this reality that has activated the ecovillage members to create an environment that would improve the city's air quality and make the environment more livable for all.

The apartment unit underneath the stairs at LAEV was turned into a room for people to store their bicycles. A monetary incentive, in the form of rent reduction, is given to members who do not own cars. However, living car-free in Los Angeles remains a huge challenge to some members, especially those with small children or whose jobs require them to travel around the city. These members try to carpool as much as they can or share one car among many members as explained below by one of the members who owns a car.

> It's almost a community car at this point. There are about maybe six different households who use it on a regular basis. And it's a station wagon, so it's very, kind of, utilitarian for people who want to get whatever – gardening supplies or moving or whatever. Finding free things on the sidewalk [laughs] that are too heavy to lug by bike.

Economic Dimensions

LAEV offers affordable housing in a city where the median house price is twice the US national average (The Economist 2016). Los Angeles ranks number 7 in North America and number 17 in the world among the cities with the highest cost of living (Expatistan 2016). Living in a community and sharing resources as opposed to a single-family house has many financial advantages.

> I feel that the model that is promoted here, and that we try to live by, is a very accessible one, economically... recognizing the links between ecology and economy. So, things from having a commitment to affordable housing, to understanding the benefits of sharing resources. So having the free table, if there are things that you don't need, somebody else can use. Also, just sharing basic tools, or things that are either expensive on their own, or just unnecessary to have. Everyone to have a vacuum cleaner, or everyone to have particular power tools [is not necessary] if you can share those things. You know, I borrow my friend's

mop or whatever. Or they borrow my drills or whatever. It just seems to both economically and ecologically makes sense. I could probably go on much longer.

It was a conscious decision of the LAEV founders to establish an urban ecovillage as opposed to a more typical rural one. Even though they were initially looking for a plot of land to build a rural ecovillage like the majority of ecovillages in Northern America, they abandoned this idea in favor of an effort to create desirable living conditions in the city.

> I think this community represented a lot of like-minded people who not only had those same concerns and some people had a lot of knowledge about one area or not...but it seemed like we had a lot of general agreement when we would have discussions. Okay, we're obviously preaching to the choir most of the time. Also, I felt like there were a number of people who were really willing to experiment with alternatives, and to do so with a consciousness...with a conscience about money. I felt like to try to do those things – part of what I thought was inspiring, which I think was...I don't know if it was necessarily what everybody thought, was that most of the global population doesn't have a lot of money. If we come up with – within an urban setting – ways to change the way that we live in a way that's sustainable with very few resources...that's replicable. Moving out to the farm and the country with a lot of money and purchasing your own land and buying solar panels like that and everything that's expensive...it didn't seem – it seems privileged from a class perspective. I felt like I wasn't as interested in that as I was for these other low-tech or low-cost alternatives. I was really interested in people who couldn't care less about the values of mainstream society, and here's what I'm going to pursue. I'm going to pursue this interest and I'm going to talk to these people that I think really need my help. I felt like there are people who are really giving a lot of themselves to change the system in a really consistent way. That was really inspiring.

Ecovillagers adopt different anticonsumerist practices to reduce their carbon footprints and to live more economically by utilizing alternative methods to money-based exchanges. Some examples include the free table at the lobby where people leave their unwanted goods for others to pick up, a street library where people can leave and pick up books from a shelf on the street outside the ecovillage, clothing swap parties, participating in trade school activities, and exchanging resources through a time bank. A time bank is a time-based exchange system where its members exchange their services for one-time credit. For example, a person can give one-hour guitar lesson in exchange for one-hour of house cleaning. No money is involved and regardless of the type of service, all services are treated equal. A trade school is similar to a barter system, but the exchange is not limited to 1 h of time. For instance, a person can teach a group candle making in exchange for food. Sharing individual skills on-site rather than paying "outsiders" to travel to the site is a key concept in cooperative living.

Socio-Cultural Dimensions

One of the main reasons people live in intentional communities is to experience the sense of being part of a larger entity, a collectivistic community, and surrounding themselves with people who share similar worldviews and values. Finding like-minded people can be a challenge in a city of four million individuals who come from all walks of life, educational backgrounds, income levels, ethnicities, and nationalities. This is also precisely the reason why LAEV becomes a meeting point for people from very diverse backgrounds but with one vision for the world: a more sustainable planet.

> So I think that when people come to us, there's so much going on when people find out about us. This idea that, 'Oh my goodness, these values.' People come and they see the common values, and I could live out these values, and it's just, when people have never seen that before – I mean, I had that experience as well, 'Oh my goodness, I could actually live my values.' I didn't know that was possible. And not only that, living in LA is very expensive, and the chance to do that more affordably here.

> Other things that I – like, I feel like living in an intentional community, and particularly an ecovillage, living with people with shared values. Feeling safe in the community that I live in. It's not necessarily – it wouldn't necessarily appear to be safe. There's certainly crime around the area, but I feel like knowing your neighbors and being. . ... When I first moved I was a single woman, so just being able to feel safe in a neighborhood I think is really good. Around the shared values, commitment and kind of curiosity and exploration around particular kinds of ecological living.

When LAEV was first founded, it had no members with children. Over the years, the members had children or families with children moved in which changed the social dynamic of the community noticeably. As the saying goes, "it takes a village to raise a child," and the parents experience this first hand. The community becomes their village in the middle of a city and acts like one big "extended family," where the availability of adults in the community at any given time helps parents with childcare, becoming a primary reason for them in choosing to live in the ecovillage.

> So, for me, my perspective is that I think it would be really hard and isolating living alone – living in a single-family home with my partner [. . .] I feel like I'd be really isolated. For me, I guess pet – cat – care and childcare are very different but similar idea of, like, if I need to go away for work for a couple of days, it's so easy to find somebody to take care of my cats, or to water the plants or whatever. And I am available to do that for other people who are going out of town. So, it's just easy to have that kind of support. Also, there are projects that are just hard to do by yourself. Like, building raised beds or whatever. It's just like you can do a work party and you have a bunch of people to do something together. It's easy to have people over.

The community life is organized around several events such as weekly meetings with a largely administrative purpose, potlucks/social events bringing members together in a more informal setting, and work parties which require manual-labor as members work together to take care of the garden or renovate the building facilities. Weekly meetings are held to discuss issues that the ecovillage needs to address. Communal decisions are made based on a *consensus decision making* process which requires a collaborative group unanimity that all members of the community actively support or at least can live with (Seeds for Change n.d.). This is significantly different from a democratic or nonconsensus voting systems where the majority rules. The meetings, facilitated by a moderator, create an arena for the members to discuss their concerns, share their viewpoints, and vote on accepting new members to the community.

The Challenges

Life in an ecovillage is not effortless, nor without its challenges. As one of the LAEV members puts it "it's kind of like a big family with all its wondrous stuff and its difficult stuff. The arguments... the love... the gossip... a little bit of testing your own patience. The way that we can have this space in the middle of a rather unforgiving city is an amazing aspect of what we can have here."

Individualistic personality traits, usually associated with Western cultures, can at times be at odds with the collectivistic demands of communal living and the notion of consensus decision-making. In fact, individuality and its contradiction with the ecovillage ideology is one of the most referred to topics in the ecovillage context (e. g., Holleman 2011). Moreover, the consensus decision-making process can be a hindrance to taking action in situations where there is no consensus, resulting in unresolved issues. Conflict resolution teams are not always successful in resolving conflicts between members, and conflicts may even cause members to leave the community.

While many members join ecovillages to escape from materialism and the dictates of capitalism, they may also continue to exhibit their old established habits of consumption. Boundaries, communication, governance, and consensus issues are reported as among the major challenges faced in ecovillages (Kirby 2003; Cunningham and Wearing 2013). As Litfin puts it "ecovillages are not utopias; they are living laboratories. Some experiments may be successful and others not, but they are all opportunities for learning" (2014, 18).

Robert Gilman, a pioneer of sustainable communal living, who was instrumental in the founding of Global Ecovillage Network (GEN) notes numerous challenges ecovillages face during the research and design, creation and implementation, and maintenance stages. He lists these challenges under the following categories of "the bio-system challenge," "the built-environment challenge," "the economic system challenge," "the governance challenge," "the glue challenge," and "the whole systems challenge" (Gilman 1991, 10).

The "bio system challenge" poses difficulties in preserving the natural habitats on the village land while producing food on site, processing the organic waste produced on site and recycling the solid waste and processing the solid waste from the village. The "built-environment challenge" requires that the ecovillage is built with ecologically friendly materials using renewable energy resources and built in ways that have a minimal impact on the land. The buildings should also encourage community interaction. The "economic challenge" includes finding answers to the questions such as "What are sustainable economic activities, both in terms of what will sustain the members of the community and what is sustainable in ecological terms?" "How can we be simultaneously economically and ecologically efficient, so as to reduce both expenses and environmental impact?" and "Are there useful alternatives and/or supplements to the money economy for facilitating economic exchange within and between ecovillages?" The "governance challenge" deals with conflict resolution, communal decision-making process, and leadership roles. Gilman views the "glue challenge" as the mechanics of holding the members together around shared values and vision. Finally, the "whole-system" challenge is described as "perhaps the biggest challenge" (Gilman 1991) given that during the formation of an ecovillage so many changes have to be made and so many factors have to be taken into consideration that the whole task becomes an insurmountable one. Therefore, he advocates that sustainability should not be a characteristic of the community, but it needs to be part of the design process from the very beginning.

Conclusion

A grim reality lies in the future of our planet and its habitants in the face of anthropogenic climate change. Urbanization plays a significant role in contributing to environmental damage and harmful impact on the quality of life for all. There is no denying the fact that the human population, especially in the Western urban areas, cannot continue to consume Earth's limited resources the way they are being consumed today.

Ecovillages are in essence born out of pure and moral intentions to offer "small-scale place-based, yet tightly networked, collective efforts" (Litfin 2009, 124). As shown in the example of an urban ecovillage in this chapter's case study, there are a number of ways in which economic, ecological, social, and cultural dimensions of sustainability can be achieved while in a city setting.

"While ecovillages may show that another world is possible on a very small scale, the question remains: can global systemic change come about through a network of communities committed to social and ecological sustainability?" asks Litfin (2009). She responds, "The short answer is we don't know. There are good reasons to doubt it can, yet also countervailing considerations to all of these good reasons" and she continues to add "we must admit at least the possibility that the ecovillage movement could play an important role in transition to a just and sustainable society" (140). Despite various challenges ecovillages face and encounter, overall they are noteworthy examples of making cities more sustainable and livable, and in the absence of more viable alternatives they pose an optimistic possibility for living a more holistic life in the cities by leaving a lower impact on the planet. As former United Nations Secretary-General Kofi Annan so aptly put it, the "future of humanity lies in cities."

Exercises in Practice

 (i) Start a communal edible garden at your school/university, workplace, or neighborhood. Organize a communal meal with the produce you harvest from the garden to gather around food to get to know your community better.
 (ii) Go to *Global Footprint Network* (http://www.footprintnetwork.org/resources/footprint-calculator/) or *World Wildlife Fund*'s "footprint calculator" (http://footprint.wwf.org.uk/) or any other similar web site where you can calculate

your ecological footprint and find out about your carbon footprint. What can you change about your consumption behavior that can lower your carbon footprint?

Engaged Sustainability Lessons

- It is possible to live sustainably in big cities.
- Living in ecovillages helps people achieve ecological, economic, social, and cultural dimensions of sustainability by making resources more readily available and shared by a close-knit community.
- Community gardens benefit the air, soil, animals, and the people around them.
- Composting helps reduce food waste that goes into landfills, and nourishes the soil compost is applied to.
- Car-dependency can be reduced by carpooling, car sharing, bicycling, and using public transportation more frequently.
- There are alternative methods of exchange such as time banks, trade schools, street libraries, swap parties, and bartering.

Chapter-End Reflection Questions

(i) What are some of the sustainable consumption practices with which you engage? Does your household generally support those practices?
(ii) Do you know where your clothes were made and who made them? Does the label provide information about the manufacturing process? Do you know if the workers had fair labor conditions?
(iii) Do you know where your food comes from? Is it locally produced/grown or shipped from across the globe? Does it matter to you whether it is locally sourced or not? Is an organic tomato shipped from 8000 miles/kilometers away better than a nonorganic locally grown one?
(iv) Do you frequently throw away food that has gone bad, that was bought but never eaten? What do you do to not waste food? What do you do with food scrapings?
(v) If your city has a recycling center, visit the center and find out what type of materials are recycled, where these materials go after recycling.
(vi) If you live in a car-dependent big city, are there ways in which you can still reduce your car consumption? Can you use public transportation or bicycles as your main mode of transportation? If that is not a viable option, can you carpool?

Cross-References

▶ Application of Big Data to Smart Cities for a Sustainable Future
▶ Community Engagement in Energy Transition
▶ Environmental Stewardship

▶ Smart Cities
▶ Supermarket and Green Wave
▶ The LOHAS Lifestyle and Marketplace Behavior
▶ The Theology of Sustainability Practice
▶ Transformative Solutions for Sustainable Well-Being
▶ Urban Green Spaces as a Component of an Ecosystem

References

About Los Angeles Eco-Village. (n.d.). https://laecovillage.wordpress.com/about/. Accessed 12 May 2017.
Andreas, M. (2013). Must Utopia be an island? Positioning an ecovillage within its region. *Social Sciences Directory, 2*(4), 9–18.
Bang, J. (2005). *Ecovillages: A practical guide to sustainable communities*. Boston: New Society Publishers.
Bossy, S. (2014). The utopias of political consumerism: The search of alternatives to mass consumption. *Journal of Consumer Culture, 14*, 179–198.
Breton, P. E. (2009a). *Organizing for sustainability at a small scale. A case study of an ecovillage*. Master's thesis, Heritage Branch: University of Northern British Columbia, Department of Natural Resources and Environmental Studies, Ottawa.
Breton, P. E. (2009b). Ecovillages: How ecological are you? *Communities, 143*, 22–24.
Chitewere, T. (2006). *Constructing a green lifestyle: Consumption and environmentalism in an ecovillage*. Ph.D. dissertation, New York.
Chitewere, T., & Taylor, D. E. (2010). Sustainable living and community building in Ecovillage at Ithaca: The challenges of incorporating social justice concerns into the practices of an ecological cohousing community. In D. E. Taylor (Ed.), *Environment and social justice: An international perspective*, Research in social problems and public policy (Vol. 18, pp. 141–176). Bingley: Emerald Group Publishing Limited.
Cunningham, P. A., & Wearing, S. L. (2013). The politics of consensus: An exploration of the Cloughjordan ecovillage, Ireland. *Cosmopolitan Civil Societies Journal, 5*(2), 1–28.
Dawson, J. (2006). *Ecovillages: New frontiers for sustainability*. Devon: Green Books.
EU Science Hub. (2015, November 30). Urban air pollution – What are the main sources across the world? Joint Research Center. https://ec.europa.eu/jrc/en/news/what-are-main-sources-urban-air-pollution. Accessed 10 May 2017.
Expatistan Cost of Living Index. (2016). https://www.expatistan.com/cost-of-living/index/north-america. Accessed 10 May 2017.
Gesota, B. (2008). *Ecovillages as models for sustainable development: A case study approach*. Master's thesis, Albert-Ludwigs-Universitat, Freiburg.
Gilman, R. (1991). The eco-village challenge. *In Context: A Quarterly of Humane Sustainable Culture, 29*. http://www.context.org/iclib/ic29/gilman1/. Accessed 18 Nov 2017.
Global Ecovillage Network (GEN). (n.d.). What is an ecovillage? http://gen.ecovillage.org/en/article/what-ecovillage. Accessed 10 May 2017.
Hedrén, J., & Linnér, B. (2009). Utopian thought and sustainable development. *Futures, 41*(4), 197–200.
Holleman, M. (2011). *Individuality in community at the EVI*. Master's thesis, Vrije Universitet, Amsterdam.
Hong, S., & Vicdan, H. (2016). Re-imagining the utopian: Transformation of a sustainable lifestyle in ecovillages. *Journal of Business Research, 69*, 120–136.
Kirby, A. (2003). Redefining social and environmental relations at the ecovillage at Ithaca: A case study. *Journal of Environmental Psychology, 23*, 323–332.

Kirby, A. (2004). Domestic protest: The ecovillage movement as a space of resistance. *Bad Subjects, 65*.

Kunze, I., & Avelino, F. (2017). Research literature on ecovillages. https://ecovillage.org/sites/default/files/files/Literature-onecovillages_2017_03_13_IK_FA.pdf?x52178. Accessed 18 Nov 2017.

Litfin, K. (2009). Reinventing the future: The global ecovillage movement as a holistic knowledge community. In G. Kütting & R. Lipschutz (Eds.), *Environmental governance: Knowledge and power in a local-global world*. London: Routledge.

Litfin, K. (2014). *Ecovillages: Lessons for sustainable community*. Cambridge, UK: Polity.

Meijering, L. (2012). Ideals and practices of European ecovillages. In *Realizing utopia: Ecovillage endeavors and academic approaches* (pp. 31–41). Munich: RCC Perspectives.

Miller, E., & Bentley, K. (2012). Leading a sustainable lifestyle in a 'non-sustainable world': Reflections from Australian ecovillage and suburban residents. *Journal of Education for Sustainable Development, 6*(1), 137–147.

Permaculture Design Principles. (n.d.) https://permacultureprinciples.com/principles/ Accessed 18 Nov 2017.

Sargisson, L. (2012). Second-wave cohousing: A modern utopia? *Utopian Studies, 23*(1), 28–56.

Seeds for Change. (n.d.). Consensus decision making. http://www.seedsforchange.org.uk/consensus. Accessed 12 May 2017.

Sherry, J. (2014). *Community supported sustainability: How ecovillages model more sustainable community*. Ph.D. dissertation in Planning and Public Policy, Rutgers University, New Brunswick.

Standen, S. (2014). *Re-examining "community" in the modern world: The role of boundaries in the intentional community construction of "community" at an ecovillage in Ithaca*. Social Anthropology B.A. dissertation, Queen's University, Belfast.

Texas A&M Transportation Institute. (2011). 2011 Congested Corridors Report. https://mobility.tamu.edu/wp-content/uploads/2011/11/ccr-all-table-rankings.pdf#page=36. Accessed 10 May 2017.

The American Lung Association. (2016). State of the Air 2016 Report. http://www.lung.org/assets/documents/healthy-air/state-of-the-air/sota-2016-full.pdf. Accessed 10 May 2017.

The Economist. (2016, August 24). American house prices: Realty check. http://www.economist.com/blogs/graphicdetail/2016/08/daily-chart-20. Accessed 10 May 2017.

United Nations. (2014). World's population increasingly urban with more than half living in urban areas. http://www.un.org/en/development/desa/news/population/world-urbanization-prospects-2014.html. Accessed 10 May 2017.

United Nations. (2015). Sustainable development goals. https://sustainabledevelopment.un.org/sdgs. Accessed 10 May 2017.

Unites Nations-Habitat. (2016). Sustainable Urbanization in the Paris Agreement Comparative review for urban content in the Nationally Determined Contributions (NDCs). https://unhabitat.org/books/sustainable-urbanization-in-the-paris-agreement/. Accessed 10 May 2017.

Van Schyndel Kasper, D. (2008). Redefining community in the ecovillage. *Human Ecology Review, 15*(1), 12–24.

Whitfield, J. (2001). *Understanding the barriers encountered by residents of ecovillages*. Master's thesis, School of Planning, University of Waterloo, Ontario.

Winston, N. (2012). Sustainable housing: A case study of the Cloughjordan eco-village. In N. Winston (Ed.), *Enterprising communities: Grassroots sustainability innovations*. Bingley: Emerald.

World Bank. (2009). World development report 2009: Reshaping economic geography. World Bank. https://openknowledge.worldbank.org/handle/10986/5991. License: CC BY 3.0 IGO.

Wu, J. (2010). Urban sustainability: An inevitable goal of landscape research. *Landscape Ecology, 25*(1), 1–4. https://doi.org/10.1007/s10980-009-9444-7.

Urban Green Spaces as a Component of an Ecosystem

José G. Vargas-Hernández, Karina Pallagst, and
Justyna Zdunek-Wielgołaska

Contents

Introduction	886
Components of Urban Green Spaces	887
Ecosystems, Functions, and Services of Urban Green Spaces	888
Methodological Considerations	893
Users of Urban Green Spaces	894
Factors of Successful Community Involvement in Urban Green Spaces	898
Challenges and Opportunities	904
Conclusion: Public Initiatives and Actions	906
Research Gaps	908
Cross-References	909
References	909

Abstract

This chapter is aimed at analyzing a review of the empirical literature on some important features of urban green spaces such as the components, functions, services, community involvement, initiatives, and actions from an ecosystem

---

J. G. Vargas-Hernández (✉)
University Center for Economic and Managerial Sciences, University of Guadalajara, Guadalajara, Jalisco, Mexico

Núcleo Universitario Los Belenes, Zapopan, Jalisco, Mexico
e-mail: josevargas@cucea.udg.mx; jvargas2006@gmail.com; jgvh0811@yahoo.com

K. Pallagst
IPS Department International Planning Systems, Faculty of Spatial and Environmental Planning, Technische Universität Kaiserslautern, Kaiserslautern, Germany
e-mail: karina.pallagst@ru.uni-kl.de

J. Zdunek-Wielgołaska
Faculty of Architecture, University of Technology, Otwock, Poland
e-mail: justyzdu@wp.pl; justynazdunek@yahoo.com

© Springer International Publishing AG, part of Springer Nature 2018
S. Dhiman, J. Marques (eds.), *Handbook of Engaged Sustainability*,
https://doi.org/10.1007/978-3-319-71312-0_49

perspective. The analysis begins from the assumption that urban green spaces are ecosystems of vital importance in enhancing the quality of life in an urban environment and supplying ecosystem services, such as biodiversity, climate regulation. Thus, the urban green space ecosystem is an important component of an ecosystem in any community development. Meeting the needs of users is related to the functions and services that urban green spaces provide to communities. Community involvement, engagement, and development require a methodology to ensure that the needs and aspirations of local users in the community are met. The methods employed in this analysis are a review of the empirical literature and documents, analysis of existing data on uses and users, interviews with authorities, and a more detailed examination of case-specific data. Also, as concluding remarks, some of the wider environmental, economic and social initiatives for local authorities and communities are suggested that might justify any funding for the all the stakeholders that are represented and involved. Finally, the chapter proposes some of the opportunities, challenges, and further research.

Keywords
Community involvement · Ecosystem · Functions · Services · Urban green spaces · Users

Introduction

The history of life on earth is one of the living things surrounded by a natural environment that supplies water, fresh air, minerals, plants, vegetation, animals, and all the fruits of nature, etc., for enjoyment of everyday life. However, these natural and environmental resources are not lasting forever, most of them are either polluted and corrupted or extinguished by consumption. Urban populations are facing ecologically vital threats from over-urbanization, such as water and air pollution, agricultural and forest lands with vegetation removal and ground water overdraft. The ecological dimension of urban green spaces considers the objective and subjective components of a place providing a supportive habitat of biological diversity.

The renewal and increasing of urban green spaces considering the fast demographic growth and agglomeration should be accompanied by inhabitants' participation in environmental, social, cultural and economic actions, and objectives to promote the bio-economy in urban biodiversity and sustainable development. Inhabitants should be informed and motivated to participate in environmental, cultural, and educational activities and become active in the designing and planning of urban green spaces. Public support and political involvement of citizens for urban green space development requires various actions. Public urban green spaces are accessible to and used by all the citizens. Arrangements of public activities and actions on planned urban green spaces raise awareness amongst inhabitants of the city. For example, a public event can have the purpose of making users aware and educating them through experiencing and enjoying different activities organized by the urban green space.

In 2014, as many as 54% of the world's population were living in urban settings and this is projected to reach 70% by 2050 (United Nations Department of Economic and Social Affairs – Population Division 2014) By 2020, around 62% of world's population will live in urban areas, covering 2% of the world's land space and consuming 75% of natural resources. By 2030, two thirds of the urban area that will exist will have to have been built in sustainable urban environments.

In 1953, the seminal report on Park Life recognized urban green spaces as a vital component of urban environment and their role in social renewal (Comedia and Demos 1995). The Urban Parks Programme was launched by the Heritage Lottery Fund and marked the change in attitude toward creating policy initiatives such as the Urban White Paper (Department of the Environment 1996).

Components of Urban Green Spaces

An urban green space system is an important component in any community development, in housing, business, leisure areas, etc. (Baycan-Levent 2002). Components of urban green areas are vegetation, water, accessibility, services of shelters, toilets, seating, playgrounds and sport areas, events and activities, environmental quality conditions and resources such as lighting, safety, litter bins, friendly staff, artistic features and artefacts such as sculptures, etc. The quality assessment of green spaces is measured by factors such as infrastructure, vegetation, accessibility, security, and equipment. Conditions that favor the use of urban green spaces are the distance walking time (Herzele and Wiedeman 2003), location and distribution, easy access, and proximity. Environmental enhancement makes better and more attractive urban green spaces by promoting inward investments, increasing the land value, and stimulating the economy of the community.

Urban green spaces are urban areas that were natural or semi-natural ecosystems that were converted into urban spaces by human influence (Bilgili and Gökyer 2012). Urban green spaces are public and private open spaces in urban areas primarily covered by vegetation, which inspire active or passive recreational and sports activities or have an indirectly positive influence on the urban environment available for the users (Tuzin et al. 2002). Urban green spaces provide sustainable, diverse places where, according to the classic report Park Life "people will find a sense of continuity, of relief from the pressure of urban living, places to be in touch with the natural cycle of the seasons and of wildlife and also places to meet and celebrate with others" (Comedia and Demos 1995).

The mixed community green space is defined as the mix of overall community-level green spaces that significantly affect land surface temperature. However, there is inequitable distribution of heat and thermal discomfort (Huang et al. 2011). Inequitable distributions of community green space are related to socioeconomic status. There are no health impacts documented on the thermal comfort provided by green spaces. A mixed neighborhood green space is a mixed area of grass, trees, and vegetation.

Urban green spaces have different forms and types of open spaces, community parks and gardens, and landscaped areas. Some types of urban green spaces are formal green space, informal green space, natural green space, children's space, public participation, active sports space, recreation activities, and further land management policies. Urban green spaces exist in a variety of types, structures, and shapes. Urban green spaces include public parks, reserves, sports fields, streams, river banks and other riparian areas, greenways, walkways and trails, communal shared gardens, street trees and bushes, nature conservation areas, and less conventional spaces such as green walls, green alleyways, and cemeteries (Roy et al. 2012).

The broader notion of green space connotes turf grass-related residential, commercial and institutional surfaces and public facilities such as parks and playing fields. Turf grass is associated with the notion of green space that connotes turf-related surfaces as residential, commercial and institutional lawns and turf surfaces.

Urban green spaces connect the urban and the natural, while caring for the environment, social, and economic elements. Public forests and green roofs in public and community buildings, and vacant and derelict land also provide ecosystem services. Productive land use ensures a long-term regeneration initiative to use green spaces properly for economic revenue by implementing sustainable urban initiatives such as drainage schemes. Green spaces include wilder, woodland-type, and untamed elements, with more areas for child development.

An urban green space can be considered a continuum without fences, hierarchies and horizontally maintained at the same level of community-oriented service and a use-oriented approach. Green spaces are the spirit of the community. Urban green spaces are a focal point for communities (Greenspace 2007). Urban green spaces contribute to building a sense of community among residents who are more likely to enjoy strong social ties. Green spaces promote interaction between people, developing social ties and community cohesion. Greenness in a neighborhood is one of the most important predictors of neighborhood satisfaction (Van Herzele and de Vries 2011).

Ecosystems, Functions, and Services of Urban Green Spaces

Urban green spaces are ecosystems of vital importance in enhancing the quality of life in an urban environment. Urban green spaces supply ecosystem services such as biodiversity and climate regulation. Urban green spaces are essential for the quality of life, health, and well-being of citizens. Urban green spaces are critical for protecting wildlife, watersheds, meads, vegetation, providing air quality for a dense urban environment and recreational activities. Cool islands in dense urban areas can be provided between green spaces. A dense green space is more effective at preventing nitrogen run-off, untreated human and industrial waste, toxic materials, and debris. Urban green spaces provide ecosystem services that can improve the conditions of the environment, pollution, and congestion of the large metropolitan area of Guadalajara.

Green spatial connectivity and density are associated with the cooling and pollution-mitigating capacity of the diversity of urban green space types and connected green path corridors. Green space density is described as the tree canopy cover (Feyisa et al. 2014), the relative percentage of vegetation (Ng et al. 2012). The cover patterns, densities, and balance of green spaces affect the urban heat island (Dobrovolný 2013; Kong et al. 2014; Li et al. 2011). The density and size of the green space are highly interrelated and multi-scale, dependent on configuration. Greener spaces are cooler than non-green spaces and contribute to lower ambient temperatures (Srivanit and Hokao 2013).

There is a significant association between the increased density and cooling effects of greenspaces (Dobrovolný 2013; Feyisa et al. 2014; Hart and Sailor 2009; Ng et al. 2012; Perini and Magliocco 2014; Vidrih and Medved 2013; Weber et al. 2014; Zhang et al. 2013). The cooling range of green spaces into surroundings is influenced by the building density, arrangements and heights (Li et al. 2012, 2013; Zoulia et al. 2009). Urban greening density is suggested to become optimal at $\geq 50\%$ coverage (Ng et al. 2012). Urban greening initiatives are insufficient for achieving air quality and climate. Greening reduces heat stress and related illnesses (Bassil et al. 2010).

The reasons for visiting urban green spaces are mostly to enjoy a wide range of environmental elements such as flowers, trees, nature, fresh air, wildlife, watching cascades, educational opportunities, social activities, taking children to play, social interaction, to meet friends, picnics, meeting people, getting away from it all, passive walking and activities, shelter and sitting, etc. Shared parks and gardens may be set up to facilitate social links, collective participatory projects and collective cultural interventions, well-being recreational areas and walkways connecting attractions and facilities.

Other reasons why people visit urban green spaces are to walk a dog, walk by the lake, river or creek, walk socially as part of a group, hanging out, passive enjoyment, sitting either on the grass or on seats, photography, messing about on swings, watching sport, reading, watching life go, smoking, sunbathing, an informal pursuit such as flying kites, fishing, etc. Walkable green spaces in urban areas are associated with a healthy environment and increasing green exercise. Walkable green spaces influence the longevity of urban senior citizens (Wolf 2010).

Urban green spaces have beneficial physical, psychological, and health effects through physical activities and green exercises. Environmental determinants affect the use of green spaces, physical activities, and leisure. The amount of green spaces available to users in the living environment correlates with socioeconomic, demographic, and self-perceived health. Higher levels of greenness have been positively associated with lower stroke mortality. Perceived neighborhood greenness is positively associated with physical and mental health. Socioeconomic and cross-cultural variations may result from the unequal distribution of green spaces.

Another important reason to visit urban green spaces is the use of facilities such as cafes, restaurants, environmental centers, libraries, museums, sports such as football, tennis, etc., biking, skateboarding, cycling, and other forms of active enjoyment. Events are other important motive to visit urban green spaces, such as group music performance, concerts, Christmas carol concerts, orchestral performance, craft fairs, fun fairs, opera, circus, firework displays, bands playing, dancing, etc.

Making inhabitants aware of the existence of urban green spaces and using values contributing to urban inhabitants' lives in the form of a more balanced quality of life and lifestyle, encouraging physical and mental fitness, reduces tensions and conflicts, relieving the harshness of the urban environment, providing places for social and cultural interaction in informal contacts and more formal participation in social events, social inclusion, recreation, aesthetic pleasure, wildlife, and fostering community development.

Urban green spaces are natural meeting points for local inhabitants facilitating social inclusion and integration, community cohesion, social capital, civic society, supported by an increasing sense of identity and belonging (Konijnendijk et al. 2013a; Abraham et al. 2010). By providing a meeting place for social interaction and integration between community users, green spaces influence social capital. Inhabitants living near urban green spaces reduce health inequalities and have lower circulatory diseases (Mitchell and Popham 2008). Inequitable distribution of green spaces correlates with the distribution of disadvantaged inhabitants.

Passive activities are the main reason why users visit urban green spaces such as passive or informal enjoyment of the environment, social activities and attending events, getting away from it all, walking activities including dog walking, active enjoyment including sport and specific activities. Surveys have shown that people are less stressed, communicate better and make sensible decisions when surrounded by green spaces.

Urban green spaces can be linked as wildlife corridors to facilitate the movement of fauna, preventing fragmentation and isolation of wildlife (Rouquette et al. 2013; Hale et al. 2012). Urban green spaces are home to many species, including those that are rare and threatened and the habitat for pollinators. More urban green spaces sustain more wildlife and biodiversity, providing a more favorable habitat, therefore requiring more protection from human interference (Cornelis and Hermy 2004; Fuller et al. 2009; Schwartz et al. 2002; Baldock et al. 2015). Creation, protection, and development of urban green spaces are relevant elements of sustainable urban development.

Urban green spaces have an impact on human thermal comfort and air quality on human health (Cohen et al. 2012; Weber et al. 2014; Nowak et al. 2014). Comparisons of the impact of green space air quality and heat show that a greater predominance of trees mitigates urban heat islands, provides thermal comfort, and improves air quality. Green spaces reduce urban heat islands and air pollution, improving air quality in urban settings (Bowler et al. 2010). Community-level air quality is dependent on the tree population (Morani et al. 2011).

A community's green spaces are associated with reduced household pollution material exposure (Dadvand et al. 2012). Vegetation density in green spaces is associated with pollution mitigation (Yin et al. 2011; Dzierżanowski et al. 2011; Escobedo and Nowak 2009; Nowak et al. 2013, 2014, Tallis et al. 2011; Tiwary et al. 2009; Tsiros et al. 2009). Bushes instead of trees may retain more pollution particles and reduce concentrations (Wania et al. 2012). A diversity of tree species including evergreen, conifer, and deciduous tree species have complementary air-pollution uptake patterns and provide maximum air-quality improvements (Manes et al. 2012). Trees and shrubs are more effective at removing pollutants than herbaceous perennials (Rowe 2011).

A diversity of evergreen and conifer tree species provides complementary air-pollution mitigation. Coniferous trees are the best for capturing pollutant material (Tallis et al. 2011; Tiwary et al. 2009) and the evergreen, rather than deciduous trees in green spaces, provides more cooling and below comfort conditions in winter (Cohen et al. 2012; Zhang et al. 2013). Evergreen and deciduous trees remove more atmospheric 03 (Alonso et al. 2011) than conifers.

The cooling capacity of green spaces is affected by multiple variables such as density, size and shape associated with an increase in air quality. Urban green spaces reduce heat, ozone, and ultraviolet (UV) radiation and improve air quality (Roy et al. 2012; Konijnendijk et al. 2013b; Bowler et al. 2010). Absorbing pollutants improve air quality. Research based on modeling has weak evidence that capturing pollutants and particles by urban green spaces improves air quality (Konijnendijk et al. 2013a). Wong et al. (2013) reviewed the evidence for the relationship between green spaces, heat and air quality considering variables such as green space type, climate, method, etc. Building orientation and heights affect cooling and air quality from green spaces.

Urban green spaces reduce the UHI effect by cooling the air on average by 1°C and by providing shade. Cooling is influenced by plant type, green patch size and density, temperature, and wind (Armson et al. 2012; Cao et al. 2010; Feyisa et al. 2014; Fintikakis et al. 2011 Fröhlich and Matzarakis 2013; Gaitani et al. 2011; Konijnendijk et al. 2013a; Lafortezza et al. 2009; Oliveira et al. 2011; Onishi et al. 2010; Vidrih and Medved 2013). Many characteristics of green spaces affect the cooling capacity such as size, cover, shape, density, and spacing.

Green space scale is the area or size of green space including a single or multiple sites. The green area impact scale includes the site and the adjacent nongreen areas. The percentage of covered greenspace (PLAND) equals the sum of the areas (m^2) of a specific land-cover class divided by a total landscape area, multiplied by 100 (Herold et al. 2003). There is a strong association between the size of green space and the cooling effects (Cao et al. 2010; Chen et al. 2014; Dobrovolný 2013; Feyisa et al. 2014; Hart and Sailor 2009; Li et al. 2012; Onishi et al. 2010; Susca et al. 2011; Weber et al. 2014). The size of the green space affects the urban cooling island because the cool air builds up and is emitted from the center (Vidrih and Medved 2013) and it is stronger during the summer (Chen et al. 2014; Li et al. 2012; Onishi et al. 2010; Susca et al. 2011).

Increased community green space is related to lower surface and air temperatures and reduced air pollution (Bassil et al. 2010). Studies are consistent in finding low temperatures and reduced air temperature in urban green spaces (Bowler et al. 2010). Average temperatures are lower inside the urban green spaces, confirming their impact on urban heat (Yu and Hien 2005). Mature trees remain relatively cool in an urban climate, in contrast to nongreen impervious surfaces, by providing shade, thermal comfort, reduction of air temperature, and relief from the effects of heat islands (Meier and Scherer 2012; Roy et al. 2012; Hwang et al. 2011; Lin et al. 2010; Lynn et al. 2009; Park et al. 2012; Shashua-Bar et al. 2012).

Temperature differences between green and nongreen spaces are greater during the hot periods of the day (Hamada et al. 2013; Doick et al. 2014). The cooling effects are greater during the times of the hottest temperatures (Bowler et al. 2010;

Cao et al. 2010; Cohen et al. 2012; Hamada et al. 2010; Hwang et al. 2011; Meier and Scherer 2012; Li et al. 2012; Oliveira et al. 2011; Park et al. 2012; Sung 2013; Zhang et al. 2013).

Changes in surface temperatures from green space are related to urban heat islands, but are not an indicator of thermal comfort improvement and heat stress reduction. Higher land surface temperature is significantly associated with lower income communities with larger ethnic minorities and older adults (Huang et al. 2011). Increased green spaces increase energy flows, while decreasing land surface temperatures (Li et al. 2012, 2013; Zhou et al. 2011).

Air temperatures in warm humid climates are significantly cooler within the urban green spaces (Oliveira et al. 2011) compared with nongreen areas (Vidrih and Medved 2013; Armson et al. 2012). Humidity tends to be higher in urban green areas than in inhabited zones. Densely inhabited areas without green spaces usually have an inadequate climate. Green spaces mitigate the effect of climate warming by providing shade. Replacing paved yards with urban green spaces reduces the heat island effect during the summer by moderating temperatures expected with climate change. Increased cover of community-level green spaces is associated with reduced air temperatures. The configuration and patch area of a community's green space have a relationship with personal exposure to air pollution at the household level, with cooler air temperatures and reduced urban heat island effects (Steeneveld et al. 2011; Li et al. 2012).

The main role and function of urban green spaces and gardens are to improve climate and reduce air pollution. The role and behavior of urban green spaces and gardens in improving climate and reducing air pollution. Pollution in urban areas is dependent on the type of architecture and proximity to green spaces. An avenue with a green space is less polluted because dispersion is better, whereas narrow streets tend to be more polluted. Walkways with large green spaces are better protected from pollution (Eliasson 2000). Planting more trees on street canyons may not be a good prescription where it may increase the concentration of pollutants (Escobedo and Nowak 2009; McPherson et al. 2011).

The predominance of trees has the greatest cooling effects, provides thermal comfort and heat stress relief (Chen et al. 2014; Cohen, Ng et al. 2012; Perini and Magliocco 2014; Cohen et al. 2012; Zhang et al. 2013). Different scales and types of green spaces have diverse cooling effects on heat mitigation. A comparison of green space types and scales may cause the effects to overlap. Green space scales have differential scales (Cohen et al. 2012). Green spaces with trees provide greater cooling than spaces with grass (Chen et al. 2014). Higher concentrations of green spaces are associated with greater cooling (Rinner and Hussain 2011). Connectivity between urban green spaces maximizes the cooling effects (Doick et al. 2014).

Other different types of green spaces are the green buildings that have a vegetated roof or wall serving for pollution, heat stress and urban heat island mitigation. Green roofs and walls provide heat island and pollution mitigation services.

A green roof is a roof of a building covered with vegetation planted over a growing medium and a waterproof dispositive. Green roofs combined with insulation provide heat mitigation (Coutts et al. 2013) and good irrigation provides cooling

(Zinzi and Agnoli 2012). Green roofs affect air quality by removal of air pollution comparable with mitigation effects of urban forests (Speak et al. 2012; Baik et al. 2012). Green roofs provide cooling effects and reduce the heat island in the urban environment (Smith and Roebber 2011; Susca et al. 2011). Green roofs and walls are an alternative in high-density urban areas for cooling and pollution mitigation. Green roofs maximize air quality by plant selection such as creeping bent grass and red fescue that have a higher level of particle capture (Speak et al. 2012).

Green roofs do not have an effect on the street level temperature, but decrease the cooling load of buildings (Perini and Magliocco 2014). Green spaces with trees are more effective than grass surfacing and green roofs planted with grass at reducing temperatures and improving thermal comfort (Ng et al. 2012; Chen et al. 2009). Green roofs reduce storm water run-off (Mackey et al. 2012). Green roofs for heat mitigation cost more (Mackey et al. 2012; Coutts et al. 2013; Smith and Roebber 2011; Zinzi and Agnoli 2012). The impact of wind on pollution, mitigating the effects of urban green spaces, is complex, but green roofs located downwind of prevailing winds have significant mitigation effects (Baik et al. 2012; Speak et al. 2012). Wind increases the cooling and pollution-mitigating effects of green space.

Green walls have positive cooling effects (Speak et al. 2012; Baik et al. 2012) and mitigate urban heat island effects through the evapotranspiration of plants (Susca et al. 2011; Smith and Roebber 2011). The cooling capacity of green walls increases with increased temperatures (Hamada and Ohta 2010; Koyama et al. 2013). Green walls with low wind speeds reduce air pollution in the street canyon (Amorim et al. 2013). Green walls are more effective than green roofs at mitigating in-canyon air pollution (Amorim et al. 2013; Buccolieri et al. 2011; Koyama et al. 2013).

Trees in an urban green infrastructure capture and sequester carbon, mitigating the negative effects of emissions. Carbon sequestration is the removal of the greenhouse gas carbon dioxide and its incorporation into plants. Any green spaces balance carbon, taking more than they return to the atmosphere (Nowak et al. 2013; Nowak and Crane 2002). A forest in a green space maximizes carbon sequestration (Strohbach et al. 2012).

Methodological Considerations

Urban green spaces reflect the need for natural and landscaped areas within the cities. Cities have mixed land use such as residential areas, industrial areas, forest and agricultural areas, but most are man-made environments such as built-up areas and urban green areas, and water. Large cities have lost natural resources and invest more than medium cities that have more natural green areas (Tuzin et al. 2002). Urban green spaces have a critical value for planning and developing sustainable eco-cities. In cities with a higher rate of growth population density, urban green spaces tend to be reduced at the expense of the urbanization process. There are variations in areas given over to urban green spaces in cities, for example, Singapore has 47% and Sydney 46%.

Community involvement, engagement, and development require a methodology to ensure that local authorities meet the needs and aspirations of local users in the community. Some of the methods employed are the literature review, the survey of local authorities, structured interviews with authorities and a review of their documents, analysis of existing data on uses and users, and more detailed examination of a case study.

An analysis of the green space deals with the physical and quantitative, functional, ecological, environmental, economic, and quality aspects. Economic aspects are the expenses of development, costs of maintenance, financing, and budget sources. The quality of the urban green space experience needs to be studied from an interdisciplinary perspective, drawing on both natural and social sciences. Some of the physical quantitative indicators are the supply and distribution of natural and landscape resources of public green areas as a percentage of the city area, the square meter per capita, and structural and morphological characteristics. Quantitative evaluation of the relationship between urban population and urban green spaces takes into account functionality, the green space ratio, green space coverage, and green space area per capita (Xion-Jun 2009). The quality aspects of urban green spaces are the suitability and quality of the site structure, design and provision, and the quality of the conditions.

Finding meaningful information on the uses and users of urban green spaces is hampered by inconsistencies with regard to information from local authorities. The use of model surveys to collect information from users of green spaces regarding the satisfaction of needs and aspirations needs to be developed through pilot studies and consultation by researchers and local authorities. Consultation on and involvement in environmental issues identify the community's needs. Also, the results of research find evidence for the differentiation of the need to have a green space close to where they live, as opposed to where they work (Greenspace 2007).

The user's perceptions of urban green areas matter to the community's image and decisions on uses. Perceptions of the image of urban green spaces affect uses and user aspirations and the value to the community of creating, designing, meeting the needs, and sustainably managing. The issues more related to designing are the variety, activities, spaces, sensory stimulation, vegetation, water, birds, animals, etc.

A sound basis for the collection and analysis of data is the means of finding out the priorities. A pool of data should be collected and analyzed to find out priorities in terms of type, quantity and quality, location, and accessibility. The amount of green space has been reduced by the trend toward more a compact urban environment (Burton 2003). The observed current trends on urban green spaces suggest increasing degradation and without support, the process is not likely to be reversed. An expert study that is already available may help to compare and check the planning context and legislation.

Users of Urban Green Spaces

Meeting the needs of users is related to issues of awareness of needs, the nature of the facilities and their conditions, the opportunities for activities, events and playing, the provisions of conveniences such as toilets, shelters, seating, and refreshments. Users of urban green areas develop patterns of informal and passive activities, with peaks in the afternoons, weekends, and holidays on a daily basis. Involvement in urban green spaces leads to the creation of facilities to meet the needs of users users who have more experience and demand higher quality. Facilities of urban green spaces must meet the environmental, socioeconomic, and psychological needs and attitudes of the user (Balram and Dragicevic 2005). Meeting the users' needs at local environmental, social, and economic levels requires the development of local standards, such as the provision of urban green space per head.

In one piece of research, users of urban green spaces manifested psychological effects (Dunnett et al. 2002). A significant relationship was found between the use of green spaces and levels of stress (Grahn and Stigsdotter 2003). Green spaces in the living environments also positively affect stress and quality of life. Urban green spaces reduce stresses for users and provide them with a pleasant positive distraction (Ulrich et al. 2010). The use of green spaces is associated with less stress. Viewing nature and urban green spaces ameliorates stress levels (Ulrich 2002).

Natural green environments have restorative effects and pleasing stimuli promoting soft fascination (Forest Research 2010). Also, users are happier and have greater well-being when they live in an urban area with large green spaces (White et al. 2013). The evidence between green spaces and physical activity is strong; although beneficial links have been reported between urban green spaces and emotional, psychological and mental health, and well-being, the evidence is weak. Large urban green spaces contribute to the physical and mental health and well-being of users.

There also appears to be seasonal patterns affected by the weather. Other reasons for using urban green spaces are for walking, including dog walking, passive and active enjoyment of the environment and sports, social encounters, and activities.

Accessibility to urban green spaces is related to ease of access by proximity with no physical barriers, transportation, open gates at an early hour, accessibility for disabled people, information on cues and path-finding features, maps, information at the entrance, path junctions, slopes, and cambers, an attendants for those with disabilities and visual impairment. Improving safety issues requires changes in the use of fencing, lighting, staff or rangers, removal of cars, restriction of cycling, roller-skating, and roller-blading, etc. Urban green areas are safer gathering places for children and young people, at least safer than being on the street.

User determinants such as gender and age affect accessibility and quality of urban green spaces and other environmental factors. Access to green spaces facilitates their use and increases the level of physical activities. Accessibility to green spaces has an impact on urban socioeconomic health inequalities. There are links between access to urban green spaces and social integration among older adults (Forest Research 2010). The availability of green spaces is associated with increased survival of elderly people.

Unequal distribution of green spaces and reduced access to green environments are related to health inequalities and increases in pollution and intense heat (Alberti and Marzluff 2004; Cohen et al. 2012; Girardet 1996; Gregg et al. 2003; Grimm et al. 2008; Hough 2004; Moore et al. 2003; Newman and Jennings 2008). Deprivation levels are linked to access to green spaces. Distance from the green spaces is related to physical activity; thus, users living nearby report higher levels of physical activities, although there is no correlation with accessibility to green spaces. Proximity to green spaces is associated with self-reported health.

Increasing green spaces and optimizing spatial configuration mitigate urban heat (Choi et al. 2012; Rinner and Hussain 2011). The ratio between the urban heat area and the urban cooling area increases with the distance from the urban green space (Choi et al. 2012). There is a negative correlation between the percentage cover of urban green spaces with a land surface temperature related to the distance where the closer it is, the stronger the cool island effects. Modifying variables that affect the relationship between green spaces and heat include density, distance, wind, temperature/season, the surrounding built-up environment, and precipitation.

Urban green space distribution inequities and neighborhood quality affect urban health inequalities. Inequalities in green space quality may affect urban health inequalities. There is evidence for the relationships among green space, heat, air pollution, and health (Lachowycz and Jones 2011; Lee and Maheswaran 2011). Heat and air pollution-related health inequalities are associated with green spaces. Urban green space distribution is related to health inequalities. There is evidence in the relationship between air pollution and heat mitigation from green space having positive effects on human health. Disparities and inequalities in distribution lead to pollution "hot spots" and green deserts (Escobedo and Nowak 2009; Huang et al. 2011; Jesdale et al. 2013; Su et al. 2011).

All types of green space are associated with reductions in heat stress, urban heat islands, and air pollution. Green space density as the relative tree cover affects the relationship between green space and the mitigation of air pollution (Baik et al. 2012; Doick et al. 2014; Tsiros et al. 2009). Community green spaces are associated with lower exposure of air pollution at the household level (Dadvand et al. 2012). Greening has different impacts on heat and air pollution (Alonso et al. 2011; Nowak et al. 2014), on individual and household-level exposure to air pollution (Dadvand et al. 2012; Maher et al. 2013). Reductions of air pollution from green spaces are insignificant relative to urban-based emissions (Baró et al. 2014). Wind increases heat and air pollution, mitigating the effects of green spaces.

Unequal distribution of green spaces are related to health inequalities derived from heat and air pollution (Escobedo and Nowak 2009; Huang et al. 2011; Jesdale et al. 2013; Su et al. 2011). The uneven distribution and quality of green spaces related to mitigation of heat and air pollution is associated with health inequalities. Green spaces have differential scales of health impacts associated with reductions in air pollution and heat (Bowler et al. 2010; Roy et al. 2012). A relationship has been identified between urban green spaces, air pollution, and health inequality (Su et al. 2011). Mitigation of pollution and heat from green spaces has a direct impact on health (Nowak et al. 2014).

Access to urban green spaces for the elderly, the disabled, children, women, and minority ethnic groups concerns issues such as ease of entrance, proximity, social inclusion, provision for the visually impaired, public transport, parking, moving safely, and surface design. Awareness and understanding of social inclusion in urban green areas is a recognition of the particular social and cultural needs and aspirations of users who are most likely to be excluded in society.

Some users of urban green areas are concerned about environmental quality issues such as litter, dog mess, graffiti and vandalism, lack of garbage cans; garbage and items such as condoms, food put out for birds that has been left lying around (Gregory and Baillie 1998); smashed bottles and broken glass. Psychological issues related to the use of urban green areas prevent users from going alone because of the feeling of vulnerability, fears, safety concerns, laziness, loneliness, lack of confidence, inertia, etc.

Some negative economic impacts of urban green spaces may be a fall in profits and potential problems derived from gentrification of the area as a result of increasing the added value of land and increases in housing prices. Negative impacts identified with green spaces are the increased green density, which increases street canyon air pollution detrimental to health (Amorim et al. 2013; Morani et al. 2011). Other negative impacts of green spaces is the tree emissions of biogenic volatile organic compounds that increase the levels of ground-level ozone (Escobedo and Nowak 2009; Roy et al. 2012). Green spaces with high BVOC-emitting tree species reduce ground-level ozone. Some negative impacts and trade-offs of green spaces are exposure to pollen and physical injuries.

Some of the personal issues that deter people from using urban green spaces are factors such as not having enough time, working unsocial hours, poor health and mobility, preferences for visiting other places, issues related to the location of urban green spaces, accessibility, user experience, and environmental quality. Other personal issues can deter users from going to urban green spaces, such as having their own park, changing circumstances, family and parental restrictions. To increase parental responsibility, training sports sessions for children and young people, encourage the active participation of parents. Users of urban green spaces are deterred by a lack of or deficient facilities, the influence of undesirable people, safety issues and psychological concerns, dog mess, litter, graffiti, and vandalism.

The deterrent effects of other users are related to conflicts between children and young people, teenagers, with adults, drug users, undesirable characters, users drinking alcohol, verbal abuse, gay men, bikes and skateboards, gamblers, noisy people, and being crowded, etc. The study of the urban environment combines the sound levels, biodiversity and green spaces. The results of this study confirm that the planning and designing of urban green spaces are enhanced by the ecological quality in issues such as noise levels of livable and sustainable communities (Williams et al. 2000; Girardet 2004). The soundscapes of green urban spaces have been less well-studied.

The declining quality of urban green spaces contributing to a decline in the urban quality of life has been studied by Irvine et al. (2009). Dog mess is a critical concern in urban green spaces and requires special attention such as dog-free areas and areas for dogs, good positions of dog bins in suitable locations, dogs on the lead and controlled, dog toilets, proper use of fines, etc.

The most relevant emerging barriers are the resource issues rather than personal concerns regarding the lack of facilities, the lack of maintenance, including play opportunities for children; not enough to do, the negative influence of other green space users; dog mess and dogs not being on a lead; physical safety and other psychological concerns such as fears concerning environmental quality, including litter, vandalism and graffiti, accessibility, poor public transport, distance, a lack of or poor facilities, a neglect of spaces and facilities, the conditions of play areas and play equipment, the lack of playing opportunities, inefficient staffing, poor conditions or a lack of toilets, seating, poor lighting, lack of provisions in spaces for children, the elderly, and for women.

Non-users of urban green spaces are people who have used them once in the last year or never. Infrequent users are those who have used these spaces only once in the last 6 months. Some of the reasons for non-use and infrequent use of urban green spaces are public drinking, vandalism and policies of care in the community, dog mess, perception of an unsafe environment, concern for personal safety and security, fear of violence, fear of bullying and racist attacks, dark passages, lack of lighting, poorly lit paths, emergency assistance and telephones, predominance of playing fields, lack of attractive activities and facilities, failure to provide activities and experiences demanded by users, the lack of character of many parks, unfamiliarity with landscapes and open space cultures, an uncomfortable feeling of "otherness" (DETR 1996; MacFarlane et al. 2000; Thomas 1999; McAllister 2000).

Factors of Successful Community Involvement in Urban Green Spaces

Some relevant factors to improve the use of urban green areas are less dog mess, improved safety, better maintenance, better facilities, more events and activities, more staff and easier access to sites, a wider variety of things to do, dogs being on a lead, dog-free areas, more staff, provision of more seats, no motor vehicles, no cycling, no roller skating and roller blading, play areas, lower planting near paths, accessibility to an urban green space and being close to home, with parking facilities and good public transportation, an information center and information boards, displays boards, braille signs, maps signing posts with directions, etc.

Urban green spaces provide opportunities for all people to meet, regardless of their cultural, religious, ethnic origin, political ideology, sexuality, profession, etc. Urban green spaces are sites for community spirit, although different friends and users' groups have different levels of involvement, engagement, and development that range from the frankly adversarial to those that have already formed partnerships. Some community activities for children and young people could spring out around the urban green spaces, providing a free open gathering place, usually at weekends, during the summer time and holidays. Members of the friends' group can clean-up the urban green space every weekend.

As a result of a consultative process, local authorities committed to partnerships must consider demonstrations of the willingness of community groups and residents to get involved in some of the initiatives and take responsibility for tasks at the urban green spaces. In fact, some participants are willing to volunteer and engage in the green space proposals, providing a unique experience.

Partnerships offer an opportunity for the coordination of environmental regeneration programs at a potentially low financial cost. For this purpose, a priority proposal is to establish a user community group to include representative volunteer local members to work as a charity in partnership. Provision of other facilities that encourage the use of urban green areas are cafés, information centers, boards, toilets, sports areas, sporting events, dog litter bins, seating, parking, staffing including attendants, park wardens, and keepers, first aid facilities, boating, water features, and water plants.

Different models of partnerships between urban green spaces and communities require cultural change to move the emphasis onto community involvement and sense of ownership which results in caring, resourcing, involvement, creativity, and innovation. Some factors contributing to a successful involvement are the institutional culture of local authorities, community groups and users, resources and capabilities, sense of funding, investing, and ownership, voluntary commitment, and communication between stakeholders. The responsibility and ownership of urban green spaces should not be fragmented between different authorities and different structures to achieve more innovation, efficiency, and community involvement. Local authorities develop approaches to engaging and involving users through discussion groups, consultations, artistic events, sports activities, ethnic minority background activities, leisure programs, environmental and horticultural activities, community gardens, organic food growing projects, etc.

Community involvement and engagement in urban green spaces leads to enhanced quality of experiences and uses meeting the needs of users and long-term sustainability, giving access to additional funding and expertise. User groups usually set up priorities for urban green spaces where funding is available and consultation from local communities is required, tied as they are into yielding tangible results. Groups should be active to complement the capabilities of local authorities. One way to motivate is to provide grants for urban green space projects available to all groups. Urban green spaces are a catalyst for community projects because they revolve around the most relevant community issues and the potential for environmental, social, and economic change is derived from adequate funding and managing of resources. Therefore, the prominence of urban green spaces in communities is in the promotion for funding in community regeneration initiatives.

The development of friends' and users' groups needs to be managed by requiring commitment from local authorities, but also from the community, moving from the concept of the local authority's duty to provide services because they are already funded by taxes. Well-managed urban green spaces have an impact on the urban fabric in benefiting urban environment and wildlife, promoting healthier lifestyles, increasing urban attractiveness and the urban value of land and infrastructure. Nature has beneficial effects on health and wellbeing and mood improvement (Hull and Michaels 1995; Kaplan and Kaplan 1989; Irvine and Warber 2002), reducing stress (Ulrich 1981), managing mental fatigue (Hartig et al. 1991) and opportunities for reflection (Kuo 2001; Fuller et al. 2007).

Creative and innovative approaches to funding and resourcing of urban green spaces are required if designing the appropriate arrangement to make the best with the available resources (DTR 1999). An innovative process is not exempt from conflicts. Conflicts arise among users and community organizations and groups who set up the trust. More innovative and creative local authorities are able to achieve more and better resources with less financial investment and spending. There are different methods of allocating, administering, and using the funding to be spent according to creative approaches aimed at enhancing the quality of life.

Access to additional funding by increasing the ownership of voluntary community involvement and group engagement with the local authorities. The multidisciplinary and multi-agency team enable the sense of ownership by local groups of the community, increasing their capacity building, and partner agencies to take risks using the grant, focusing on long-term regeneration and renewal objectives.

Local authorities should ensure that the backgrounds, culture and environmental resources, new expertise, skills, and interests brought in are in harmony so that the potential to develop is self-fulfilled without leaving aside the commitment and voluntary efforts of traditional users. Urban green spaces are focal for community volunteer groups to achieve change, providing facilities and activities to local users and involving and engaging other users. Activities developed by community groups in urban green spaces are most essentially voluntary actions such as conservation and maintenance tasks, although volunteer maintenance is coordinated by rangers and the feeling of ownership of upkeep is the responsibility of local authority.

Volunteers and trainees can be in charge of maintenance. Usually, voluntary community groups get involved in some of the routine operations and maintenance such as planting, grass cutting, cleaning, etc. Volunteering activities are more common in business groups conducted through staff initiatives. Active volunteers in the community need to be more motivated and negativity managed to achieve more active involvement, engagement and collaboration in a task-orientation approach with local authorities. A green space watch scheme run by volunteers can be set up in partnership with the police.

These arrangements can incur some risks, but can improve the facilities, infrastructure, maintenance, etc. However, after the initial investments, it is difficult to sustain the pace of change. Siting housing and business areas in green landscape environments is a means of promoting a green image, enhancing quality of life, and encouraging investment and economic activities. An eco-village can mix community development with providing good-quality accommodations for local residents.

Urban green space service delivery from environmental authorities may have a more holistic approach to policy and budget implementation. Resources available to local authorities and their efficient use make better provision of quality service delivery. One of the main problems facing the urban green spaces is the capital and financial resources and budget decline in real terms by the spending per head of population for funding urban green space projects. Spending per head does not necessarily take into account the area of the green space. Comparison can be made with regard to the spending per head and per hectare of green space, despite the fact that there is no consistent methodology. Urban green space officers must have expertise in community involvement and engagement, with environmental training.

Community engagement and involvement occur with a change of institutional culture of the local government and changes in the users' culture. Sensory stimulation experiences, such as gardens to smell and touch, statues, warning sounds and colored items such as seats, litter bins, lights shining on the water at night stimulation through running water, scented plants, quacking ducks, calm flowery areas, etc., should be provided. Planting programs with volunteering users from a community not only inject color, but also the sense of identity.

Diversity of vegetation and greenery is one of the most important elements of urban green spaces, color for aesthetic appreciation, grass, trees, flowers, natural trails, wild plants, an arboretum with labeled plants, plant names in Braille, and all varieties of plants such as tropical species. Animals within the urban green spaces are also an important element for children, such as birds, ducks, etc.

Determining the economic value of urban green spaces considers their natural resources. Some economic factors of urban green spaces include the production of wood, the supply of fruits, the economic value of the area, job creation, tourist attractions. Urban green spaces are of ecological value (Bilgili and Gökyer 2012) (which has become a necessity together with aesthetic and recreational value). Evidence for the value of green ecological networks to wildlife is limited, although it has become an element of urban planning (Tzoulas et al. 2007). Ecological and environmental aspects are the biodiversity and ecological values, urban climate, and natural corridors.

This vision must be agreed and shared with all the users and stakeholders and local authorities. A vision can develop and protect the quality standards of using urban green spaces, in healthy and pleasant environments and improving new uses and ensuring sustainability with high ecological and environmental value for healthy living, offering well-designed and -maintained green space, meeting the demands of users, ensuring participative action and accessibility, stimulating socioeconomic development and quality of lifestyle in the community, and contributing to the spatial identity. The concept of economic development linked with the environment is one of the principles.

The spatial concept of urban green space incorporates green into the urban structure and is related to the concept of a green system, a network of corridors. A spatial concept for urban green space development describes and incorporates green issues, interconnects the existing urban spaces and the future desired network and their relationships with the hinterland of the city. Green spaces are related to and connected to green networks and green corridors, defining preservation, improvement and development areas, the neighboring countryside, the regional green network, pedestrian and cycling paths, etc.

The quality standard measures the amount of urban green spaces per inhabitant for each type based on providing appropriate sizes for different activities, security and protection, distance and accessibility based on the travel time and the willingness to walk. Regulations and standards ensure the quality standards of accessibility of users to urban green spaces. Guidelines and standards for the provision of quality services are set out. Some standards related to urban green spaces are recreation areas near to residential districts, larger recreation areas with multifunctional uses, protection for open spaces, nature protection, local climate, land use, and soil sealing (Stadt Leipzig 2003). Combining various factors results in rendering a standardized method of classifying urban green spaces virtually impossible.

Planning and design of urban green spaces consider the recreational and visual attraction, residential and business areas, leisure and tourism development (Dole 1994). International tourism is attracted by the creativity, innovation and quality of urban green spaces, but also rolling programs of garden festivals. A scheme can include a festival space, market, fairs, etc., once the greens have been invigorated and their uses revitalized.

Urban green areas can provide countryside activities and educational activities to children outside school hours and to adults through training programs, workshops, and cultural events on urban regeneration initiatives ranging from horticulture, maintenance, school education visits on nature, art activities, lectures and training on environmental education, vocational qualifications in horticulture, animal husbandry, and a 4-week summer play scheme. Also, children are supported by their schools in some activities related to the environment, ecology, tree planting, etc., for example, providing an eLearning module to increase awareness and knowledge. Educational institutions can benefit from making use of urban green spaces for educational, sporting programs and community-based education activities for children, young people, and adults. Urban green spaces offer children the development of a social environment to improve cognitive and motor skills, and to achieve higher levels of creative play, socialization, more collaboration, and emotional resilience (Forest Research 2010).

A partnership structure that enables a crosscutting integration of community group initiatives with officers of local authorities and the urban green space in a network to coordinate responsibilities, developing action plans and activities to improve biodiversity and improve the environment as a whole. The action plan describes the specific tasks for implementing and achieving each type and each issue, actions, timescale, potential funding sources, and partners. Local authorities of urban green spaces, acting as the eyes and ears of the friends and user's groups work as a partnership shaped by community orientation. Unintended consequences of urban green spaces are avoided with community-based decision-making (Jesdale et al. 2013; Su et al. 2011). Partnerships raise the quality of urban green space.

Partnerships among business, agencies, and communities with the local authorities make base-line funding available to achieve higher and better added value, far more than can be achieved by a local authority alone. Effective partnerships among local government, business, agencies, neighborhood organizations, and community groups can add financial and quality value to the green spaces. The identification of spatial, organizational, and financial problems with regard to the planning and management of urban green spaces, such as distribution, change of use, green corridors, and networks. Among the organizational problems are communication and cooperation problems. Financial problems are related to funding. Other important arrangements for increasing and making more efficient financial resources are among others, partnerships with grant-making foundations, private financial initiatives, community and business groups, targeted grant funding, and creative initiatives to increase revenue spending.

External funding and resources from externally funded capital programs amount to a small proportion of the budget required to maintain quality standards, although they are essentially crucial for capital works. Other forms of external funding are the so-called landfill tax credit scheme and private and business sponsorship that enable creation and operation of facilities and a wide range of financial private initiatives as a means of injecting private capital. Partnering to achieve external funding and expertise from community and business involvement is a formula for raising quality standards. An active sports program of events can attract funding so that it can be financially self-supporting.

Creative and innovative approaches for external funding from community and business groups are usually selective in their applications such as tackling deprivation. Local authorities have to change radically to find and make use of the best opportunities available for external funding through partnerships. Other relevant factors important for success are political support and networking support. Some factors contributing to external funding are the political will of the local authorities to match funding to urban green spaces by embracing an entrepreneurial culture and creativity of external funding officers to investigate sources and resources through partnership opportunities. Voluntary activities enable volunteers with creative, innovative and entrepreneurial capabilities and skills to contribute to urban renewal by pursuing personal development. Bringing the necessary external resources to the urban green spaces by managing change through the involvement and engagement of local residents requires professional input expertise to discuss and accept the evolving structure.

Private sponsorship should make more significant contributions in budgeting and enabling more facilities. The bulk of urban green space is transferred by contract to a private contractor, but accountability and quality monitoring roles are retained essentially through consultation mechanisms and to ensure public accountability and quality of service delivery.

Financial values result in increasing land prices, attracting more inward investments, economic growth and development, community economic spin-offs, etc. Urban green space is one of the main drives to attracting investments and multinational corporations that usually choose to build facilities taking into consideration the urban environment and landscape (Baycan-Levent and Nijkamp 2009; Wuqiang et al. 2012).

Urban green spaces are the catalyst for social economic spin-offs in the community, such as sports activities, community centers, training programs, job creation, etc. Urban green space-based groups counting on the right individuals involved, have the potential for spin-off effects in the community. Quality values are more intangible and may result in community strengthening and environmental quality. These programs and projects can be carried out in partnership with local business, industry, companies, and financial organizations in a continuing involvement with local schools, universities, research centers, museums, heritage organizations, local authorities, local community, neighborhoods and people, green and environmental societies and organizations, etc.

Partnerships among local authorities, funding agencies and institutions, community groups, and business can contribute time and resources to adding value and quality. Partnerships can be with voluntary sector support, voluntary sector led and managed, environmental/regeneration projects, and finally partnerships around a hub. Ground work trusts are locally based partnerships committed to national organizations as an area-wide player and as a network with local operators, although sometimes they have difficulties securing a long-term commitment and leave the community with aspirations to continue the project.

Trusts are an alternative for recreation and amenity facilities, environmental and wildlife, potential new business, and urban opportunities. Urban green spaces have the capacity to be attractive to local, national, and international leisure visitors, while playing a beneficial role for the brand of the city. Thus, they indirectly play a role in location business decisions. Research has found a positive correlation between urban green spaces and business location decisions (Woolley and Rose undated for CABE), although there is little reliable evidence for the effect of green spaces on the decision to locate in a certain area and on economic growth and investments (Forest Research 2010). Trusts and private finance initiatives are partnerships with communities in different situations, with different roles of partners, including appropriate safeguards. Trust partnerships provide assistance at the level of the friends' groups.

Value-added benefits that essentially result from community involvement and engagement are contradicted by the costs and problems derived from involving nonparticipative groups, because they requires capacity building and development. The costs of urban green spaces for local authorities include all kinds of resources such as human capital, financial investments, material, knowledge, etc. Other costs are conflicting demands. Communities face the costs of responsibilities, the skills balance in services, the commitment of volunteers, etc.

Some costs associated with the involvement of local authorities in community development are a lack of long-term vision, the increase in workload without a complementary resource, a major demand on resources, greater expectations, motivating and maintaining momentum in capacity and supporting groups, over-reliance on volunteers and jealousy, identification of good leaders and representativeness in the community, lack of appropriate capabilities and skills, a hard learning process, maintaining volunteer commitment and responsibility, community development and maintenance, managing demands that conflict with and contradict constructive engagement, extending and delaying the process, job security, and successful community development. Volunteers receive training and are hired when there is funding, thus building capacities and promoting employment and ensuring commitment to the project.

Challenges and Opportunities

Some of the important challenges faced by urban green spaces are the structural and organizational changes in service delivery, re-engineering the staffing, redesigning and refurbishment, involving communities and agencies to promote the cause, providing guidance and sharing experience in good practice, avoiding duplication of responsibilities and stimulating active cooperation. Good practices are successful initiatives that are transferred and utilized in similar situations.

One of the most important challenges that mankind faces are sustainable urban green spaces where more than half of the world's population live and need to improve their lifestyle. Urban green spaces provide environmental, economic, social, political, cultural, and psychological services for the wellbeing and have an environmental sustainability impact on human activities, a role that cannot be ignored by policy makers. Use of urban green spaces improves wellbeing, reducing anxiety and depression. These impacts raise awareness of the rational use of natural resources. Environmental sustainability of any urban agenda must be considered when designing, planning, managing, and maintaining the distribution and qualitative improvements of urban green spaces as an integrated approach to providing a quality service and delivering it to users.

A variety of social and physical opportunities are available in urban green spaces, a particular recreational activity, getting away from the demands of daily life and relieving stress. Some social factors of urban green spaces to be included are the biodiversity of land uses, healthy and active lifestyles, inclusiveness and social justice (Thomas 1999) and opportunities (Scottish Executive 2001). Urban green spaces offer opportunities for aesthetic experiences with a positive psychosomatic effect that reduces attention deficits and other cognitive disorders, stress and mental fatigue, and blood pressure (Tzoulas et al. 2007).

Urban green spaces are opportunities for community dwellers to be surrounded by nature and biodiversity and gaining more enjoyment from spending time (Dallimer et al. 2012). Users are willing to pay for access to green spaces if they receive in return the opportunity to see the richness of nature in the various species (Dallimer et al. 2014). Urban green spaces provide opportunities for physical exercise and better air quality. Opportunities and potential for green urban spaces included the spatial, structural and morphological, functional, and ecological. The ecological aspects of urban green spaces comprise the maintenance of a healthy environment with water, air and soil, diversity of wildlife and resources, reduction of the impact of human activities, preservation of natural and cultural heritage, and promotion of private investments to conserve cultural heritage.

The spatial system of the urban green space requires conservation, restoration, maintenance, improvements, protection of existing and developments of new spatial forms (Kong et al. 2010), taking into consideration the impedance in the habitats posed by land use and landscape. A green infrastructure in a green network provides an overall value by offering opportunities for wildlife including various species of animals, birds, and insects. The urban green space infrastructure provides natural drainage, water interception, infiltration, storage, pollutant removal, surface flow and rainwater runoff reduction, and water quality. The storage of water in green areas is superior in quality to the runoff from streets, roads, and roofs (Hou et al. 2006; cited in Zhang et al. 2012).

Evaluation of the current conditions and economic perspectives is important for achieving results in relation to costs. Urban green spaces combine both monetary and nonmonetary valuations to assess the value. Green spaces are linked to residential and commercial property values. Cost-benefit analysis of urban green spaces is a critical factor in policy action (McPherson et al. 2011). The cost-benefit analysis of green spaces is a trade-off. A cost-benefit analysis of green spaces should take into consideration all the significant benefits such as energy consumption, reduction of pollutant materials, effects of heat stress island, reduction of greenhouse gas emissions, storm water runoff, mortality rates, etc. (Mackey et al. 2012).

Conclusion: Public Initiatives and Actions

Urban green spaces are wider initiatives of local authorities and communities with environmental, social, and economic objectives that can justify any funding for all the represented and involved stakeholders. The institutional structural framework of urban green spaces is a design concern of local authorities in response to providing services for the satisfaction of users' needs. Urban green spaces must be large enough to satisfy the urban users' needs and aspirations and distributed throughout the total urban area in such a way that better relationships with the environment can be sustained.

Therefore, promoting cooperation relationships through networks between urban green spaces and groups of friends is usually a political issue of high priority and commitment for local authorities. Some of the driving forces of urban green space initiatives behind community development are to improve the poor state of development, to generate employment, and return to the greenery as it was in the past.

Public initiatives and actions supported by local authorities aimed at inhabitants with regard to urban green spaces, parks and gardens in public spaces should demonstrate their attachment to sustainable development and the environment. Grass roots initiatives usually form community groups to work toward achieving better provision of services. Local and community initiatives in a green space develop as a result of inadequate provision for users' needs and aspirations by the local authorities, or had not been developed because of a lack of resources.

Beyond this, there is also economic stimulation with the regeneration of the community. Urban green space stimulates social and economic regeneration of communities in a multi-agency area, beyond landscaping, which can be considered only a cosmetic change. The link between environmental regeneration and economic stimulation provides substantial infrastructure in housing, job creation, etc. Usually, policymakers underestimate the role of urban green spaces beyond the greenery or landscaping in urban regeneration to reinvigorate communities and neighborhoods, by improving lifestyles, making them more pleasant and attractive, increasing the land value, strengthening the community spirit and social networks, economic stimulation, etc. Urban green spaces play a relevant environmental, economic, social, and cultural role.

Green objectives must be integrated into spatial planning. Planning for distances between urban green spaces is required to provide climate cooling to communities and neighborhoods (Doick et al. 2014). Designing, planning and sustainable management should address the resources for improving urban green areas focusing on meeting the users' needs in location, access and environmental quality, lighting, security and safety, playgrounds for children, the elderly, the disabled, and young people, dog-free areas, no vehicles, cycling or roller skating. Urban green spaces should be accessible, uniformly distributed, optimal in quality and quantity, large enough to accommodate the population's needs (Haq 2011), sustainable, and livable. These features should be considered at the stages of designing, managing, maintaining, and protecting.

The designing of urban green spaces results in good quality and variety of activities in open spacious areas, trees and spaces, exit points, quiet areas, a good network of paths, meadows, water, formal areas, meeting places, monuments, hills, mazes, etc. Designing play areas with community engagement provides opportunities for skill development and major satisfaction of the final users. External experts in design and mediation are very welcome to the task.

Designing and planning of urban green spaces must move from traditional park railings and interconnected webs to planning land uses for multiple purposes, that is, recreational and conservation uses with other purposes such as wildlife corridors beside streams and roads, public gardens on top of buildings, reservoirs and waterparks; flood prevention with canoe courses, hides and ornithological habitats in conjunction with sewage farms. Conservation planning of urban green spaces should ensure natural flora, fauna, landforms, water, air, soil, etc., and protect them from other land uses. Urban green spaces comprise habitats supporting a wide range of species, some of them with a conservation concern (Park and Lee 2000; Mörtberg and Wallentinus 2000).

Clean outdoor green spaces should provide facilities for pedestrian and cycle routes to promote well-being and health and encourage physical activities such as walking, jogging, trim trails, running, and cycling. Walking in green spaces, self-guided or led by guides for health walks on prescription, is a service that can be provided. Other offers are to promote healthy living by providing safe routes to school or business, facilitating journeys among home, the school, and the community. A green space surrounding schools lowers the levels of traffic-related pollution (Dadvand et al. 2015). Also, it is recommended to promote healthy lifestyles by growing vegetables and fruits in community urban green spaces.

Creating a sense of shared ownership for all users and stakeholders is crucial for communities using urban green spaces and may result in innovativeness, creativity, resourcing, funding, and care. Maintenance of facilities such as cafés involving community engagement recognizes the value of the services provided. The focus is on innovative and creative approaches to delivering urban green space services in the local communities. Innovation lies in applying principles with a more holistic approach to the urban green space as part of a wider network to meet the needs and aspirations of urban users. A green network improves environmental quality and safety by providing green routes and adjacent buildings, which can be business units, chapels, youth and children's facilities and diversification of activities to take advantage of further funding opportunities. There is no correlation between innovation practices and levels of spending.

This approach sets the framework for the greater potential of a range of creativity and innovation with the sense of empowerment and ownership streaming from the partnership collaboration between the local authorities and the users. To investigate creativity and innovation in creating alternative models and developing funding partnerships may increase budgeting, and the use and spending of resources. On a plot scheme, green spaces can be adopted by groups or individuals, residents' and tenants' associations in agreements with local authorities. Any plot of any size across from a large area of green space, taking over parts of a street by the participants of the program adopting a plot, for example, who have the responsibility for clearing the site and maintenance. Despite residents maintaining the green area, the local authorities have the responsibility for the bulk of planting and landscaping and the community may use it for public events.

Urban green space renewal action plans require political attention and have a wide range of environmental, economic, and social regeneration, an increase in resources and investment and educational benefits, and contributes to improving the urban landscape and its use for recreation and enjoyment. Planned activities in urban green spaces for development operations may include: plant trails, biological corridors, botanical gardens, shared gardens, play areas for recreational use.

Urban spaces that could potentially receive biodiversity should be inventoried, including open spaces, abandoned wastelands, etc. Regarding water as an element of urban green spaces, fountains and waterfalls with sound, water for children's play, ponds with wildlife, rivers, streams, boating lakes, etc, should be included. A skate park, wheelchair activities, a graffiti wall, outdoor chess and draughts, and a community fishing space could also be provided.

Sport facilities in urban green spaces require changing facilities in good condition, so that users can dress correctly, play on football pitches, access to bowling greens, tennis, basketball courts, facilities for organized clubs and for casual teams, and the option to join in activities without being a member. Programs involving the elderly, such as playing bowls, children and young people, such as football, and women, etc., contribute to better social cohesion.

Spaces for musical performances provide opportunities for better cultural awareness in the community, such as active event programs concerts, theater, fairs, bandstands, etc.

An inventory should be made of resources and maintenance and building facilities for which the urban green space is responsible, finding new ways of reducing the costs of maintenance without a loss of green assets, such as using woodland spaces for recreation and protection from climatic conditions, also involving the extensive agricultural and agro-ecological use of urban green spaces and afforestation of derelict land green spaces.

Research Gaps

There is a research gap in the empirical literature, including a lack of data on the optimal size, characteristics, distribution, and the influence on health effects of green spaces (Bowler et al. 2010). Research into inequalities in urban greener environments is necessary to improve health equity. However, there is little research on the effects of green areas on carbon capture, although study on green spaces and pollution looks at the link between carbon capture in green spaces' capacity for pollution and the absorption of particles. Research into the impact of urban green spaces on air pollution has been limited, showing moderate evidence that mitigates sulfur oxides, nitric oxides, carbon oxides, and particulate matter (Konijnendijk et al.

2013a; Yin et al. 2011). The urban green infrastructure contributes to carbon capture by building up carbon reserves in the soil (Forest Research 2010).

Green barriers are useful in protection from traffic emissions, but require further research to clarify the effects of green street canyon geometries, wind speeds, velocity of air pollutants velocity, types of vegetation, etc. Furthermore, a research gap has been reported by Bowler et al. (2010) on the cooling effect on nongreen areas adjacent to urban green spaces. The impact of small green urban areas on heat have also been less frequently explored (Oliveira et al. 2011; Bowler et al. 2010).A final gap in the research on urban green spaces is the reduction of a habitat to one independent variable, such as levels of vegetation (Kuo et al. 1998a, b), overlooking the structural complexity of biodiversity patterns interacting with social and psychological benefits and bypassing the intangible benefits associated with socioeconomic factors (Pickett et al. 2001; Hope et al. 2003; Martin et al. 2004; Kinzig et al. 2005).

Cross-References

▶ Collaboration for Regional Sustainable Circular Economy Innovation
▶ Environmental Stewardship
▶ Just Conservation
▶ Responsible Investing and Environmental Economics
▶ Sustainable Living in the City
▶ Smart Cities
▶ Strategic Management Innovation of Urban Green Spaces for Sustainable Community Development
▶ The Theology of Sustainability Practice

References

Abraham, A., Sommerhalder, K., & Abel, T. (2010). Landscape and wellbeing: A scoping study on the health-promoting impact of outdoor environments. *International Journal of Public Health, 55*(1), 59–69.

Alberti, M., & Marzluff, J. M. (2004). Ecological resilience in urban ecosystems: Linking urban patterns to human and ecological functions. *Urban Ecosystems, 7*(3), 241–265.

Alonso, R., Vivanco, M. G., González-Fernández, I., Bermejo, V., Palomino, I., Garrido, J. L., . . ., & Artínano, B. (2011). Modelling the influence of peri-urban trees in the air quality of Madrid region (Spain). *Environmental Pollution, 159*(8–9), 2138–2147.

Amorim, J. H., Rodrigues, V., Tavares, R., Valente, J., & Borrego, C. (2013). CFD modelling of the aerodynamic effect of trees on urban air pollution dispersion. *Science of the Total Environment, 461*, 541–551.

Armson, D., Stringer, P., & Ennos, A. R. (2012). The effect of tree shade and grass on surface and globe temperatures in an urban area. *Urban Forestry & Urban Greening, 11*(3), 245–255.

Baró, F., Chaparro, L., Gómez-Baggethun, E., Langemeyer, J., Nowak, D. J., & Terradas, J. (2014). Contribution of ecosystem services to air quality and climate change mitigation policies: the case of urban forests in Barcelona, Spain. *Ambio, 43*(4), 466–79.

Baik, J. J., Kwak, K. H., Park, S. B., & Ryu, Y. H. (2012). Effects of building roof greening on air quality in street canyons. *Atmospheric Environment, 61*, 48–55.

Baldock KCR et al. (2015) Where is the UK's pollinator biodiversity? The importance of urban areas for flower-visiting insects. *Proceedings of the Royal Society B, 282*, 20142849. https://doi.org/10.1098/rspb.2014.2849

Balram, S., & Dragicevic, S. (2005). Attitude towards urban green spaces; integrated questionnaire survey and Collaborative GIS techniques to improve attitude measurement. *Elsevier: Landscape and Urban Planning, 71*(2–4), 147–162.

Bassil, K. L., Cole, D. C., Moineddin, R., Lou, W., Craig, A. M., Schwartz, B., & Rea, E. (2010). The relationship between temperature and ambulance response calls for heat-related illness in Toronto, Ontario, 2005. *Journal of Epidemiology and Community Health, 65*, jech-2009.

Baycan-Levent, T. (2002). Demographic Transition and Urbanization Dynamics in Turkey in International Textbook of Urban Systems: Studies of Urbanization and Migration in Advanced and Developing Countries. In H.S. Geyer (ed), Edward Elgar Publishing, 329–361.

Baycan-Levent, T., & Nijkamp, P. (2009). Planning and management of urban green spaces in Europe: Comparative analysis. *Journal of Urban Planning and Development, 135*, 1.

Bilgili, B. C., & Gökyer, E. (2012). Urban green space system planning. In Dr. M. Ozyavuz (Ed.). landscape planning. InTech Publisher.

Bowler, D., Buyung-Ali, L., Knight, T., & Pullin, A. (2010). Urban greening to cool towns and cities: A systematic review of the empirical evidence. *Landscape and Urban Planning, 97*(2010), 147–155.

Buccolieri, R., Salim, S.M., Leo, L.S., Di Sabatino, S., Chan, A., Ielpo, P., ..., & Gromke, C. (2011). Analysis of local scale tree–atmosphere interaction on pollutant concentration in idealized street canyons and application to a real urban junction. *Atmospheric Environment, 45*(9), 1702–1713.

Burton, E. (2003). Housing for an urban renaissance: Implications for social equity. *Housing Studies, 18*(4), 537–562.

Cao, X., Onishi, A., Chen, J., & Imura, H. (2010). Quantifying the cool island intensity of urban parks using ASTER and IKONOS data. *Landscape and Urban Planning, 96*(4), 224–231.

Chen, H., Ooka, R., Huang, H., Tsuchiya, T. (2009). Study on mitigation measures for outdoor thermal environment on present urban blocks in Tokyo using coupled simulation. *Building and Environment, 44*, 2290–2299. https://doi.org/10.1016/j.buildenv.2009.03.012

Chen, A., Yao, X. A., Sun, R., & Chen, L. (2014). Effect of urban green patterns on surface urban cool islands and its seasonal variations. *Urban Forestry & Urban Greening, 13*.

Choi, H. A., Lee, W. K., & Byun, W. H. (2012). Determining the effect of green spaces on urban heat distribution using satellite imagery. *Asian Journal of Atmospheric Environment, 6*(2), 127–135.

Cohen, P., Potchter, O., & Matzarakis, A. (2012). Daily and seasonal climatic conditions of green urban open spaces in the Mediterranean climate and their impact on human comfort. *Building and Environment, 51*, 285–295.

Comedia and Demos. (1995). *Park life: Urban parks and social renewal*. Stroud: Comedia.

Cornelis, J., & Hermy, M. (2004). *Landscape and Urban Planning, 69*, 385–401.

Coutts, A. M., Daly, E., Beringer, J., & Tapper, N. J. (2013). Assessing practical measures to reduce urban heat: Green and cool roofs. *Building and Environment, 70*, 266.

Dadvand, P., De Nazelle, A., Triguero-Mas, M., Schembari, A., Cirach, M., Amoly, E., Figueras, F., Basagana, X., Ostro, B., & Nieuwenhuijsen, M. (2012). Surrounding greenness and exposure to air pollution during pregnancy: An analysis of personal monitoring data. *Epidemiology, 120*(9), 1286–1290.

Dadvand P, et al. (2015) The association between greenness and traffic-related air pollution at schools. Sci Total Environ 523:59–63

Dallimer, M., et al. (2012). Biodiversity and the feel-good factor: Understanding associations between self-reported human well-being and species richness. *Bioscience, 62*, 47–55.

Dallimer, M., et al. (2014). Quantifying preferences for the natural world using monetary and nonmonetary assessments of value. *Conservation Biology, 28*, 404–413.

Department of the Environment. (1996). *People, parks and cities – A guide to current good practice in urban parks*. London: HMSO.
DETR. (1996). English House Conditions Survey. DETR London.
DETR. (1999). By design, urban design in the planning system: Towards better practice. Available from www.odpm.gov.uk.
Dobrovolný, P. (2013). The surface urban heat island in the city of Brno (Czech Republic) derived from land surface temperatures and selected reasons for its spatial variability. *Theoretical and Applied Climatology, 112*(1–2), 89–98.
Doick, K. J., Peace, A., & Hutchings, T. R. (2014). The role of one large greenspace in mitigating London's nocturnal urban heat island. *The Science of the Total Environment, 493*, 662–671.
Dole, J. (1994). Greenscape 5: Green cities, architects' journal. In G. Haughton & C. Hunterm (Eds.), *Sustainable cities* (pp. 61–69). London: JKP.
Dunnett, N., Swanwick, C., & Woolley, H. (2002). Improving urban parks, play areas and green spaces. Department for Transport, Local Government and the Regions. [pdf] Available at: http://www.ocs.polito.it/biblioteca/verde/improving_full.pdf. Accessed Apr 2013.
Dzierżanowski, K., Popek, R., Gawrońska, H., Sabo, A., & Gawroński, S. W. (2011). Deposition of particulate matter of different size fractions on leaf surfaces and in waxes of urban forest species. *International Journal of Phytoremediation, 13*(10), 1037–1046.
Eliasson, I. (2000). The use of climate knowledge in urban planning. *Landscape and Urban Planning, 48*, 31–44.
Escobedo, F. J., & Nowak, D. J. (2009). Spatial heterogeneity and air pollution removal by an urban forest. *Landscape and Urban Planning, 90*(3–4), 102–110.
Feyisa, G. L., Dons, K., & Meilby, H. (2014). Efficiency of parks in mitigating urban heat island effect: An example from Addis Ababa. *Landscape and Urban Planning, 123*, 87–95.
Fintikakis, N., Gaitani, N., Santamouris, M., Assimakopoulos, M., Assimakopoulos, D. N., Fintikaki, M., ..., & Doumas, P. (2011). Bioclimatic design of open public spaces in the historic centre of Tirana, Albania. *Sustainable Cities and Society, 1*(1), 54–62.
Forest Research. (2010). Benefits of green infrastructure. Report to Defra and CLG. [pdf] Available at: http://www.forestry.gov.uk/pdf/urgp_benefits_of_green_infrastructure_main_report.pdf/$ file/urgp_benefits_of_green_infrastructure_main_report.pdf. Accessed Apr 2013.
Fröhlich, D., & Matzarakis, A. (2013). Modeling of changes in thermal bioclimate: Examples based on urban spaces in Freiburg, Germany. *Theoretical and Applied Climatology, 111*(3–4), 547–558.
Fuller, R.A., Irvine, K. N., Devine-Wright, P., Warren, P.H. & Gaston, K. J. (2007). Psychological benefits of greenspace increase with biodiversity. Biology Letters, 3 (4), 390–394. https//doi.org/10.1098/rsbl.2007.0149.
Fuller, R. A. et al. (2009). Diversity and Distributions, 15, 328–337.
Gaitani, N., Spanou, A., Saliari, M., Synnefa, A., Vassilakopoulou, K., Papadopoulou, K., ..., & Lagoudaki, A. (2011). Improving the microclimate in urban areas: A case study in the centre of Athens. *Building Services Engineering Research and Technology, 32*(1), 53–71.
Girardet, H. (1996). *The Gaia atlas of cities: New directions for sustainable urban living*. London: Gaia. UN-HABITAT.
Girardet, H. (2004). *Cities people planet: Liveable cities for a sustainable world*. London: Wiley.
Grahn, P., & Stigsdotter, U. A. (2003). Landscape planning and stress. *Urban Forestry & Urban Greening, 2*(1), 1–18.
GreenSpace. (2007). The park life report. The first ever public satisfaction survey green spaces: The benefits for London bibliography of Britain's parks and green spaces. [pdf] Available at: http://www.greenspace.org.uk/downloads/.
Gregg, J. W., Jones, C. G., & Dawson, T. E. (2003). Urbanization effects on tree growth in the vicinity of New York City. *Nature, 424*(6945), 183–187.
Gregory, R. D., & Baillie, S. R. (1998). Large-scale habitat use of some declining British birds. *Journal of Applied Ecology, 35*(5), 785–799.
Grimm, N. B., Faeth, S. H., Golubiewski, N. E., Redman, C. L., Wu, J., Bai, X., et al. (2008). Global change and the ecology of cities. *Science, 319*(5864), 756–760.

Hale, J. D., et al. (2012). *PLoS ONE, 7*, e33300.

Hamada, S., & Ohta, T. (2010). Seasonal variations in the cooling effect of urban green areas on surrounding urban areas. *Urban Forestry & Urban Greening, 9*(1), 15–24.

Hamada, S., Tanaka, T., & Ohta, T. (2013). Impacts of land use and topography on the cooling effect of green areas on surrounding urban areas. *Urban Forestry & Urban Greening, 12*(4), 426–434.

Haq, S. M. A. (2011). Urban green spaces and an integrative approach to sustainable environment. *Journal of Environmental Protection, 2*(5), 601–608.

Hart, M. A., & Sailor, D. J. (2009). Quantifying the influence of land-use and surface characteristics on spatial variability in the urban heat island. *Theoretical and Applied Climatology, 95*(3–4), 397–406.

Hartig, T., Mang, M., & Evans, G. W. (1991). Restorative effects of natural environment experiences. *Environment and Behavior, 23*(1), 3–26.

Herold, M., Liu, X., & Clarke, K. C. (2003). Spatial metrics and image texture for mapping urban land use. *Photogrammetric Engineering & Remote Sensing, 69*(9), 991–1001.

Herzele, V., & Wiedeman, T. (2003). A monitoring tool for the provision for accessible and attractive green spaces. *Elsevier Sciences: Landscape and Urban Planning, 63*(2), 109–126. https://doi.org/10.1016/S0169-2046(02)00192-5.

Hope, D., et al. (2003). Socioeconomics drive urban plant diversity. *Proceedings of the National Academy of Sciences, 100*(15), 8788–8792.

Hou, L. Z., Ding, Y. Y., Feng, S. Y. , Zhang, S. H., Chen, J. G. & Liao, R. H. (2006). Comparison of water quality of rainwater runoff from different underlying surface in Beijing city, China. *Water and Wastewater, 22*(23), 35–38 (in Chinese).

Hough, M. (2004). *Cities and natural process*. London: Routledge.

Huang, G., Zhou, W., & Cadenasso, M. L. (2011). Is everyone hot in the city? Spatial pattern of land surface temperatures, land cover and neighborhood socioeconomic characteristics in Baltimore, MD. *Journal of Environmental Management, 92*, 1753–1759.

Hull, R. B., & Michaels, S. E. (1995). Nature-based recreation, mood change and stress restoration. *Leisure Sciences, 17*(1), 1–14.

Hwang, R. L., Lin, T. P., & Matzarakis, A. (2011). Seasonal effects of urban street shading on long-term outdoor thermal comfort. *Building and Environment, 46*(4), 863–870.

Irvine, K. N., & Warber, S. L. (2002). Greening healthcare: Practicing as if the natural environment really mattered. *Alternative Therapies in Health and Medicine, 8*(5), 76–83.

Irvine, K. N., Devine-Wright, P., Payne, S. R., Fuller, R. A., Painter, B., & Gaston, K. J. (2009). Green space, soundscape and urban sustainability: An interdisciplinary, empirical study. *Local Environment, 14*(2), 155–172.

Jesdale, B. M., Morello-Frosch, R., & Cushing, L. (2013). The racial/ethnic distribution of heat risk-related land cover in relation to residential segregation. *Environmental Health Perspectives, 121*, 811–817.

Kaplan, R., & Kaplan, S. (1989). *The experience of nature: A psychological perspective*. Cambridge: Cambridge University Press. (Republished by Ann Arbor, MI: Ulrich's, 1995.)

Kinzig, A., et al. (2005). The effects of human socioeconomic status and cultural characteristics on urban patterns of biodiversity. *Ecology and Society, 10*(1), 23.

Kong, F., Yin, H., Nakagoshi, N., & Zong, Y. (2010). Urban green space network development for biodiversity conservation: Identification based on graph theory and gravity modeling. *Landscape and Urban Planning, 95*(1–2), 16–27.

Kong, F., Yin, H., James, P., Hutyra, L. R., & He, H. S. (2014). Effects of spatial pattern of greenspace on urban cooling in a large metropolitan area of eastern China. *Landscape and Urban Planning, 128*, 35–47.

Konijnendijk, C. C., Annerstedt, M., Nielsen, A. B., & Maruthaveeran, S. (2013a). Benefits of urban parks: A systematic review. *IFPRA World, 2012*(6), 10–12.

Konijnendijk, C. C., Annerstedt, M., Nielsen, A. B., & Maruthaveeran, S. (2013b). Benefits of urban parks: A systematic review. A report for IPFRA, Canada.

Koyama, T., Yoshinga, M., Hayashi, H., Maeda, K., & Yamauchi, A. (2013). Identification of key plant traits contributing to the cooling effects of green façades using free-standing walls. *Building and Environment, 66*, 96–103.

Kuo, F. E. (2001). Coping with poverty: Impacts of environment and attention in the inner city. *Environment and Behavior, 33*(1), 5–34.

Kuo, F. E., Bocacia, M., & Sullivan, W. C. (1998a). Transforming inner-city neighbourhoods: Trees, sense of safety, and preference. *Environment and Behavior, 30*(1), 28–59.

Kuo, F. E., et al. (1998b). Fertile ground for community: Inner-city neighbourhood common spaces. *American Journal of Community Psychology, 26*(6), 823–851.

Lachowycz, K., & Jones, A. P. (2011). Green space and obesity: A systematic review of the evidence. *Obesity Reviews, 12*(5), e183–e189.

Lafortezza, R., Carrus, G., Sanesi, G., & Davies, C. (2009). Benefits and well-being perceived by people visiting green spaces in periods of heat stress. *Urban Forestry & Urban Greening, 8*(2), 97–108.

Lee, A. C. K., & Maheswaran, R. (2011). The health benefits of urban green spaces: A review of the evidence. *Journal of Public Health, 33*(2), 212–222.

Li, X., Zhou, W., Ouyang, Z., Xu, W., & Zheng, H. (2012). Spatial pattern of greenspace affects land surface temperature: Evidence from the heavily urbanized Beijing metropolitan area, China. *Landscape Ecology, 27*(6), 887–898.

Li, J., Song, C., Cao, L., Zhu, F., Meng, X., & Wu, J. (2011). Impacts of landscape structure on surface urban heat islands: a case study of Shanghai, China. *Remote Sensing of Environment, 115*(12), 3249–3263.

Lynn, B.H., Carlson, T.N., Rosenzweig, C., Goldberg, R., Druyan, L., Cox, J., …, & Civerolo, K. (2009). A modification to the NOAH LSM to simulate heat mitigation strategies in the New York City metropolitan area. *Journal of Applied Meteorology and Climatology, 48*(2), 199–216.

MacFarlane, R., Fuller, D., & Jeffries, M. (2000). Outsiders in the urban landscape? An analysis of ethnic minority landscape projects. In J. F. Benson & M. H. Roe (Eds.), *Urban lifestyles: Spaces, places, people*. Rotterdam: Balkema.

Mackey, C. W., Lee, X., & Smith, R. B. (2012). Remotely sensing the cooling effects of city scale efforts to reduce urban heat island. *Building and Environment, 49*, 348–358.

Maher, B. A., Ahmed, I. A., Davison, B., Karloukovski, V., & Clarke, R. (2013). Impact of roadside tree lines on indoor concentrations of traffic-derived particulate matter. *Environmental Science & Technology, 47*(23), 13737–13744.

Manes, F., Incerti, G., Salvatori, E., Vitale, M., Ricotta, C., & Costanza, R. (2012). Urban ecosystem services: Tree diversity and stability of tropospheric ozone removal. *Ecological Applications, 22*(1), 349–360.

Martin, C. A., Warren, P. S., & Kinzig, A. P. (2004). Neighborhood socioeconomic status is a useful predictor of perennial landscape vegetation in residential neighborhoods and embedded small parks of Phoenix, AZ. *Landscape and Urban Planning, 69*(4), 355–368.

McAllister, S. (2000). Institutionalised racism in the landscape: The exclusion of ethnic minorities from landscape processes. Unpublished manuscript, University of Sheffield, Department of Landscape.

McPherson, E. G., Simpson, J. R., Xiao, Q., & Wu, C. (2011). Million trees Los Angeles canopy cover and benefit assessment. *Landscape and Urban Planning, 99*(1), 40–50.

Meier, F., & Scherer, D. (2012). Spatial and temporal variability of urban tree canopy temperature during summer 2010 in Berlin, Germany. *Theoretical and Applied Climatology, 110*(3), 373–384.

Mitchell, R., & Popham, F. (2008). Effect of exposure to natural environment on health inequalities: An observational population study. *Lancet, 372*, 1655–1660. Retrieved from: http://www.thelancet.com/journals/lancet/article/PIIS0140-6736(08)61689-X/fulltext.

Moore, M., Gould, P., & Keary, B. S. (2003). Global urbanization and impact on health. *International Journal of Hygiene and Environmental Health, 206*(4), 269–278.

Morani, A., Nowak, D. J., Hirabayashi, S., & Calfapietra, C. (2011). How to select the best tree planting locations to enhance air pollution removal in the MillionTreesNYC initiative. *Environmental Pollution, 159*(5), 1040–1047.

Mörtberg, U., & Wallentinus, H.-G. (2000). Red-listed forest bird species in an urban environment – Assessment of green space corridors. *Landscape and Urban Planning, 50*(4), 215–226.
Newman, P., & Jennings, I. (2008). *Cities as sustainable ecosystems: Principles and practices*. Washington, DC: Island Press.
Ng, E., Chen, L., Wang, Y., & Yuan, C. (2012). A study on the cooling effects of greening in a high-density city: An experience from Hong Kong. *Building and Environment, 47*, 256–271.
Nowak, D. J., & Crane, D. E. (2002). Carbon storage and sequestration by urban trees in the USA. *Environmental Pollution, 116*, 381–389.
Nowak, D. J., et al. (2013). Carbon storage and sequestration by trees in urban and community areas of the United States. *Environmental Pollution, 178*, 229–236.
Nowak, D. J., Hirabayashi, S., Bodine, A., & Greenfield, E. (2014). Tree and forest effects on air quality and human health in the United States. *Environmental Pollution, 193*, 119–129.
Oliveira, S., Andrade, H., & Vaz, T. (2011). The cooling effect of green spaces as a contribution to the mitigation of urban heat: A case study in Lisbon. *Building and Environment, 46*(11), 2186–2194.
Onishi, A., Cao, X., Ito, T., Shi, F., & Imura, H. (2010). Evaluating the potential for urban heat-island mitigation by greening parking lots. *Urban Forestry & Urban Greening, 9*(4), 323–332.
Park, C. R., & Lee, W. S. (2000). Relationship between species composition and area in breeding birds of urban woods in Seoul, Korea. *Landscape and Urban Planning, 51*(1), 29–36.
Park, M., Hagishima, A., Tanimoto, J., & Narita, K. I. (2012). Effect of urban vegetation on outdoor thermal environment: Field measurement at a scale model site. *Building and Environment, 56*, 38–46.
Perini, K., & Magliocco, A. (2014). Effects of vegetation, urban density, building height, and atmospheric conditions on local temperatures and thermal comfort. *Urban Forestry & Urban Greening, 13*(3), 495–506.
Pickett, S. T. A., et al. (2001). Urban ecological systems: Linking terrestrial, ecological, physical, and socioeconomic components of metropolitan areas. *Annual Review of Ecology and Systematics, 32*, 127–157.
Rinner, C., & Hussain, M. (2011). Toronto's urban heat island – Exploring the relationship between land use and surface temperature. *Remote Sensing, 3*(6), 1251–1265.
Rouquette, J. R., et al. (2013). *Diversity and Distributions, 19*, 1429–1439.
Rowe, D. B. (2011). Green roofs as a means of pollution abatement. *Environmental Pollution, 159*(8), 2100–2110.
Roy, S., Byrne, J., & Pickering, C. (2012). A systematic quantitative review of urban tree benefits, costs, and assessment methods across cities in different climatic zones. *Urban Forestry and Urban Greening, 11*(4), 351–363.
Schwartz, M. W., et al. (2002). Conservation's disenfranchised urban poor. *Bioscience, 52*, 601–606.
Scottish Executive. (2001). *Rethinking open space, 1 stationery office*. Edinburgh: Kit Campbell Associates.
Shashua-Bar, L., Tsiros, I. X., & Hoffman, M. (2012). Passive cooling design options to ameliorate thermal comfort in urban streets of a Mediterranean climate (Athens) under hot summer conditions. *Building and Environment, 57*, 110–119.
Smith, K. R., & Roebber, P. J. (2011). Green roof mitigation potential for a proxy future climate scenario in Chicago, Illinois. *Journal of Applied Meteorology and Climatology, 50*(3), 507–522.
Speak, A. F., Rothwell, J. J., Lindley, S. J., & Smith, C. L. (2012). Urban particulate pollution reduction by four species of green roof vegetation in a UK city. *Atmospheric Environment, 61*, 283–293.
Srivanit, M., & Hokao, K. (2013). Evaluating the cooling effects of greening for improving the outdoor thermal environment at an institutional campus in the summer. *Building and Environment, 66*, 158–172.
Stadt Leipzig. (2003). *Umweltqualitatsziele und -standards fur die Stadt Leipzig*. Leipzig: Amt fur Umweltschutz. http://www.leipzig.de/imperia/md/content/36_amt_fuer_umweltschutz/umweltziele.pdf. Accessed July 2008.

Steeneveld, G. J., Koopmans, S., Heusinkveld, B. G., Van Hove, L. W. A., & Holtslag, A. A. M. (2011). Quantifying urban heat island effects and human comfort for cities of variable size and urban morphology in the Netherlands. *Journal of Geophysical Research Atmospheres (1984–2012), 116*(D20), 2156.

Strohbach, M. W., et al. (2012). The carbon footprint of urban green space – A life cycle approach. *Landscape and Urban Planning, 104*, 220–229.

Su, J. G., Jerrett, M., de Nazelle, A., & Wolch, J. (2011). Does exposure to air pollution in urban parks have socioeconomic, racial or ethnic gradients? *Environmental Research, 111*(3), 319–328.

Sung, C. Y. (2013). Mitigating surface urban heat island by a tree protection policy: A case study of the woodland, Texas, USA. *Urban Forestry & Urban Greening, 12*(4), 474–480.

Susca, T., Gaffin, S. R., & Dell'Osso, G. R. (2011). Positive effects of vegetation: Urban heat island and green roofs. *Environmental Pollution, 159*(8), 2119–2126.

Tallis, M., Taylor, G., Sinnett, D., & Freer-Smith, P. (2011). Estimating the removal of atmospheric particulate pollution by the urban tree canopy of London, under current and future environments. *Landscape and Urban Planning, 103*(2), 129–138.

Thomas, H. (1999). Urban renaissance and social justice. *Town and Country Planning, 68*(11), 332–333.

Tiwary, A., Sinnett, D., Peachey, C., Chalabi, Z., Vardoulakis, S., Fletcher, T., ..., & Hutchings, T. R. (2009). An integrated tool to assess the role of new planting in PM10 capture and the human health benefits: A case study in London. *Environmental Pollution, 157*(10), 2645–2653.

Tsiros, I. X., Dimopoulos, I. F., Chronopoulos, K. I., & Chronopoulos, G. (2009). Estimating airborne pollutant concentrations in vegetated urban sites using statistical models with microclimate and urban geometry parameters as predictor variables: A case study in the city of Athens Greece. *Journal of Environmental Science and Health, Part A, 44*(14), 1496–1502.

Tuzin, B., Leeuwen, E., Rodenburg, C., & Peter, N. (2002). Paper presented at the 38th International Planning Congress on "The Pulsar Effect" Planning with Peaks, Glifada, Athens, 21–26 September 2002.

Tzoulas, K., Korpela, K., Venn, S., Yli-Pelkonen, V., Kazmierczak, A., Niemela, J., & James, P. (2007). Promoting ecosystem and human health in urban areas using green infrastructure: A literature review. *Landscape and Urban Planning, 81*(2007), 167–178.

Tzu-Ping Lin, Andreas Matzarakis, Ruey-Lung Hwang. (2010). Shading effect on long-term outdoor thermal comfort. *Building and Environment, 45*(1):213–221

Ulrich, R. S. (1981). Natural versus urban scenes: Some psychophysiological effects. *Environment and Behavior, 13*(5), 523–556.

Ulrich, R.S. (2002). Health benefits of gardens in hospitals. International Exhibition Floriade. 2002. Retrieved from: http://www.planterra.com/research/research_3.php.

Ulrich, Quan, X., & Zimring, C. (2010). The role of the physical environment in the hospital of the 21st century: A once-in-a-lifetime opportunity. Report prepared for TriPoint Hospital Center. Retrieved from: tinyurl.com/healthdesignstudy as cited in http://www.cleveland.com/healthfit/index.ssf/2010/09/blueprint_for_healing_–_hospi.html.

United Nations (Department of Economic and Social Affairs – Population Division). (2014). World Urbanization Prospects: The 2014 revision, highlights (ST/ESA/SER.A/352).

Van Herzele, A., & de Vries, S. (2011). Linking green space to health: A comparative study of two urban neighbourhoods in Ghent, Belgium. *Population and Environment, 34*(2), 171–193.

Vidrih, B., & Medved, S. (2013). Multiparametric model of urban park cooling island. *Urban Forestry & Urban Greening, 12*(2), 220–229.

Wania, A., Bruse, M., Blond, N., & Weber, C. (2012). Analysing the influence of different street vegetation on traffic-induced particle dispersion using microscale simulations. *Journal of Environmental Management, 94*(1), 91–101.

Weber, N., Haase, D., & Franck, U. (2014). Zooming into temperature conditions in the city of Leipzig: How do urban built and green structures influence earth surface temperatures in the city? *The Science of the Total Environment, 496*, 289–298.

White, M., Alcock, I., Wheeler, B., & Depledge, M. (2013). Would you be happier living in a greener urban area? A fixed effects analysis of panel data. European Centre for Environment and Human Health. [online] Available at: http://www.ecehh.org/publication/would-you-be-happier-living-greenerurban-area. Accessed Apr 2013.

Williams, K., Burton, E., & Jenks, M. (2000). *Achieving sustainable urban form*. London: E&FN Spon.

Wolf, K.L. (2010). Active living – A literature review. In *Green cities: Good health*. College of the Environment, University of Washington. Retrieved from: http://depts.washington.edu/hhwb/Thm_ActiveLiving.html.

Wong, G., Greenhalgh, T., Westhorp, G., Buckingham, J., & Pawson, R. (2013). RAMESES publication standards: Meta-narrative reviews. *BMC Medicine, 11*(1), 20.

Woolley, H. & Rose, S. (undated). The value of public space. How high quality parks and public spaces create economic, social and environmental value. CABEspace. [pdf] Available at: http://webarchive.nationalarchives.gov.uk/20110118095356/ http://www.cabe.org.uk/files/the-value-of-publicspace.pdf. Accessed Apr 2013.

Wuqiang, L., Song, S., & Wei, L. (2012). Urban spatial patterns based on the urban green space system: A strategic plan for Wuhan City, P. R. China Shi Song. www.intechopen.com.

Xiaoma Li, Weiqi Zhou, Zhiyun Ouyang. (2013). Relationship between land surface temperature and spatial pattern of greenspace: What are the effects of spatial resolution?. *Landscape and Urban Planning, 114*, 1–8.

Xion-Jun, W. (2009). Analysis of problems in urban green space system planning in China. *Journal of Forestry Research, 20*(1), 79–82.

Yin, S., Shen, Z., Zhou, P., Zou, X., Che, S., & Wang, W. (2011). Quantifying air pollution attenuation within urban parks: An experimental approach in Shanghai, China. *Environmental Pollution, 159*(8), 2155–2163.

Yu, C., & Hien, W. (2005). Thermal benefits of city parks. *Energy and Buildings, 38*((2006), 105–120.

Zhang, B., Xie, G., Zhang, C., & Zhang, J. (2012). The economic benefits of rainwater runoff reduction by urban green spaces: A case study in Beijing, China. *Journal of Environmental Management, 100*(2012)), 65–71.

Zhang, Z., Lv, Y., & Pan, H. (2013). Cooling and humidifying effect of plant communities in subtropical urban parks. *Urban Forestry & Urban Greening, 12*(3), 323–329.

Zhou, W., Huang, G., & Cadenasso, M. L. (2011). Does spatial configuration matter? Understanding the effects of land cover pattern on land surface temperature in urban landscapes. *Landscape and Urban Planning, 102*(1), 54–63.

Zinzi, M., & Agnoli, S. (2012). Cool and green roofs. An energy and comfort comparison between passive cooling and mitigation urban heat island techniques for residential buildings in the Mediterranean region. *Energy and Buildings, 55*, 66–76.

Zoulia, I., Santamouris, M., & Dimoudi, A. (2009). Monitoring the effect of urban green areas on the heat island in Athens. *Environmental Monitoring and Assessment, 156*(1–4), 275–292.

Strategic Management Innovation of Urban Green Spaces for Sustainable Community Development

José G. Vargas-Hernández, Karina Pallagst, and Patricia Hammer

Contents

Introduction	918
Functions of Urban Green Spaces	919
Typology of Green Spaces	919
Benefits of Urban Green Spaces in Community Development	922
Sustainable Management for Community Development	927
Strategic Management Innovation	933
Some Research Gaps Detected for Future Research	940
Some Concluding Remarks	940
Cross-References	941
References	941

Abstract

This chapter aims to analyze the strategic management innovation in sustainable management of urban green spaces for neighborhood and community development. The report is intended to review the available theoretical and empirical literature on urban green spaces in the main related topics of community and neighborhood development, sustainable management, and strategic management innovation. The research methods employed are the analytical from a functionalist approach moving later into the critical analysis finally from a holistic or integrative

---

J. G. Vargas-Hernández (✉)
University Center for Economic and Managerial Sciences, University of Guadalajara, Guadalajara, Jalisco, Mexico

Núcleo Universitario Los Belenes, Zapopan, Jalisco, Mexico
e-mail: josevargas@cucea.udg.mx; jvargas2006@gmail.com

K. Pallagst · P. Hammer
IPS Department International Planning Systems, Faculty of Spatial and Environmental Planning, Technische Universität Kaiserslautern, Kaiserslautern, Germany
e-mail: karina.pallagst@ru.uni-kl.de; patricia.hammer@ru.uni-kl.de

© Springer International Publishing AG, part of Springer Nature 2018
S. Dhiman, J. Marques (eds.), *Handbook of Engaged Sustainability*,
https://doi.org/10.1007/978-3-319-71312-0_50

point of view. It begins with the identification and review of cases, initiatives, and projects to demonstrate detailed examination of good practice, innovation, and creativity. However, the results of this review on strategic management innovation of urban green spaces are not conclusive, particularly because of the different local development circumstances and the nature of communities. Finally, this chapter adopts a prepositive and prescriptive strategic management approach of urban green spaces by presenting some research gaps and suggesting future research.

Keywords
Community development · Strategic management innovation · Sustainable development · Urban green spaces

Introduction

In the beginning, God created the earth, God made the earth to be a garden-like home for mankind, and He pronounced all the work should be well cultivated and taken care of. Some important topics of urban green spaces to take into account are the sustainable management, maintenance, community commitment, enhancement and involvement, participative culture, and spatial planning with green objectives.

The term green space has its origins in the green space planning and in the urban conservation movements to describe the green environment of urban areas. Urban green space is considered a long-term comprehensive tool for protection and maintenance of environmental sustainability by providing ecosystem services to users. Urban green space is based on the protection and optimization of natural ecological system as the base of the ecological balance of the city, its communities, and neighborhoods. Urban green space is defined as land of unsealed and permeable "soft" surfaces such as soil, grass, shrubs, and trees as the predominant character publicly accessible and managed.

An open space is "a mixture of civic spaces and green spaces" (Chesterton and Pedestrian Market Research Services Ltd 1997) encompassing "a mixture of public (or civic) and green space, where public spaces are mainly 'hard' spaces such as squares, street frontages and paved areas" (Kit Campbell Associates 2001; Scottish Executive 2001). Research Report for the Scottish Executive Central Research Unit 2001). Therefore, public open space is an open space having both green spaces and hard civic spaces with public access. Public open space may not have recreational facilities (Ilam 1999). Urban green spaces refer to land uses and land covered with natural or man-made vegetation in the city and planning areas.

Green space is any form of natural urban forests and seminatural environments with plant species (Bowler et al. 2010). Urban green space embraces all types of parks, green spaces, and play areas intended for recreational uses. The nature of urban green spaces is determined by some relevant factors such as quantity in the urban area, activities, experiences, perceived benefits to the users, (Oguz 2000; Herzele and Wiedeman 2003) location, and accessibility or distribution (Grahn and Stigsdotter 2003; Neuvonen et al. 2007).

This chapter aims to review the strategic management innovation of urban green spaces for sustainable community development. In doing so, this chapter begins with a functional analysis of urban green spaces by reviewing the main functions and deriving a typology. An analysis of benefits of urban green spaces is conducted to understand the relationship and association with the quality of lifestyle by providing a linkage between nature and the social and economics of people living in the community. Later, the analysis leads to review the sustainable management of community development and the association with the strategic management innovation as an approach to improve the quality of life of users. Finally, this analysis leads to present some research gaps and suggests some future research in this field.

Functions of Urban Green Spaces

Urban green spaces may have several functions and different forms of access, such as amenity, recreation and enjoyment, farming, horticulture, burial grounds, educational, institutional, domestic gardens, incidental green space, play areas, outdoor sports areas, informal recreation areas, parks, and gardens. Every urban green space can have special functions and uses, from a kite-flying center, herbaceous plants, different types of recreational areas, tennis to forests at urban green spaces which can supply wild food and firewood and infiltrate rainwater into the ground. Street markets and fairs can fit into the urban green spaces (Turner 1998).

Urban green space functions as linkage to the urban area with a natural world for reproduction of species and conservation of plants, soil, and water quality. Urban green spaces have the function to preserve, protect, and stabilize soil against wind and water erosion. Urban green spaces absorb rainfall and reduce runoff and trap and remove pollutants.

Urban green spaces provide better quality of life by fulfilling relevant functions. Some designed landscape of urban green spaces has historic and cultural values due to the historic period and the designer either for the purposes of amenity, as a functional green space, archeological remains, plant collections, etc.

Typology of Green Spaces

Biodiversity in urban spaces varies among the different types of city green space projects, an aspect to consider since it is the first action to take for the design, such as the diversity, biotope, creation, species of plants, running water, etc. Thus, it is important to investigate if the different types of urban green spaces meet the needs and expectations of users; the social, economic, and environmental contributions and benefits; and barriers and encouraging factors to use them, in such a way that a typology of users can be developed. Resourcing urban green space management is more than an increase in funding to become a creative and innovative orientation of available resources, the associated costs, partnerships, and comparing the spending between the different types.

The structure of a typology may vary among categories and types, such as formal and informal open space; outdoor recreational parks; amenity areas and allotments; gardens and green squares; children's play; public parks; forest park; heritage historical and archaeological parks; nature conservation parks; district, community, and neighborhood parks; green street; cemeteries; greenbelts; and green networks. A site categorization system and typology of urban green spaces at city level, defined as sites that have attraction potential for visitors, accessibility, facilities in good conditions, and staff, can be classified in park, garden, sports site, playing field, playground/open space, woodland, moorland/heathland, allotment, and closed churchyard/open space. This site categorization system assesses the levels of provision and quality standards for urban green space strategies.

There have been identified different types of users and use of urban green spaces, to determine the frequency, locations, nature of activities, and facilities required, among them the children, youngers, older people, disabled, and ethnic minority groups. There are different categories of users of urban green spaces with a range of type of uses of facilities and different types of walks and events taking place across different types of urban green spaces such as cultural, educational, sports, etc. The typology of the nature of urban green areas is dependent on the use. A well-known typology of urban green spaces characterizes four categories and subdivided into 26:

1. Amenity green spaces: recreational, incidental, and private
2. Recreational: parks and gardens, informal recreation areas, outdoor sport areas, and play areas

A consistent typology is one that categorizes planning designations as physical green space resources on which can be based a quantitative inventory combined with information on recreational functions an environmental value. A typology of urban green spaces classifies categories of types as a framework to facilitate consistent reporting existing data to avoid overlap and double counting. A consistent typology should consider type of space, its extent, size, the nature of the resource, and facilities provided.

Urban green space can be categorized as institutional grounds, school grounds, burial grounds, allotments, farmland, linear green spaces, other linear features, transport corridors, river and canal banks, disturbed ground, grassland, moor and heath, woodland, wetland, and other seminatural green spaces. The Institute of Leisure and Amenity Management (ILAM 1999) defines four types of parks:

1. **Principal/city/metropolitan parks** of more than 8.0 ha, with a town-/city-wide catchment, a varied physical resource, a wide range of facilities, and recognized as a visitor attraction
2. **District parks** with an extension of less than 8.0 ha serving a catchment area from 1500 to 2000 m, with a mixture of landscape features and a variety of facilities such as sports field/playing fields and play areas
3. **Neighborhood park has** up to 4.0 ha in extent, a catchment area of between 1000 and 1500 m, landscape features, and a variety of facilities

4. **Local park** up to 1.2 ha in extent, a catchment area of between 500 and 1000 m, consisting of a play area, informal green area, and landscape

A Typology of Urban Green Spaces: Main Types of Green Spaces

Urban green spaces are categorized according to their typology, location, size, function, etc. for management and development. To identify the elements to develop the typology of urban green spaces according to the urban green space strategy, consider their options:

1. **Recreation green space.** Parks and gardens. Informal recreation areas. Outdoor sports areas. Play areas.
2. **Incidental green space.** Housing green space. Other incidental space.
3. **Roof green.**
4. **Wall green.**
5. **Private green space.** Domestic gardens.
6. **Productive green space.** Remnant farmland. City farms. Allotments.
7. **Burial grounds.** Cemeteries. Churchyards.
8. **Institutional grounds.** School grounds (including school farms and growing areas). Other institutional grounds.
9. **Wetland.** Open/running water marsh. Fen.
10. **Woodland.** Deciduous woodland. Coniferous woodland. Mixed woodland.
11. **Other habitats.** Moor/Heath. Grassland. Disturbed ground.
12. **Linear green space.** River and canal banks. Transport corridors (road, rail, cycle ways, and walking routes). Other linear features (e.g., cliffs) but lacking other facilities.
13. **Urban green space network.** An urban green space network has some functions that include provision of nature, fresh air to breathe, recreation, water, food, etc. The integration of several green spaces to form an urban green space network connecting sites with urban habitats is a trend in nature conservation value.

Some other designations embrace different types of green space to facilitate some strategic approaches for designing, planning, and sustainable management, therefore recognizing a wide range of green spaces within urban environments.

However, this information needs improvement using a typology of users of urban green spaces by person and activity as a framework for any analysis based on consistent recorded estimates. A social typology of users of urban green spaces is based on demographic characteristics such as age, gender, ethnicity, and physical and mental abilities. Users usually change the motives for using urban green spaces at different times and occasions. Also the analysis of these categories can be combined with a social typology to categorize users regarding age, gender, ethnicity, and physical and mental abilities. Elderly age group over 65 years old, disable people, women, ethnic minorities, and young people are among the nonusers and infrequent users.

Benefits of Urban Green Spaces in Community Development

Human beings need green spaces and more who live in urban settlements where, more than the environment and social benefits, the economic returns should become more visible. Urban green spaces bring environmental, health, educational, social, and economic benefits. There exists a large body of theoretical and empirical literature related to the benefits of urban green spaces. A large proportion comes from Asian countries. Urban green spaces have important functions, meanings and benefit the quality of lifestyle by providing a linkage between nature and the social and economics of people living in the community (Alm 2007). Environmental, health, social, and economic benefits are dependent upon the underlying physical characteristics of green spaces. Urban green spaces contribute to rain water storage, pollutant capture, space for social events, environmental and health benefits, relaxation, and restoration.

The lack or low awareness and recognition of the benefits and functions of urban green spaces for the cities may experience temporary uncertainties. Sometimes the authorities cannot understand the real value of urban green spaces beyond the recreational, environmental, social cohesion, regeneration, and economic benefits. Investigation, diagnosis, and recognition of all needs and aspirations of all users enhance awareness of the benefits of urban green spaces. Each urban green space provides different benefits, although a key feature is its capacity for multi-functionality with multiple benefits for a wide range of uses and users. The benefits provided by urban green spaces to urban communities are multiple and categorized as social, environmental, economic, etc.

Some **environmental and ecological benefits** of urban green space are the biodiversity in urban habitats, reduction of carbon dioxide, enhancement of landscape, improvement of urban climate, sustainable drainage systems, and cultural heritage. The benefits of urban green space are categorized into environmental and health. Urban green spaces provide substantial benefits to the environment in terms of runoff-induced soil erosion control, oxygen generation, carbon absorption and sequestration, water purification, and temperature/energy cost saving. Air pollution mitigation from urban green spaces is associated to direct health benefits.

Urban green spaces have health and environmental benefits beyond the ornamental and aesthetic. Health disparities in income, health, and in all-cause mortality are narrower between people living near green spaces by enabling them to become more physically active (Mitchell and Popham 2008; Hartig 2008). However, there is no clear evidence for physical and nonphysical health benefits, and there is a weak link between physical, mental, well-being, and urban green spaces.

Other environmental benefits of urban green space are conservation, maintenance and enhancement of biodiversity; reduction of pollution, landscape, and cultural heritage; moderation of urban weather and climate and carbon dioxide; and more cost-effective and sustainable urban drainage system. The environmental benefits that urban green spaces provide to users are to meet objectives in urban climate, biodiversity, wildlife, habitat conservation, landscape and cultural heritage, air quality, and noise levels. Green spaces filter water, clean air, and heat pollution and ground temperature in urban communities. These ecological benefits providing natural protection are related to the quality, size, and density of green spaces.

Ecological benefits of urban green spaces range from conserving, protecting, and maintaining biodiversity and wildlife, improving air quality, and reducing energy costs of cooling systems. Some ecological benefits of an urban green space ecosystem are the environmental sustainability, maintenance of biodiversity and natural ecological network, sustainable urban landscape, solar input and radiation, urban heat island, rainfall pattern and relative humidity, wind speed and air temperature, connectivity of urban forests as wildlife corridors with populations of species (Haq 2011; Byrne and Sipe 2010), air pollution control, and noise pollution.

Among other ecological benefits, green spaces preserve biodiversity and nature conservation; environment; gardening; urban climate; air cleaning; moderation of noise; energy plant cultivation; social, economic, and cultural values; citizenship and education; aesthetics and attractiveness; cultural heritage; and allotments. The benefits of urban green spaces include biodiversity and providing ecosystem services such as carbon sequestration, temperature regulation, etc. (Bolund and Hunhammar 1999; Nowak and Crane 2002; Crane and Kinzig 2005; Gaston et al. 2005; Smith et al. 2005).

The environmental benefits are categorized in biodiversity, landscape and cultural heritage, physical environment, and sustainable practices. Urban green areas benefit biodiversity in wildlife natural areas and spontaneous urban vegetation that colonizes brownfield sites which have greater biodiversity mixed of native and introduced species. Some environmental benefits related to water and air quality are the protection of soil against erosion and prevention of nutrient runoff, water purification, removal of pollutants, air purification, oxygen generation, energy cost saving and temperature modification, and carbon sequestration.

Some benefits to physical environment of urban green areas are the reduction of air pollution and heavy metals and provision of clean air and oxygen. Vegetation potentially sinks for carbon dioxide, a major contributor of global warming. Trees potentially counteract the rise of greenhouse gases. Urban green spaces create microclimates that have cooling and humidity effects in ameliorating climatic factors such as higher temperatures, restrictions of winds and dispersal of pollutants, and increased runoff of rainfall. Runoff from roofs fed wetlands and water bodies that purify and sustain wildlife while being aesthetically pleasant. Also, sustainable urban drainage, biological filtration of runoff, and water storage systems are elements of urban green spaces such as ponds, water bodies, lakes, etc. Managing water in urban green spaces improves the environmental performance. Woodland offers benefits in carbon displacement.

Urban green spaces benefit landscape and cultural heritage by providing the green fabric to the environment. When urban green spaces and high-quality historic buildings occur, they create landscape and cultural heritage value in urban conservation areas. Urban green spaces bring ecological benefits to meet the need and demand of urban inhabitants such as spending spare time, relaxing from work and study pressures, improving communication, etc. (Wuqiang et al. 2012).

Social benefits. Urban green space has social benefits which provide the existence value in a neutral ground for all users, social inclusion for interaction, community spirit and positive influence behavior, physical and psychological health, child and educational development, engagement in healthy outdoor exercise, and a more relaxing and less stressful environment. Social and psychological benefits of urban green spaces provide resources for recreation and relaxation, easy access, and quantity and quality. Social benefits of urban green spaces are, among others, promotion of social interaction and sense of community and crime reduction (Kuo and Sullivan 2001; Sullivan et al. 2004; Kim and Kaplan 2004).

Social benefits have an existence value of community spirit for different activities including education and health and contribute to social inclusion through different opportunities, such as child development, healthy outdoor exercise as an escape to relaxing environment. Social benefits provide opportunities for users to make contact with nature and passive and active recreation and involvement in social, community, and cultural activities and events. Social and health benefits are linked with nature and trees.

Social benefits of urban green spaces extend beyond gathering or meeting. Urban green spaces strengthen community capacity building and neighborhood quality when offering free unlimited services. Free accessibility to open and available urban green spaces to all encourages inclusiveness. Dog walking is also a social activity if owners and their dogs meet each other. Of course, there should be some neutral areas for mutual tolerance between different users, considering that urban green spaces are for everybody. Urban green spaces are the hub and the spirit of urban communities, for gathering together whole families, for social and cultural events such as picnics, and for keeping teenagers out of the streets. Urban green spaces influence the behavior of users by providing opportunities for outdoor activities in contact with nature, such as taking children and the young for playing and socializing with others, reducing crime rates, etc.

Urban green spaces enhance the quality of the city environment and the life of inhabitants and provide benefits for the users subject to innovative approaches involving and engaging community to creating, managing, and maintaining the services. Urban green spaces play a vital role for the wide range of benefits provided to users in their urban living (Dunnett et al. 2002). Local authorities can adopt a quality of life capital approach as a framework for reporting on the benefits of urban green spaces.

The quality of life capital approach provides a framework for presenting and evaluating multiple benefits offered by urban green spaces (CAG Consultants and Land Use 2001). This approach is designed to analyze the values that people have on urban green spaces and benefits of environmental services in such areas as biodiversity, environmental health, and sense of place, recreational and educational benefits, historical landscape and heritage, and economic value.

Social and psychological benefits of urban green spaces provide resources for recreation and relaxation and emotional and psychological healing.

Health and psychological benefits of urban green spaces are vital motives of users who wish to escape from urban life and people to get restorative effects while needing to get away from all the hustle and bustle, relaxing, and avoiding the stress and sound of the city. Contact with nature, trees, woodlands, and greenery in the living environment brings health benefits in convergence with leisure, thus reducing healthcare costs (DeVries et al. 2000). Health benefits resulting from social and environmental benefits of urban green spaces provide physical, psychological, and mental health to users and the community as a whole. Nature has value in itself, but its effects on humans. Peace of mind also comes from something about trees, birds, and animals.

Urban green spaces provide benefits for human urban populations enhancing the health and well-being, reducing depression, improving physical fitness, reducing urban temperatures, increasing air quality, storing carbon, mitigating climate change, reducing the likelihood of flooding, enhancing patterns of brain behavior with low stress and positive blood pressure, increasing performance at attention-demanding tasks (Ward Thompson et al. 2012; Hartig et al. 1991, 2003; Roe and Aspinall 2011; Tennessen and Cimprich 1995), and encouraging physical activity by exercising, walking, and cycling. Green spaces benefit human health by providing opportunities for recreational and physical activities and reducing the risks of obesity and stress. Estimates of human health and air quality benefits of trees of green spaces integrate pollution and population data costs (Nowak et al. 2014) for monitoring purposes (Tallis et al. 2011).

Green spaces benefit health due to the capacity to promote physical activity. Landscapes in urban green areas promote physical activity and well-being (Abraham et al. 2010). However, green spaces in living environment are not related to meeting health recommendations for physical activities (Richardson and Parker 2011). Although green spaces offer opportunities for exercise, this causal relationship between physical activity and urban green spaces is still uncertain. Being physically active is high for inhabitants living in residential green environments (Ellaway et al. 2005). Physical activity in urban green spaces has strong mental health benefits.

Socioeconomic position of users does not affect frequency of using green space, reducing socioeconomic inequalities (Brown et al. 2014; Coombes et al. 2010. Grahn and Stigsdotter 2003; Mitchell and Popham 2008; Mitchell et al. 2015), facilitating socialization and community relationships and pleasant areas to relax, and maintaining high quality of life for elderly (Sullivan et al. 2004; Sugiyama et al. 2009; Kweon et al. 1998; Sugiyama and Ward Thompson 2007).

Biological quality of green spaces has psychological benefits for users (Fuller et al. 2007) throughout the park allowing the researchers to identify different soundscape zones and design implications. Urban green spaces provide psychological benefits in the environments of peaceful areas, being quiet, calm, relax, and tranquil, allowing users to unwind; get away from it all; reduce stress; be free of pollution; look for comfort and peace after a quarrel, hot discussion, and arguments; alleviate pains; relieve boredom; enjoy birds singing; and escape from traffic noise, smells, and fumes and from electronic devices such as radios and television.

Social and psychological benefits of urban green spaces are the recreation, relaxation, well-being, physical and mental health, emotional warmth, and decrease in respiratory illness.

Aesthetic and amenity benefits of urban green spaces are to make them more attractive and increase the value of property. Aesthetic benefits of urban green spaces provide attractiveness and pleasing view to residents attracting investments.

Some **economic benefits** of urban green spaces are direct employment, revenue creation, and attraction of tourism, increase of property prices, retainment and creation of business in the community, and contribution to sustain local economy opportunities for commercial operations with bioproducts such as orchards. As economic benefits, urban green spaces and landscaping increase property values and financial returns, save energy consumption, and reduce the costs of energy in cooling buildings by improving air circulation and lowering air temperatures. Economic benefits are associated with green spaces and the effects on air pollution mitigation.

The creation, conservation, maintenance, and management of green space benefit local economies by generating employment opportunities and encouraging further investment and property development. It has been estimated that any investment in urban green spaces is more than double counting all the economic, environmental, and well-being benefits such as tourism attraction, creation of related jobs, social savings, etc. Some other economic benefits contributing to sustained urban economies are creating revenue and employment, attracting tourists, creating business, increasing the value of nearby properties, and meeting agendas in education, health, environment, etc. The economic benefits of urban green spaces create employment, attract inward investments, create tourism attractions, and increase value of nearby property, community enterprise café, and garden center.

Green spaces provide direct and indirect economic benefits to communities. Green spaces provide economic valuable amenities for different stakeholders such as direct environmental, health, and social benefits, contributing to reduce the costs incurred by governments. Governments and stakeholders of urban green spaces can establish the monetary value of the benefits provided (Esteban 2012) such as in the case of green spaces on the well-being (Zhang et al. 2012). Urban green spaces have a weak evidence on the economic benefits. The evidence that green spaces have an impact on business location decisions cannot be substantiated.

There is a strong evidence in the premium on property values nearby the urban green spaces. Government gains an indirect economic benefit from cost savings linked to green spaces. Government meets the challenge to reduce costs to environmental and socioeconomic benefits of green spaces which have a positive effect. Residents may be willing to pay the costs of living in areas close to green spaces, resulting in the increase of property land value. Research has found a strong correlation between property value and proximity to green spaces, depending on how close the property is to the facilities of green spaces.

Regarding **educational benefits,** urban green spaces provide opportunities of both formal and informal education to users. Studies in nature such as biology should be encouraged as an educational opportunity for users in educational programs, among others, watch groups, and urban nature reserves having open recreational and educational spaces and art and event programs.

Joint programs in strategic alliances with educational institutions of elementary, secondary, and higher education and labor volunteers can offer a variety of educational programs involving the whole community in the development by providing informational recreation, educational resources, voluntary employment learning new skills on conservation tasks, supervised area for children educational programs, adult further educational programs, recycling schemes, wildlife reserves, educational events and meetings, adult education classes in areas such as alternative technology backyard gardening, fruit growing, composting, a work experience, training and day care program for people with special needs, programs on work experience from schools.

Any improvement in an urban green space is the result of designing and managing social vision, focusing on meeting the needs of users, and overcoming the barriers to provide the best benefits in services to the users. Irrigation for green roofs provides a microclimate benefit (Coutts et al. 2013; Zinzi and Agnoli 2012), but relying on rainfall limits the cooling capacity. The potential benefits of green wall infrastructure for air quality are undervalued (Pugh et al. 2012).

Sustainable Management for Community Development

In 1964 was published "People, Parks and Cities" to examine the management practices of urban green spaces, although it was very limited, it began to consider and important factor of community development. Management issues are related to users of urban green spaces, the attendants, staffing, quality, and varied experiences. Urban sustainable development of communities requires urban green spaces considered under an integrative and interdisciplinary framework, in order to improve and optimize the application of economic, social, political, cultural, planning, and management principles to formulate policies and strategies to provide high-quality facilities and services.

Sustainable management and expansion of urban natural green spaces contribute to support the sustainable community development, ecological system integrity, livability and general public health, psychological well-being, physical activities, gentrification, and environmental justice. Urban ecosystems applying differentiated environmental sustainable management based on green spaces may optimize the landscape related to ecological diversity and biological, economic, social, and cultural variables contributing toward a sustainable development. To improve biodiversity in urban areas is necessary to practice differentiated environmental sustainable management in green spaces to varying degrees to host diversified flora and grass protected areas with other man-made and natural amenities.

In urban green space, the sustainable management of urban forest cares for tree populations, vegetation, grass, and droves for the purpose of improving our urban environment. Sustainable maintenance and management requires tools and stewardship to identify priorities for action and design and implement a biodiversity strategy to improve the quantity and quality of wildlife habitats that can live alongside humans. Improvement of wildlife value can be achieved by creation, maintenance, and management techniques of urban nature techniques.

Local authorities, who are managing and delivering services in urban green services, must provide a more holistic approach for more efficient policy implementation and budget protection that other institutional governance structures benefiting the community involvement and responsibility by promoting diversification of urban green spaces in sustainable management and institutional governance structures, forms, contents, users, creativity, and innovation. The locally based steering group evolves to take more responsibility and an active role in the green space management.

A community-led initiative encounters problems when transferring control from local authorities to self-management gaining advantages if financial and management costs associated with community involvement with an increase in sense of ownership are reduced. Management and delegation of responsibilities on community groups can take on real ownership of urban green spaces and enable varying degrees of self-management.

The decline in quality of service delivery in urban green parks is usually linked to declining or reduced budgets. Also, management of urban green space is linked to organizational culture, and structure of spending management calls to increase funding requires budgeting the resources to be applied and considering the policies and strategies of providers in such areas as environmental benefits. External creative redirection of funding and revenue spent can be placed on partnerships, rather than cutting internal budgets, and can result in higher benefits by pooling all the resources into a single investment coordinated by the management structure of local authorities. Using targeted grant funding ensures achievement of strategic plans such as landscape restoration management.

Some responsibilities under this type of arrangement can be arboricultural and horticultural services, clinical waste, etc. Quality of service delivery and user satisfaction can be monitored by feedback mechanisms from user groups. This approach has the advantage to work with community groups to develop and implement a plan. Local authorities developing a community engagement culture create links networking between friends and groups to promote wildlife, to approach the green space management and share experiences, but also to increase awareness and understanding of financial restrictions.

Community involvement has different levels measured by the degree of active participation in decision-making and management devolved away by local authorities. Community involvement is dependent upon the culture, resources, and capacities of local authorities. Diversification of urban green spaces to meet the user's needs in terms of services provided, contents, forms, etc. is associated with creativity and innovation driven by management and local authorities in partnership with other stakeholders such as local communities, local business, etc.

Management and maintenance quality is under the pressure of budgeting and revenue spending resulting in a decline in facilities and infrastructure of green spaces. To stimulate involving the community in environmental and horticultural activity projects, each individual engaged has a small development budget. To be more successful in operation of the groups, the funding can be used to bring more resources in partnership working within cross-cutting teams and at a distance management with the officers involved. Officers should work closely with nonuser's groups of urban green spaces to get access and acceptance and develop relationships.

Improvement of management and maintenance of urban green spaces is vital to give quality services to the users. Appropriate staff's activities are vital for the functioning and operation of any urban green center, staffing needing good supervision, gardeners taking care of flowered areas, park keepers and ranger's service deterring crime and vandalism, to ask for directions, wardens to clean up the space, give first aid, support, confidence, and help users, introducing and guiding to new locations, providing company and giving.

Responsibilities of management and maintenance of urban green space are based on the users' needs by facilitation and consultation processes supported by site-based groups. Planning and management of an urban green space is improved by site-based decision-making with the involvement, engagement, active participation, and collaboration of community groups. Active participation of different stakeholders and partners involves active participation in decision-making process, planning, and management of different actions such as organizing events for fund-raising and income generation.

Urban green space audits incorporate quantitative and qualitative data to formulate a typology of a system as the base for holistic green space policies and strategies to be implemented in an action plan of a larger environmental network. Local authorities must maintain a database of sustainable management of urban green spaces, ecological values, environmental sustainability, landscapes, etc. (Jim 2004).

The aim of the audit is to have an assessment of the urban green space according to the standards of quality, quantity, and needs, weighting the user opinion and the professional expertise. Examples of setting local standards of quality, quantity and accessibility, principles, regulations, guidelines describing the condition of green spaces, etc. are important for strategic management. These can be detailed and specific or general. The quantitative standard sets the minimum amount of space per capita.

Assessing urban green spaces considered as assets in a community is to find economic value subordinated to economic development from environmental actions and green infrastructure (Walker 2004). Policy makers, budget holders, foundations, and other organizations implied should be aware of this when calling for investments and management of urban green spaces in the promotion of green communities.

Added economic value of green spaces as the result of community regeneration has an impact on increasing local inward investment and raising land values, resources, and political awareness. However, urban green spaces may be a political priority, and citizens are well informed of the potentials and benefits and get involved and engaged participating in all phases of the strategic management process. These benefits are difficult to quantify. Green spaces support other initiatives to spin off into the whole community generating contacts and cohesiveness.

Also, audits of urban green spaces provide information on deficiencies of a framework for decision-making. This approach to urban green spaces is centered around the needs of local communities and organizations. Audits produce a baseline information to feed planning and a set of recommendations and perceptions of how management practices need to change.

A short questionnaire is used to collect records on the nature of the urban green space resources, uses and users from local authorities, organizational structure, approaches to community involvement, engagement and participation, strategic planning and designing, sustainable management and maintenance practices, finance, partnerships and strategic alliances, innovative practices, and awards and other recognitions.

Local authorities have responsibility for the development and implementation of biodiversity action plans of urban green spaces. These action plans are on policy and strategy community-focused operations and maintenance, all linked to responsibilities of management, transferring from policy making to community engagement. Thus, the emphasis changes from direct delivery to developing supporting groups reflecting the real needs of community group users.

Focusing on community group consultation such as environmental organizations; educational and health institutions; associations of elderly, disable, children, young people, sports clubs, etc.; and the delivery of services from the urban green spaces will be expedite and according to their needs, in such a way that the administrative structure matches the service delivery. Community engagement is the responsibility for all the services by area and site-based that enables a structure with direct input of service delivery to users with the managerial support and community consultation.

To increase the level of community development, engagement and involvement in decision-making process and management of the urban green spaces must participate in terms of the quality of facilities, use, maintenance, etc. considering the costs and benefits associated. Among the benefits are the ownership and empowerment of communities, individuals that want to put something back into the community, access to funding sources and grants such as hanging basket projects in horticultural or gardening activities, and allotment clubs for children partly involved in environmental activities, increasing the use of the green space and increasing the involvement of users.

Other benefits include improving communication and understanding of operation constraints, additional funding, political profile to secure funding, providing safe and secure environment, issues of safety, security and rapid reporting incidents, crime reporting and vandalism reduction, additional expertise, spreading the voluntary work, ability to respond to local users' needs and aspirations, bringing the community together in a community spirit, long-term sustainability and viability, developing partnerships to complement resources and to achieve the best value, and personal satisfaction and development.

A community development approach based upon a geographical location requires a structure and skills based on community consultation, site-based staff for sustainable management, and delivery of the services received by the site users. Community engagement in designing, planning, sustainable management, and maintenance of urban green spaces involves information and communication exchange, active consultation, and collaboration in decision-making. Engagement of all represented stakeholders; participation in funding; awareness of physical, emotional, psychological, and cultural needs; and involvement in sustainable management and maintenance encourage social inclusion of vulnerable community groups such as children, elderly, disable, young people, etc.

A whole systems application to urban green space conference to produce a shared vision can bring interested representatives of community groups, businesses, clubs, organizations, and agencies. Public support and political will and commitment are necessary to develop a strategic management plan with a shared vision, to implement it, and to achieve real success.

An urban green space strategy is a long-term collective vision and perspective integrated within the planning and management system of the city to provide guidelines for development to meet the needs, aspirations, and priorities of urban dwellers and to provide objectives and the ways to achieve them.

Those interested groups to participate in decision-making may continue to develop detailed planning of actions and develop some proposals inviting members to engage in such responsibilities as financing, landscape design, management, maintenance, community activities, etc. Afterward, interested members of the extended group can discuss the initiatives and proposals.

Community engagement and involvement has the objectives of improving information exchange, communication, consultation and understanding, active collaboration in design, planning, decision-making, maintenance, and development of self-management. Services of urban green spaces support developing community groups. There are some mechanisms to stimulate groups of users to get involved and engaged on decision-making processes.

The ethos of working with user groups of the community in a more self-management orientation means that the delivery service moves to users' motivated services. Delivery services require to engage the users by meeting their needs and aspirations and send the information back to management for planning and development. The involvement of friends and user groups demonstrates the responsibilities and commitment of communities to raise the standards of quality service provided by urban green spaces in their neighborhoods.

Self-management of the urban green space involving voluntary activities from the community groups, business, and school's participation is an ultimate goal requiring financial control and delegation of planning and budgeting of resources, facilities, and activities. Self-management of facilities in urban green spaces owned by local authorities where revenue can be invested in the site enhances the sense of ownership and the entrepreneurial spirit, such as in sports events and activities. Maintenance is in charge of volunteers.

An ethical dilemma emerges when voluntary labor from the community groups participate in the operations of running the urban green spaces that have to be provided by the local authorities. Therefore, voluntary management and local community engagement cannot be acceptable if the volunteers already pay taxes for the provision of public services. However, it can be argued that community involvement and engagement with voluntary efforts in management of the green space can complement and enrich the local authority while gaining relevant benefits from their input. Engaging and developing community groups involve individual negotiations and renewing contracts in an equitable approach for setting standards and goals between local authorities of the urban green spaces and the voluntary groups.

Self-management of allotments through social meetings can create a sense of community. Under the regime of the compulsory competitive tendering system, the client functions of management, strategizing, and policing are separated from the contractor functions of maintenance and provision of service delivery to the users. Some urban green spaces have moved from a management by officers, policy, management contract role, and direct delivery service toward larger community development-based teams having the responsibilities by involving existing staff and new skills mix.

Allotments can be self-managed by residents and community involvement and support, and some plots can be set aside for community green space and other plots for productive use, which give incentive to the tenants to create income. Trusts enable urban green spaces to be self-managed in tune with community and local needs and aspirations while creating links with users, neighborhood groups, and friend's groups associated also with historic landscapes, nature conservation spaces, etc., in response to development and funding initiatives.

To make these changes, it is necessary to take into consideration many factors for the implementation of friend and user group development, involvement, and engagement to evolve from consultation functions to a more active commitment, participation, and collaboration. However, the client and contractor functions can be merged by combining strategic management and policy development for improved performance and better service delivery. Co-management with other community and business organizations and collaborators is active in planning, organizing, and staffing resources to achieve common goals. Some assistant providers help with volunteers, education, health, and programming, such as the catalyst groups involved in advocacy, consultation, etc.

Sustainable management practices in urban green spaces provide a framework for balanced management, provision and use of green environment, put in place environmental policies, resources self-sufficiency use and waste, measures to reduce energy consumption, reduce use of pesticides and peat, recycling by composting waste plant material, high standards on horticulture and arboriculture, encouraging biomass planting, new ways of working with sustainable development policies, center for local community recycling and composting schemes.

Overcoming and tackling barriers to urban green spaces increase the willingness of users to share good practices in designing, sustainable management, and maintenance. Most of these barriers and issues could be overcome if addressed correctly by planning, designing, and sustainable management of these urban green spaces. Responsibility fragmentation of urban green space management is one of the main barriers to achieve efficient community involvement under a holistic approach. Breaking down barriers between the different roles of workers in an urban green space creates a sense of ownership in all the involved encouraging creativity and innovation. Changing roles in urban green space require new skills and a new vision to cut across the traditional professional profiles emphasizing entrepreneurial skills, community consultation and facilitation, design awareness, creativity, and innovation.

Local authorities can combine policy and operations with responsibilities in a unified structure approach. Grassroots' initiatives as result of politics are important to consider during the process of integration under a functional structure with flexible management in charge of strategic planning of the whole urban green space. The new agency in charge of the public green spaces should design and implement strategies and public policies for sustainable management and care to promote standards and stimulate innovative practices in sustainable planning, design, and integrated management of urban green areas.

Strategic Management Innovation

Local authorities of cities should design and implement strategies and public policies to develop strategic management innovation of urban green spaces as an instrument of addressing environmental justice to poor green areas, such as the greening of urban land and underutilized land infrastructure to improve quality of life, make the healthier living, social, economic, ecological, and environmental sustainable and attractive. The strategic public policies should promote restoration of green space services providing leisure, culture, and sports services.

Conventional strategic and policy approaches to urban green spaces are very limited to achieving effective results in community engagement and development instead of taking a more integrative approach through addressing the full range of issues required to meet the needs and aspirations of the community. The development strategic process of urban green spaces must identify the needs, values, aspirations, priorities, spatial problems, opportunities, and potentials to improve the quality of life of the users. Strategic issues are linked with strategic priorities such as enhancement, protection, access, community, and stakeholders' involvement.

A green system or green network is a spatial concept for strategic issues. All urban green spaces open to the users and regardless of accessibility and functions are included in the strategy that contributes to the sustainable development of the city and well-being of citizens. A strategy to increase density is the establishment of greenbelts and networks of urban green spaces, minimizing distance between them and increasing the cooling (Tallis et al. 2011) of urban settlements, requiring green spaces for new urban developments (Rinner and Hussain 2011), and taking advantage of other green space alternatives (Cameron et al. 2014; Koyama et al. 2013).

The identification of strategic priority issues helps to analyze demands, problems, challenges, opportunities, potentials, etc. focusing on the resulting measures for urban green space development. Priorities of existing and new green spaces are set for types derived from specific aims of the strategy. The identification of the strategic issues related to the quality, accessibility, connectivity, etc. of green space has an impact on the implementation of a strategic action. Defining priorities of strategic issues regarding green spaces derive from the identification of strategic problems and evaluation of green spaces on relation to the vision and strategic objectives. The strategic vision statement sets out the future direction of the urban green space describing the aims to be achieved.

Opportunities in urban green space are related to the public good. Strategic development requires public support and cooperation from communities, involving them in a shared vision on the possibilities. Vision and strategic objectives of urban green space strategy have to be set down in implementing and achieving them together with an action plan with the final priorities, the strategic route, and the means. A strategic aim of urban green spaces can be to provide attractions for balanced lifestyle patterns.

Integrative approach to overcome the challenges faced by urban spaces and contributions to solutions to environmental sustainability, including the number, size, and land allocation to urban green spaces based on population, visitors, and accessible facilities but also considering other factors such as environmental, economic, social, population growth, migration, inefficient management, and lack of proper implementation of environmental policies.

The integrative approach is a transdisciplinary framework to study and analyze variables, complexities, and challenges of urban green spaces such as environmental conservation and sustainability, climate change adaptation, etc. An integrative approach facilitates the dialogue between local authorities, users, communities, business, social and non-governmental organizations and other stakeholders, academics and researchers, policy makers, societal actors, representatives of management and the citizens in general, etc. (Tress et al. 2005). An integrative approach using transdisciplinary approach analyzes urban green spaces and their implications and challenges to achieve environmental sustainability and climate change while obtaining economic efficiency and social well-being.

An urban green space strategy strengthens long-term sustainability of facilities and environment; develops a shared vision of needs, values, and priorities for the provision of quality service delivery; and generates public participation and internal and external cooperation of stakeholders for management and funding opportunities. The design, planning, management, and policy and strategy implementation of urban green spaces are related with the quality of city development with a sustainable environment with contributions to economic, social, cultural, and psychological well-being.

A basic framework for building up the green space strategy has three stages: preliminary activities, analytical or information gathering and evaluation, and strategy formulation. The building up strategy stage can be initiated by political will and governmental mandate by creating a supportive environment and the cooperation of a variety. Presentation for approval of the urban green space strategy requires clear criteria and transparent evaluation process of aspects such as the priorities, quantity and quality, design, distribution, accessibility, security, etc. An efficient implementation of an urban green strategy has the evaluation results and presented for discussion to boost participation and involvement. The urban green space strategy subject to approval can be used as a reference for making decisions of the local authority related to planning, design, development, management, and maintenance.

Systematic information gathering on strategic issues for urban green spaces includes provision and quality service delivery, internal and external administrative variables, deficiencies and gaps, priorities for action, etc. The geographic information system (GIS) enables the integration of strategies into the urban green space strategy. Records on user information collected by local authorities on urban green spaces are inconsistent despite introducing survey and monitoring schemes to determine users' needs and aspirations, impacts of development strategies, etc. One tool for monitoring is the strategy developing table used for city administrators with the purpose to evaluate the process of building. Urban green space strategy building, development, implementation, and evaluation processes must be monitored to be effective, using different methods such as mapping and observation.

Innovation in a wider network of urban green spaces lies in the best practices in a more holistic approach on strategic management to meet the needs and aspirations of all users. The pool of strategies examines possible solutions to obstacles and problems delivering best practices and new thinking to face the challenges. An integrative and holistic green space plans and strategies consider the different structural components of the whole green space network of any metropolitan area. An integrative and holistic view of urban green spaces should be the starting point of designing by the local government, considering the resources available and the environmental, social, and economic benefits contributing to the quality of life of urban inhabitants.

Less intensive management of green space network with native forest offers the greatest benefits. An innovative approach for designing of urban green spaces is critical for an integral ongoing management to solve problems and to find the right balance on funding to achieve the highest quality for creation, restoration, or renewal. A comprehensive strategic management process requires the cooperation of actors, decision-makers, and stakeholders involved. Involving participants ensures long-lasting partnerships. A strategic management building process must be flexible and supportive on the development of urban green space strategy. Strategic partnerships with cross-cutting responsibilities provide coordination and active cooperation within the local authority's structure which is in charge of community renewal and regeneration bringing funding and resources from public, private, and voluntary sectors.

Urban green spaces as public goods have to be planned and managed for more sustainable urban development. An urban green space strategic management confronts current and future needs, problems, conflicts, and potentials with a shared collective vision of the future providing objectives, proposals, tasks, and actions. A cooperative environment of an urban green space allows the integration of strategic analysis and management and development in strategic issues related to improvement of quality service delivery, accessibility and social inclusion, community involvement and engaging, and partnerships for funding.

A strategic partnership has at its core environmental and green space demonstrating real benefits of cross-agency and multi-discipline interconnected on community-based initiatives that are centered on economic, social, and community regeneration. Sustainable long-term regeneration strategy needs to meet sources of funding and professional expertise and make available to communities either through partnerships, employment, or a trust system on a basis of equality, communication, and without conflicts of interests.

Flexible initiatives of investment approaches such as partnerships with environmental organizations and trusts to achieve strategic plans by targeting funding and resources at facilities in need are important assumptions to design and implement a strategic management plan. While forming the trust, also a conflict may arise when there are opposing interests between the concentration on methods to build capacity and coordination on both long-term strategic vision and ground action that are vital for local authorities and users of community groups. Most of these partnerships are far-reaching involving cross-cutting work within local authorities, although the results do not have the right impact in site-specific outcomes.

Horticultural planting and floral displays beyond the decorative motives are forms of high-quality environmental enhancement and attracting inward investments into town by creating outdoor recreational facilities, attracting tourists and shoppers, and encouraging spending. Under this scheme of community renewal and regeneration, some important private sponsorship is attracted essentially from business groups. Trees and flowers are a means of economic investment to create an environment contributing to the quality of life and attracting tourism and shopping. One relevant activity is to chase funds from business sponsorship. One way that has been done in this strategy is to organize garden competitions for gardens in private, public, and commercial premises with the community involvement from local schools and allotment holders.

Strategic management and support from a community development ethos and friend's groups are essential elements to achieve excellence. Some important issues contributing to this excellence are the institutional culture of local authorities and managers at the urban green spaces sharing a philosophy of community engagement and a sense of ownership, empowerment and delegation of responsibilities, high degree of self-management and community involvement, the nature and type of site supported by volunteer labor, the available resources, local capacity building, high standards of quality maintenance and service delivery meeting the users' needs and aspirations, self-management encouraging outsources, and trust status and commitment.

A trust created between the different stakeholders and partners may have a social entrepreneurial approach on urban green spaces and aims more on action plans learned through experiences at a certain level of risks to focus on improvement of facilities and infrastructure. A strategy adopted may be to take advantage of some regenerating funding sources aimed to develop the site that later will support itself. This enables the trust to spend and invest freely to benefit the urban green space and the local community.

The green space strategic management considers the existing complementary policies, strategies, plans, and official documents to determine the relevance and relationships to urban green space management. Structure is adapted to the strategic management of urban green spaces, based on the city characteristics such as demographic, environmental, economic, financial, historical, spatial, social, cultural, institutional, and organizational data.

Developing urban green spaces with sustainable development objectives requires a strategic planning and management approach to improve the quality service delivery, sustain better use making them more attractive to attract investments, and enhance the well-being of users and tourism. Strategic planning and high-level innovative, creative, and quality designing of urban green space to suit the needs and aspirations of users is the most relevant issue to overcome barriers including the process of landscape management practices. An agenda of a strategic planning process for an urban green space strategy must identify the key strategic issues and challenges in addressing linkages between internal and external interactions. Participative strategic planning involves internal and external stakeholders, users, communities, funding agencies, etc. in a changing environment bringing diversity of problems, views, values, and opportunities.

Planning and designing process must incorporate participation of users and stakeholder providing their needs, values, and attitudes toward uses of urban green spaces. Planning urban green spaces requires the users' needs integration such as evergreen plants, grass, pleasant landscape, walking and running facilities, sitting, peaceful atmosphere, etc. Urban green space development strategies must include stakeholders' participation and consider the environmental, economic, and social functions (Oguz 2000).

Strategic planning and management is suitable for the development of urban green spaces taking into consideration the nature characteristics and the public good provision. Urban green space strategic planning and management is embedded in a complex system integrated in environmental, natural, economic, social, and cultural components. Urban green space strategic planning and management reacts to complexities and uncertainties. The strategic development of urban green spaces should consider the larger environmental network as a system in the planning structure. Urban green spaces must meet green policies and strategies of local authorities linked to other public policies in biodiversity and environment, economic development, education, health, etc.

As a community resource, green spaces need creative and innovative designing according to the needs of the community and to meet them and with the involvement of the community groups of users. A strategic approach adopted by the planning and management authorities provides opportunities for development of urban green areas and ensures accessibility and provides quality service delivery and recreational facilities (Laing et al. 2006).

Gathering of information, dialogue, and consultation with the groups of the community of common interest are important input activities for strategic management in directing service delivery to meet the current needs of local users. A formal structure considers the function of a consultation body established by representatives of communities of interest to discuss and exchange information and participate in making decisions and developing policies and strategies. A community development orientation of management of urban green spaces takes into consideration the information from consultation to inform the service delivery.

The development of urban green spaces requires strategic thinking to design the policies and management actions centered around an inventory of resources available to develop policies, strategies, and plan of action according to the satisfaction of user needs. An audit carried out proceeds to the development of a strategic management that needs to be based on knowledge extent and quality of the urban green space. Green space audits require to incorporate quantitative and qualitative data to categorize the impact of typologies and systems driven by policies and strategies. Green space audits involving community groups in its assessment categorize systems and typologies result in green structure strategic plan of an urban green space to consider it as part of a larger green space network.

Green space audits incorporate quantitative and qualitative information and data to characterize typologies and systems of urban green spaces to be driven by the policy and strategy design and implementation. Thus, the benefit of the audit is to provide the information basis in strategic planning. Urban green space audits require quantitative and qualitative information based on space green policies and strategies; economic, social, and environmental local standards; typologies of green spaces and users; and planning of green structures and resources.

An analytical summary of the strategic management document must explain the vision, mission, values, strategic objectives, obstacles, problems, opportunities, potentials, and challenges for urban green spaces. The strategic plan must have a good description and analysis of the urban green space situation, identification of trends and tendencies, the demographic structure, attitudes and values of the community, and identification of demands, needs and aspirations of users, priorities, objectives and goals, and the use and a set of thematic maps explaining different characteristics, considering the size of the document. Exchanging information and transferring knowledge provides an analysis of the benefits.

A green estate program of urban renewal, an area-based approach, has creative and innovative practices such as the holistic environmental initiatives in economic and social regeneration that enables resources to reallocate housing, community, education, etc. The environmental strategy has the same status than the other areas. The regeneration process is cross-cutting integrated to bridge environment with communities and its economic, social, environmental, and recreational potential. Integration of land use and sustainable management provides sustainable long-term green business income generating such as a café, fishing ponds, etc.

Urban green spaces are inextricably bound as a core element in urban renewal and regeneration integrated by a strategic synergy element between the environment, economy, and community regeneration processes. Some benefits derived of strategic urban regeneration are the attraction of inward investments, spin-offs from green space initiatives, multi-agency area, and educational, social, and economic benefits difficult to quantify.

A regenerating program of green spaces is an innovative design approach with emphasis on major investment for children and youth facilities involving principles of good practices in consultation, conflict resolution, design innovation, and collaborative partnership to long-term management and maintenance. Building, developing, and supporting strategic alliances and partnerships involving local government institutions, local communities, businesses, non-governmental organizations, and other private and public agencies may increase the quality of facilities and services provided by urban green spaces ensuring to meet the needs and aspirations of final users. Friends and user groups should be actively managed by building capacities to achieve greater levels of partnerships and strategic alliances between local authorities and communities.

Urban green spaces can also contribute to urban renewal and environmental enhancement identified by levels of integration and characterized by the strategic synergy of the interrelationship between the community, the economy, and the environment in terms of investment attraction by urban landscapes, spin-offs from green spaces initiatives, neighborhood renewal, and strategic regeneration.

An innovation can come from management by local authorities to change toward self-management changing toward community engagement and development or at least in between linking local authorities and community engagement and development. Innovation and creativity in urban green spaces can consider new forms of funding including external for new facilities; restructuration of service delivery; strategic alliances and partnerships with local businesses, voluntary sector, civil society, etc.; creation of consultative bodies and network of sustainable greenspaces; developing a program of events; delivering of biodiversity and wildlife benefits to local users; forming friend groups to be engaged in regenerating the local urban green spaces and promote community action to organize programs and networks for support for the elderly, teenagers, women, etc.; and designing and managing to provide events to experience the countryside landscapes, arboriculture, horticulture, etc. and to penetrate into the urban core.

Individual initiatives based on strategic innovation are required to promote new roles and functions of future urban green areas. The value of urban green spaces toward designing and developing eco-cities provides relevant functions that benefit the quality of life, protecting and maintaining the biodiversity, improving air quality, reducing the energy costs, and increasing property value. However, greening strategies to mitigate heat and air pollution should be applied as a multi-scale approach (Baró et al. 2014). A change in organizational and management culture in the urban green spaces is leading from contract management with direct service delivery toward user groups in a community involvement and engagement approach.

Self-management and trust status are trends to create and manage urban green spaces. A strategy in self-management in urban green spaces should be to move from groups of users to friend groups. Group users provide communication and user consultation in a top-down approach, while friend groups have more formal structure with larger membership with active participation, involvement, and collaboration.

The strategy for green space to avoid fragmented responsibility should be to bring together responsibilities for management and maintenance by creating a social entrepreneurial culture centered on a sense of ownership in all the users and stakeholders involved. A public relations and marketing strategy to increase political and public awareness and commitment for managing urban green spaces should include press releases, site events, etc. Citizens should participate in planning and maintenance of urban public spaces.

Merged structures between clients and contractors of urban green spaces can allow the integration of policy development and operations with management involving maintenance and managerial roles and functions. This integrated approach of policy development, operations, and management of practices results in a progressive engagement strategy between the urban green parks and the community.

A sustainable management practice of urban green spaces is to manage strategic partnerships with communities and user groups, being representatives of the community according to their objectives. To reach out and engage with communities and local businesses, it is required to identify the opportunities for supporting and building partnerships. A representative community group of local community according to the objectives through active participation in management ensures that any developments benefit the community.

Some Research Gaps Detected for Future Research

There are several research gaps that are derived from the lack of study if the benefits derived from small inner-city green spaces, the benefits in driving tourism. There is evidence based on the relationship between green space and health, heat, and air quality, but there is a gap on the greening strategies to maximize the health benefits and to reduce urban heat and air pollution. There is no comparative analysis between cities exploring different approaches. Also it is necessary to explore the benefits of blue spaces such as rivers, lakes, and ponds out of the study of green spaces.

There is a need for more research on strategic management innovation addressed to community green space and its relationship with heat-related stress and illness and into health benefits of green spaces at local level. Future research should be on green space program, and strategic planning and policies impact on physical, psychological, economic, and social benefits from reduction of air pollution and heat.

Some Concluding Remarks

Urban green spaces are an important issue for research. Despite that there is a growing concern for the study and analysis, the field is still in its infancy. Moreover, the research on the implications and associations between strategic management of innovation of urban green spaces for community development is very scarce despite the relevance it has under the pressures of climate change.

The results of this review and analysis of strategic management innovation of urban green spaces and their associations and implications are not conclusive, particularly because of the different local and neighborhood development circumstances and the nature of communities. However, the analysis has been useful to identify innovative models for sustainable designing, creating, planning, implementing, improving, managing, and maintaining urban green spaces with the involvement, participation, and capacity development of users, residents, and business community, among other stakeholders.

Cross-References

▶ Collaboration for Regional Sustainable Circular Economy Innovation
▶ Community Engagement in Energy Transition
▶ Environmental Stewardship
▶ Just Conservation
▶ Responsible Investing and Environmental Economics
▶ Smart Cities
▶ Sustainable Living in the City
▶ The Spirit of Sustainability
▶ The Theology of Sustainability Practice
▶ Transformative Solutions for Sustainable Well-Being
▶ Urban Green Spaces as a Component of an Ecosystem
▶ Utilizing Gamification to Promote Sustainable Practices

References

Abraham, A., Sommerhalder, K. Abel, T. (2010) Landscape and well-being: a scoping study on the health-promoting impact of outdoor environments. *International Journal of Public Health, 55* (1):59–69

Alm, L.E. (2007). Urban green structure A hidden resource, Baltic University Urban Forum Urban Management Guidebook V. In D. Wlodarczyk (Ed.), *Green structures in the sustainable city.* Chalmers University of Technology. Project part financed by the European Union (European Regional Development Fund) within the BSR INTERREG III B Neighbourhood Programme.

Baró, F., Chaparro, L., Gómez-Baggethun, E., Langemeyer, J., Nowak, D. J., & Terradas, J. (2014). Contribution of ecosystem services to air quality and climate change mitigation policies: The case of urban forests in Barcelona, Spain. *Ambio, 43*(4), 466–479.

Bärbel Tress, Gunther Tress, Gary Fry, (2005) Integrative studies on rural landscapes: policy expectations and research practice. *Landscape and Urban Planning, 70*(1-2), 177–191.

Biao Zhang, Gaodi Xie, Canqiang Zhang, Jing Zhang, (2012) The economic benefits of rainwater-runoff reduction by urban green spaces: A case study in Beijing, China. *Journal of Environmental Management, 100,* 65–71.

Bolund, P., & Hunhammar, S. (1999). Ecosystem services in urban areas. *Ecological Economics, 29* (2), 293–301.

Bowler, D., Buyung-Ali, L., Knight, T., & Pullin, A. (2010). Urban greening to cool towns and cities: A systematic review of the empirical evidence. *Landscape and Urban Planning, 97,* 147–155.

Brown, G., Schebella, M. F., & Weber, D. (2014). Using participatory GIS to measurephysical activity and urban park benefits. *Landscape and Urban Planning, 121,* 34–44.

Byrne, J., & Sipe, N. (2010). *Green and open space planning for urban consolidation – A review of the literature and best practice.* Brisbane: Urban Research Program. ISBN 978-1-921291-96-8.

CAG Consultants and Land Use. (2001). *Quality of life capital: Managing environmental, social and economic benefits-overview report.* Consultants for Countryside Agency, English Heritage, English Nature, Environment Agency Department of Landscape, University of Sheffield.

(2001). *Improving parks, play areas and green spaces – interim literature review.* London: Department for Transport, Local Government and the Regions.

Cameron, R. W. F., Taylor, J. E., & Emmett, M. R. (2014). What's 'cool' in the world of green façades? How plant choice influences the cooling properties of green walls. *Building and Environment, 73*, 198–207. (Complete).

Chesterton and Pedestrian Market Research Services Ltd. (1997). *Managing urban spaces in town centres, Department of the Environment and the Association of Town Centre Managers.* London: The Stationery Office.

Coombes, E., Jones, A. P., & Hillsdon, M. (2010). The relationship of physical activity and overweight to objectively measured green space accessibility and use. *Social science & medicine, 70*(6), 816–822.

Coutts, A. M., Daly, E., Beringer, J., & Tapper, N. J. (2013). Assessing practical measures to reduce urban heat: Green and cool roofs. *Building and Environment, 70*, 266–276.

Crane, P., & Kinzig, A. (2005). Nature in the metropolis. *Science, 308*(5726), 1225.

DeVries, S., Verheij, R. A., & Groenewegen, P. P. (2000). 'Nature and health: An exploratory investigation of the relationship between health and green space in the living environment' ('Natuur en gezondheid een verkennend onderzoek naar de relatie tussen volksgezondheid en groen in de leefomgeving'). *Mens En Maatschappij, 75*(4), 320–339.

Dunnett, N., Swanwick, C., and Woolley, H. (2002). Improving parks, play areas and green spaces. London: ODPM.

Ellaway, A., Macintyre, S., & Bonnefoy, X. (2005). Graffiti, greenery, and obesity in adults: Secondary analysis of European cross sectional survey. *British Medical Journal, 331*(7517), 611–612.

Esteban, A. (2012). *Natural solutions. Nature's role in delivering well-being and key policy goals – Opportunities for the third sector.* New economics foundation. [pdf] Available at: http://www.greenspace.org.uk/downloads/Publications/Natural_solutions_nef.pdf. Accessed Apr 2013.

Fuller, R. A., et al. (2007). Psychological benefits of greenspace increase with biodiversity. *Biology Letters, 3*(4), 390–394. 170 K.N. Irvine et al. By: [University of Queensland] At: 03:04 19 January 2009.

Gaston, K. J., et al. (2005). Urban domestic gardens (IV): The extent of the resource and its associated features. *Biodiversity and Conservation, 14*(4), 3327–3349.

Grahn, P., & Stigsdotter, U. A. (2003). Landscape planning and stress. *Urban Forestry & Urban Greening, 2*(1), 1–18.

Haq, S. M. A. (2011). Urban green spaces and an integrative approach to sustainable environment. *Journal of Environmental Protection, 2*(5), 601–608.

Hartig, T. (2008). Green space, psychological restoration, and health inequality. *Lancet, 372*, 1614–1615. Retrieved from: http://www.thelancet.com/journals/lancet/article/PIIS0140-6736 (08)61669-4/fulltext.

Hartig, T., Mang, M., & Evans, G. W. (1991). Restorative effectsof natural environment experience. *Environment and Behavior, 23,* 3–26.

Hartig, T., Evans, G., Jamner, L., Davis, D., Garling, T., (2003). Tracking restorationin natural and urban field settings. *Journal of Environmental Psychology 23,* 109–123.

Herzele, V., & Wiedeman, T. (2003). A monitoring tool for the provision for accessible and attractive green spaces. *Elsevier Sciences: Landscape and Urban Planning, 63*(2), 109–126. https://doi.org/10.1016/S0169-2046(02)00192-5.

Ilam. (1999). Open space terminology. *ILAM Fact Sheet* 00/99 (1999).

Jim, C. Y., (2004) Green-space preservation and allocation for sustainable greening of compact cities. *Cities, 21*(4):311–320.

Kim, J., & Kaplan, R. (2004). Physical and psychological factors in sense of community. *Environment and Behavior, 36*(3), 313–340.

Kit Campbell Associates (2001) Rethinking Open Space: Open Space Provision and Management: A Way Forward. A research report prepared for the Scottish Executive Central Research Unit. Edinburgh: Scottish Executive Central Research Unit.

Koyama, T., Yoshina, M., Hayashi, H., Maeda, K., & Yamauchi, A. (2013). Identification of key plant traits contributing to the cooling effects of green façades using free-standing walls. *Building and Environment, 66*, 96–103.

Kuo, F. E., & Sullivan, W. C. (2001). Environment and crime in the inner city: Does vegetation reduce crime? *Environment and Behavior, 33*(3), 343–367.

Kweon, B. S., Sullivan, W. C., & Wiley, A. W. (1998). Green common spaces and the social integration of inner-city older adults. *Environment and Behavior, 30,* 832–858. https://doi.org/10.1177/001391659803000605.

Laing, R., Miller, D., Davies, A.-M., & Scott, S. (2006). Urban green spaces; the incorporation of environmental VALUES in a decision support system, 2006. http://www.itcom.org/2006/14/.

Mitchell, R., & Popham, F. (2008). Effect of exposure to natural environment on health inequalities: An observational population study. *Lancet, 372*, 1655–1660. Retrieved from: http://www.thelancet.com/journals/lancet/article/PIIS0140-6736(08)61689-X/fulltext.

Mitchell R. J., Richardson E. A., Shortt N. K., et al. (2015) Neighborhood environments and socioeconomic inequalities in mental well-being. *American Journal of Preventive Medicine, 49*, 80–88.

Neuvonen, M., Sievanen, T., Susan, T., & Terhi, K. (2007). Access to green areas and the frequency of visits: A case study in Helsinki. *Elsevier: Urban Forestry and Urban Greening, 6*(4), 235–247.

Nowak, D. J., & Crane, D. E. (2002). Carbon storage and sequestration by urban trees in the USA. *Environmental Pollution, 116*(3), 381–389.

Nowak, D. J., Hirabayashi, S., Bodine, A., & Greenfield, E. (2014). Tree and forest effects on air quality and human health in the United States. *Environmental Pollution, 193*, 119–129.

Oguz, D. (2000). User survey of Ankara's parks. *Elsevier Science: Landscape and Urban Planning, 52*(2), 165–171.

Pugh, T. A., MacKenzie, A. R., Whyatt, J. D., & Hewitt, C. N. (2012). Effectiveness of green infrastructure for improvement of air quality in urban street canyons. *Environmental Science & Technology, 46*(14), 7692–7699.

Richardson, D., & Parker, M. (2011). *A rapid review of the evidence base in relation to physical activity and green space and health.* HM Partnerships for NHS Ashton Leigh and Wigan Shah, H. and Peck, J., 2005.

Rinner, C., & Hussain, M. (2011). Toronto's urban heat island – Exploring the relationship between land use and surface temperature. *Remote Sensing, 3*(6), 1251–1265.

Roe, J., Aspinall, P. (2011). The restorative benefits of walking in urban and rural setting in adults with good and poor mental health. *Health & Place, 17,* 103–113.

Scottish Executive. (2001). *Rethinking open space*. Edinburgh: The Stationery Office, Kit Campbell Associates.

Smith, R. M., et al. (2005). Urban domestic gardens (V): Relationships between landcover composition, housing and landscape. *Landscape Ecology, 20*(2), 235–253.

Sugiyama, T., & Ward Thompson, C. (2007). *Environment and Planning A, 39,* 1943–1960.

Sugiyama, T., Leslie, E., Giles-Corti, B., & Owen, N. (2009). Physical activity for recreation or exercise on neighbourhood streets: Associations with perceivedenvironmental attributes. *Health & Place, 15,* 1058–1063. https://doi.org/10.1016/j.healthplace.2009.05.001.

Sullivan, W.C., Kuo, F.E., and DePooter, S.F. (2004). The Fruit of Urban Nature: Vital Neighborhood Spaces. *Environment and Behavior, 36*(5): 678–700.

Tallis, M., Taylor, G., Sinnett, D., & Freer-Smith, P. (2011). Estimating the removal of atmospheric particulate pollution by the urban tree canopy of London, under current and future environments. *Landscape and Urban Planning, 103*(2), 129–138.

Tennessen, C. H., & Cimprich, B. (1995). Views to nature: Effects on attention. *Journal of Environmental Psychology, 15,* 77–85.

Turner, T. (1998). *Landscape planning and environmental impact design*. Florence: Routledge.

Walker, C. (2004). The public value of urban parks. Retrieved 20th November 2010 from http://www.urban.org/uploadedPDF/311011_urban_parks.pdf.

Ward Thompson, C., Roe, J., Aspinall, P., Mitchell, R., Clow, A., & Miller, D. (2012). More green space is linked to less stress in deprived communities: Evidencefrom salivary cortisol patterns. *Landscape and Urban Planning, 105*(3), 221–229.

Wuqiang, L., Song, S., & Wei, L. (2012). *Urban spatial patterns based on the urban green space system: A strategic plan for Wuhan City*. P. R. China Shi Song. www.intechopen.com.

Zinzi, M., & Agnoli, S. (2012). Cool and green roofs. An energy and comfort comparison between passive cooling and mitigation urban heat island techniques for residential buildings in the Mediterranean region. *Energy and Buildings, 55*, 66–76.

Application of Big Data to Smart Cities for a Sustainable Future

Anil K. Maheshwari

Contents

Introduction	946
Modern City	947
Smart and Flourishing	948
Smart City	949
Flourishing City	950
Smart Citizens	950
Smart Services for a Smart City	952
Smart Information and Communications Technology (ICT) Infrastructure	952
Smart Health	953
Smart Education	954
Smart Energy	955
Smart Transportation	956
Smart Spaces	956
Smart Sanitation	957
Smart Connectivity	958
Smart Water	958
Smart Work	958
Smart Governance	959
Smart City Examples	959
Barcelona	959
Amsterdam	961
Big Data	961
Chicago's Bike Rental Data Analysis	963
Dangers of Big Data in Smart Cities	963
Implications and Conclusion	967
Cross-References	967
References	967

A. K. Maheshwari (✉)
Maharishi University of Management, Fairfield, IA, USA
e-mail: Akm2030@gmail.com

© Springer International Publishing AG, part of Springer Nature 2018
S. Dhiman, J. Marques (eds.), *Handbook of Engaged Sustainability*,
https://doi.org/10.1007/978-3-319-71312-0_36

Abstract

A smart city is a way of organizing and facilitating secure, sustainable, and flourishing life for large numbers of people. Ideally, smart cities will enable a blissful life for all, in complete alignment with all the laws of nature. A city can make itself smarter by making its key infrastructure components – such as health, education, transportation, utilities, and other essential services – more efficient and responsive through integrated design and electronic governance (e-governance) using information and communications technologies (ICT). It can also become smarter by offering environments to flourish in and thus attract more smart people. Smart engaged citizens would actively monitor that the technology works for them, and not the other way around, as they seek a free and flourishing life. This chapter offers a smart services framework and discusses about the ten types of smart services. It will then demonstrate examples of successful smart city transitions, in particular Barcelona and Amsterdam, along with the solutions they used to become smarter. We will finally include specific examples of the use of big data to analyze the receptivity of bike rental services in Chicago. We will also examine why the use of big data can invoke the specter of Big Brother, and thus there is a need for sensitivity to human needs and privacy. Finally, we will examine a few implications of big data for smart cities and citizens.

Keywords

Sustainability · Flourishing · Natural living · Smart city · Big data · Information and communications technologies · Internet of Things

Introduction

Smart cities ideally mean enabling the things that make human life flourish. Smart city has however taken an Internet of Things (IoT) image, where collecting and analyzing large amounts of data helps uncover service usage patterns to design better services and experiences (Dameri 2017). Smart cities are thus digital governance and service delivery platforms that responsively and interactively serve a superior living experience to their citizens (Peris-Ortiz et al. 2016). However, there is an imperative need to understand the smart city phenomenon from multiple points of view, including freedom and privacy. The smart city must be created not only for citizens but also in collaboration with them.

Smart cities are emerging around the world. Barcelona is the poster child for a well-done smart city. India has targeted many cities to become smarter through electronic governance (e-governance). Many American cities have smart city initiatives well underway. McKinsey estimates that the worldwide market for smart city technologies will reach $400 billion by 2020. Other estimates go as high as $1.5 trillion (Singh 2014). Through innovative design of living and public spaces, smart cities can bring together ecosystems of skills to enable new waves of innovation and creation. Many communities devastated by natural disasters are rebuilding their infrastructure using the best principles of sustainable and smarter living.

Bicycle highways in Denmark reduce congestion, energy consumption, and air pollution while at the same time enhance citizen health through exercise and fresh air. Chicago has developed a bicycle rental system for commuting and leisure. This chapter will include a brief analysis of Chicago bike rental data to discover patterns and deeper insights into the needs and usage patterns of the smart citizens. A instrumented and integrated city with pervasive data gathering and analysis, however, also raises the *specter* of Big Brother, the perceived loss of individual liberty.

Modern City

A city is an old concept of a place to live, to work, to earn, and to connect. The production of an agricultural surplus created the context for the creation of cities in the Mesopotamia region around 4500 BCE. Many cities lay claim to the title of "first city," including currently thriving cities such as Jericho, Damascus, Aleppo, Jerusalem, Athens, and Varanasi. Most ancient cities had only modest populations, often under 5,000 persons. Probably no city in antiquity had a population of much more than a million inhabitants, not even Rome. There is a near-perfect correlation between urbanization and prosperity across societies (Gleaser 2012). The rise of technologies and engineering continues to increase the attraction of cities.

Cities offer resources and services to attract people. Cities differentiate themselves on the basis of their unique activities and services, for which they often develop a reputation (Scientific American Editors 2014). Around the world there has been a mass migration to large cities. Companies find skilled people, and people find enough good employers. Smaller cities and towns find it difficult to retain their young and ambitious people. They have been losing young people in large droves to the larger cities for employment and excitement. Large cities thus keep on getting even bigger. The number of people living in large cities will rise from 55% now to about 80% by mid-century. Larger cities offer bigger problems though. Tokyo is a home to 36 million people, and there are many cities with populations of more than 10 million each. Large cities are congested, polluted, and expensive. The cost of urban living typically rises about 15% for every doubling of the size of the city. The contemporary urban challenge is to make the quality of life similar to that of small town living while still enjoying the higher standard of living of larger cities. This is the major driver for the concept of smart cities.

There are many books for the new science of the smart cities. They agree that achieving triple bottom line of profit, people, and planet is the chief objective of smart cities (Song et al. 2017). Winkless (2016) describes how new building materials help to construct the tallest skyscrapers in Dubai, how New Yorkers use light to treat their drinking water, and how Tokyo commuters' footsteps power gates in train stations. Batty (2013) suggests that cities should be viewed not simply as places in space but as complex systems of networks and flows designed for collective action., i.e., the relations between objects that comprise the system of the city. Defining the central flows and their networks can help understand different aspects of city structure. For example, this could help optimize land-use policies, the size of cities, their internal order, and their transport routes.

These days people and cities survive based upon their ability to innovate. Cities need to continually evolve, or they can decline and die due to certain exigencies such as natural calamities or man-made policies and events. In the USA, coastal cities of Seattle and Portland have been growing, while Midwestern cities like Cleveland and St Louis have been declining. New York City saw a decline from being a manufacturing hub and reemerged as a financial and fashion capital (Jacobs 1985). New York is considered more open to immigrants and ideas. Young people pay exorbitant rents for housing just to be in New York City, with greater hopes of fulfilling career ambitions and finding desirable life partners.

Large cities speed up innovation and collaboration by connecting their smart inhabitants to one another. However, the alienation produced from large commutes and distances within the suburbs of the cities has reduced civic cooperation and participation. Also, the wealth and thus the greater availability of resources under personal control have promoted self-absorption and egotistic behavior. Thus, there is a need for a smarter city where people can feel more connected and can flourish.

Smart and Flourishing

The word smart has many meanings. Smart can mean being rational and intelligent. Smart can also mean being wise and flourishing, aligned with long-term, and supporting what truly matters.

From an intelligent, problem-solving approach to life, smartness means an empirical resource-based view of the world and connotes a sense of efficiency to accomplish goals. It means being able to do more with less, using an optimizing mindset. It means making judicious use of scarce resources and competing to acquire those resources. Smart can thus mean tech-oriented and with greater use of information and communications technologies. A trillion connected sensors and instruments can monitor and control every house, office, store, street, and the underlying infrastructure. This is often called the Internet of Things (IoT). These huge volume, velocity, and variety of data contribute to what has been called big data. Analyzing this data can help discover patterns of use and can help design better service experiences to citizens. Figure 1 shows an urban planner's imagination of a highly integrated and connected large city, about 100 years from now.

To flourish means to live a joyful, productive, prosperous, and fulfilling life. The purpose of life is the expansion of happiness, says Maharishi Mahesh Yogi (1963). It means waking up to one's true nature and also growing up to live up to one's full potential (Wilber 2000). From a flourishing perspective, smart means being wise, in tune with nature, and being self-sufficient and happy in the long term. From an urban planning perspective, it can mean creating a longer-term plan that anticipates growth, new needs, and opportunities. A smart city thus naturally evolves with the times and in tune with what makes people happy. It means being creative, productive, and playful. A flourishing life means being present every moment and open to possibilities. A flourishing city would focus on higher-level work, which is creative and fulfilling. It could also be different kinds of evolutionary work, as new artificially intelligence

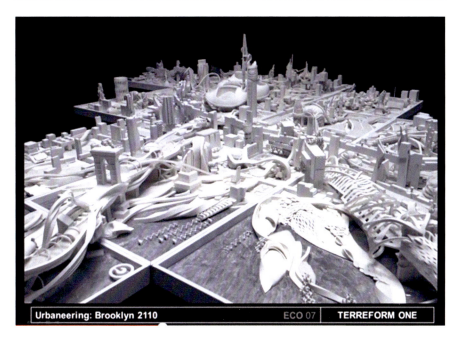

Fig. 1 A futuristic view of a modern city (Source: Terreform.org)

technologies become available in the marketplace. The intellect would be valued, but so also would be other forms of being and doing. Creative design and expression will be of the essence. Powers (2014) offers a delightful yearlong experiment of learning to live with less in New York City. Through smart work, with long weekends, he could free up time to pursue his creative passions. Permaculture is another positive approach to seek solutions for environmental and social challenges confronting us. The nature-based approach for growing local healthy nourishing food – can help meet many of our other needs including a close-knit community (Hemenway 2015).

Smart City

The concept of a smart city has largely been the creation of marketing campaigns by global ICT companies. In 2005, Cisco invented the concept of connected communities. In 2009, IBM coined the term smarter planet. The concept of a smart city has resonated with citizens, governments, and businesses alike. This moniker came in handy for cities that were devastated by the financial crisis of 2008 and wanted a new label to redefine and reenergize themselves. The emerging ICT technologies offered a new basis to reimagine and reinvent cities. ICT industry promotes ubiquitous connectivity and data-analytic approach to smart city. This view represents the goal of high efficiency in resource usage and delivery of services, thereby also maximizing profit for the service delivery organizations.

Smart city thus relies mostly on a materialist, object-oriented view of the world, where everything is a unique resource or service that has a technological and economic value associated with it. A smart city system could gather data about the demands and supply of every type of resources and services, from all providers and consumers, and then help deliver them in the most efficient way. Data would be continually and automatically gathered from instruments and sensors placed everywhere in the city, to track every object and activity. Optimizing mechanisms will then be used to deliver the needed services in a prioritized manner. A smart city thus conceived would deliver the largest quantity and range of services speedily and at low cost. The ideal citizen of such a smart city will be tech-savvy and nerdy.

Flourishing City

Not everyone agrees with the technological Utopian view of a smart city though. Anderson and Pold (2012) highlight and critique the profit motive of the ICT companies in pushing the idea of more technology, more data, and more cybernetic control. These companies have a history of working with the powers-that-be, and do not always care how the technology is used, for the good or the bad of the citizens. The narrative of smart cities has been marketed as value-neutral and promoting greater happiness, though a city has always been a contested space for ideas and resources. A homogenized and standardized view of life in such a smart city takes away some of the richness and variety of the human experience.

The ideal of a flourishing city is where one can lead a life at one's own terms, i.e., of a person feeling special, safe, creative, productive, and fulfilled. Access to nature such as woods, rivers, oceans, and mountains would be highly valued in a flourishing city. The city would incrementally grow to adapt to new demands from new citizens. The city will be judicious in the use of natural resources and services. Everyone will get essential services in terms of access to food, water, air, and information and detoxification in terms of disposal of sewer and garbage. Additional services such as energy, transportation, communication, and financial services would be needed as the city size increases. Figure 2 shows an artist's imagination of a flourishing integrated community. People will share and connect with one another and help one another in a gentle and trustworthy relationships.

The views of a smart city from the two different points of view can be summarized as follows (Table 1):

Many cities have launched projects to become incrementally smarter. They use public-private partnerships to try to get the best of both worlds – individual ownership and contribution and state support and reliability.

Smart Citizens

Smart citizens are self-aware, educated, motivated, informed, and active owners of their life. Smart citizens take their civic responsibilities seriously even as they

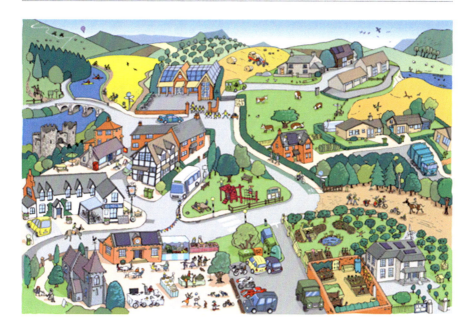

Fig. 2 A conceptual sustainable flourishing community (Source: clipartkid.com)

Table 1 Two viewpoints on a smart city

Dimension	Tech-smart	Flourish-smart
Dominant paradigm	Techno-economic	Sociocultural
Goal	Efficiency, scale, speed	Joy, fulfilling life
Modus operandi	Instruments, rational logic	Human spirit, creativity
Actions	Resources, deliver services	Safe and easy access to nature
Measures	Big data, instrumented, integrated, intelligent	Openness, freedom, spontaneity
Institutions	Corporations, governments	Communities, governments,
Smart citizen	Nerd	Activist

want to flourish in areas of their own personal interests. Smart citizens will need to make trade-offs, such as between the centrality and size of smart spaces, with the cost and time of smart transportation, and affordability. Smart cities need smart citizens to smartly guide the deployment of smart technologies! It is important to nurture the aspirations and growth of smart citizens who don't simply use the smart technology but play an active part in developing and implementing these tools for their city. One can imagine a new noncommercial organization which provides the same basic capabilities as Facebook, but with automatic shared ownership, where all have a stake. Smart people will connect and share extensively to create sharing cities (McLaren and Agyeman 2015). People willingly contribute their time, connections, knowledge, pictures, etc. to Facebook, for example, for free. The unified ocean of human consciousness is infinite and can accomplish anything. Existing technologies of consciousness can help align the spiritual forces of people around the world.

Smart Services for a Smart City

From a systems approach, a smart city can be examined as an interlinked set of services (Fig. 3). There are many services that the city provides. The list of services can be long. The most highly valued city services would be air, water, sanitation, energy, health, education, transportation, spaces, connectivity, and management. Underlying all these services is an interconnected information and communication infrastructure. These services enable smart living and flourishing for smart citizens.

These subsystems will have interdependencies, e.g., energy is critical for delivering almost all other services. Here, we will examine some of the key services. Each service can be seen from the viewpoint of its technological infrastructure needs and what it enables (Table 2).

Smart Information and Communications Technology (ICT) Infrastructure

Smart cities have been fundamentally defined by the ICT infrastructure. The entire concept was popularized by the ICT companies as a framework to sell their wares. For example, Dameri (2017) analyzes the concept and theory of the recent smart city phenomenon and highlights the role of ICT as a prime enabler for smart cities. ICT provides computational power, data storage, network connectivity, mobile access, data analytics, and visualization capability, for billions of transactions and activities happening simultaneously. For example, cloud computing services provide efficient ICT services to individuals, corporations, and governments to deploy their own specific solutions. Publicly available privately owned smart networks such as Facebook connect two billion people into one giant network and facilitate community building, commerce, and more. ICT will continue to be enhanced with related technological advances in quantum computing and nanotechnology. There will be embedded nanodevices almost everywhere including inside the human body. Artificial intelligence systems will improve the capability of systems, from regular

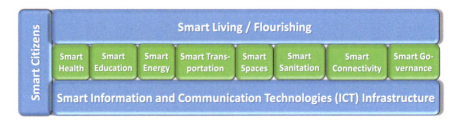

Fig. 3 A smart services framework for a smart city

Table 2 Smart services framework – infrastructure and flourishing

Smart city services	Tech-smart vision	Flourish-smart vision
Smart health	Disease-free, long life, pain-free body	Healthy, whole, joyful
Smart education	Efficient, on-demand, online, just-in-time learning	Self-awareness, broad-based, consciousness development
Smart energy	Multimodal, smart grid	Renewable sources, local
Smart transportation	Fast, efficient, integrated, multimodal, low pollution, nonpolluting, inexpensive	Bicycles, walking trails, healthy
Smart spaces	Efficient, hygienic, maximum packing	Open spaces, encourage community, aligned with nature
Smart sanitation	Efficient garbage collection and processing	Reduce, reuse, recycle
Smart connectivity	Fast, ubiquitous, secure Internet access	Mix of face-to-face and remote meetings
Smart water	Quality of treated water, monitoring consumption	Natural flowing spring water
Smart work	Intellectual work, integrated and efficient	Creative design and expression, aligned with nature
Smart governance	E-governance, surveillance	Autonomy, evolutionary

monitoring to analysis of needs, and deliver the relevant services. Intelligent voice-enabled user interfaces such as Siri by Apple and Alexa by Amazon will be integrated in all aspects of the ICT infrastructure. Robotic systems will improve efficiency and will save humans from hazards of highly inconvenient work.

Smart Health

Smart health means delivering healthcare services efficiently and reducing pain and suffering. This means optimizing the use of available skilled healthcare resources to match the needs and demands for those services. A smart health system would free up the healthcare practitioners to focus on patient care, while the administrative system will take care of the rest. Optimizing the location and staffing of clinical centers would be important. Telemedicine would be an efficient delivery method. Web-based delivery of essential medication information would be an important part. Electronic health record (EHR) systems will store all health history and make relevant data instantly securely available to health service providers. The system will be designed to open standards such as FHIR (Fast Healthcare Interoperability Resource) so any provider can access those resources in a secure manner. Flexible health insurance solutions could be customized to the needs of every person. Smart health also means a range of personalized self-monitoring devices such as a smart watch that would gather data and provide reminders and preliminary advice. Efficiently gathering data for e-clinical trials would help quickly discover or invent pharmaceutical solutions for emerging public diseases. Genetic approaches to disease identification, prevention, and curation would also be included for personalized care. Smart health would allow a combination of private and public ownership for reliability, excellence, and profit. Smart public health would include tracking using systems like Google Flu Trends system. This would enable public health initiatives to prevent diseases from happening.

Smart health would include preventive healthcare, including diet and exercise suggestions. It would be holistic, and inclusive of western and the holistic eastern medical systems, such as Ayurveda and acupuncture, to provide the right set of preventive and curative health services to every person. It would include building running and hiking trails alongside natural beauty to encourage healthy living. It would include incentives for exercise and meditation and for shedding unhealthy habits.

Smart Education

Smart education describes learning in the digital age. Education literally means "to draw out" or develop. Smart education means rethinking and reimagining the content and process of education. The idea of a classroom-based and grade-based teaching system is an outdated, industrial-era invention. Educators now need to think of how best to truly produce lifelong learners and children in the information and cognitive era where artificial intelligence and robots will do amazing range of work. Human creativity and imagination are at least as, if not more, important as rote learning. For example, Albert Einstein was a laggard and hated his schools and could not get a good job; yet he produced four remarkable research papers in 1 year at the age of 25. Formal teaching is likely to diminish in the coming years, and self-motivated, self-paced learning will take its place. A few interesting initiatives are listed below.

Stress Management and Relaxation Techniques in Education (SMART) is a program to support teachers and staff for personal renewal. It helps reconnect to personal and professional meaning and purpose and develop emotional intelligence which has been found to be a great predictor of leadership capabilities.

SMART Education is the abbreviation for self-directed, motivated, adaptive, resource-enriched, and technology-embedded education. It can be considered an educational paradigm shift for digital natives. Teachers cannot imagine what the future will be like after 20–30 years; they should simply facilitate the students inventing the future.

"Flipped classrooms" is an increasing popular educational technique. Students listen to video lectures at home and then use the classroom time to do exercises and solve problems in collaboration with teachers and fellow students.

Massively open online courses (MOOCs) are an on-demand, self-paced, and achieved at a fraction of the cost of in-person training. Most universities have already started offering MOOCs to supplement on-campus training. Many companies such as Coursera provide technological platforms for universities to deliver such education.

Teachers need to develop a new Twenty-First-Century Learner Skills Framework. They need to analyze their schools' and students' SWOT (strengths, weaknesses, opportunities, threats) and select the best future course of action. A smart pedagogical framework will include a mix of class-based instruction, group-based collaboration, individual-based generativity, and mass-based learning. A vibrant public-private-community partnership is likely to help achieve a smartly educated populace.

Finally, smart education would develop students' creativity. Teaching for creativity requires encouraging and supporting spontaneity, intrinsic motivation, and autonomy. Thus, students will work on what, when, and however they will choose to work (Runco 2014).

Smart Energy

Smart energy is nonpolluting, renewable, inexpensive, abundant, and efficient. Smart energy is distributed through a smart grid that can match energy production to energy consumption and include energy storage systems.

Energy is the backbone of the modern economy; it is central to reducing drudgery. Energy is used for agriculture, manufacturing, homes, offices, shops, transportation, data centers, and almost all aspects of life. Fossil fuels (coal, oil), hydro, and nuclear energy have been the mainstays of current energy production systems. Control of fossil fuel resources is the cause for many of the nastiest wars in the world. The shift to clean, renewable, low-cost energy production has immense geopolitical consequences.

Solar and wind energy are two of the fastest-growing renewable means of producing energy. Large solar farms of as much as 1 GW capacity are coming up fast. The cost of solar installation per unit is falling exponentially, and the cost of solar energy is already competitive with the cost of fossil fuel sources of energy production. Wind turbines are coming up in many parts of the world. China has about 15 GW worth of wind energy installed capacity, double that of the USA. A smart city can set up a policy and incentive structure to encourage the move to renewable sources of energy. It can encourage installation of solar panels, windmills, and other renewable modes of producing energy. Tesla Motors has been producing electric cars that reduce noise and pollution. Solar recharging stations for electric car batteries reduce the cost of transportation. Both residential and commercial buildings in smart cities would be more efficient in using energy, and their energy consumption data would be gathered and analyzed. A smart home can be energy-efficient through better insulation and need-based usage. Providing real-time access to energy consumption data alone can motivate consumers to reduce their consumption by about 10%.

Smart street lighting is an easy entry point for cities since LED lights save money and pay for themselves within a few years. By implementing sensor-equipped streetlights, urban centers can save money on energy efficiency, and they can also able to capture useful data on how citizens use public spaces and where more (or less) lighting is needed. Smart lighting also makes the streets more secure.

Smart Transportation

Smart transportation will be fast, efficient, safe, multimodal, pollution-free, and personalized. A smart city will support multimodal transportation, smart traffic management, and smart parking. Sensors and surveys can help collect data for demand and supply data to feed into transportation planning. Integrated transport hubs can seamlessly connect multiple modes of transportation like private cars, bus system, metro system, boat, helicopter, bike, walking, etc. Multimodal coordination and fare integration can help citizens reach their destinations faster and without hassle. Smart traffic management can reduce traffic jams and speed up traffic flow. Smart parking can allow a clear picture of where available parking spaces can be found while providing an overview of improper usage of any non-parking areas. Traffic management and adaptive light management can light up the roads and sidewalks for safe movement of people and vehicles. Smart cities would install surveillance mechanisms for monitoring of traffic flows and for providing security, especially at dark hours or locations.

Cars brought mass-produced affordable opulent housing to middle class in cities around the world. Car-based transport however imposes great environmental costs. Smart carpooling and car sharing services can link drivers and passengers in real time, thus enabling shared rides and lower costs and pollution. Electric cars can reduce pollution. Self-driving cars can reduce driver stress on the road and thus result in fewer accidents and happier people. Smart parking can save time and improve parking space utilization. Dynamic road-use charges can optimize the use of scarce road resources and reduce congestion on the roads. Smart toll systems can reduce delays and errors. Traffic lights can be coordinated to ensure smooth flow of traffic. Smart GPS-based systems can provide congestion reports along with best rerouting options. Emergency vehicle notification systems can provide information about any accidents and send help promptly.

Special bicycle lanes can improve security and comfort of travel. A bike rental system helps commuters and leisure users use the bicycle on a short-term basis for a small fee. It reduces congestion and pollution. Smart citizens can use mobile access features to quickly notify the correct department about roadway damage, injured wildlife, or illegal dumping. This would help foster a sense of collaboration and cooperation among citizens, in turn creating cleaner and safer places to live.

Smart Spaces

Smart spaces for living and working would be health- and wealth-giving; they would maximize openness and liveliness. Smart common spaces will help build vibrant communities. Smart playgrounds will bring joy and fun to life. Proper planning and design gives home and office spaces a professional and aesthetic look and feel. Smart spaces are designed using special pattern languages (Alexander 1977) in tune with natural law. Smart buildings will be functional, efficient, and yet aesthetic and encouraging community formation.

Smart healthy spaces should be designed around Vastu principles. Maharishi Vastu (maharishivastu.org) is a set of architectural and planning principles assembled by Maharishi Mahesh Yogi based on ancient Vedic texts. It infuses the living spaces with solar energy. The most important factor is the entrance, which must be either due east or due north. The slope and shape of the plot, exposure to the rising sun, location of nearby bodies of water, and the other buildings or activities in the nearby environment are also important considerations. Vastu also emphasizes the use of natural or green building materials such as wood, bricks, adobe, rammed earth, clay, stucco, and marble. Other natural fibers such as wood, paper, cotton, and wool are used in the interior. Nontoxic materials reduce the exposure to chemicals and allergens that can impact the joy of living.

A sustainable or green building design and construction is a method of wisely using resources to create high-quality, healthier, and more energy-efficient spaces for homes and offices. An environmentally friendly design is about finding that balance between high-quality construction and low environmental impact. Green building combines both materials and processes to maximize efficiency, durability, and savings. Leadership in Energy and Environmental Design (LEED) is a voluntary national certification process that helps industry experts develop high-performance, sustainable residential and commercial buildings.

Smart cities will need to balance expandability of spaces with maintaining the aesthetic character of the city. Smart cities may need permit increase in heights of buildings to generate enough smart spaces for newcomers to move into the city.

Smart Sanitation

Smart sanitation will be efficient, nonpolluting, health-giving, and cost-effective. The absence of efficient waste management can cause serious environmental problems and cost issues. Incinerating the waste can release toxic gases in the air. Leeching of garbage in landfills also releases toxic chemicals that can mix with water supply. There are two major functions of smart waste management: operational efficiency and waste reduction, i.e., reduce the amount of time and energy required to provide waste management services and reduce the amount of waste created. Recycling is a way of waste reduction. Many smart cities such as Barcelona have already deployed smart sanitation systems.

Automated bins can be located at regular street corners of the city, and they would be controlled by a centralized control room. The IoT-embedded bins would notify the levels of garbage in the dumpsters, and the municipal control room would be automatically informed after 80% filling, thus triggering a collection request. The bins would also have sensors that would sense the temperature of the bin and the presence of any harmful, poisonous, flammable articles that should be removed immediately. Self- cleaning technology would ensure that the bin is cleaned itself after being emptied every time. A smart dumpster (with sensors and software) has shown to cut waste management costs by up to 50%.

In the USA, consumers waste about 30% or 133 billion pounds of food each year! An experiment in Korea using an IoT-based smart garbage system showed that the average amount of food waste could be reduced by 33%. Food is wasted only if one cannot do something smart about it. Smart recycling carts provide an efficient solution for storing and transporting wet waste. Finished compost can be returned to local farms and community organizations that support a local, pesticide-free food culture.

For more efficient recycling, there could be smart garbage sorting systems where consumers dispose their recyclable garbage into a single recycling bin, and the smart garbage bin sorts them into appropriate categories automatically at the point of disposal. Every recyclable product can be tagged with a low-cost noninvasive RFID tag at the manufacturing stage. These tags will be noninvasive and could be printed on any surface and can hold little but enough information about the type and recyclability of each product. This will prevent human error and contamination of wastes. Education programs can bring awareness of the importance of garbage reduction and reducing their landfill footprint. For example, by reinventing their packaging system, companies can significantly reduce their cardboard waste footprint.

Smart Connectivity

Smart connectivity means fast, ubiquitous, mobile, secure connectivity over the Internet and other means. A smart planet is predicated upon the digital era. Without online access and the use of information and communications technologies, managing cities can become even more complex. Connectivity is a precondition of the smart cities and the smarter planet. A vast wireless Internet access network is being deployed in many cities, and a minimum level of access is being provided free of charge to all citizens. Data collection and analysis is a key component in predicting future demand for services. A multi-tier IoT architecture would gather and store data from all devices and then process it to generate actionable information in real time.

Smart Water

Water is essential for life. Water is considered the blue gold. In developing countries the availability of fresh water for drinking and cooking purposes is limited. It is said that by 2075, major wars will be fought over fresh water. Smart water would not be contaminated by effluents from all sources, be they nitrates from agricultural lands or excess antibiotics from human metabolism. Evaporation of water could be controlled by shielding the water reservoirs through solar panels which can double up as energy sources. Desalination plants could work from the solar energy generated at location. When the quality of water improves everywhere, there will be little need for bottled water to be brought from places as far as Fiji and the Alps.

Smart Work

There is a big debate going on about the future of work. Work is a source of income and also a source of meaning in life. The concept of a job as we know it is essentially a creation of the industrial era when people were needed to work around machines. A lot of the routine work could now be done by machines, including robots.

Traditionally, work and management has been centered around the human desires to acquire and defend resources. However, smart humanistic work will support the additional needs to bond with others and comprehend the world and their own place in it. Smart work should support the human dignity by providing opportunity for all four drives (Pirson 2017).

Smart work is that which produces the most amount of joy for the least amount of work. Smart work will be creative and fulfilling and in tune with the rapidly evolving needs of the community. Many people are experimenting with slow life, including a 2-day workday and a 5-day weekend, to live joyfully and stress-free.

Smart Governance

Smart governance implies being responsive, innovative, and efficient in serving the needs and concerns of the citizens (Goldsmith and Crawford 2014). Smart governance will include electronic governance (or e-governance) which is the application of information and communications technologies for delivering government services, exchange of information, and business transactions, between government, business, and citizens. E-governance is also predicated upon gathering of data and dissemination of information electronically. It will require modifying and transforming existing rules and processes. Many developing countries are taking advantage of mobile and Internet connectivity to completely leapfrog generations of old systems and processes and deliver responsive governance of the city. Through e-governance, government services will be made available to citizens in a convenient, efficient, and transparent manner.

Smart City Examples

Dozens of cities have done self-transformation to make themselves smarter. We examine a couple of shining examples.

Barcelona

This large European city was hit very hard by the global financial crisis of 2008. They had to do something different to get out of the funk. Barcelona clearly understood the huge potential of the Internet of Things. Starting in 2012, the city deployed responsive technologies across urban systems including public transit, parking, street lighting, and waste management. These innovations yielded significant cost savings, improved the quality of life for residents, and made the city a center for the young IoT industry. "Smart City Barcelona" initiative identified 12 areas for intervention, including transportation, water, energy, waste, and open government, and initiated 22 programs, encompassing 83 distinct projects across urban systems.

Their fiber optic network now provides 90% fiber-to-the-home coverage and serves as a backbone for integrated city systems. The fiber network serves as a direct link to the Internet for Barcelona's residents and visitors. The city draws on the fiber infrastructure to provide citywide Wi-Fi, at a maximum distance of 100 m from point to point.

The city installed smart meters that monitor and optimize energy consumption in targeted areas of the city. In waste management, households deposit waste in municipal smart bins that monitor waste levels and optimize collection routes. These sensors can be further enhanced, and plans have been developed to integrate sensing for hazardous or offensive waste material. In transportation, Barcelona has pursued a multimodal strategy, advancing the use of electric cars and bike sharing while investing heavily in improving the bus and parking systems. For drivers, Barcelona has implemented a sensor system that guides them to available parking spaces. The sensors, embedded in the asphalt, can sense whether or not a vehicle is parked in a given location. By directing drivers to open spaces, the program has reduced congestion and emissions.

Barcelona uses smart technologies to enhance the efficiency and utility of city lampposts. Lampposts had been transitioned to LED, reducing energy consumption. The lampposts sense when pedestrians are in close proximity; when the streets are empty, lights automatically dim to further conserve energy. Cumulatively, the improvements produced 30% energy savings across the urban lighting system. The lampposts are also part of the city's Wi-Fi network, providing consistent, free Internet access throughout the city. Moreover, the lampposts are equipped with sensors that collect data on air quality, relaying information to city agencies and to the public. The new digital bus stops turn waiting for buses into an interactive experience, with updates on bus location, USB charging stations, free Wi-Fi, and tools to help riders download apps to help them learn more about the city. Kiosks facilitate citizen access information and services and make requests to the government.

The city's parks remotely sense and control irrigation and water levels in public fountains. Using sensors to monitor rain and humidity, park workers can determine how much irrigation is needed in each area. A system of electro-valves is then remotely controlled to deliver necessary water across the city. The program helped the city achieve a 25% increase in water conservation.

Together, these systems constitute a "network of networks" generating data that can be used by city agencies to improve city operations and by citizens seeking to better understand their local environment. Barcelona's integrated sensor network is relayed through what is now open source and available for reuse by other governments. Through this platform, data is managed and shared with citizens and city workers. Already, these improvements have saved the city money and reduced the consumption of valuable energy and water.

Through investment in IoT for urban systems, Barcelona has achieved a wide array of benefits. The city's commitment to producing smarter urban infrastructure is changing the quality of governance and the quality of life for residents, workers, and visitors.

Amsterdam

Amsterdam is one of the smartest cities in the world (Fitzgerald 2016). One of the main challenges right now is the balance between economic growth and sustainability. Smart citizen project helps people monitor the air pollution, noise, and light intensity in different neighborhoods to let the community contribute to the city's open data program. Throughout Amsterdam there are "living labs" or communities that act as petri dishes for ideas and initiatives to be tested before scaling them across the city. Projects like free Wi-Fi and a new fiber network, personalized television and transportation services, and a coworking space allow residents to experiment and test city projects to improve healthcare, environment, and energy programs in the city.

A smart work project created an alternative workspace for people to work remotely rather than commuting into the city center on a regular basis. The facility offers telepresence technology, an inexpensive option for employers who want to enable colleagues to work remotely. An automated traffic density system that can tell residents what the traffic will be like at any point throughout the day. By using traffic data from private and public organizations, the application can automatically tell drivers the quickest route to whatever's on their mobile calendar. Residents can use such technologies to make managing their homes easier, and with the help of the city's sustainable energy programs, they are able to use more solar power to do it.

Amsterdam is one of the two European pilot sites for City-Zen, an energy-saving program that will significantly lower the amount of carbon emissions and improve the city's energy infrastructure. City-Zen stands for "city-zero carbon energy," and through projects like smart, future-proof energy grids, and retrofitting buildings to be more sustainable, it is expected that Amsterdam will save 60,000 metric tons per year in carbon dioxide (CO_2), which is the same impact as removing about 12,000 cards from the road. The Amsterdam Energy Atlas maps the energy use and potential in the city, based on real data. The city can figure out where there is a lot of potential for solar energy on roofs.

The city's vehicle-to-grid project is attempting to balance solar energy use and consumption by using the batteries of electric vehicles and boats to store energy generated during the day to be used during peak hours in the evening. Through this program, people could be entirely self-sustainable, relying on stored solar energy to power their homes and appliances. It also helps to reduce investments in the local electricity grid.

Big Data

Big data heralds a paradigm shift into a digital world. Smart devices and machines (Brynjolfsson and McAfee 2014; Zuboff 1989) have a democratizing influence on the generation and consumption of data, which fuels the generation of big data. At one level, the digital economy is all about data, about the life cycle from generation to consumption of data. Big data can be examined on two levels

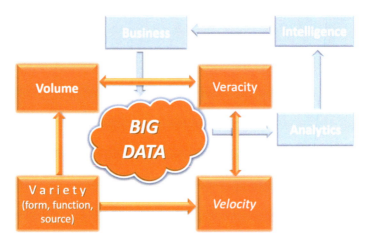

Fig. 4 Big data has special features that influence data analytics (Source: Maheshwari 2016)

Table 3 Comparing big data with conventional data (Adapted from Maheshwari 2016)

Feature	Conventional data	Big data
Metaphor	Steady and defined – tank/pool	Flowing and inclusive – stream/ocean
Benefits	Control and manage operations	Real-time monitoring and analytics, data-based products
Sources	Business transactions, documents	Social media, weblogs, sensors
Volume	Terabytes	Exabytes (million times more)
Velocity	Controlled generation	High and unpredictable data ingest
Variety	Data from business operations	Audio, video, graphs, weblogs, etc.
Veracity	Clean, trustworthy	Varies widely (e.g., fake news)
Structure	Well structured	Unstructured
Technologies	Commercial databases, SQL	Open-source (Hadoop, Spark), NoSQL
Processing	Conventional processing	Parallel processing

(Fig. 4). At a fundamental level, it is just another collection of data that can be gathered, organized, analyzed, and consumed for the benefit of the business. On another level, it is an extremely large and fast, diverse, and uncertain data that poses unique challenges and offers unique benefits. These challenges are seen often as a set of Vs or volume, velocity, variety, and veracity (Maheshwari 2016). The business benefits of big data include pervasive monitoring, real-time analytics, and data-based products and services. Big data posits now as an asset on an organization's balance sheet.

Big data and conventional data differ in the sources and types of data, the process of storage and processing, and the resulting applications and benefits (Table 3).

Chicago's Bike Rental Data Analysis

A few years ago, Chicago used a public-private partnership to set up a bike rental system for commuters and for tourists. This organization released millions of records of their rental data to the public to crowdsource the analysis and visualization of this data, for new ideas and insights. The analysis below is based on almost a million rental records and is a small example of the kind of insights that can be derived from big data.

Figure 5 shows the analysis of bike trips by time: by day of week and the hour of renting. The chart shows interesting patterns. There are definitely weekday patterns, when rentals peak at commute hours in the morning and the evening. There are also different weekend rental patterns, when rentals peak at the middle of the day.

Figure 6 continues the time analysis, but instead of the day of the week, the month is included along with the hour of rental. There is again a clear pattern. More rentals happen in summer months when the weather is good and sunny. The rentals continue to peak on commute hours in the morning and the evening.

Figure 7 shows bike trips by the day of the week and renter type. There are two types of renters: the subscribers have an account, while customers pay for each rental. The chart shows that there are contrasting trends between subscriber and customer. The subscribers rent most on weekdays and less on weekends. The occasional customers rent most on weekends and less often on weekdays.

These are among the many kinds of analyses that can ultimately lead to actionable insights. Table 4 shows that one could discern a clear segmentation of bike renters and their renting pattern. Regular subscribers were younger people who rented for short duration, mostly at peak hours, and from near metro train stations. The casual customers were older, who rented for longer durations, mostly on the weekends. Different fare plans can be designed for subscribers and customers.

Smart cities can become single points of control in the hands of malicious elements. Cybercriminals can shut down a city's infrastructure or seriously impair it. Who should own the data? How can the data be protected? Who can mine the data? What kind of patterns will be discovered? How will that knowledge be shared and used? Who will benefit from that knowledge?

Dangers of Big Data in Smart Cities

Data sabotage is a serious security risk for smart cities (Townsend 2014). For example, what would happen if a metropolis running on instrumented, integrated technology suddenly finds itself under the control of cybercriminals? If the criminals chose to suddenly make every light in a city green at once, there will be a barrage of accidents that would clog up roads. They could also clog up network bandwidth so that other more critical systems could be left unguarded.

Corporations lust after the big data to discover patterns and grow themselves. There is a fear of concentration of power, of Big Brother watching. What kind of foolproof antidotes can be designed for it? All these are very important questions that

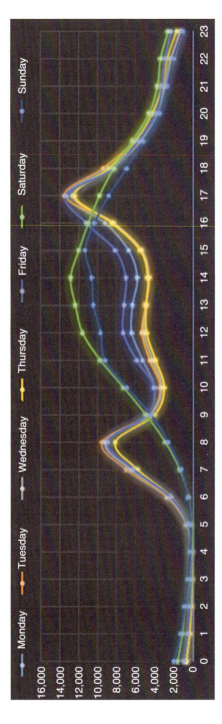

Fig. 5 Analysis of bike usage by day and hour

Application of Big Data to Smart Cities for a Sustainable Future 965

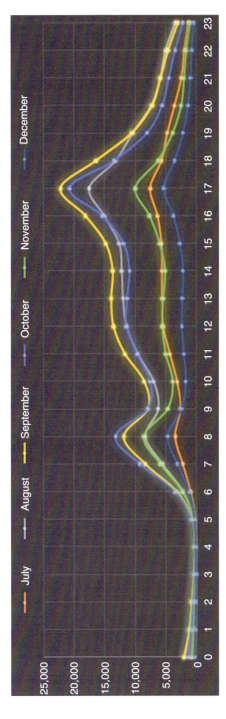

Fig. 6 Analysis of bike usage by month and hour

Fig. 7 Bike trips by subscribers and casual customers

Table 4 Data analysis-based profile of subscribers and casual customers

Feature	Subscribers	Customers
Age	Younger	Older
Day of week	Weekdays	Weekends
Time of day	Peak-hours, morning & evening	Middle of the day
Length of rental	Short (modal rental is 10 min)	Long duration
Total number of trips	High	Fewer
Location	Near train stations	Wider variety of stations

can have life-and-death consequences for cities and their citizens. Citizens can be manipulated politically, socially, and economically based on an analysis of their data. Politicians can also learn a great deal about their supporters and mobilize them to their advantage. They can also exact vendetta from their perceived enemies using the data.

There are health implications for citizens as connected smart devices such as smart meters for electricity and water emit strong radio signals. These emissions have been known to cause sleep and other health disorders. These devices are software operated and their capabilities can be enhanced without notification to the citizens. These devices can be used to collect personal data about residents of houses. These devices also may break down more often and the costs will be added to the bills of the citizens.

For example, Chicago is a relatively newcomer in the so-called smart city game. The city has begun installing sensor boxes on municipal light poles to monitor air quality, rainwater pools, air temperatures. However, the city's nascent efforts to collect environmental data are sparking concerns about further erosion of individual privacy in a city already outfitted with police cameras, red light cameras, in-store cameras, and public transit cameras. Many wonder if the collection and analysis of data will lead to meaningful improvements to urban life or just enrich big ICT vendors. It is also a challenge to promote these smart solutions when the city is struggling with funding basic needs such as schools and with creating trust between the police and the citizens.

Implications and Conclusion

Smart cities are a planned way of building intelligence into the infrastructure and services that the city provides to its citizens. These services attract skilled and ambitious people who collaborate to innovate and create prosperity. Cities will differentiate themselves on the basis of quality of services provided to the citizens. The cities however have downsides in terms of their ecological footprint and social fragmentation and anomie. There is also a danger of smart cities being hijacked by cybercriminals and of the citizenry being manipulated by unscrupulous corporations or governments.

Sleepy industrial towns will need to provide smart work opportunities to attract and retain smart work forces. Similarly, smaller towns in rural areas also need to smarten up and be able to set themselves apart on some account or else face decline. In particular, smart cities will need to become places for smart work, so that creative talent may assemble and stay there.

Citizens will need to continually smarten up and use smart education tools to continually improve their skills. They would need to become aware of their consumption patterns. They also need to move between cities based on the match between the evolution of cities and their own evolution of interests and skills. People may need to move across cities more often than in the past.

Cross-References

▶ Smart Cities
▶ The LOHAS Lifestyle and Marketplace Behavior
▶ The Theology of Sustainability Practice

References

Alexander, C. (1977). *A pattern language: Towns, buildings, construction*. New York: Oxford University Press.
Anderson, C. U., & Pold, S. B. (2012). Occupation of the open city. *ACM Media Architecture Biennial Conference*, 1–4. ACM.
Batty, M. (2013). *The new science of cities*. Cambridge, MA: MIT Press.

Brynjolfsson, E., & McAfee, A. (2014). *The second machine age: Work, progress, and prosperity in a time of brilliant technologies*. WW Norton & Company.

Dameri, R. P. (2017). Using ICT in smart city. In Smart City Implementation (pp. 45–65). Springer, Cham.

Fitzgerald, M. (2016). *Data-driven city management: A close look at Amsterdam's smart city initiative*. MIT Sloan Management Review, 57(4).

Glaeser, E. L. (2012). *Triumph of the city: How our greatest invention makes us richer, smarter, greener, healthier, and happier*. Penguin.

Goldsmith, S., & Crawford, S. (2014). *The responsive city: Engaging communities through data-smart governance*. San Francisco: Jossey-Bass.

Hemenway, T. (2015). *The permaculture city: Regenerative design for urban, suburban, and town resilience*. White River Junction: Chelsea Green Publishing.

http://datasmart.ash.harvard.edu/news/article/how-smart-city-barcelona-brought-the-internet-of-things-to-life-789.

Jacobs, J. (1985). *Cities and the wealth of nations: Principles of economic life*. Vintage.

Maheshwari, A. (2016). *Big data made accessible*. Kindle Books. ASIN: B01HPFZRBY.

McLaren, D., & Agyeman, J. (2015). *Sharing cities: A case for truly smart and sustainable cities*. Cambridge, MA: MIT Press.

Peris-Ortiz, M., Bennett, D. R., & Yábar, D. P.-B. (2016). *Sustainable smart cities: Creating spaces for technological, social and business development*. Cham: Springer.

Pirson, M. (2017). *Humanistic management: Protecting dignity and promoting well-being*. Cambridge: Cambridge University Press.

Powers, W. (2014). *New slow city: Living simply in the world's fastest city*. Novato: New World Press.

Runco, M. A. (2014). *Creativity: Theories and themes: Research, development, and practice*. San Diego: Elsevier.

Scientific American Editors (2014). Designing the urban future: Smart cities. Kindle.

Singh, S. (2014). Smart cities – A $1.5 trillion market opportunity. https://www.forbes.com/sites/sarwantsingh/2014/06/19/smart-cities-a-1-5-trillion-market-opportunity/.

Song, H., Srinivasan, R., Jeschke, S., & Sookoor, T. (2017). *Smart cities: foundations, principles and applications*. Hoboken: Wiley.

Townsend, A. M. (2014). *Smart cities: Big data, civic hackers, and the quest for a new utopia*. New York: W.W.Norton.

Wilber, K. (2000). *Integral psychology*. Boston: Shambala.

Winkless, L. (2016). *Science and the city: The mechanics behind the metropolis*. London: Bloomsbury Publishing.

Yogi, M. M. (1963). *Science of being and the art of living*. New York: Penguin.

People, Planet, and Profit

Training Sustainable Entrepreneurs at the University Level

Dolors Gil-Doménech and Jasmina Berbegal-Mirabent

Contents

Introduction	970
Sustainability in Higher Education	971
Why Is It Important to Teach Sustainability at Universities?	971
Methods for Teaching Sustainability	974
Sustainable Enterpreneurship	977
Case Study	979
Academic Programs on Sustainable Entrepreneurship	979
Service Management	980
Conclusions	984
Reflection Questions	986
Exercises in Practice	986
Engaged Sustainability Lessons	987
Chapter-End Reflection Questions	987
Cross-References	988
References	988

Abstract

The dynamic and highly volatile environment that characterizes today's society demands professionals who are capable of engaging in the process of continuous innovation, adaptation, and learning. It is no longer enough to create and place new products and services in the marketplace; it is also necessary to recognize and relentlessly pursue new opportunities that create and sustain social value. Within this context, sustainable entrepreneurs are upon called to play a major role by identifying unique opportunities that help society resurge from the economic downturn and promote a prolific, inclusive, and economically sustainable

---

D. Gil-Doménech (✉) · J. Berbegal-Mirabent
Department of Economy and Business Organization, Universitat Internacional de Catalunya, Barcelona, Spain
e-mail: mdgil@uic.es; jberbegal@uic.es

© Springer International Publishing AG, part of Springer Nature 2018
S. Dhiman, J. Marques (eds.), *Handbook of Engaged Sustainability*,
https://doi.org/10.1007/978-3-319-71312-0_38

development. A correct alignment among environmental, social, and economic issues is therefore paramount. Increasingly these values are permeating many domains of society, including the academic sphere. As a result, universities are now looking for how best to instill these values in their students by generating sustainability awareness and boosting the development of sustainable skills. This chapter reflects on the key role that sustainability plays in entrepreneurial education and reviews different ways for fostering these skills. For illustrative purposes, the chapter includes a case study which presents an example on how to train nonbusiness-related master's students to become sustainable entrepreneurs.

Keywords
Sustainable skills · Entrepreneurship · Sustainable entrepreneurship · Higher education · Teaching sustainability

Introduction

Nowadays, society demands professionals who are capable of continuously innovating, adapting, and learning. The increasingly globalized and industrialized business environment requires flexible business models that allow companies to readapt and update on a regular basis. As a result, professionals cannot limit their contribution to the design and launching of new products or services but have to do so in a sustainable way.

Within this framework, the role of sustainable entrepreneurs is paramount. By designing and developing new products and projects in which environmental and societal objectives are fully incorporated into feasible and profitable business models, entrepreneurs can promote and pursue sustainability. By aligning social (people), environmental (planet), and economic (profit) aspects, sustainable entrepreneurs can improve, promote, and pursue more balanced initiatives that bring value to the whole society.

Universities play a key role in this process. They are key actors in transmitting the importance of sustainability and educating entrepreneurs-to-be in such values. Higher education institutions are thus expected to instill sustainable principles and attitudes in their students, while promoting research projects in the field of sustainable business development.

This book chapter elaborates on the benefits of teaching students the fundamentals of both entrepreneurship and sustainability in a combined formula. To do this, we first review the role that sustainability plays in higher education and question whether current existing programs place enough importance on generating sustainability awareness among students. Next, the chapter focuses on active learning methods that have been proved to help students develop sustainable skills. The section that follows reviews the specific literature on sustainable entrepreneurship. Several examples from around the world are used to illustrate how universities are adapting their traditional entrepreneurship programs by adding a social component. Finally, the case of the Service Management course is presented as an example of a subject that trains nonbusiness-related master's students to become sustainable entrepreneurs. The chapter ends with some reflection questions and practical exercises.

Sustainability in Higher Education

Why Is It Important to Teach Sustainability at Universities?

In 2004, the United Nations Educational, Scientific and Cultural Organizations (UNESCO) signaled universities as one of the key drivers to promote sustainability. As a result, higher education institutions are expected to be "*places of research and learning for sustainable development*" and act as leaders "*by practicing what they teach through sustainable purchasing, investments and facilities that are integrated with teaching and learning*" (UNESCO 2004, p. 22). In the manifest, UNESCO highlighted the need for the development of soft (transversal) skills linked to sustainability, such as problem-solving skills or critical thinking. Some years later, in 2009, UNESCO re-emphasized the central role of universities in pursuing sustainability, encouraging them as part of their mission to promote interdisciplinary and cooperative education and research programs related to sustainable development (UNESCO 2009).

By its nature, sustainability cannot be limited to a skill fostered in the classroom; instead, it needs to be conceived and practiced inside and outside it. Thereby, sustainability education has meant a fundamental change in the traditional operating of universities. They are now required to rethink and adapt not only the curricula and pedagogies used but also their policies and institutional structures (Corcoran 2010).

As mentioned before, sustainability is a transversal skill that can and should be promoted by universities. At the same time, it encompasses the acquisition of many other skills such as critical thinking, problem- and project-solving, personal responsibility, and the capacity of having a broader view of a given problem. When education is conceived in its holistic view (soft + hard skills), it provides students a lifelong form of learning, that is, an invaluable asset that they acquire through practice and formative assessment and that they will be able to apply in all kind of disciplines throughout their entire lives (Kember 2009; Star and Hammer 2008).

In order to test how much students know about sustainability, a survey was designed and administered to students enrolled in the bachelor's degree program in Business at the Universitat Internacional de Catalunya (Barcelona, Spain). Responses were collected during May 2017. Responses from 112 students were collected, 65 (58%) of them male, with an average age of 19.6 years old. The survey combined different types of questions, including items in the form of statements to which respondents had to indicate their level of agreement/disagreement on a 5-point Likert scale ranging from (1) completely disagree to (5) completely agree, short questions to which respondents should indicate agreement or disagreement (yes/no), and lastly a set of open questions that invited students to self-reflect about their sustainability-oriented habits.

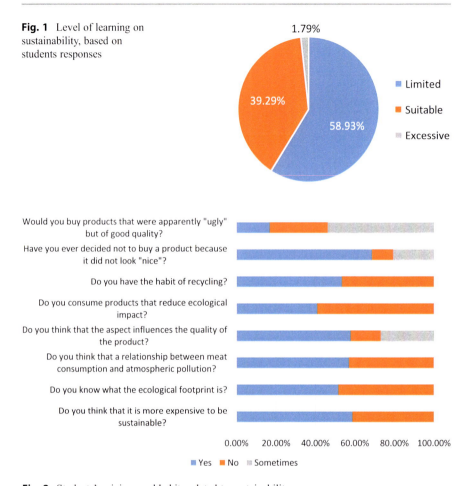

Fig. 1 Level of learning on sustainability, based on students responses

Fig. 2 Students' opinions and habits related to sustainability

After analyzing the results, one of the key findings is that almost 60% of the students interviewed said that the level of learning on sustainability was limited (see Fig. 1). Only a small percentage of students (1.79%) indicated that they had received excessive training on this topic.

The survey also included questions where respondents should answer with yes/no/sometimes. Figure 2 shows students' responses with respect to a set of habits related to different practical aspects of sustainability. As is shown in Fig. 2, the first four questions refer to students' sustainability behaviors, while the last four ask about specific knowledge on different aspects concerning sustainability. Based on the responses, it can be observed that the majority of students converge in saying that their consumption decisions are highly tied to the physical appearance of the products to be purchased. As for recycling habits, students tend to separate the rubbish; however, they do not consume products with reduced ecologic impact. As for the second group of questions, a significant proportion of students believe that

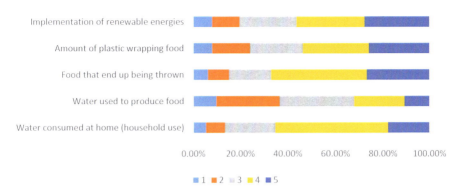

Fig. 3 Level of concern about different sustainability topics

how the product looks has an effect on its quality. This thought is consistent with their consumption decisions. Moreover, a great proportion of students indicate that some habits of consumption (such as eating meat) have a direct effect on pollution, and many of them are familiar with the concept of ecological footprint. Finally, it is remarkable that, in general, students consider being more sustainable and more expensive. This way of thinking might have a negative and direct effect on their sustainability-related decisions.

The surveys also gathered information on critical issues related to sustainable practices. In this respect, students were asked to indicate on a scale from 1 (the lowest) to 5 (the highest) how important were a set of topics. Responses are reported in Fig. 3. According to the results, hot topics on sustainability (with a score equal to 4 or superior) include food waste, followed by the water consumed in households. In contrast, the topic that scored the lower mark is the quantity of water used to produce food.

Finally, with open questions students were given the opportunity to indicate their sustainability-oriented habits. Their responses were later classified into six different categories: water consumption, energy consumption, recycling, means of transportation, food consumption, re-use of goods, and care of the environment. Results are presented in Figs. 4 and 5. While the former shows the habits already adopted by students, the latter shows the ones students said they still have to acquire.

According to Fig. 4, students' most common routines are those aimed at reducing water and energy consumption. Some examples are turning off the water while brushing their teeth, taking shorter showers, and turning off the lights when they are not strictly necessary. Two of these practices – recycling and sustainable water consumption – are, at the same time, the ones students would like to acquire, followed by reducing food waste, mainly by making rational food purchases and buying local products. On the other hand, few students consider re-utilizing goods, or taking care of the environment by avoiding the use of sprays or plastic bags. The sustainable habit of re-using goods is also the one with the lowest likelihood of being acquired in the short term.

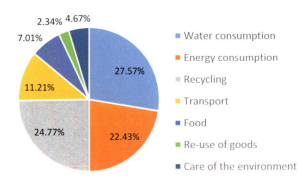

Fig. 4 Sustainable habits adopted by students

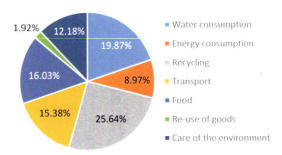

Fig. 5 Sustainable habits students want to adopt

Taken altogether, these results indicate that sustainability is a concept present in students' daily lives, meaning that students are aware of sustainability practices and acknowledge the importance of being sustainably responsible. Notwithstanding, it is worth recalling that a significant proportion of students consider that their sustainability education was limited. This result reinforces the need for integrating sustainability in higher education.

Methods for Teaching Sustainability

Most of the methods that can be used to teach sustainability are based on active learning, which has been demonstrated to facilitate the acquisition of skills (Prince 2004). Active learning consists of placing students at the center of their learning process while they participate in activities and reflect on what they are doing and how (Bonwell and Eison 1991). This way, students become more independent and responsible, and improve their performance not only while doing the activity but also after completing it (Bell and Kozlowski 2008).

Active learning methods imply changing the traditional role of lecturers: instead of acting as "expert authorities" they act as facilitators or guides of the educational process (Phillips 2004). When applied correctly, these teaching methods motivate learners while simultaneously providing them with a deeper knowledge and positive attitudes (Michael 2006).

Given that sustainability is a soft skill that should be promoted in the educational process, active learning methods are suitable for this purpose. In fact, the majority of methods used for teaching sustainability are based on active learning. The following pages summarize some of the most commonly used active learning methods that have been found to be sound in boosting sustainable skills among students.

Problem-Based Learning

Problem-based learning is a "*collaborative and participatory student centered approach to teaching and learning, based on work and problem exploration*" (Bessant et al. 2013, p. 2). When learning through problems, students are collectively involved in a shared process of acquiring new knowledge and skills. This teaching method requires an inductive and contextual approach to the hypotheses raised in a complex and structured problem with prompts from lecturers. When solving the problem, students construct knowledge by building and testing a greater understanding of the learning outcomes pursued (Wiek et al. 2014).

Problem-based learning in the context of sustainability education has been used as a method for exploring authentic problems related to sustainability. Students develop a deeper understanding of the nature of the problem proposed and acquire transferrable sustainability competences (Bessant et al. 2013). By enhancing skills for sustainable development using real-life problems within this framework, higher education institutions are able to become student-centered learning environments that encourage sustainability (Cörvers et al. 2016).

Project-Based Learning

According to Friedman (2000), project-based learning is one of the most common and relevant active learning methods. It can be defined as a "*comprehensive perspective designed to engage students in investigation to solve real problems by using multiple cognitive instruments and various sources of information, while working in a social context*" (Berbegal-Mirabent et al. 2017). The main aim is to make students responsible for their own learning while they apply competences and knowledge to real projects. In this context, the facilitator offers students the organized structure of contents and requirements of the project to solve. They do so by guiding students to the acquisition of the necessary skills and accompanying them through the autonomous educational process (Doppelt 2003).

Given that project-based learning encourages students to find solutions for projects in a real situation, it can serve as a link between classroom experiences and real-life events (Blumenfeld et al. 1991). For a proper sequencing of the project, students have to organize themselves from the start, gather and manage information, communicate and discuss their ideas and findings, follow the instructions and accomplish the assigned tasks, and collaborate with others (Laffey et al. 1998). This way, project-based learning establishes a model of authentic instruction in which students plan, implement, and evaluate projects with applications beyond the classroom walls, that is, in the real world. While developing the project students also become more responsible for their learning and increase their sense of pride (Barron et al. 1998; Kubiatko and Vaculová 2011).

Problem-based learning is particularly suitable for teaching sustainability. This method can be used by lecturers who want to integrate real situations in the classroom. Under this approach, students typically work in groups and have to collaborate with each other in order to create a project or plan an event.

Living Labs

Living or campus labs consist of using the university infrastructures as an experimental laboratory where different stakeholders are able to develop and test new technologies (Evans et al. 2015). Living labs combine the use of both academic and campus facilities to supply students with transversal skills.

Living labs are becoming increasingly popular to promote sustainability in the higher education context (Evans et al. 2015). In fact, universities' campuses offer real locations which are ideal for performing practical research. Some examples of the application of such strategies can be found in the use of solar panels or wind turbines at universities. Thus, while allowing institutions to reduce their carbon footprint, these green practices also provide students with real-world, hands-on learning experiences (Cohen and Lovell 2011). This way, campus labs arise as an initiative that enables institutions to meet their sustainability objectives and, at the same time, encourage students to adopt the attitudes and skills to think critically about different and relevant challenges related to sustainability.

Echoing Evans et al. (2015, p. 3), living labs "*provide a framework for students and academics to engage with the opportunities to work with Estates staff and their environmental consultants on applied sustainability challenges.*" However, it is important to note that proper implementation of such labs requires close collaboration among students, academics, and technical staff. It also encompasses a redesign of the role of universities as promoters of experimental courses for the generation of new ideas towards sustainability (Vezzoli and Penin 2006). Universities should thus take into account all these considerations – boosting cooperation and offering their facilities – when designing and implementing campus labs.

Transdisciplinary Activities

Transdisciplinary teaching involves academics, students, and practitioners who collaborate to solve real-world problems together. In doing so, participants obtain a superior benefit from the one they could have obtained if working independently (Scholz et al. 2006).

Following Lang et al. (2012, p. 27) and Vilsmaier and Lang (2015), transdisciplinarity can be understood as a type of research that encompasses the following characteristics: (i) it focuses on important real-world problems, (ii) it empowers mutual learning among people participating from inside and outside academia, and (iii) it aims at generating new knowledge: "*solution-oriented, socially robust, and transferable to both the scientific and societal practice.*"

Sustainability is a multidimensional, transdisciplinary concept, as a result of the three pillars on which it is based: environment, society, and economy. These three pillars, informally referred to as the "triple P" (3P), stand for people, planet, and profit (Epstein and Buhovac 2014). For this reason, in order to teach sustainability, it is highly recommended to use transdisciplinary techniques, which integrate different perspectives of a specific problem. Additionally, applying transdisciplinary teaching in sustainability programs through an active imparting of practice-based knowledge has been shown to have a positive impact on learners' motivations (Merck and Beermann 2015). When the teaching is addressed in this sense, students are able to create, integrate, and manage sustainability knowledge in their day-to-day decisions (Scholz et al. 2006).

When designing transdisciplinary activities with sustainability teaching purposes, it is thus essential to create a multidisciplinary team, bringing specialists from diverse disciplines. This diversity in backgrounds can contribute to the creation of activities that integrate all of the fundamental principles of sustainability.

Sustainable Entrepreneurship

Companies are usually perceived as responsible for causing the majority of ecological and social problems. However, depending on the way they carry out their production, while it is true that they can work against sustainability, it is also true that they can do just the opposite. By designing and developing new projects in which environmental and societal objectives are fully incorporated into feasible and lucrative business models, businesses can promote and pursue sustainability (Lans et al. 2014). Actually, entrepreneurship is increasingly being considered as an essential channel for carrying out a transformation toward sustainable processes and products (Hall et al. 2010).

Actors integrating environmental, economic, and social development in their core businesses are called "sustainable entrepreneurs" (Schaltegger and Wagner 2011). Sustainable entrepreneurs are thus said to combine the "best of both worlds": by promoting new and profitable processes and products (entrepreneurship), they foster social and ecological progress (sustainability) (Lans et al. 2014). Thereby, they "*generate new products, services, techniques and organizational modes that substantially reduce environmental impacts and increase the quality of life*" (Schaltegger and Wagner 2011, p. 223). Following this rationale, Kuckertz and Wagner (2010) assert that "*sustainable entrepreneurs manage to the 'triple bottom line' by balancing economic health, social equity and environmental resilience through their entrepreneurial behavior*" (p. 524).

The relevance of sustainable entrepreneurship has been highlighted by several authors in recent years (Hall et al. 2010; Kuckertz and Wagner 2010; Lans et al. 2014; Lourenço et al. 2012; Schaltegger and Wagner 2011). Given this remarkable importance, the role that universities play in teaching the necessary skills for developing sustainable entrepreneurship is paramount. By doing so, universities promote the acquisition of abilities that in the future will help students make business decisions that result in a socially, economically, and environmentally better world. As Lourenço et al. (2012) argue, *"entrepreneurship education for sustainable development is a pluralistic approach that can bridge the two paradigms: doing well (instrumental stakeholder perspective) while doing good (normative stakeholder perspective)"* (Lourenço et al. 2012, p. 858). The two paradigms to which the authors refer are the instrumental perspective, which maintains that both social and environmental viewpoints should be integrated in the business strategies of all companies in order to generate mutual benefits, and the normative perspective, which affirms that companies do not have the right to affect social well-being negatively through externalities (Donaldson and Preston 1995; Jones 1995; Lourenço et al. 2012).

In the framework of sustainable entrepreneurship education, there has been a bulk of recent studies determining the necessary skills that higher education should foster for this purpose. In a similar vein, some other works report experiences on different practices for teaching sustainable entrepreneurship at the university level. Along these lines, Schaltegger and Wagner (2011), pioneers in coining the term "sustainable entrepreneurship," analyzed the links between sustainability entrepreneurship and innovation and specified the conditions necessary to companies in order to move towards sustainability. In their study, these authors recommended that firms be ready to adapt rapidly to market changes and also to establish partnerships with different stakeholders in order to pursue a sustainable innovation. Similarly, Lans et al. (2014) focused on the figure of the entrepreneur, specifying the skills, knowledge, and attitudes that are indispensable for integrating entrepreneurship and sustainability. Basing their study on focus group discussions with professors and structured questionnaires addressed to higher education students, these authors concluded that enhancing systems thinking and normative and interpersonal competences, as well as integrating diversity and interdisciplinarity and promoting foresighted thinking, are key aspects for imparting a satisfactory sustainable entrepreneurship education.

With respect to teaching practices, Bonnet et al. (2006) proposed an activity that combined entrepreneurship, sustainability, and project education in an undergraduate engineering course titled Sustainable Entrepreneurship and Technology at Delft University of Technology (The Netherlands). Specifically, they examined how to integrate the triple P (People + Planet + Profit) in a business plan which students had to develop. Their results suggest that the activity contributed, on the one hand, to fostering skills such as entrepreneurship, presentation, problem-based learning, teamwork, and cooperation skills, and, on the other hand, it helped students to be aware of their capability for being entrepreneurs and doing so in a sustainability context. Lourenço et al. (2012) also focused their attention on this area by analyzing the extent to which sustainability education enlightens the principles of emerging entrepreneurs. As a contribution of their study, these authors identified internal factors with potential to operate as barriers (profit-first mentality) or drivers (perception of benefit, learning, ease of use) which influence nascent entrepreneurs to exploit their learning associated with sustainable entrepreneurship.

Case Study

Academic Programs on Sustainable Entrepreneurship

The scarcity of natural resources, the exponential human population growth, and the current inequalities demand people able to apply business principles to solving social, environmental, and economic problems. Sustainable or eco- entrepreneurs are those entrepreneurs who aim at tackling these issues. They do so by proactively introducing environmentally and socially friendly innovations to a large group of stakeholders (Dean and McMullen 2007).

Such a profile is in great demand. However, there is a lack of studies examining how these two disciplines – entrepreneurship and sustainability – can reinforce each other. Traditionally, scholars have focused either on entrepreneurship education or on education for sustainability. Notwithstanding, we argue that higher education institutions should adopt a more holistic role, laying the foundation for sustainable entrepreneurs.

Although the number of programs exploring the borderland between these two disciplines is significantly low, it is possible to identify some academic programs in which students enrolled will be able to explore areas of opportunity in green businesses. In March 2011, the electronic magazine *Entrepreneur* published a list of some of the top business schools for eco-entrepreneurs (Daley 2011). It was headed by Babson College's F.W. Olin Graduate School of Business, one of the leading centers in the field of entrepreneurship that has now enlarged its academic portfolio, offering courses and certificate programs that bring together the fields of innovation, engineering, and sustainability. This idea was not new. In 2008, the McGill University Desautels Faculty of Management launched a new curriculum, integrating social, environmental, and ethical issues into business education.

Universities from all over the world have followed these steps and, as of today, it is possible to find different programs that prepare individuals to create profitable and sustainable business opportunities in a world undergoing transformational change. To reach this goal, students can participate in research projects and help local companies to develop sustainability plans. The Graduate School of Business at Stanford University, the University of Wisconsin-Madison School of Business, New York University, the University of Vermont, the Centre for Enterprise and Entrepreneurship Studies at University of Leeds, the McCallum Graduate School of Business at Bentley University, and the International Business School at Brandeis University exemplify the emergence of graduate schools that are increasingly adapting their traditional MBA programs in order to teach the next generation of leaders who are expected to think, reinvent, disrupt, and build sustainable enterprises.

Other initiatives include Doing Good Doing Well (DGDW), the largest student-run conference in Europe that takes place each year at IESE Business School; the annual Duke Conference on Sustainable Business and Social Impact, held at the Fuqua School of Business (Duke University), which attracts 350 participants each year; and the 11-day program "Doing Business in a Culture of Sustainability" in

Costa Rica, promoted by the Drucker School of Management at Claremont Graduate University. It is worth mentioning that in 2009, a team of students from this university won the Net Impact Case Competition (NICC), the premier MBA case competition focused on solving real-world social and environmental business challenges.

Going a step further, some universities have created support infrastructures such as business incubators to accelerate socially responsible businesses. Examples include the Social Innovation Incubator located in the Center for Global Leadership in Sustainability at Portland State University and the Kenan-Flagler's Business Accelerator for Sustainable Entrepreneurship at the University of North Carolina at Chapel Hill. This latter university hosts a Sustainable Venture Capital Investment Competition, in which students present sustainable business plans in an attempt to win seed money.

From the examples underlined above, it can be inferred that entrepreneurship and sustainability cannot be set off to one side, but rather need to be fully integrated in the curriculum. Businesses are constantly changing the way they think about their operations, business models, and delivery of products and services. What is more, they are expected to do so while ensuring that they respect the environment, contribute to social welfare, and obtain financial benefits.

Individuals can address social and environmental problems by means of creating and developing new businesses and innovations. In this respect, educational programs play an important role, as they should infuse students with the right skills and confidence to develop and launch their own ideas and plans for new and sustainable ventures.

Service Management

This section describes the experience of a course on Service Management taught at Universitat Politècnica de Catalunya (UPC Barcelona Tech) in Spain. This course is part of the master's program in Innovation and Research in Informatics, and its main aim is to provide students a complete and comprehensive picture of the management of service industries, offering them an integrated view. As economies all over the world are increasingly becoming more service-oriented, there is a need to go further in the study of all aspects related to the management of service industries.

One of the main outcomes of the course is the final project. To achieve this goal, the course adopts an entrepreneurial approach. In groups of four, students are assigned a project-based activity. Specifically, they are required to design a new service business. Throughout the course, students gain theoretical and practical insights on how to manage this particular type of firm and the specific features they need to consider.

Compared to other entrepreneurship courses, the value added of this course is that students should come up with an idea that, besides having the potential for being profitable, will be respectful of the environment and/or contribute to reducing social inequalities. As there are few courses exploring and/or crossing the boundaries between these two disciplines (entrepreneurship and sustainability), the ultimate goal of this course is threefold: (i) to foster sustainable entrepreneurship, (ii) to influence students' behavior by dealing with the challenge of sustainability, and (iii) to prepare students to thrive in and contribute to sustainable development.

The theoretical foundations are imparted in the form of explanatory classes. These lectures are complemented by reading specific articles, class exercises, and seminars. Students are expected to actively participate. In order to make sure that students possess both the required knowledge and skills for conducting the final project (business plan), two prior assignments are included in the course program. The first one is called "Innovation Awareness & Opportunity Recognition." For this activity, students have to subscribe to a blog, newsletter, or similar source from which to obtain information about new start-ups. Among the new businesses, they have to select two start-ups operating in the service sector that are at an early stage of development. One of them should address a social need. For each idea, they have to prepare a 10-min presentation covering these following points: the business idea and need being addressed, the developer, and the target audience. Presentations should end with a critical assessment. During the first month of the course, students present their work at the end of each session. The underlying rationale behind this activity is to train students to adopt a critical perspective when evaluating business ideas.

Once this step is reached, students can move on and assess how business models evolve over time. In this respect, they are asked to compare the initial business model of a start-up (also operating in the service sector) with the current one. To do this they should make use of the business model canvas. What students learn from this activity is that firms, if they want to survive, need to innovate in response to changing customer demands and lifestyles.

As for the final project, the lean start-up method is used. This method is particularly convenient as it favors experimentation and iterative design (Blank 2013). Students learn how to develop and test ideas rapidly. Specifically, hypotheses are summarized in a business model canvas. By getting "out of the building," students test their hypotheses with prospective customers and validate/reject their assumptions (Trimi and Berbegal-Mirabent 2012). This circle is performed in an iterative basis until no more adjustments are necessary.

The following pages deal with some of the business ideas students came up with during the first semester of academic year 2016–2017. For this year, the focus was on companies operating in the educational sector.

Mediafy

The debate between textbooks vs. computers has been going on for a long time. Increasingly, technology is invading the classrooms, providing alternative learning tools for students. Despite many generations of students' having used textbooks to learn, textbooks present some pitfalls that e-books can overcome. First, printed books can sometimes have outdated information if the latest edition is not available. Second, additional material might be useful to complement what is written. Third, some textbooks cannot be reutilized (e.g., by younger brothers or sisters) because blank spaces for responses have already been filled in. Fourth, approximately

30 million trees per year are used to make books sold in the United States. A large proportion of these trees are sourced from endangered forests with devastating impacts on the people and wildlife that rely on them. Therefore, e-books' environmental impact is lower.

In this context Mediafy emerges as a disruptive start-up in the publishing sector. Nowadays digital products and services are more important than they have ever been before. Everybody is using different kinds of devices to go online and consume digital content. Traditional media is being replaced by new and innovative solutions. The idea of Mediafy is, instead of replacing school books with their e-versions, to enrich the material with digital content and make it more interactive, entertaining, and appealing to the younger generations. Current online solutions are focused on specific functions such as visualization. However, there is no real media provider for physical books. Mediafy will provide a Content Management System (CMS) that will enable publishing houses to publish digital content online in a secure environment. As the application will be tailored to the publishing houses' needs, the publishing process will be fast and easy to fulfill.

School Pooling

School Pooling is a carpool sharing application service that provides different routes to transport children to school. The underlying objective is to diminish the number of commutes using private transportation and consequently diminish pollution and energy consumption.

With this system, parents who usually drive their children to school will be able to "sell" their spare seats by picking up other students. They will get paid for the service. For security reasons, an individual platform will be created for each school. Using an app, driving parents will be able to publish the route they follow. Other parents from the same school who are searching for a driver to pick-up their child (ren) can browse among the available routes and request a pick-up. Within the following 24 h they will be notified if accepted or not. If so, a chat for all the parents with children in the same route will be created in order to arrange the place to meet the following day. Once students arrive safely at the school, a notification will be shared among the nondriving parents. In case of an emergency, which means that the driver is not able to drive, a private driver or taxi driver will be called by the company and s/he will be in charge of bringing children to the school.

The service differentiates from that of competitors in several ways. First, compared to public transportation, School Pooling is safer. Depending on the town, young students might feel insecure taking public transportation on their own. Also commutes from home to a metro/bus/tram station might be long. This system allows pick-ups from homes. Second, one of the main advantages of School Pooling over private school buses will be the price. School Pooling will cost less than half the price of a private bus. Parents acting as drivers will receive free credit for future services. The payment will be processed through the app. As the application will be extremely user-friendly, the service will be easy to use.

Revid

Students attending private academies want to review the content of the lectures before exams and whenever they miss a class. On the other hand, the owners of academies want a service that allows them to be more competitive by providing recorded lectures to their students, hence improving their performance. Nevertheless, they do not want to invest high amounts to make this improvement or to increase the workload of their teachers.

Revid is a service that provides private academies with all they need to record their lectures and make them available to their students. Revid will take care of the setup (choosing and installing the cameras and microphones), the training (how to use the platform), and the maintenance. The service will be contracted via either the website, phone calls, or on-site visits. A team of technical staff will be in charge of installing the cameras and the required material. Likewise, in case of a technical problem, a technician will go to the academy and fix it. For doubts and queries, a call center will be available 24/7. The online platform will provide useful tips for making the most of the equipment installed.

Following the spirit of work integration social enterprises, Revid aims at improving the employment prospects of people farthest from the labor market. Reasons for potential exclusion might be multiple (e.g., low education attainment, discrimination related to ethnicity, gender, homelessness, migration or asylum seekers, illness or disability, and unemployment). Specifically, Revid will work closely with foundations and associations in order to train people at risk of exclusion on how to install the technical equipment at academies and provide maintenance services. The ultimate goal is to integrate or reintegrate these individuals into the workplace permanently.

Tutoria

Tutoria is a web platform that brings together tutors and students with the right tools to enable a digital environment that simulates a real-time tutoring session. The tutor provides input to the process in the form of information, guidance, and tutoring. The student provides input to the process in the form of feedback and questions. The output of the process is an increase of knowledge in the specified course of the student.

Tutoria's service differs from that of its competitors in three main aspects. First, tutors' qualifications would be verified and publically available for registered users; this way, prospective students would be able to select which tutor they prefer based on the ratings and recommendations from other users. Second, all tutors working in Tutoria will be using a graphical tablet that will act as a digital notebook or whiteboard for the student. Lastly, inspired by the social enterprise Glovico.org, Tutoria combines the business aspect (providing students the necessary skills to pass their exams), with a social mission. It does so by offering people from developing countries the opportunity to earn additional income by acting as tutors.

Conclusions

Education is a decisive factor of change. Students – tomorrow's leaders – need to take an active role in creating a better, sustainable, and safe future. Educational institutions play a paramount role in this process. This topic is not new. The term "education for sustainable development" was coined in the mid-1990s and emerged as a need for changing the traditional focus of "environmental education" (Keeble 1988). The main implications of this approach entailed broadening its scope. That is, education for sustainable development is a lifetime process that should create the conditions for the development of environmental consciousness and formation of ecological culture. Citizens need to better respond to socioeconomic – rather than only environmental – challenges at the local, regional, and global level (Filho 2015). Said differently, as today's resources need to be used with care, so that they are available to future generations, education for sustainable development should foster awareness about the issues pertaining to sustainable development in all its different dimensions – social, political, economic, and ecological.

Since the 1990s, educational initiatives oriented towards generating sustainability awareness have significantly increased, and education for sustainable development is now a relevant part of the curriculum in different educational settings. Higher education institutions are not an exception to this reality. Universities are the perfect place where to teach students how to identify and implement solutions to environmental challenges. This ability is highly tied to the development of entrepreneurial mindsets. Indeed, entrepreneurship has been found to help economies spring up and create new jobs. Moreover, if entrepreneurial activities are oriented towards the offering of creative solutions to complex and persistent social problems, they would not only report benefits to the entrepreneurs but they will also boost social wealth creation (Zahra et al. 2009).

This book chapter has reflected on the benefits of teaching students at universities the fundamentals of sustainable entrepreneurship. To do so, we have first highlighted the need for promoting sustainability values among entrepreneurs and the central role that universities play in this context.

In order to illustrate students' perceptions of sustainable practices and their current knowledge on some key topics dealing with sustainability issues, the results of a survey – specifically designed for this purpose – are presented. Overall, it can be concluded that, although students are aware of sustainability practices and acknowledge their importance, most of them consider the education for sustainable development they received to be limited. This result is consistent with the existing literature and the different calls for an urgent need to better integrate sustainability in the context of higher education.

Sustainable entrepreneurs promote new and profitable processes and products while fostering social and ecological progress. Thus, given the importance of boosting sustainable entrepreneurial attitudes among students, we have next reviewed different active learning methods that have been found to help in this process. Specifically, we have focused on problem- and project-based learning, living labs, and transdisciplinary activities.

Sustainability and entrepreneurship can complement each other, and certainly, higher education institutions are expected to take a leading role in hastening these abilities among students. A number of initiatives worldwide have been presented, thus evidencing how higher education institutions have enlarged their academic offerings and created academic programs addressing the 3Ps – people, planet, and profit. These programs assist students in developing the appropriate skills that are necessary to explore new opportunities in green businesses, participate in research projects, and help local companies to develop sustainability plans. Because universities should not lose sight of their responsibility and only adopt a passive role – e.g., enlarging the academic offer – but an active one, we have also reviewed examples of universities that have gone a step further and implemented educational sustainable development using different shapes and formats. Some examples include the establishment of institutional guidelines on how to foster sustainable development within the institution, the creation of a broaden support for sustainability initiatives, the adoption of innovative practices to make universities' footprint more sustainable, or the establishment of local and regional partnerships so as to yield results in the communities surrounding universities.

The chapter ends with an example – course on Service Management – of how to train nonbusiness-related students to become sustainable entrepreneurs. We believe this example to be worth as the original purpose of the course is to teach neither entrepreneurship nor sustainability but how to manage service firms. Thus, this example shows that the fundamentals of a course can be complemented with other relevant skills students need to develop – in this case, entrepreneurship and sustainability – before graduating. Using a project-based approach, students learn how to run a business operating in the service sector, while developing the ability of identifying social needs and creating a solution. This approach is much more enriching, and it does not require reducing content on service management. However, we acknowledge that such an approach is much more demanding for the instructor, as s/he needs to think outside of the box in order to come up with class activities that combine different topics, all of them, of interest for the student. Likewise, such a holistic approach presupposes important challenges and changes in conceptualizing the role of university lecturers, their educational profiles, and the preparation of the teaching material (Llorens et al. 2017). Universities – and by extension, professors – are responsible for educating students not only in terms of specific knowledge but also in terms of skills. As such, although much more demanding, we have in our hands the choice of helping students making business decisions that result in a socially, economically, and environmentally better world.

Although in this chapter we have seen that entrepreneurship and sustainability have been found to reinforce each other, literature examining this connection is scarce. We therefore encourage researchers to conduct further studies in this direction. The authors hope this chapter inspires universities to think beyond the approach of merely teaching students, but to embark on activities that seek to create impact and induce large-scale positive change within society or for the environment (Bradach 2010).

Reflection Questions

- Use the SWOT matrix (strengths, weaknesses, opportunities, and threats) to analyze the case of Medifay. Do you think this business to be salable?
- Which are the main competitors of School Pooling? Would you be willing to use this service?
- In the case of Revid, which specific profile of people at risk of exclusion would you consider easier to reincorporate into the job market?
- Search for similar services to the one provided by Tutoria. What actions and strategies would you suggest in order to make this company more sustainable?

Exercises in Practice

In this section are some examples of activities designed for teaching sustainable entrepreneurship. As typical entrepreneurship courses are designed around the generation of a business idea (Berbegal-Mirabent et al. 2016), the activities listed below converge in the writing of a business plan. The statements for these activities as well as the acquisition of skills pursued are as follows.

Exercise 1: The 3P Mission
List five companies with a sustainable mission. When doing so, consider their main objectives and values, and check if the members of the companies are involved with this mission.

Skills to develop: search for information, critical thinking.

Hint: This TED Talks video can be helpful to understand why companies can solve social problems: https://www.youtube.com/watch?v=0iIh5YYDR2o

Exercise 2: 3P Products and Processes
List five enterprises with high sustainability principles related to their products and management. Also specify, if possible, which of them are known for generating opportunities for new sustainable business.

Skills to develop: search for information, critical thinking.

Exercise 3: Analyzing the Market: Sustainable Companies' Hunters
Working in teams of three, perform an analysis of a market formed by social start-ups. Consider also the elements that can let this start-up be profitable and indicate the stakeholders that can help these firms to achieve their goals.

Skills to develop: search for information, teamwork, critical thinking.

Hint: This TED Talks video can be helpful to understand the role of entrepreneurs for fostering sustainability: https://www.youtube.com/watch?v=Dqza5Uo1cFE

Exercise 4: 3P Product Designers
Design a sustainable product and a marketing strategy associated with it. You will have to present your ideas in front of the class. Your classmates will give you

feedback, taking into account the triple P (effects on society, environment, and profits) and the originality and feasibility of your project.

Skills to develop: critical thinking, communication and presentation, project-development, peer-assessment.

Exercise 5: InnoGame

List as many innovative production practices as possible in 15 min. For each correct answer, you will get 1 point, 3 points if the practice can be considered sustainable. If a production practice on your list is not innovative, you will lose 0.5 points. The jury (formed by classmates and the lecturer) will decide the points imputable to each participant. The one who ranks in first place will be the InnoWinner.

Skills to develop: critical thinking, thinking speed, peer-assessment.

Exercise 6: Sustainable Business Plan Development

Write a business plan for creating a new company or developing and launching a new product that fulfills sustainability standards. It is of the utmost importance that you integrate People + Planet + Profit in your project. Thus, the business plan must consider and promote aspects that cause an improvement for sustainability and also include the design of the product; an analysis of the context; a plan for the production, management, and marketing; and a financial study. The social, ecological, and economic aspects should be clearly integrated in the work. You will have to do both an oral (30 min) and written presentation (maximum length 30 pages) of your business plan. Two lecturers and a practitioner from a sustainable start-up will be involved in the evaluation, considering the originality and the correct realization of the task.

Skills to develop: search for information, critical thinking, communication and presentation, project-development.

Engaged Sustainability Lessons

– The role of sustainable entrepreneurs is paramount, as they are responsible for designing and developing new products and projects aligning social, environmental and economic objectives.
– Universities must transmit the importance of sustainability, instill these values in their students, and promote research programs related to sustainable development.
– Many methods exist to teach students about sustainable practices. Some of the most commonly used methods include problem-based learning activities, project-based learning activities, living labs, and transdisciplinary activities.

Chapter-End Reflection Questions

After reading this chapter:

– Do you think it is feasible to become a sustainable entrepreneur, looking to the common good for both society and environment, without giving up profits? If you

were thinking of creating a new business, would you try to be a sustainable entrepreneur?
- Taking into consideration the results obtained from students' feedback, would you consider changing some of your habits in order to contribute to a more sustainable world?
- Have you become more aware of the importance of educating for sustainability? If you are in a higher education setting, are you willing to apply teaching methodologies or participate in activities that promote the acquisition of sustainable skills?
- If you ever have to make a business plan, will you follow the patterns proposed in this chapter in order to make the project sustainable, integrating People + Planet + Profit in the design of it?

If the answers to the previous questions are "yes," it means that the objectives pursued with this chapter have been fulfilled: highlighting the importance of sustainability and providing tools to instill this skill in students in a higher education context.

Cross-References

▶ Education in Human Values
▶ Environmental Intrapreneurship for Engaged Sustainability
▶ Expanding Sustainable Business Education Beyond Business Schools
▶ Sustainable Higher Education Teaching Approaches

References

Barron, B., Schwartz, D., Vye, N., Moore, A., Petrosino, A., Zech, L., & Bransford, J. (1998). Doing with understanding: Lessons from research on problem- and project-based learning. *Journal of the Learning Sciences, 7*(3), 271–311.

Bell, B. S., & Kozlowski, S. W. J. (2008). Active learning: Effects of core training design elements on self-regulatory processes, learning, and adaptability. *The Journal of Applied Psychology, 93*(2), 296–316.

Berbegal-Mirabent, J., Gil-Doménech, D., & Alegre, I. (2016). Improving business plan development and entrepreneurial skills through a project-based activity. *Journal of Entrepreneurship Education, 19*(2), 89–97.

Berbegal-Mirabent, J., Gil-Doménech, D., & Alegre, I. (2017). Where to locate? A project-based learning activity for a graduate-level course on operations management. *International Journal of Engineering Education, 33*(5), 1586–1597.

Bessant, S., Bailey, P., Robinson, Z., Tomkinson, C. B., Tomkinson, R., Ormerod, R. M., & Boast, R. (2013). Problem-based learning: A case study of sustainability education. A toolkit for university educators. https://www.keele.ac.uk/media/keeleuniversity/group/hybridpbl/PBL_ESD_CaseStudy_Bessant_et al._2013.pdf.

Blank, S. (2013). Why the lean start-up changes everything. *Harvard Business Review, 91*(5), 63–72.

Blumenfeld, P. C., Soloway, E., Marx, R. W., Krajcik, J. S., Guzdial, M., & Palincsar, A. (1991). Motivating project-based learning: Sustaining the doing, supporting the learning. *Educational Psychologist, 26*(3&4), 369–398.

Bonnet, H., Quist, J., Hoogwater, D., Spaans, J., & Wehrmann, C. (2006). Teaching sustainable entrepreneurship to engineering students: The case of Delft University of Technology. *European Journal of Engineering Education, 31*(2), 155–167.

Bonwell, C. C., & Eison, J. A. (1991). *Active learning: Creating excitement in the classroom* (J. D. Fife, Ed.) ASHE-ERIC higher education report no. 1. Washington, DC: The George Washington University, School of Education and Human Development.

Bradach, J. (2010). Scaling impact. *Stanford Social Innovation Review, 6*, 27–28.

Cohen, T., & Lovell, B. (2011). *The campus as a living laboratory: Using the built environment to revitalize college education. A guide for community colleges.* Washington, DC: Center for Sustainability Education and Economic Development, American Association of Community Colleges.

Corcoran, P. B. (2010). Sustainability education in higher education: Perspectives and practices across the curriculum. In P. Jones, D. Selby, & S. Sterling (Eds.), *Sustainability education: Perspectives and practices across higher education* (pp. xiii–xxiv). London/New York: Taylor & Francis.

Cörvers, R., Wiek, A., de Kraker, J., Lang, D. J., & Martens, P. (2016). Problem-based and project-based learning for sustainable development. In *Sustainability science* (pp. 349–358). Dordrecht: Springer Netherlands.

Daley, J. (2011). The top business schools for eco-entrepreneurs. Published online 14 Mar 2011. Available on www.entrepreneur.com/article/219236. Last retrieved 26 June 2017.

Dean, T. J., & McMullen, J. S. (2007). Toward a theory of sustainable entrepreneurship: Reducing environmental degradation through entrepreneurial action. *Journal of Business Venturing, 22*(1), 50–76.

Donaldson, T., & Preston, L. E. (1995). The stakeholder theory of the corporation: Concepts, evidence, and implications. *Academy of Management Review, 20*(1), 65–91.

Doppelt, Y. (2003). Implementation and assessment of project-based learning in a flexible environment. *International Journal of Technology and Design Education, 13*(3), 255–272.

Epstein, M. J., & Buhovac, A. R. (2014). *Making sustainability work: Best practices in managing and measuring corporate social, environmental, and economic impacts* (2nd ed.). San Francisco, CA: Berrett-Koehler Publishers.

Evans, J., Jones, R., Karvonen, A., Millard, L., & Wendler, J. (2015). Living labs and co-production: University campuses as platforms for sustainability science. *Current Opinion in Environmental Sustainability, 16*, 1–6.

Filho, W. L. (2015). Education for sustainable development in higher education: Reviewing needs. In W. Leal Filho (Ed.), *Transformative approaches to sustainable development at universities* (pp. 3–12). Cham: Springer International Publishing. https://doi.org/10.1007/978-3-319-08837-2.

Friedman, K. (2000). Creating design knowledge: From research into practice. *IDATER Conference, 1*, 5–32.

Hall, J. K., Daneke, G. A., & Lenox, M. J. (2010). Sustainable development and entrepreneurship: Past contributions and future directions. *Journal of Business Venturing, 25*(5), 439–448.

Jones, T. M. (1995). Instrumental stakeholder theory: A synthesis of ethics and economics. *Academy of Management Review, 20*(2), 404–437.

Keeble, B. R. (1988). The Brundtland report:'Our common future'. *Medicine and War, 4*(1), 17–25.

Kember, D. (2009). Nurturing generic capabilities through a teaching and learning environment which provides practise in their use. *Higher Education, 57*(1), 37–55.

Kubiatko, M., & Vaculová, I. (2011). Project-based learning: Characteristic and the experiences with application in the science subjects. *Energy Education Science and Technology Part B: Social and Educational Studies, 3*(1), 65–74.

Kuckertz, A., & Wagner, M. (2010). The influence of sustainability orientation on entrepreneurial intentions – Investigating the role of business experience. *Journal of Business Venturing, 25*(5), 524–539.

Laffey, J., Tupper, T., Musser, D., & Wedman, J. (1998). A computer-mediated support system for project-based learning. *Educational Technology Research and Development, 46*(1), 73–86.

Lang, D. J., Wiek, A., Bergmann, M., Stauffacher, M., Martens, P., Moll, P., et al. (2012). Transdisciplinary research in sustainability science: Practice, principles, and challenges. *Sustainability Science, 7*(Suppl 1), 25–43.

Lans, T., Blok, V., & Wesselink, R. (2014). Learning apart and together: Towards an integrated competence framework for sustainable entrepreneurship in higher education. *Journal of Cleaner Production, 62*, 37–47.

Llorens, A., Berbegal-Mirabent, J., & Llinàs-Audet, X. (2017). Aligning professional skills and active learning methods: An application for information and communications technology engineering. *European Journal of Engineering Education, 42*(4), 382–395.

Lourenço, F., Jones, O., & Jayawarna, D. (2012). Promoting sustainable development: The role of entrepreneurship education. *International Small Business Journal, 31*(8), 841–865.

Merck, J., & Beermann, M. (2015). The relevance of transdisciplinary teaching and learning for the successful integration of sustainability issues into higher education development. In W. Leal Filho, L. Brandli, O. Kuznetsova, & A. M. F. do Paço (Eds.), *Integrative approaches to sustainable development at university level* (World sustainability series, pp. 19–26). Cham: Springer International Publishing. https://doi.org/10.1007/978-3-319-10690-8.

Michael, J. (2006). Where's the evidence that active learning works? *Advances in Physiology Education, 30*(4), 159–167.

Phillips, J. M. (2004). Strategies for active learning in online continuing education. *Journal of Continuing Education in Nursing, 36*(2), 77–83.

Prince, M. (2004). Does active learning work? A review of the research. *Journal of Engineering Education, 93*, 223–231.

Schaltegger, S., & Wagner, M. (2011). Sustainable entrepreneurship and sustainability innovation: Categories and interactions. *Business Strategy and the Environment, 20*, 222–237.

Scholz, R. W., Lang, D. J., Wiek, A., Walter, A. I., & Stauffacher, M. (2006). Transdisciplinary case studies as a means of sustainability learning. *International Journal of Sustainability in Higher Education, 7*(3), 226–251.

Star, C., & Hammer, S. (2008). Teaching generic skills: Eroding the higher purpose of universities, or an opportunity for renewal? *Oxford Review of Education, 34*(2), 237–251.

Trimi, S., & Berbegal-Mirabent, J. (2012). Business model innovation in entrepreneurship. *International Entrepreneurship and Management Journal, 8*(4), 449–465.

UNESCO. (2004). UN decade of education for sustainable development (2005–2014): Draft international implementation scheme. Paris.

UNESCO. (2009). Education for sustainable development: United Nations decade (2005–2014). Paris.

Vezzoli, C., & Penin, L. (2006). Campus: "lab" and "window" for sustainable design research and education. *International Journal of Sustainability in Higher Education, 7*(1), 69–80.

Vilsmaier, U., & Lang, D. J. (2015). Making a difference by marking the difference: Constituting in-between spaces for sustainability learning. *Current Opinion in Environmental Sustainability, 16*, 51–55.

Wiek, A., Xiong, A., Brundiers, K., & van der Leeuw, S. (2014). Integrating problem- and project-based learning into sustainability programs: A case study on the school of sustainability. *International Journal of Sustainability in Higher Education, 15*(4), 431–449.

Zahra, S. A., Gedajlovic, E., Neubaum, D. O., & Shulman, J. M. (2009). A typology of social entrepreneurs: Motives, search processes and ethical challenges. *Journal of Business Venturing, 24*(5), 519–532.

Ecopreneurship for Sustainable Development

The Bricolage Solution

Parag Rastogi and Radha Sharma

Contents

Introduction	992
The Context	993
Bricolage: An Introduction	994
Sustainable Entrepreneurship	994
Ecopreneurship	995
How Entrepreneurs Can Contribute to Sustainability	996
Entrepreneurship and Sustainability	996
Profit-Seeking Entrepreneur	996
Institutional Entrepreneur	996
Resource-Oriented Entrepreneur	997
How the Concepts of Sustainability and Entrepreneurship Can Be Linked to Business Models	997
How Entrepreneurial Actions Can Address Sustainable Requirements in Construction Industry	998
Case Study 1: Vikram Varma	999
Case Analysis	1000
Case Study 2: Chitra Vishwanath	1003
Case Analysis	1007
Case Study 3: Sanjay Prakash	1008
Case Analysis	1011
Bricolage: A Feasible Option in Construction Industry that Can Address Sustainability Concerns	1012
Proposed Model for Bricoleur Entrepreneurs	1012
Cross-References	1014
References	1014

P. Rastogi (✉) · R. Sharma
Management Development Institute, Gurgaon, India
e-mail: parag.rastogi@gmail.com; radha@mdi.ac.in; radhasharma308@gmail.com

© Springer International Publishing AG, part of Springer Nature 2018
S. Dhiman, J. Marques (eds.), *Handbook of Engaged Sustainability*,
https://doi.org/10.1007/978-3-319-71312-0_46

Abstract
Sustainability is a critical issue across the globe, and the debate on sustainable development is still in its incipient phase. The purpose of this chapter is to add another perspective in this ongoing debate. Sustainability concerns have led several countries to introduce policy interventions to reduce carbon footprint. However, these policies are not enough to deal with the challenge. The chapter addresses critical sustainability issues through a detailed discussion on key concepts, methods, and lessons learned in operationalizing ecopreneurship as a solution for solving environmental concerns.

Using construction industry as an example, the chapter discusses ecopreneurship as a solution using the twin lenses of sustainability and environmental concerns. Buildings contribute to 24% of the world's CO_2 emissions. A developing country like India with resource constraints which added to the need to reduce the CO_2 footprint poses a significant challenge to building architects. This chapter encapsulates case studies of some Indian architects in sustainable innovation and development. Concerned with sustainability in their work, they face challenges and considerable risks. The chapter analyzes the barriers they encounter, how they view these risks, and the methods they adopt to mitigate the risks. The chapter examines in detail the existing definitions and typologies in dealing with sustainability, explores the motivational aspects for ecopreneurs, and researches the various parameters of operationalizing sustainable solutions. The chapter introduces a novel entrepreneur typology – a *bricoleur ecopreneur* – who uses bricolage as a solution to the challenges of sustainable development.

The chapter can be helpful for researchers in designing research in the overall context of sustainable development debate. Apart from researchers, this would also be useful for international institutions, NGOs, architects, developers, and building users.

Keywords
Sustainability · Entrepreneurship · Ecopreneurs · Bricolage · Environmental concerns · Building architecture · Bricoleur · Disruptive innovators · Construction industry

Introduction

Globally sustainability has become a critical issue, and researchers and industry practitioners are trying innovative solutions that are sustainable and protect the environment. Many countries have brought in policy interventions for reducing carbon footprint, waste material management, rainwater harvesting, and promoting the use of alternative energy resources and materials. United Nations Conference on Environment and Development (UNCED) in 1992 brought the focus on climate change and uneconomic growth (Elkington 1994).

At the global level, cities consume 75% of the world's energy and are responsible for 80% of greenhouse gas emissions. Buildings are responsible for 24% of the world's CO_2 emissions (International Energy Agency 2008). Building industry has a high level of impact on the economic growth on one hand and environmental impact on the other. According to United Nations Environment Program – Sustainable Buildings and Climate Initiative (UNEP-SBCI) report (2009), "the building sector contributes up to 30% of global annual greenhouse gas emissions and consumes up to 40% of all energy... if nothing is done, greenhouse gas emissions from buildings will more than double in the next 20 years."

Moreover, most of the energy from buildings (57% of domestic consumption) is used for space heating or cooling (Association for the Conservation of Energy 2004). This highlights the critical need for sustainability-driven designs of buildings. Given the criticality and the impact on environment it plays, there is an urgent need to study the construction industry from sustainability perspective. According to Jong and Muizer (2005), a study in which 58 different business sectors in Netherlands were studied based on their levels of innovative input, output, and future investing plans, the construction industry ranked 55.

Several studies have shown that many businesses view sustainable agenda as detrimental to their businesses and they are resistant to improve their environmental performance (Revell and Rutherfoord 2003; Tilley 2000). This is truer for most owner-managers (Revell and Blackburn 2007). This chapter, in contrast, builds on the current knowledge about entrepreneurs who are working toward addressing the challenges of sustainable development. In this chapter, we discuss entrepreneurs who take considerable risk to aggressively pursue sustainability agenda. These entrepreneurs use bricolage as a solution for overcoming the challenges of sustainable development.

The Context

There is an increasing interest in linking entrepreneurship and sustainability. However, theoretical understanding of the link is only emergent. The study by York and Venkataraman (2010) posited "although there has been recent interest in the concept of environmental or sustainable entrepreneurship, under what conditions and how entrepreneurial action can address such problems remains unclear."

Entrepreneurs have the unique ability to combine self and circumstance (Anderson 2000). There has been a focus on understanding entrepreneurship as a potential mechanism for sustainable development (Harbi et al. 2010). We have used multiple-case approach to develop deeper understanding and build theoretical knowledge of environmentally sustainable entrepreneurship. These case studies seek to expand our understanding of how building architects can play a role in environmentally sustainable entrepreneurship.

The case studies encapsulate experiences of few architects/entrepreneurs from India who are concerned about sustainability in their work, the barriers they encounter, how they view these risks, and innovative methods they employ in mitigating these risks. These cases help in understanding the concept of sustainable entrepreneurship from a resource-based perspective in a country like India, where resources are limited. The focus is on the actions taken by the ecopreneurs in the process of sustainable innovation and development.

These case studies seek to answer the following questions:

1. How can building entrepreneurs contribute to sustainability?
2. How can the concepts of sustainability and entrepreneurship be linked with resources?
3. How can entrepreneurial actions address sustainable requirements in construction industry?
4. Is bricolage a feasible option in building (construction) industry that can address sustainability concerns?

Bricolage: An Introduction

Bricolage was first introduced by Levi-Strauss (1967) as a theory of how meanings are assigned to objects (or resources) and the institutional rules of when and what combinations can be applied in society. Levi Strauss proposed a bricoleur (a person that engages in bricolage) as one who is "trying to make his way out of and to go beyond the constraints imposed by a particular state of civilization."

Since then, bricolage has been used from its structuralist anthropology origins to cognitive science, information technology, innovation, and organization theory (Duymedjian and Rulings 2010). We find that bricolage can be a viable method to relate entrepreneurship to sustainability.

Sustainable Entrepreneurship

Sustainable entrepreneurs are those entrepreneurs who combine various aspects of sustainability and have a different organizing logic to more conventional entrepreneurs (Tilley and Parrish 2006). These entrepreneurs operate their businesses in ways that are counter to conventional entrepreneurial behavior (Hart 2006). According to Parrish (2006), these entrepreneurs seek to perpetuate resources focused on sustainable development. They play a critical role in bringing a paradigm shift to a new form of development. According to Beveridge and Guy (2005), such development can alleviate global warming and other negative environmental impacts.

There have been several ecological advancements that have happened in the recent years. Modernization efforts like hybrid cars, solar power, and windmills combine new technologies and even trigger changing institutions. Such ecological innovations are getting into the mainstream now through commercial ventures. Dean and McMullen (2007) have used this evidence to argue that there is an increasing importance for environmental entrepreneurship. Willis et al. (2007) have called such entrepreneurs "disruptive innovators" – implying that established business models and user expectations are transformed by such innovations. Further, Schlange (2009) has posited that "sustainability-driven entrepreneurs view their ventures as integral parts of a larger societal context in which they are able to contribute to the improvement of life conditions in the most general sense."

Ecopreneurship

The term "ecopreneurship" is a combination of two words, "ecological" ("eco") and "entrepreneurship." Ecopreneurship can thus be roughly defined as "entrepreneurship through an environmental lens." According to Dean and McMullen (2007), ecopreneurship is a "sustainable and environmental entrepreneurship which detail how entrepreneurs seize the opportunities that are inherent in environmentally relevant market failures." Schlange (2006) has suggested that "ecologically driven entrepreneurship has *sustainability* as a key element to *motivate* its basic approach."

Hence, ecopreneurship is different from other forms of corporate environmental development by the company's commitment to environmental progress and its desire for business growth. Ecopreneurship has its own characteristics and the associated business risks which are different.

Post and Altman (1994) have identified three main drivers of change from an external perspective:

1. *Compliance-based*, which emerge as an outcome of government regulation and legislation
2. *Market-driven*, with environmentally beneficial behavior coming as a result of profit motivation
3. *Value-driven*, with environmental change coming in response to end-user demands

Schaltegger (2002) has posited ecopreneurs do more than just a change and that they "destroy existing conventional production methods, products, market structures and consumption patterns and replace them with superior environmental products and services." In a similar context, Walley and Taylor (2002) see ecopreneurs as "crucial change agents" or "champions" of "collective learning process."

For ecopreneurs, "market creation is even more difficult for environmental business ideas than it is for non-environmental business ideas" (Linnanen 2002). Randjelovic et al. (2003) have pointed out that ecopreneurial development may require longer gestation periods to achieve market breakthrough than conventional entrepreneurial activity, and this long period can deter investors ("Green VC") looking for a quick return on their investment. Further, Randjelovic et al. (2003) have proposed that ecopreneurs "need to develop competences on environment-related strategies and practices, which can create economic value *and* reduce environmental impacts/risks."

How Entrepreneurs Can Contribute to Sustainability

There are certain types of entrepreneurial actions that are particularly conducive to the creation of sustainable solutions (Dean and McMullen 2007). Sustainable entrepreneurship research has primarily focused on how entrepreneurs can work on profit opportunities within a market which are environmentally friendly.

Ecopreneurship explores various forms of entrepreneurial action that are resource oriented and seek to make the most of the resources at hand (Baker and Nelson 2005). Ecopreneurship stresses on parsimony in usage of resources. Application of bricolage in ecopreneurship is premised on the *frugal use of resources* as well as *usage of resources which are available currently* to the entrepreneur.

Entrepreneurship and Sustainability

For environmentally sustainable entrepreneurial action, Pacheco et al. (2010) identify two types of actions. The first type of actions is taken by the entrepreneurs who identify and exploit opportunities that are environmentally sensitive. This entrepreneur could be a profit-seeking entrepreneur (Kirzner 1973) or in the nexus perspective of entrepreneurship (Shane and Venkataraman 2000). Second-type actions are taken by the entrepreneurs who create the opportunities for sustainable entrepreneurship through institutional entrepreneurship (Aldrich and Fiol 1994).

Profit-Seeking Entrepreneur

Pacheco et al. (2010) have used the metaphor of "green prison" and using the prisoner's dilemma construct argued that there are situations where actors necessarily take the less environmentally unsustainable option, e.g., because it is cheapest or easiest, even though in the long run it may be less rational. An example of this could be over-exploitation of natural resources where over-exploitation in short term can result in long-term scarcity of the resource.

Institutional Entrepreneur

The term institutional entrepreneurship refers to the "activities of actors who have an interest in particular institutional arrangement and who leverage resources to create new institutions or to transform existing ones" (Maguire et al. 2004). Moreover, such an entrepreneur can "create a whole new system of meaning that ties the functioning of disparate sets of institutions together" (Garud et al. 2002). "Institutional entrepreneurs act to change the incentives of the institutional setting by reducing the benefits of environmentally unsustainable strategies and/or enhancing the benefits of the environmentally sound strategies" (Pacheco et al. 2010). Dean and McMullen (2007) further suggest that the elimination of environmentally damaging government subsidies is perhaps the most important form of sustainable institutional entrepreneurship.

Resource-Oriented Entrepreneur

According to Stevenson and Jarillo (1990), entrepreneurship is "a process by which individuals – either on their own or inside organizations – pursue opportunities without regard to the resources they currently control." Can entrepreneurial action also be resource oriented? The concept of bricolage suggests that entrepreneurs take their starting point in currently available resources and create effects from these resources. Such entrepreneurs create opportunities from the resources at hand (Baker and Nelson 2005). This implies a bias toward parsimony in utilization of resources. Such entrepreneurs can focus on creative utilization of resources at hand so that action can take place in circumstances where the interest is in improving the environment.

How the Concepts of Sustainability and Entrepreneurship Can Be Linked to Business Models

Ecopreneurship deals with entrepreneurs focused on sustainability. How do we relate ecopreneurship to business models? Eastwood et al. (2001) proposed a business classification based on the types of green businesses. There are other proposed typologies that focus on the characteristics of the ecopreneurs. Linnanen (2002) has based the typologies on internal motivations and desires. Schaltegger (2002) classifies ecopreneurs based on the priority given by them to environmental issues as a business goal and their focus on the market effect of the business. Taylor and Walley (2002) have used influence of structural drivers and orientation toward sustainability or financial outcomes. Summary of these typologies is given in Table 1.

The ecopreneur typologies are not without criticism. Gibbs (2009) has extensively reviewed the ecopreneur typologies and critiqued ecopreneur literature that "it is heavy on speculation and extremely light on empirical evidence." Similarly, Beveridge and Guy (2005) have commented that these typologies "fail to fully engage with the processes involved in the emergence of eco-preneurship." Harbi et al. (2010) have concurred and concluded that very little empirical work has been done on ecopreneur typologies.

Harbi et al. (2010) did an empirical research based on Taylor and Walley typology on 56 Tunisian entrepreneurs and concluded that only "ethical maverick" type of ecopreneur is present in their sample. They found a profile resembling "innovative opportunists" and no presence of "ad hoc environpreneurs" and "visionary champions." This implies that Taylor and Walley typology does not capture all the types of ecopreneurs.

Table 1 Ecopreneur typologies

Reference	Parameters considered	Characteristics	Type of ecopreneur
Eastwood et al. (2001)	Types of green businesses	Owners in 11 types of green businesses	Green producer
Liannen (2002)	1. Desire to change the world 2. Financial drive	High on first and low on second parameter	Nonprofit businessman
		Low on first and low on second parameter	Self-employer
		Low on first and high on second parameter	Opportunist
		High on first and high on second parameter	Successful idealist
Schaltegger (2002)	1. Environmental performance goals are core to business	Combined with alternate scene away from market orientation	Alternative actors
		Combined with eco-niche	Bioneers
		Combined with mass market	Ecopreneur
Taylor and Walley (2002)	1. Influenced by structural drivers – hard (regulation, incentives, etc.) or soft (past experience, family, etc.) 2. Financial orientation or commitment to sustainability	Hard structural drivers and financial orientation	Innovative opportunists
		Hard structural drivers and commitment to sustainability	Visionary champions
		Soft structural drivers and sustainability orientations	Ethical mavericks
		Soft structural drivers and financial orientation	Ad hoc environpreneurs

How Entrepreneurial Actions Can Address Sustainable Requirements in Construction Industry

Yin (2009) points out that case studies are the preferred strategy when "how" and "why" questions are posed. Hence to explore the question "how do entrepreneurs evaluate risk?" we propose to use case study method – using the practitioners' experience/actions.

Building theory/model from case studies is a research strategy that involves using one or more cases to create theoretical constructs, propositions, and/or midrange theory from case-based empirical evidence (Eisenhardt 1989). The theory-building process occurs via recursive cycling among the case data, emerging theory, and, later, extant literature (Eisenhardt and Graebner 2007). For phenomenon-driven research questions, a researcher has to frame the research in terms of the importance of the phenomenon and the lack of plausible existing theory (Eisenhardt and Graebner 2007).

Multiple-case studies typically provide a stronger base for theory building. They are chosen for theoretical reasons such as replication, extension of theory, contrary replication, and elimination of alternative explanations (Yin 2009). Adding three cases to a single-case study is modest in terms of numbers but offers four times the analytic power (Eisenhardt and Graebner 2007).

The cases to study have been selected using a critical case strategy (Flyvbjerg 2004). The value of a critical case is determined, not by how it is representative of a larger population but by the amount of information and learning that can be derived from the case (Flyvbjerg 2004).

The critical cases have been selected on the basis of the following criteria:

1. The entrepreneur is focused on sustainable architecture.
2. Project designs and actions are focused on creating value from natural environment.
3. Architects have the freedom to practice non-sustainable designs (no commercial or policy constraints).

This research evaluates three case studies of building architects, who are motivated by sustainable designs and use bricolage extensively.

Case Study 1: Vikram Varma

Vikram's definition of sustainability is "living for less." "Sustainability is different from green. Many green buildings are technology-based. They do not care about energy cost. They are more about efficiency and better technology."

Eye Hospital in Goa, India

One of Vikram's key projects was in Goa, in the western part of India. Here he experimented and learnt much about sustainable architecture. He was tasked with building a 30,000 sq. ft eye clinic.

The client had taken proposals from various architects. One of the proposals was based on futuristic material and used an exterior envelope of aluminum and steel. The facade looked stunning. The second brief the client got was a design based on modular furniture and clutter-breaking decor. Both were modern architectural proposals.

Vikram's own proposal was starkly different. He proposed that he would do an 80,000 sq. ft eye hospital with inpatient facilities as well. He would also build it so that it would be a very energy-efficient building from an air-conditioning perspective

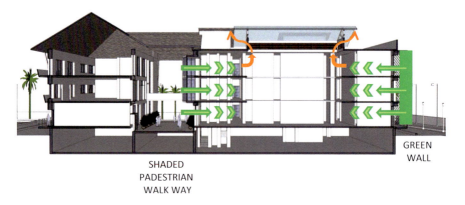

Fig. 1 Schematic of airflow for natural air conditioning

(Fig. 1). As the operational cost would be low, it would be a facility with low operating costs. He also suggested that he would do the project in phases.

As an experiment, he wanted to use coconut wood. He had heard of some architects using coconut wood in Tamil Nadu, a southern coastal state in India. Coconut wood does not need seasoning. His reasoning was that coconut wood would be widely available in Goa as well as it is also a coastal state. He surveyed the local constructions. While there were a few domestic users, there were no commercial buildings that had used coconut wood. He made up his mind on using coconut wood as it would be a novelty for Goa. The wood would be drawn from trees which had become non-fruit giving.

For the cooling solution without air conditioning, he proposed a novel vertical garden which has double-sided ventilating capability – a vertical garden with a "double skin." The hospital waste would be used for irrigation of the vertical garden. The construction material would be coconut waste material (Figs. 2 and 3).

Case Analysis

Vikram made systematic and extensive use of the resources at hand – what Baker and Nelson (2005) refer to as bricolage. The understanding of how resource-oriented entrepreneurship leads to the creation of environmentally sustainable practices and solutions is discussed below.

Scavenging (Baker and Nelson 2005) is understood as the collecting of resources that others have discarded or do not intend to use. Vikram's use of coconut wood which others had not intended to use is an example of scavenging.

Scavenging provides Vikram the confidence that he can derive resources for a new project. This broadens his bricolage capabilities. Scavenging practices are also evident in Vikram's use of resources that are by-products or waste products from the hospital. Such practices reduce strain on the environment, as they substitute other

Fig. 2 Vertical garden setup

products that would have incurred production costs. It also makes local waste management convenient.

Use of waste material in irrigating the vertical garden is a form of recycling as well. Recycling, as practiced by Vikram, where the resources to be recycled are acquired at no cost, represents a form of bricolage that reduces the strain placed on the natural environment as the environmental costs of production have already been incurred. Moreover, it reduced the financial burden for the client.

In some cases where recycled materials were not available, Vikram made an effort to use locally available resources, i.e., resources that are naturally available in the local setting in which the practice takes place. The attempt at using coconut wood was in this direction.

Using locally available resources ideally reduces strain on the environment, mainly due to the transportation costs.

Another practice is the use of unprocessed materials. Coconut wood does not require seasoning and hence can be used directly.

Vikram extended this further and used another form of bricolage in his closed/semi-closed systems – using the hospital waste is the best example there.

Vikram not only reduced his use of water but also derived value from the system.

Deriving value from such closed system constitutes a "making do," as value is created with resources that are available and that under normal circumstances would be fed into the public waste management system.

The above mentioned six forms of bricolage used by Vikram are summarized in Table 2.

The case analysis and various forms of bricolage demonstrate that resource-oriented environmentally sustainable entrepreneurship exists as a phenomenon in

Fig. 3 (a) and (b) Vertical garden close-up

Table 2 Forms of bricolage used in Case 1

Form of bricolage	Examples of use
Scavenging	Visible from the many resources used by scavenging
Recycling	In the water recycling systems
Using local materials	Intent to use local material like coconut wood
Using unprocessed materials	Coconut wood which does not require seasoning
Establishing semi-closed systems	In the waste systems reusing the hospital wastes
Extracting value from semi-closed systems	In the use of novel air-conditioning solution

building industry. It indicates the importance of resources in sustainability architecture.

This type of entrepreneurship is innovative and disruptive – with focus on recombinations of resources. The actions performed by the entrepreneur are distinct from the classical view of entrepreneurs.

Baker and Nelson (2005) cite three examples of bricolage that have harmful environmental consequences and where new resources – from an environmental perspective – should have been acquired.

The six forms of bricolage practiced by Vikram need to be more sustainable than other forms of architectural practices. For example, scavenging is sustainable only if the resources scavenged and stored do not lead to the emission of toxic or otherwise harmful substances. The same is the case for recycling, which is sustainable if the potential environmental damages offset those caused by the production and use of a newer alternative.

Hence, it cannot be said that all forms of resource-oriented entrepreneurial actions are environmentally sustainable.

In terms of inspiring others, Vikram's most valuable contribution to sustainability is providing a response to two key challenges in terms of sustainable resource consumption:

(i) Making visible the flow of resources (such as airflow system and recycled waste)
(ii) Anchoring these flows in a closed system

Vikram accomplished both tasks successfully.

Case Study 2: Chitra Vishwanath

Chitra runs her firm Biome Environmental Solutions Pvt. Ltd. in Bangalore, India. She has 21 years of experience in the field, though she has not had any formal training in sustainability.

She believes that sustainable construction is the only way forward. She is getting more and more entrenched into the concept that we see social unrest due to unsustainable lifestyle choices being made. "Lack of water and the race for energy are the top-most unsustainable lifestyle choices that people are making."

She started freelancing in her name in 1990 in Bangalore. She built their own home in 1995 with stabilized mud blocks (or compressed and stabilized earth blocks), rainwater harvesting, and solar installations which eventually became a very easy product to convince the clients. "We walked the talk and that made a lot of difference to our prospective clients."

Chitra's House ("San Souci")

The house was built to conserve and enhance biodiversity on-site and also make use of all natural resources available on it. In the design, a basement was put in which provided all the earth required to make the compressed stabilized earth blocks for the home. The basement is a thermally balanced space with temperatures never going above 23° C and is used as a study and playroom (Fig. 4a). The house has been designed to work with rainwater harvesting and organic waste disposal, and gray-water reuse are practiced.

Fig. 4 (**a**) and (**b**) Chitra's house

The roof is another part of the house designed to capture rainwater and use solar energy for cooking, water heating, and for lighting. An "ecosan" toilet placed here provides nutrients to grow rice in a rooftop paddy patch which is irrigated using wastewater from the washing machine which is treated using a reed drum system. Rice and millet grown on the rooftop provide food, while vegetables are grown in pots. The mulch layer rests upon silpaulin lining sheets to prevent dampness. The rooftop garden also helps to keep the house cool (Fig. 4b).

The house inside has no plaster or paint and works with mezzanines to use the volume available to the maximum. One mezzanine area is a sleeping loft, and the other is a study room for the son.

No artificial cooling systems such as fans or ACs have been used in this home. The green walls through creepers bring 32 different types of birds plus the occasional snake on to the rooftop. Sans Souci is an open house to study the various ecological ideas used and has been frequented by kids, college students, students of architecture, and others of the curious variety.

Bio-pool (Fig. 5a–d)
Built as part of the Native Village Eco Resort at Hessarghatta, a town in Karnataka state in India, this is a natural swimming pool that tries to recreate nature's mechanisms for purifying and maintaining water quality. Rather than using chemicals, the bio-pool incorporates reeds and other living organisms that replicate the kind of system existing in a lake.

Broadly there are two sections – one for swimming and another for purifying the water. Reeds and plants help to oxygenate the water while microorganisms and other creatures like frogs help keep it clean. Water circulation is achieved using a difference in levels thereby eliminating the need to artificially pump and circulate the water.

Playing Parks (Fig. 6a, b)
Kilikili is an NGO in India which works with disabled children (www.kilikili.org) with a focus on creating playspaces for such kids in public parks. Their concern stems from the fact that one never really finds a disabled child playing in a park – mainly due to lack of access and equipment. Kilikili first approached Biome in 2006 to help out with the design of such a play area in Coles Park, Bangalore, India. Chitra's aim is not to have separate spaces for disabled kids but rather a play area where all children – regardless of their abilities – can play together. Since all public parks are under the purview of Bruhat Bengaluru Mahanagar Palike (Municipal Corporation of Bangalore), India, Chitra and her team have been constantly engaging with them to sensitize the authorities and get their support for these activities. She has done five more similar projects since then – all in Bangalore.

Her innovation approach is based on the use of sustainable materials. She believes that there needs to be a lot of focus on sustainable materials, water treatment systems, and energy. "We want to always be able to give the client a salad bowl of various choices and we need innovations in them." In doing so, she foresees that there are sometime "end of the pipe" risks for such novel innovation. She mitigates such risks by first testing them on projects where the client is agreeable to test new ideas.

Most of their innovations can be easily implementable and are replicable. Most of the innovations are adopted in their designs. For example, they were one of the first to suggest adding a basement in the home designs in Bangalore. It was not a common practice to have it in homes. They have been also insisting that the basement be a habitable space since it is suited climatically too. The earth excavated from the basement is used for making the walling and roofing units.

She has found that sustainable designs are not more expensive than the conventional ones and in the longer run cheaper to maintain. It is sometimes the lack of conviction on the part of the future owner to do things which is not in-line with what

Fig. 5 (continued)

Fig. 5 (a–d) Phases in construction of the bio-pool

is seen commonly. It is also partly not enough knowledge about the consequences of one's actions as well as total disregard of the same in some cases.

The strategy she uses to convince her prospective clients are the following:

1. Walking the talk
2. Engaging them with her existing clients
3. Gently making them aware on what is unsustainable and that we will leave a sorry earth for our kids

There are times when this gentle cajoling and sharing of best practices do not help. In about 10% of the cases, the clients do say no to such sustainable ideas. Most of the resistance comes on the use of mud, bare walls without paint, treatment of gray/black water and its reuse, and the usage of ecosan options.

When this happens, she considers it as a new challenge and strives to make the design a lot more appealing while still incorporating what she believes is the best for the environment.

Case Analysis

Chitra has made use of bricolage to a great extent. Some of the uses of bricolage by Chitra are given in Table 3.

The concept of bio-pool is of significant value as there are many places in India where it can be used.

Fig. 6 (**a**) and (**b**) Playing parks

Case Study 3: Sanjay Prakash

Sanjay Prakash is an architect with a commitment to energy-conscious architecture, eco-friendly design, people's participation in planning, music, and production design. Over the years, he has integrated all his work with the practice of new urbanism and sustainability in his professional and personal life.

His area of practice and research over the last 33 years includes passive and low energy architecture and planning, hybrid air conditioning, autonomous energy and water systems, bamboo and earth construction, community-based design of common property, and computer-aided design. Under his guidance, hundreds of persons have developed capabilities in performing design, conceptual, or management work in these areas.

Table 3 Forms of bricolage used in Case 2

Form of bricolage	Examples of use
Scavenging	Visible from the many resources used by scavenging, e.g., use of truck tires in children parks
Recycling	In the water recycling systems, e.g., ecosan solutions and bio-pool
Using local materials	Intent to use local material like reeds and plants in bio-pool
Using unprocessed materials	Reeds, plants, etc. to create water filtration solution
Establishing semi-closed systems	In the bio-pool system
Extracting value from semi-closed systems	In the use of novel bio-pool system

He is the principal consultant of his design firm, SHiFt: Studio for Habitat Futures (formerly Sanjay Prakash and Associates) which has 15 associates. He is a senior advisor at the Indian Institute for Human Settlements (IIHS). His name and work are mentioned in the twentieth edition of one of the main reference works in architectural history, *A History of Architecture*, by Sir Bannister Fletcher.

Sanjay knows no other way of working – but only the sustainable way. He believes that "it is authentic, unself-conscious and natural for me." He just cannot follow non-sustainable practices – even at the cost of losing work, though his reputation precludes that. His interest in sustainability was triggered by reading Schumacher's *Small is Beautiful*. His first projects were in designing and developing solar houses in Srinagar, Jodhpur, and Delhi (all cities in India). The most important innovations for him are the ones that are resource-conserving and which are integral to the architecture.

Some of the risks that he takes cognizance of are as follows:

1. Innovations that do not work or do not work as well as planned.
2. There are few vendors and contractors who are conversant with the best sustainable methods, and hence the vendors would need to be developed from a scratch.

The biggest barrier in adoption of sustainable innovations and technologies is the misunderstanding of green versus sustainable versus innovative. "Going green" does not necessarily imply being sustainable.

He has seen demand for sustainable buildings increase over the years. As his reputation has also grown over the years, he has seen that his clients self-select on the basis of his reputation. He has also seen that now there are a lot of "charlatans in the garb of green" offering their services.

He does not anymore need to convince clients to accept sustainable solutions. The most often discussion he has now is the "sufficiency discussion." The clients want to know how much is enough.

Based on his own experiences, his thoughts are centered around the following six tactics to manage physical stock and flows in a community:

1. Use less with Factor 4 technologies for supply and social limits of sufficiency and equity on demand.
2. Grow your own tapping harvestable yields as autonomously as possible.
3. Build two-way networks for security: every consumer is also a producer.
4. Store a lot because renewable resource yields are often diurnal and seasonal.
5. Transport less over shorter distances using least life-cycle cost technologies.
6. Exchange using intelligent wireless networks to enable real-time trade and delivery of goods.

He then explained the approach to mainstream green and sustainability for large development and settlements and how to go beyond (bringing in equity, sufficiency, resilience).

Residence in Delhi, India

This was one of Sanjay's early projects in Delhi and is of EASE (environmentally appealing and saving energy) type (Fig. 7).

The following were the design features:

- The house was oriented to face south so that every habitable room has liberal solar exposure. Shading was designed to prevent solar gains during summer and allow the same during winter.
- External wall and roof insulation help the thermal mass to act in tandem with the inner space.
- Reflective finishes of wall and roof for heat reflection.

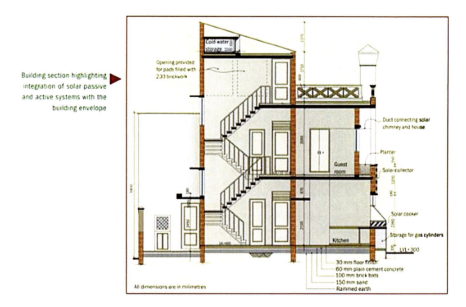

Fig. 7 Schematic of a residence designed in Delhi

- Multidirectional evaporative wind tower atop stairwell, for cooling of house during dry summer.
- Monsoon comfort dependent on strategically opening and closing of windows.
- South-facing solar chimneys assist air exhaust during summer.
- Architecturally integrated solar water heating system and solar cooker.
- Compost pits in garden for composting of kitchen wastes.
- Low water flush valves without cisterns.

Mirambika, New Delhi

The building was conceived as a place to hold an innovative program of education, research, training, and scholarship. Mirambika is a "free progress" school where the stress is on child-centric value-oriented education. The spaces were designed to be organic, amorphous, and such that the boundary between inside space and outside space would vanish.

The building is planned as a space-filling modular grid on a basic structural dimension of 7.2 m, with concrete ribs subdividing the slab into eight or four equal panels. Though they are very few, walls can be shifted to any of these locations on the ribs, and the floor which is made under them will need to be ground, while the electrical services in the ceiling are redundantly arranged so that they can just drop down according to flexible arrangements of the spaces below. Water-based services (sinks, toilets, water play areas) are placed in defined, relatively inflexible, locations.

The building is furnished with white terrazzo and china mosaic roofs, extensive courtyards and vegetation for cooling and ventilation, and integrated solar hot water.

The design team was apprehensive that the type of building detailed – almost like a verandah, with few walls, therefore open to dust – might prove a liability due to heat and dust, but as it turns out, the dust-free environment of the ashram around has allowed this building to actually be comfortable in all seasons.

Case Analysis

Some of the uses of bricolage by Sanjay are given in Table 4.

Table 4 Forms of bricolage used in Case 3

Form of bricolage	Examples of use
Scavenging	South-facing construction so that every habitable room has liberal solar exposure
Recycling	Compost pits in garden for composting of waste materials
Using local materials	Use of natural sunlight for lighting/heating, etc.
Using unprocessed materials	Use of natural sunlight
Establishing semi-closed systems	Use of wind tower/solar chimneys/solar systems
Extracting value from semi-closed systems	Multidirectional evaporative wind tower atop stairwell, for cooling of construction during dry summer

Bricolage: A Feasible Option in Construction Industry that Can Address Sustainability Concerns

Schaltegger (2002) suggested ecopreneurial types: alternative actors, bioneers, and ecopreneurs do not fall in the above cases of *bricoleur entrepreneurs*. It is suggested that there is a novel type of ecopreneur: bricoleur sustainable entrepreneurs. They are inventors, with high R&D focus, *and* use the material that is readily available. They interact with the market in a large way.

Responses from the three ecopreneurs are summarized in Table 5.

Some of the themes which emerge from the above three cases are given in Table 6.

Proposed Model for Bricoleur Entrepreneurs

Resource-based entrepreneurship as a phenomenon exists in the building architecture industry, not highly reputed for innovation. Bricoleur ecopreneurship exists as a phenomenon as evident from the above three cases. The forms of bricolage as identified from the analysis are given in Table 7.

Ecopreneurs exist in building architecture industry. They play an important role in catalyzing the sustainable designs. Ecopreneurs see themselves differently from others – the others who are "charlatans in green." Ecopreneurs innovate – however, they are also concerned about the effectiveness of their innovations ("end of the pipe risks," "innovations do not work as well as desired").

Some of them are concerned about what customers want and make designs suitable to customer needs ("we then look at how we can strive to make the designs a lot more appealing"). However there are others who do not change their design style ("I do not need to convince them. They come to me").

Ecopreneurs practice bricolage extensively. The model for such entrepreneurs is presented in Fig. 8.

Table 5 Responses from ecopreneurs

	Sanjay Prakash	Vikram Varma	Chitra Vishwanathan
Special training in sustainability	No	Yes	No
Importance of what customer wants	Emphatic no	Yes	Emphatic yes
Understanding of implications of long-term use of non-sustainable designs	Yes	Yes	Yes
Importance of latest technology/innovation	Yes	Yes	Emphatic yes
Importance of vendor base to succeed	Yes	Yes	Yes
Lifestyle of customers as a factor in sustainable designs	No	No	Yes
Walking the talk	No need	Ambiguous	Strongly needed

Table 6 Themes emerging from the case analyses

Repetitions	1. Designs to address climate/environment issues 2. Vendor base for executing sustainable designs is weak 3. Growing demand for sustainable designs 4. Innovations do not work as well as desired 5. Lifestyle needs of customers is a deterrent
Indigenous typologies	1. Ecosan designs 2. Bricolage applications in various forms
Metaphors used	1. Millimeters make a meter and meters make kilometers 2. Small is beautiful 3. Walk the talk 4. End of the pipe risks 5. New avatar of innovations 6. Charlatans in green
Linguistic connectors	1. … "we then look at how we can strive to make the designs a lot more appealing"

Table 7 Forms of bricolage

Form of bricolage
Scavenging
Recycling
Using local materials
Using unprocessed materials
Establishing semi-closed systems
Extracting value from semi-closed systems

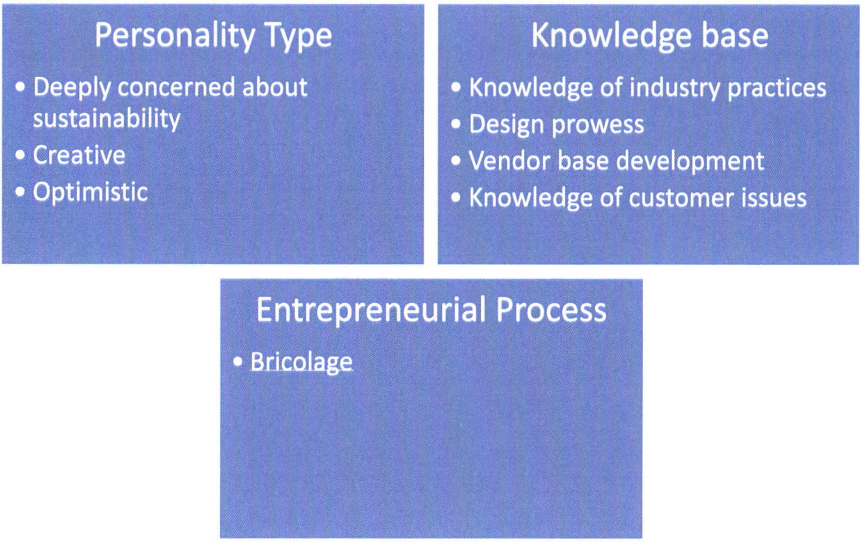

Personality Type
- Deeply concerned about sustainability
- Creative
- Optimistic

Knowledge base
- Knowledge of industry practices
- Design prowess
- Vendor base development
- Knowledge of customer issues

Entrepreneurial Process
- Bricolage

Fig. 8 Model for bricoleur entrepreneurs

The contributions within entrepreneurship literature relate bricolage use of knowledge domains (Baker et al. 2003) and creation of "something from nothing" in resource-poor environments (Baker and Nelson 2005). Ferneley and Bell (2006) have applied bricolage to information systems and term it as "a useful concept as it deals with the need for SMEs to learn about the possibilities of IS in situ, simultaneously exploiting the can-do approach." Wu et al. (2017) have concluded that "bricolage hastens new-product development." Weick (1993) identified bricolage as one of the sources of resilience and useful in creating order out of whatever is available impacting the firm's resilience in resource-constrained environments.

The three cases discussed here add to the knowledge on the understanding of *bricoleur ecopreneurs*. This is a new generation of ecopreneurs seeking to combine environmental awareness with business success and conventional entrepreneurial activity and creating new forms of business models. In ecopreneurship paradigm, such entrepreneurial activity may become increasingly central to market success as environment concerns become more prominent.

Cross-References

▶ Low-Carbon Economies (LCEs)
▶ People, Planet, and Profit
▶ Sustainable Living in the City
▶ The Spirit of Sustainability
▶ The Theology of Sustainability Practice

References

Aldrich, H. E., & Fiol, C. M. (1994). Fools rush in – The institutional context of industry creation. *The Academy of Management Review, 19*(4), 645–670.

Anderson, A. (2000). The protean entrepreneur: The entrepreneurial process as fitting self and circumstance. *Journal of Enterprising Culture, 08*, 201.

Association for the Conservation of Energy (ACE). (2004). (http://www.ukace.org/category/ace-research/). Briefing on the energy performance of buildings directive.

Baker, T., & Nelson, R. E. (2005). Creating something from nothing: Resource construction through entrepreneurial bricolage. *Administrative Science Quarterly, 50*(3), 329–366.

Baker, T., Miner, A. S., & Eesley, D. T. (2003). Improvising firms: Bricolage, account giving and improvisational competencies in the founding process. *Research Policy, 32*(2), 255–276.

Beveridge, R., & Guy, S. (2005). The rise of the eco-preneur and the messy world of environmental innovation. *Local Environment, 10*(6), 665–676.

Dean, T., & McMullen, J. (2007). Toward a theory of sustainable entrepreneurship: Reducing environmental degradation through entrepreneurial action. *Journal of Business Venturing, 22*(1), 50–76.

Duymedjian, R., & Rulings, C. C. (2010). Towards a foundation of bricolage in organization and management theory. *Organization Studies, 31*(2), 133–151.

Eastwood, D., Eaton, M., Guyer, C., & Stark, T. (2001). An examination of employment change in Northern Ireland's environmental industry 1993–2003. *European Environment II, 11*, 197–210.

Eisenhardt, K. M. (1989). Building theories from case study research. *The Academy of Management Review, 14*(4), 532–551.

Eisenhardt, K. M., & Graebner, M. E. (2007). Theory building from cases: Opportunities and challenges. *Academy of Management Journal, 50*(1), 25–32.

Elkington, J. (1994). Towards the sustainable corporation: Win-win-win business strategies for sustainable development. *California Management Review, 36*(2), 90, Berkeley, Calif.

Ferneley, E., & Bell, F. (2006). Using bricolage to integrate business and information technology innovation in SMEs. *Technovation, 26*(2), 232–241.

Flyvbjerg, B. (2004). Five misunderstandings about case-study research. In C. Seale, G. Gobo, J. F. Gubrium, & D. Silverman (Eds.), *Qualitative research practice* (pp. 420–434). London: Sage.

Garud, R., Jain, S., & Kumaraswamy, A. (2002). Institutional entrepreneurship in the sponsorship of common technological standards: The case of Sun Microsystems and Java. *Academy of Management Journal, 45*, 196–214.

Gibbs, D. (2009). Sustainability entrepreneurs, ecopreneurs and the development of a sustainable economy. *Greener Management International, 55*, 63–78.

Harbi, S. E., Anderson, A. R., & Ammar, S. H. (2010). Entrepreneurs and the environment: Towards a typology of Tunisian ecopreneurs. *International Journal of Entrepreneurship and Small Business, 10*(2), 181–204.

Hart, J. (2006). The new capitalists: Is it possible to make money and really make a difference? *Utne, 135*, 39–43.

International Energy Agency (www.iea.org/) IEA. (2008). *World energy outlook*. Paris: IEA.

Jong, J., & Muizer, A. (2005). *De meest innovatieve sector van Nederland, ranglijst van 58 sectoren*. Zoetermeer: EIM.

Kirzner, I. (1973). *Competition and entrepreneurship*. Chicago: University of Chicago Press.

Levi-Strauss, C. (1967). *The savage mind*. Chicago: The University of Chicago Press.

Linnanen, L. (2002). An insider's experiences with environmental entrepreneurship. *Greener Management International, 38*, 71–80.

Maguire, S., Hardy, C., & Lawrence, T. B. (2004). Institutional entrepreneurship in emerging fields: HIV/AIDS treatment advocacy in Canada. *The Academy of Management Journal, 47*(5), 657–679.

Pacheco, D. F., Dean, T. J., & Payne, D. S. (2010). Escaping the green prison: Entrepreneurship and the creation of opportunities for sustainable development. *Journal of Business Venturing, 25*(5), 464–480.

Parrish, B. D. (2006). *Beyond cleaner production: Entrepreneurship and the design of sustainable enterprise*. Paper presented to the international conference on green and sustainable innovation, Chiang Mai, p. 29.

Post, J. E., & Altman, B. W. (1994). Managing the environmental change process: Barriers and opportunities. *Journal of Organizational Change Management, 7*(4), 64–81.

Randjelovic, J., O'Rourke, A. R., & Orsato, R. J. (2003). The emergence of green venture capital. *Business Strategy and the Environment, 12*(4), 240–253.

Revell, A., & Blackburn, R. (2007). The business case for sustainability? An examination of small firms in the UK's construction and restaurant sectors. *Business Strategy and the Environment, 16*, 404–420.

Revell, A., & Rutherfoord, R. (2003). UK environmental policy and the small firm: Broadening the focus. *Business Strategy and the Environment, 12*, 26–35.

Schaltegger, S. (2002). A framework for ecopreneurship. *Greener Management International, 2002*(38), 45–58.

Schlange, L. E. (2006). *What drives sustainable entrepreneurs?* Paper presented at applied business and entrepreneurship association international conference, Kona, pp. 16–20.

Schlange, L. E. (2009). Stakeholder identification in sustainability entrepreneurship. *Greener Management International, 55*, 13–32.

Shane, S. & S. Venkataraman (2000). "The promise of entrepreneurship as a field of research. *Academy of Management Review 25*, 217–226

Stevenson, H. H., & Jarillo, J. C. (1990). A paradigm of entrepreneurship – Entrepreneurial management. *Strategic Management Journal, 11*, 17–27.

Tilley, F. (2000). Small firm environmental ethics: How deep do they go? *Business Ethics: a European Review, 9*(I), 31–41.

Tilley, F. (2007). *Conceptualising sustainability entrepreneurship*. Paper presented to the first world symposium on sustainable entrepreneurship, University of Leeds.

Tilley, F., & Parrish, B. D. (2006). From poles to wholes: Facilitating an integrated approach to sustainable entrepreneurship. *World Review of Entrepreneurship, Management and Sustainable Development, 2*(4), 281–294.

United Nations Environment Program – Sustainable Buildings & Climate Initiative (UNEP-SBCI). (2009). *Buildings and climate change summary for decision makers*. www.unep.org/sbci/pdfs/sbci-bccsummary.pdf.

Walley, E. E., & Taylor, D. W. (2002). Opportunists, champions, mavericks. . .? *Greener Management International, 38*, 31.

Weick, K. E. (1993). The collapse of sensemaking in organisations: The Mann Gulch disaster. *Administrative Science Quarterly, 38*(4), 628–652.

Willis, R., Webb, M., & Wilsdon, J. (2007). *The disrupters: Lessons for low-carbon innovation from the new wave of environmental pioneers*. London: NESTA.

Wu, L., et al. (2017). Bricolage effects on new-product development speed and creativity: The moderating role of technological turbulence. *Journal of Business Research, 70*, 127–135.

Yin, R. K. (2009). *Case study research design and methods* (4th ed.). Thousand Oaks: Sage.

York, J. G., & Venkataraman, S. (2010). The entrepreneur–environment nexus: Uncertainty, innovation, and allocation. *Journal of Business Venturing, 25*(5), 449–463.

Part VII

Contemporary Trends and Future Prospects

Responsible Investing and Environmental Economics

Green Finance and the Transition to a Green Economy

Carol Pomare

Contents

Introduction and Literature Review	1020
What Are you Getting with Socially Responsible Investments (SRIs)?	1021
Do Investments with a Sustainable Label Actually Fulfill their Ethical and Financial Promises?	1027
Summary	1029
Featured Case in Point	1029
Environmental Economics and the Case of Crude Oil Extraction Firms	1029
Environmental Economics and the Case of the Canadian Crude Oil Extraction Firms	1030
Summary	1033
Reflection Questions	1033
Theory on Investing and Economics	1033
Theory on Responsible Investing and Environmental Economics	1034
Theory on Responsible Investing and Passive Versus Active Investing Strategies	1036
Exercises in Practice	1037
The World's Most Sustainable Companies	1037
Engaged Sustainability Lessons	1039
Chapter-End Reflection Questions	1039
Cross-References	1040
References	1040

Abstract

This chapter aims at discussing responsible investing and environmental economics within the context of climate change. It was observed that publicly traded firms are applying sustainability strategies in their operations and capital spending, because investors are increasingly interested in responsible investing. For publicly traded firms and responsible investors, such changes towards more

---

C. Pomare (✉)
Ron Joyce Center for Business Studies, Mount Allison University, Sackville, NB, Canada
e-mail: cpomare@mta.ca

© Springer International Publishing AG, part of Springer Nature 2018
S. Dhiman, J. Marques (eds.), *Handbook of Engaged Sustainability*,
https://doi.org/10.1007/978-3-319-71312-0_42

sustainability mitigate the risk for future unburnable carbon and/or bursting of the carbon bubble on capital markets because of climate change-related regulations. However, because of greenwashing, some concerns arise that material amounts of publicly traded firms' unsustainable performances may be hidden from financial statements and annual reports. As a consequence, the challenge for responsible investors is not only to: (i) evaluate the scale of environmental operational and capital investments (e.g., attempts to decrease toxic emissions) but also to (ii) evaluate publicly traded firms' environmental performance associated with the assets locked into their reserves (e.g., oil reserves for oil publicly traded firms). This chapter follows a specific structure to explore the challenge for responsible investors. Section "Introduction and Literature Review" is focused on an introduction and literature review. Section "Featured Case in Point" is focused on a featured case related to responsible investing and environmental economics. Section "Reflection Questions" refers to reflection questions with theoretical developments on: (i) investing and economics, (ii) responsible investing and environmental economics, and (iii) responsible investing with passive versus active investing strategies. Section "Exercises in Practice" presents applied exercises to be used in class. Section "Engaged Sustainability Lessons" summarizes engaged sustainability lessons. Section "Chapter-End Reflection Questions" expands on the chapter with reflection questions. The discussion in terms of the impact of responsible investing on environmental economics is of interest, since it is believed that substantial adjustments of regulating practices are required for publicly traded firms to realistically alter their financial statements and annual reports, for responsible investors to be fully informed, and for engaged sustainability to thrive.

Keywords
Socially Responsible Investing (SRI) · Divestment · Green finance · Environmental economics · Green house gas emissions · Stranded assets · Carbon bubble · Capital markets · Efficient market hypothesis

Introduction and Literature Review

This chapter aims at discussing responsible investing and environmental economics within the context of climate change. Indeed, it was observed that publicly traded firms were applying sustainability strategies in their operations and capital spending, because investors were interested in responsible investing. For publicly traded firms and responsible investors, such changes towards more sustainability mitigated the risk for stranded assets (i.e., unburnable carbon) and bursting of the carbon bubble (i.e., sudden decrease of stock prices) on capital markets.

In the next sections, responsible investing strategies will be reviewed, as well as, their link with environmental economics. The featured case in point will be related to Canadian crude oil extraction firms.

As an introduction to investing, here are some broad definitions of typical investing options (i.e., for details see Investopedia at: http://www.investopedia.com/terms/s/security.asp). First, bonds are debt securities under which the issuer owes a debt and typically is obliged to pay interest (i.e., a coupon) and to repay the debt (i.e., the principal) to the lender, at a later date (i.e., the maturity date). Second, stocks of a corporation are related to a publicly traded firm's ownership with a single share of stock representing fractional ownership of the firm. Third, Mutual Funds (MFs) and Exchanged Traded Funds (ETFs) typically are baskets of bonds and/or stocks. Fourth, a stock index or stock market index is a measurement of the value of a section of the stock market that is used to compare the return on any specific investment (i.e., it is computed from the prices of selected stocks to describe the market).

For responsible investors, being responsible means using green investment strategies to assess environmental operations and capital investments, as a way to mitigate any adverse impact of publicly traded firms' unsustainable practices. For example, responsible investors may decrease investment in industries with high levels of Green House Gas (GHG) emissions. Responsible investors may change their investment strategies for ethical reasons, but also because they expect a bursting of the carbon bubble related to traditional investors investing in fossil fuels and the sudden decrease in stock prices because of climate change-related regulations (e.g., regulations against Green House Gas, GHG, emissions) leading to fossil fuels becoming stranded or unburnable.

In a related vein, Socially Responsible Investing (SRI) is an ethical positioning that has recently been endorsed by capital markets. This endorsement by capital markets led to the development of green finance options for responsible investors, including bonds (e.g., debts of publicly traded firms investing in environmental operations and capital expenditure) and funds (e.g., basket of stocks from publicly traded firms investing in environmental operations and capital expenditure). More specifically, in the context of green finance, the focus is typically on green bonds, green Mutual Funds (MFs) and/or green Exchanged Traded Funds (ETFs).

In this section, we explore Socially Responsible Investing (SRI) and capital markets with the following questions: (i) What are you getting when you put your money into a Socially Responsible Investments (SRIs)? and (ii) How do you know which investments with a "sustainable" label actually fulfills its promise?

What Are you Getting with Socially Responsible Investments (SRIs)?

Socially Responsible Investing (SRI) is more than not buying the shares of publicly traded firms in controversial industries like tobacco, firearms, alcohol, gambling, or oil (i.e., divestment). Socially Responsible Investing (SRI) is considering a wide range of corporate behaviors as related or not to Environmental, Social and Governance (ESG) factors (Davidson 2016).

Responsible Investing Through Divestment

Divestment is the opposite of investment, as it is the process of avoiding an asset for either financial, or social, or political reasons. Assets that can be divested may include a subsidiary, business department, real estate, equipment and/or investment in stocks (i.e., for details see Investopedia at: http://www.investopedia.com/terms/d/divestment.asp). In the context of climate change, divestment happens when responsible investors withdraw from a particular industry or a specific geographic region due to political, social or environmental pressures (i.e., for details see Investopedia at: http://www.investopedia.com/terms/d/divestment.asp).

A fossil fuel equity exposure ratio is a ratio used to assess the exposure of any investment option (e.g., Mutual Fund (MF), Exchange Traded Fund (ETF), or publicly traded firm) to fossil fuel as compared to the broader capital market (Ansar et al. 2013). As a reference, the fossil fuel exposure ratio of a university endowment or a public pension fund is generally very small (Ansar et al. 2013). On average, university endowments in the United States have 2–3% of their assets invested in fossil fuel equities (Ansar et al. 2013). The proportion in the United Kingdom (UK) is higher with an average of 5%, because the Financial Times Stock Exchange (FTSE) 100 Index has been described as having greater proportion of fossil fuel companies (Ansar et al. 2013).

On the one hand, the *direct* impact of fossil fuel divestment on equity or debt is limited (Ansar et al. 2013). The maximum possible capital that might be divested from fossil fuel companies by university endowments and/or public pension funds represents a small pool of funds as per the fossil fuel equity exposure ratios described above (Ansar et al. 2013). As such, even if the maximum possible capital was divested from fossil fuel companies, these companies' share prices would be unlikely to suffer massive declines (Ansar et al. 2013). Some investors would probably even welcome the opportunity created by the divestment to increase their holdings of these fossil fuel companies, particularly if these stocks presented a short-term discount (i.e., bargain) because of other investors' divestment (Ansar et al. 2013).

On the other hand, even when divestment outflows are small and do not directly affect future cash flows of publicly traded firms, divestment triggers an *indirect* change in social norms that tends to close channels of previously available money, with a downward pressure on the stock price of the targeted publicly traded firms (Ansar et al. 2013). One indirect impact is the stigmatization resulting from a divestment campaign which may pose a real threat to firms and their value chain (Ansar et al. 2013). As with individuals, a stigma may produce negative consequences. For example, firms heavily criticized in the media may suffer from a bad image that scares suppliers, subcontractors, employees, and customers away. Governments and politicians may prefer to engage with "clean" firms to prevent adverse effects on their own reputation or potential for reelection. Another indirect impact is the restrictive legislation resulting from divestment campaigns. Indeed, in most divestment campaigns, there is intensive lobbying in favor of restrictive legislations against the targeted firms (Ansar et al. 2013).

Historically, research credits divestment by prominent American universities, as well as, by important public pension funds, as leading to a tipping point, and as paving the way for other universities or public institutions in the United States (USA) and abroad (Ansar et al. 2013). For example, a first wave of divestment started in the United States (USA) about social issues related to tobacco or Apartheid in South Africa, with divested amounts that were very small but created public awareness (Ansar et al. 2013). Both in the case of tobacco and the Apartheid, the divestment campaigns took a few years to take off, until very high profile American universities announced their divestment from these industries or geographic area, during what was known as a second wave of divestment (Ansar et al. 2013). This involvement of very high profile American universities was the tipping point paving the way to a generalized global impact.

Similarly, a withdrawal of debt finance from fossil fuel companies by some institutions or any increase in cost of capital as a consequence of a divestment campaign may not pose any serious debt financing problem in terms of short-term liquidity or capital expenditure for fossil fuel firms in highly functioning markets (i.e., high information efficiency) (Ansar et al. 2013). Changes in social norms are relevant, however, in relatively poorly functioning markets (i.e., low information efficiency). In particular, borrowers in countries with low financial depth (i.e., low informational efficiency) may experience a restricted pool of debt financing if financial institutions in the local financial network withdraw their support because of a divestment campaign (Ansar et al. 2013). While an increase in cost of capital is unlikely to have an effect on overall corporate finance of major fossil fuel firms, their ability to undertake large capital expenditure projects in difficult environments may be diminished due to divestment campaigns and lower availability of debt financing within poorly functioning markets (Ansar et al. 2013).

As a consequence, a divestment campaign may entail small outflows in the early phases of a campaign, but drastic outflows once a certain tipping point has been reached in terms of market norms (Ansar et al. 2013). As such, even if the direct impact of divestment outflows is small in the short term, a campaign may have massive long-term indirect impacts on the value of a target firm (Ansar et al. 2013).

Responsible Investing and Business with a Purpose
Sadler (2017a) uses a quote from Sir Richard Branson, the well-known CEO of Virgin, in one of his book "*Screw Business as Usual*," which is as follows: "Never has there been a more exciting time for all of us (...) where doing good really is good for business."

Richard Branson's quote highlights a phenomenon where purpose (i.e., doing good) becomes a critical part of business success, as: (i) customers demand it; (ii) employees are attracted to it; and (iii) investors embrace it through responsible investing (Sadler 2017a).

Richard Branson's quote also highlights a generational shift as, 25 years ago, entrepreneurs wanted to be successful and seldom alluded to a greater purpose (Sadler 2017b). Apparently, generational change is a big factor in this phenomenon, as Deloitte's Millennial Survey 2014 found that millennials demand that companies from which they buy endorse social and environmental responsibility (Sadler 2017b). About 51% of millennials say they would pay extra for sustainable products which are four times the number of baby boomers ready to pay extra for sustainable products (Sadler 2017b). The new generation of entrepreneurs, especially millennial entrepreneurs, also has a support network of nonprofit organizations (e.g., Green Chamber of the South's Greenhouse Accelerator, etc.) whose specific purpose is to enhance a millennial entrepreneur's chance of success when embracing a social or environmental purpose (Sadler 2017b).

Richard Branson's quote finally highlights a transition with more than 10% of professionally managed investments now focusing on Environmental, Social and Governance (ESG) factors (Sadler 2017b). The democratization of investments for social and environmental impact projects could be one of the most important trends nowadays (Sadler 2017b). First, responsible investors have now access to potentially lucrative partnerships with a focus on: (i) renewable energy; (ii) organic farmland conversion; or (iii) brownfield remediation (Sadler 2017b). Second, more than 2,000 companies in 35 countries have passed a series of "triple-bottom-line" tests to become what is known as "Certified B Corporations" (Sadler 2017b). The success of "Certified B Corporations" with investors shows that responsible investors are looking for real and measurable environmental impacts (Sadler 2017b). Third, among celebrities who made a strong case for responsible investing, Bill Gates, the well-known billionaire and philanthropist, recently launched a $1 billion fund called Breakthrough Energy Ventures (Dolan 2016). With Breakthrough Energy Ventures, Bill Gates, Mike Bloomberg, the founder of Bloomberg LP and the former mayor of New York City (i.e., for details see Bloomberg at: https://www.boomberg.com), and 20 other well-known investors, including Silicon Valley Venture capitalists, focused on investing in new and "clean" forms of energy (Dolan 2016). Fourth, the well-publicized Breakthrough Energy Venture was not an exception, since: (i) thanks to a tracking method based on energy supply and other industries estimates, it is now known that, in 2013, responsible investing and green finance amounted to an approximate US$331 billion (Campiglio 2016); and (ii) 69% of Fortune 500 companies presented higher demand for "low carbon" ventures in 2016 (Velasquez-Manoff 2016). Finally, the development of green finance was fueled by very high demands in developing regions, like China (Campiglio 2016). Indeed, China is now seemingly the main investor in renewable energy with US$56 billion invested in such type of energy (Campiglio 2016).

Responsible Investing Through Green Bonds
As described previously, green bonds are fixed income instruments which finance debt for environmentally sustainable positive net present value projects while supposedly offering competitive returns to the responsible investor (i.e., for details see Investopedia at: http://www.investopedia.com/terms/g/green-bond.asp). Green bonds are issued to generate money that supports environmentally friendly business ventures, and when one invests in green bonds, one is buying into this responsible investing philosophy (Sadler 2017b).

Fixed-Income Instruments for Socially Responsible Investing (SRI) is a relatively new concept compared to other types of financial instruments (Sadler 2017b). Green bonds are increasingly being issued to fund environmentally-oriented projects delivering similar credit quality and income characteristics as any other mainstream fixed-income issue (Sadler 2017b). Even the municipal bond market can be segmented to deliver social impact (Sadler 2017b). By focusing on green segments (e.g., green education, green health, clean water, green space, etc.), responsible investors can build a portfolio aligned with their social and environmental values (Sadler 2017b). While the idea is still in its infancy in Europe or the United States (USA), it is likely to grow rapidly as green investing becomes more and more mainstream (Sadler 2017b).

In Europe, in May 2007, the European Investment Bank issued more than one billion Euros worth of "Climate Awareness Bonds" (Sadler 2017b). It was the first bond issue made available through a public offering process in the European Union (E.U.). The funds were set to be used in funding renewable energy projects as part of Europe's commitment to produce 20% of its energy from renewable sources by 2020 (Sadler 2017b). Rather than pay a coupon, the bond was held for 5 years before being redeemed at face value plus an amount linked to the performance of an index, the Financial Times Stock Exchange (FTSE) "Good Environmental Leaders Europe 40" (i.e., an index of large-cap companies that are involved in renewable energy or energy efficiency-related businesses) (Sadler 2017b). A 5% minimum return was guaranteed (Sadler 2017b). Similarly, the Deutsche Bank was apparently one of the founding members of the green bond principles in Europe (Deutsche Bank 2017). Alexander von zur Muhlen, the Group Treasurer of Deutsche Bank, reportedly said that the green bond market had matured in 2014 and that the size and number of green bonds offerings had substantially increased since then, making green securities a viable investment option (Deutsche Bank 2017). For that reason, Deutsche Bank had already made 200 million Euros in eligible green bond investments (Deutsche Bank 2017). A Deutsche Bank report released in 2017 even announced the Bank's intention to invest one billion Euros into a portfolio of high-quality liquid assets in the form of green bonds (Deutsche Bank 2017). These high-quality liquid assets were to be held as part of the Bank's Liquidity Reserve investments. By establishing this portfolio, the Deutsche Bank, among other European financial & investing institutions, aimed at supporting the expansion of the green bond segment in Europe (Deutsche Bank 2017).

In the United States (USA), green bonds got a boost from an amendment to the America Jobs Creation Act in 2004 (i.e., officially titled the Brownfields Demonstration Program for Qualified Green Building and Sustainable Design Projects and shortened to green bonds) (Sadler 2017b). This program was designed to provide funding, in the form of AAA-rated bonds issued by the United States (USA) Treasury to finance environmentally friendly developments. The objective was to reclaim contaminated industrial and commercial land (e.g., brown fields) (Sadler 2017b). The use of bonds to fund environmentally friendly projects, in some cases, even gave tax-exempt income (i.e., green bonds are usually taxable, but recently some green bonds issued by states in the United States (USA) were made nontaxable to investors) while generating a feel-good factor from supporting environmentally friendly projects (Sadler 2017b).

As such, Europe and the United States (USA) are building on their successful green bond origination franchises and highlighting this opportunity to fund sustainable initiatives while achieving relatively attractive returns for investors.

Responsible Investing Through Green Funds

As described previously, green funds are investments with a focus on supporting positive net present value environmental projects committed to: (i) the conservation of natural resources; (ii) the production and discovery of alternative energy sources; (iii) the implementation of "clean" air and water projects; and/or (iv) Socially Responsible Investing (SRI) practices (i.e., for details see Investopedia at: http://www.investopedia.com/terms/g/green-investing.asp).

First, Wall Street now has sustainable investing divisions that create products for key demographics, like for millennials who are eager to align their investments with their values (Davidson 2015). Indeed, capital markets relate to publicly traded firms that are listed on various stock exchanges and there has been a huge increase in Environmental, Social, and Governance (ESG) indexes in 2015 to more than 150 up from 25 in 2010 (Davidson 2015).

Second, a Mutual Fund (MF) is an investment made up of a pool of assets such as stocks of publicly traded firms and similar assets (i.e., for details see Investopedia at: http://www.investopedia.com/terms/m/mutualfund.asp). In terms of green Mutual Funds (MFs), major investment firms, like Morgan Stanley (i.e., for details see Morgan Stanley at: https://www.morganstanley.com) welcome the major responsible investing business opportunity: "Clients are really starting to think differently about their portfolios" says Audrey Choi, Chief Executive of Morgan Stanley's Institute for Sustainable Investing (Davidson 2015). Jean Rogers, the Chief Executive and Founder of the nonprofit Sustainability Accounting Standards Board (SASB) (i.e., for details see the Sustainability Accounting Standards Board at: https://www.sasb.org) says that several popular funds are available for investors who want to leave all decisions to the experts (Davidson 2015).

Third, an Exchange Traded Fund (ETF) is also an investment made up of a pool of assets such as stocks of publicly traded firms and similar assets, but generally has lower fees than Mutual Fund (MF) (i.e., for details see Investopedia at: http://www.investopedia.com/university/exchange-traded-fund/). The fees are generally lesser, as Exchange Traded Funds (ETFs) are available on self-directed platforms (i.e., platforms where investors make all investing decisions without the help of a financial advisor) or robo-adviser platforms (i.e., platforms with a logarithm picking investments based on users' answers about investment preferences and assembling a "sustainable" portfolio if required to). In terms of Exchange Traded Funds (ETFs), since December 2013, assets in the sustainable category have more than doubled, with 22 new Environmental, Social and Governance (ESG) Exchange Traded Funds (ETFs) tracking Environmental, Social and Governance (ESG) indexes (Davidson 2015). As of July 2015, assets invested in Exchange Traded Funds (ETFs) tracking

Environmental, Social and Governance (ESG) indexes have grown nearly 30% to $1.8 billion (Davidson 2015).

As such, it is more and more common to use green Mutual Funds (MFs) or green Exchange Traded Funds (ETFs). The idea is to invest in a basket of "sustainable" firms with a selection based on a series of established rubrics identifying "sustainable" firms that have a focus on Environmental, Social and Governance (ESC) practices.

Do Investments with a Sustainable Label Actually Fulfill their Ethical and Financial Promises?

Investors not only expect their investments to be related to Socially Responsible Investing (SRI); they also expect a competitive return for their Socially Responsible Investing (SRI). In February 2015, the Wall Street Journal (i.e., for details see Wall Street Journal at: https://www.wsj.com) interviewed two professors, Alex Edmans (i.e., Professor of Finance at London Business School) and David J. Vogel (i.e., Professor of Ethics at the Haas School of Business) and reported their answers to the following question: Does Socially Responsible Investing (SRI) make sense?

No, Socially Responsible Investing (SRI) Does Not Make Sense

Davidson (2015) states that the main issue for Socially Responsible Investing (SRI) may be the lack of uniform information for both individual and institutional investors (e.g., low information efficiency), with "sustainable" meaning different things depending on who is using the terminology. According to David J. Vogel, Professor of Ethics, interviewed by the Wall Street Journal, Socially Responsible Investing (SRI) does not always make sense, because there is no consensus on: (i) How to define sustainability, and (ii) how to construct a portfolio based on this concept.

On the one hand, the challenge for Socially Responsible Investing (SRI) is not only to evaluate the scale of operational environmental performance (e.g., energy efficiency of operations) but also the realistic assessment of environmental performance associated with the product locked into a firm's reserves (e.g., oil for crude oil extraction firms). According to James Leaton from the Carbon Tracker Initiative (i.e., for details see Carbon Tracker at: http://www.carbontracker.org), "the imperative to tackle climate change will require an energy transition. It will require changes to our energy evaluation system and infrastructure" (Davidson 2015). Josh Ryan-Collins, from the New Economics Foundation, adds that climate change is important, because "Environmental, Social and Governance (ESG) are not always properly incorporated into the price setting mechanism" (Davidson 2015).

On the other hand, Socially Responsible Investing (SRI) and socially screened portfolios use diverse and inconsistent criteria to assess Corporate Social Responsibility (CSR). The data used to assess a company's Corporate Social Responsibility (CSR) or sustainability is limited. For example, the Global Reporting Initiative (GRI), the largest reporting guideline, contains voluntary data from only 1,295 firms (i.e., less than 2% of the world's publicly traded firms) (Davidson 2016). In the area of carbon emissions, analysts rely on a firm self-reporting, which may or may not be accurate, and which rarely includes the carbon footprint of a firm's products in the supply chain (Davidson 2016). As discussed previously, Jean Rogers, the Chief Executive and Founder of the nonprofit Sustainability Accounting Standards Board (SASB) is trying to develop uniform standards for about 80 industries across 10 sectors (Davidson 2015), in order to mitigate this lack of consistency, even if this may take a long time (i.e., for details see Investopedia at: http://www.investopedia.com/terms/s/sec.asp).

As such, there may be a role for Governments and regulatory bodies to enforce rules to establish detailed policy and regulatory frameworks domestically and correct this failure in using uniform sources of information (i.e., low information efficiency) (Davidson 2015).

Yes, Socially Responsible Investing (SRI) Does Make Sense

According to Alex Edmans, Professor of Finance, interviewed by the Wall Street Journal in 2016, Socially Responsible Investing (SRI) makes sense, because many traditional investors think that it does not, and as a consequence many investors undervalue Socially Responsible Investing (SRI).

Traditional investors ignore companies' Corporate Social Responsibility (CSR) or sustainability, because traditional investing focuses on: (i) tangible measures of a company's value (e.g., profits and sales growth), and (ii) Socially Responsible Investing (SRI) is intangible or hard to value right now (i.e., low information efficiency). Traditional investors only catch on when Socially Responsible Investing (SRI) effects show up on the bottom line, and as such when it is too late to get a bargain (i.e., since the stock price has already risen) (Davidson 2016). If climate change is here to stay in the long run, this means that Socially Responsible Investing (SRI) investors get the bargain, as long as there is such information inefficiency.

Over the past two decades, numerous studies have attempted to establish a link between a publicly traded firm's tendency to embrace sustainable business practices and their financial performance within a weaker market environment (i.e., low information efficiency) (Shank and Shockey 2016). First, returns for companies with high scores on the American Customer Satisfaction Index (ACSI) were double those of the Dow Jones Industrial Average (DJIA) from 1997 to 2003 (Davidson 2016). Second, companies with high eco-efficiency (i.e., that generate the least waste relative to the value of their products and services) outperformed companies with low eco-efficiency (Davidson 2016). Third, a look at specific market sectors provided further evidence of the financial value of Socially Responsible Investing (SRI). Ethical investors typically over-weighted health care (i.e., based on the social benefits of treating disease) and technology (i.e., e-commerce uses far fewer resources than physical stores). Both sectors have apparently far outperformed the Standard & Poor's (S&P) 500 over the past 10 years, while the energy sector (i.e., dominated by production from nonrenewable sources) has been flat overall or fallen substantially. Please note that the Standard & Poor's (S&P) 500 is an index of 500 stocks seen as a leading indicator of American equities and as a reflection of the performance of the large cap universe (i.e., for details see Investopedia at: http://www.investopedia.com/terms/s/sharperatio.asp).

As such, these findings may have some relevance for investors, fund managers, financial analysts, policymakers, and regulators (Shank and Shockey 2016). Indeed, these findings suggest that global warming has increased awareness of environmental concerns and begun to impact publicly traded firms' practices, because it is having an impact on investment profits (Sadler 2017b).

Summary

As a summary, this section aimed at discussing responsible investing within a context of climate change (e.g., green finance). Indeed, for responsible investors, being responsible means changing their investment strategies to focus on environmental operations and capital investments in order to mitigate any adverse impact of publicly traded firms' assets (e.g., unburnable carbon) on capital markets (e.g., bursting of the carbon bubble related to investors over-investing in fossil fuels publicly traded firms until their stock price falls).

Featured Case in Point

This section aims at discussing responsible investing as an incremental strategy that mitigates the adverse impact and likelihood for stranded assets (e.g.,unburnable carbon) and stock market collapse (e.g., bursting of the carbon bubble) related to the move to environmental economics.

In this section, incremental changes are explored through a case study, the case of crude oil extraction firms in Canada. In fact, it is believed that incremental changes towards a more sustainability oriented financial system mitigate the possible adverse impact of transition risks related to stranded assets and nonsustainable financial system for emission-intensive firms, like crude oil extraction firms in Canada.

Environmental Economics and the Case of Crude Oil Extraction Firms

Hawley (2015) describes fossil fuels, in general, as being on their way to becoming stranded (i.e., unburnable), but Hawley (2015) also adds that there seems to be a lag in response to the stranded asset analysis for some fossil fuels, in particular for oil extraction firms, and this despite the fact that the decreasing price of oil has had a negative impact on the price of stocks for oil extraction firms in general.

With the long-term inflation-adjusted price of oil decreasing around $40 a barrel in 2014–2016: (i) stock prices for oil extraction firms have fallen about 17% since mid-2014; and (ii) the S&P Energy Index has fallen about 32% during the same period (Olson 2016a, b, c). Decreasing oil prices led to significant capital expenditure reductions for oil extraction firms, with some companies moving to "cleaner" natural gas projects (Olson 2016a, b, c). However, according to Olson (2016a, b, c), from the Wall Street Journal, some oil extraction firms were recently under investigation, because of how they valued their oil wells (i.e., oil reserves). New York Attorney General and the Security Exchange Commission (SEC) reportedly questioned the price assumptions some oil extraction firms used to book their oil reserves (Olson 2016a, b, c). Such a scrutiny led to an additional fall in share prices for these oil extraction firms (Olson 2016a, b, c).

This example demonstrated a lag in response to the stranded asset analysis of fossil fuel, in particular for oil extraction firms (Hawley 2015). Therefore, it is believed that the case of Canadian crude oil extraction firms may shed an interesting light on the complexity of the drivers involved in environmental economics.

Environmental Economics and the Case of the Canadian Crude Oil Extraction Firms

Canada is the sixth largest crude oil producer in the world, with the third largest petroleum reserve (Canadian Geographic 2014). The production of crude oil in Canada contributed CA$18 billion in royalty and tax revenue in 2012 (Canadian Geographic 2014). Interestingly, 97% of Canada's share of crude oil production is situated in Alberta, one Province of Canada (Canadian Geographic 2014).

Canadian crude oil extraction firms have received a lot of attention in the media lately. Peritz (2014) reports that the Hollywood star, Leonardo DiCaprio, travelled to Alberta. The purpose of Leonardo DiCaprio's trip was to research for a documentary, called "Before the Flood." The Canadian Press recounted that Leonardo DiCaprio's new climate change documentary was against crude oil extraction firms in Alberta, since the actor held a high-profile environmentalist position (Peritz 2014). This event followed other celebrities' apparent attacks against crude oil extraction firms in Alberta (e.g., musician Neil Young and film director James Cameron). In January 2017, Jane Fonda (2017) also talked about the "hard truths that brought her to Alberta" in an open letter to the Globe and Mail (i.e., for details see The Globe and Mail at: https://www.theglobeandmail.com). Jane Fonda made the following statement: "Even if we do everything needed to make a managed and compassionate transition to a low carbon economy, climate change that we have already caused will have a dramatic effect."

In light of the media attention, the question of the complexity of the drivers of environmental economics may be explored through the case of crude oil extraction firms in Canada at two different levels: (i) the concept of stranded assets (i.e., unburnable carbon related to crude oil), and (ii) the concept of non sustainable financial systems (i.e., stock market collapse and bursting of the carbon bubble related to the Canadian crude oil extraction firms).

Stranded Assets and Bursting of the Carbon Bubble

The dominant crude oil asset held by some crude petroleum extraction companies in Canada is unconventional crude oil, which may be more emission-intensive to produce than other oil related assets (Charpentier et al. 2009; Heidari and Pearce 2016). On average, crude oil extraction companies in Canada (i.e., unconventional crude oil industry) are believed to be three times more Green House Gas (GHG) emission-intensive than the conventional oil industry (Charpentier et al. 2009). Indeed, the production of unconventional crude oil apparently results in emissions ranging from 99–176 compared to 27–58 kgCO$_2$eq/bbl for conventional oil production (Charpentier et al. 2009). The difference in Green House Gas (GHG) emission intensity between unconventional and conventional oil production is apparently due to higher energy requirements for extracting bitumen and upgrading bitumen from the tar sands (Charpentier et al. 2009).

The emission-intensive nature of crude oil extraction companies in Canada may challenge the industry's ability to export unconventional crude oil in the future. While the Keystone XL and additional pipeline capacity may represent the simplest path to get Canadian unconventional oil to market in the United States (USA), Canada is keen to move more of its resources to Asia (Johnson and Boersma 2015). If regulatory risks increase worldwide, the Green House Gas (GHG) regulatory framework may evolve to limit Green House Gas (GHG) to sustainable levels only. As a consequence, the reserve resources and energy security of Canada controlled by crude oil extraction companies may become stranded (i.e., unburnable), in which case these assets may no longer sustain the market price of the publicly traded firms that control them (i.e., stock prices of crude oil extraction companies may fall leading to the bursting of the carbon bubble).

Lenient Versus Constrained Regulatory Framework

Under a lenient regulatory framework, crude oil extraction companies may have continued their planned production without being restricted by any national emissions target in Canada (Mc Diamid 2015). Indeed, G20 countries were spending $452 billion a year to financially subsidize their fossil fuel industries (Mc Diamid 2015). Crude oil extraction companies in Canada have long benefited from Government financial support to reduce political and financial risks (Mc Diamid 2015).

Under a constrained regulatory framework, however, financial subsidies to emission-intensive companies have been phased out, including financial subsidies that ended in January 2015 (Mc Diamid 2015). At the Federal level, the Government of Canada want to set-up a carbon tax of $50 a ton by 2022 (Bakx 2016b). At the Provincial level, the Provinces of British Columbia, Manitoba, Ontario, and Quebec have implemented policies intended to reduce Green House Gas (GHG) emissions (Bakx 2016b). Several Provinces have joined the Western Climate Initiative (WCI), a regional cap-and-trade system (Bakx 2016b). British Columbia has introduced a carbon tax. Manitoba and Ontario have implemented subsidies, grants, and loans for the adoption of management practices that reduce Green House Gas (GHG) emissions (Bakx 2016b). Quebec has adopted Green House Gas (GHG) emission standards for transportation (Bakx 2016b). The government of Alberta has also expressed its willingness to cap emissions for big crude oil extraction firms and to implement an economy-wide "price for carbon" policy in an effort to curb their emissions. According to Bakx (2016a), since 2007, Alberta had a type of carbon tax for its large emitters which charged $15 a ton of Green House Gas (GHG) emissions. In January 2016, the rate has increased to $20 a ton. Alberta is expected to charge $30 a ton by 2018. On November 2, 2016, a new legislation was announced to make Alberta the first major oil-production jurisdiction to implement a Green House Gas (GHG) emission limit (Giovannetti 2016). For large industrial players, Alberta designed its policy to: (i) reward the most efficient when it comes to Green House Gas (GHG) emission per barrel of crude oil; and (ii) penalize those who produce more pollution to produce the same amount of crude oil (Bakx 2016a). These financial incentives/penalties were the biggest change under the new royalty system to support emission-intensive companies that were more efficient in terms of their Green House Gas (GHG) emissions.

Incremental Mitigation of Stranded Assets and Bursting of the Carbon Bubble

On the one hand, in a Green House Gas (GHG) emission constrained regulatory environment, other countries may become increasingly cognizant of the Green House Gas (GHG) emission related to their energy imports, which may place crude oil extraction companies in Canada at a significant competitive disadvantage. Drinkwater (2016) reports that the Canadian Association of Oilwell Drilling Contractors was concerned that a carbon tax was going to move international capital away from Alberta (Bakx 2016a). First, restrictive legislations on Green House Gas (GHG) emissions may prevent crude oil extraction firms from converting their reserve into marketable products. Crude oil extraction firms in Canada may then hold "stranded assets," which may have a negative impact on the valuation of crude oil extraction firms' stocks. Second, restrictive legislations on Green House Gas (GHG) emissions may lead to the bursting of a carbon bubble, since high allocation rates of emission-intensive investments by Canadian institutional investors, such as the Canada Pension Plan (CPP), may massively expose Canadians to stranded fossil fuel assets, if restrictions in the Green House Gas (GHG) emission regulatory framework are to be imposed further.

On the other hand, in a Green House Gas (GHG) emission constrained regulatory environment, the Canadian oil patch may benefit from a competitive advantage (Bakx 2016c). Some industry associations reportedly endorsed the regulatory changes in Alberta (Bakx 2016b). Some companies even seemed so enthusiastic about the new policy that they applied for early access. For instance, one companie apparently spent $25 million to drill new wells in the Duvernay-Montney basin in Northeast Alberta (Bakx 2016b) and reportedly stated that this spending would not have happened without the royalty changes (Bakx 2016b). Another example was another companie Quest project, a joint venture with the Government of Alberta (Bakx 2016b). The companie apparently did not make money from Quest, but the project reportedly reduced how much the company had to pay in carbon pricing and reportedly allowed for investments in Alberta to compete with those in the United States (USA) according to the COO at the time (Bakx 2016b).

As such, each driver or combination of drivers may or may not have a material impact on business operations for crude oil extraction firms in Canada depending on the way financial support is being used (i.e., financial subsidy to reduce emissions as opposed to financial subsidy to increase production). As well, more restrictive legislations on Green House Gas (GHG) emissions may or may not lead to the bursting of a carbon bubble.

In fact, it is believed that incremental changes towards a more sustainability oriented financial system mitigate the possible adverse impact of transition risks related to stranded assets and nonsustainable financial systems for emission-intensive firms in Canada.

Summary

As a summary, the case of the Canadian crude oil extraction firms demonstrated the complexity of the drivers for stranded assets and bursting of the carbon bubble within an environmentally oriented economy. These drivers seem to interact in highly complex ways with a long list of policy, legal, political, technological, reputational, and economic drivers that help the mitigation of transition risks, in terms of stranded assets and bursting of the carbon bubble.

Reflection Questions

This chapter aims at discussing responsible investing and environmental economics within the context of climate change at a more theoretical level. Indeed, in the previous sections, it was observed that publicly traded firms were applying sustainability oriented strategies in their operations and capital spending, because investors were interested in responsible investing to move towards environmental economics.

Theory on Investing and Economics

Broadly speaking, investing is related to economics through the "increasing role of financial motives, financial markets, financial actors and financial institutions in the operation of the domestic and international economies" (Epstein 2005, p. 3). For example, since the 2008 subprime mortgage financial crisis, academics have been increasingly interested in the link existing between investing and economics (Davis 2009; Lounsbury and Hirsch 2010; Marti and Scherer 2016), since questionable investing practices (i.e., derivatives of subprime mortgages) have been shown to damage the economy (i.e., 2008 financial crisis).

The Efficient Market Hypothesis (EMH) is a finance theory developed in the 1960s by Eugene Fama (i.e., for details see Investopedia at: http://www.investopedia.com/university/concepts/concepts6.asp). The Efficient Market Hypothesis (EMH) states that it is impossible to beat capital markets (e.g., S&P500, NSDAQ, TSX200, etc.) in terms of investment returns, because stocks of publicly traded firms already reflect all relevant information within capital markets (e.g., financial statements, quarterly financial reports, etc.).

On the one hand, supporters of this model believe that it is useless to search for undervalued stocks of publicly traded firms (i.e., for details see Investopedia at: http://www.investopedia.com/university/concepts/concepts6.asp). Indeed, if markets are efficient, prices already reflect all available information, and an investor cannot buy a stock at a bargain price (i.e., high information efficiency and passive investing as the best option).

On the other hand, fundamental and technical analysts believe that it is not useless to search for undervalued stocks of publicly traded firms, as markets are not always perfectly efficient, and prices do not always perfectly reflect all available information (i.e., for details see Investopedia at: http://www.investopedia.com/university/concepts/concepts6.asp). As such, an investor can buy a stock at a bargain price under certain conditions (e.g., low information efficiency and active investing as the best option).

Theory on Responsible Investing and Environmental Economics

In relation to responsible investing, the Efficient Market Hypothesis (EMH) supports the theoretical argument that a publicly traded firm's value increases when it incorporates environmental business strategies into its operations and capital expenditure (Shank and Shockey 2016). This is based on the connection existing between the implementation of such sustainable business strategies, and the stabilization of future cash flows (e.g., no bursting of the carbon bubble) as well as the reduction of corporate risk exposure (e.g., no assets stranded because of climate change-related regulations) (Shank and Shockey 2016) as seen in the case of Canadian crude oil extraction firms.

The theoretical question that responsible investors may consider, however, is whether or not environmental information is already reflected in capital markets and in publicly traded firms' stock prices (Shank and Shockey 2016). If not, responsible investors should identify those publicly traded firms whose equity values and stock prices do not fully reflect their sustainable efforts yet in order to get the bargain (Shank and Shockey 2016).

Green House Gas (GHG) Emissions and Environmental Reporting

Responsible investing is related to environmental economics with the question of Green House Gas (GHG) emission reporting and its impact on capital markets. Indeed, research shows that the stock price and equity value of companies reflect Green House Gas (GHG) emission information.

First, academic research shows that capital markets are taking the cost of Green House Gas (GHG) emission into consideration when valuing companies. Whether using accounting-based measures (Ameer and Othman 2012; Lopez et al. 2007) or market-based measures (Hill et al. 2007; Shank et al. 2005), some studies found that financial value was created when publicly traded firms were recognized as being more sustainability focused than their peers (Shank and Shockey 2016). Academic research also showed the impact of Green House Gas (GHG) emission on equity value, with the cost of Green House Gas (GHG) emission on equity value going from AU$17 to AU$26 in Australia (Chapple et al. 2013), and being €75 in Europe (Clarkson et al. 2008). The cost of Green House Gas (GHG) emission on equity value went from US$204 to US$348 for American companies from 2006 to 2008 (Matsumura et al. 2014). Using a model of prediction of GHG emission when not disclosed by emission-intensive firms, Griffin et al. (2012) even found that nondiscloser and discloser firms were affected to the same extent by their levels of Green House Gas (GHG) emission. This phenomenon was explained by the fact that Green House Gas (GHG) emission is value relevant independently of the level of disclosure by emission intensive firms, since investors and capital markets use multiple channels of information to be efficient (Griffin et al. 2012).

Second, academic research shows that capital markets are taking the impact of present and/or future regulations in terms of Green House Gas (GHG) emission (e. g., carbon taxes and cap-and-trade programs) into consideration when valuing companies. Indeed, Europe has pioneered the development of carbon dioxide emissions trading programs, known as Emissions Trading Schemes (ETS) (Considine and Larson 2009). Research shows that Emissions Trading Schemes (ETS) are value relevant for companies in Europe (Clarkson et al. 2008). Research also shows that the "future carbon permit price" can be estimated with the cost of Green House Gas (GHG) emission on equity value going from AU$17 per ton to AU$26 per ton of carbon dioxide emitted in Australia (Chapple et al. 2013). Griffin (2013) shows that the average S&P500 company's balance sheet and net income are to be adversely affected under several different accounting treatments for emission allowances, with the greatest impact being for emission intensive companies.

Corporate Social Responsibility (CSR) and Environmental Reporting

Responsible investing is related to environmental economics with the question of how the "good organization" may be recognizable through Corporate Social Responsibility (CSR). Some suggests that a distinct contribution to our understanding of this link can be made by: (i) analyzing the rise of corporate governance for "good organizations," and (ii) exploring how different types of shareholders/stakeholders influence corporations to be "good organizations" (Jackson 2000).

First, corporate governance follows an agency model, where "principals" (i.e., shareholders/stakeholders) are opposed to "agents" (i.e., executive managers) in the sense that agents may not communicate all information to the principals (i.e., which results in information asymmetry and moral hazard from executive managers inside the corporation who may be hiding information from shareholders/stakeholders outside the organization) (Jackson 2000). In the context of environmental economics, information asymmetry and moral hazard may be referred to as "greenwashing" (i.e., when a "green organization" spends more time and money claiming to be "green" through advertising and marketing than actually implementing business practices that minimize its environmental impact).

Second, over the past decades, a theory and practice of corporate governance came to prominence in which shareholders played a key role against information asymmetry and moral hazard (Davis 2009). The belief was that shareholders were the legitimate beneficiaries of corporations as their focus was on ensuring the flow of reliable information (i.e., no information asymmetry) and accountability of executive managers (i.e., no moral hazard) (Veldman and Willmott 2016). In this context, corporate governance through Corporate Boards (i.e., that represent shareholders) was believed to promote the flow of reliable information and accountability in different ways: (i) from direct engagement for information exchange with Board Members (McNulty and Nordberg 2015) to (ii) arms' length evaluations and responses to executive managers' strategies (Aglietta and Rebérioux 2005). In the context of environmental economics, however, corporate governance may not be as useful if Corporate Social Responsibility (CSR) is hard to contrast from "greenwashing," because Corporate Social Responsibility (CSR) is not well-defined to begin with.

Theory on Responsible Investing and Passive Versus Active Investing Strategies

In relation to responsible investing, the Efficient Market Hypothesis (EMH) supports the theoretical argument that a publicly traded firm's value increases when it incorporates environmental business strategies into its operations and capital expenditure (Shank and Shockey 2016). As a consequence, in an efficient market, a passive investment approach of responsible investing and environmental economics seems to be the best option (Shank and Shockey 2016). As a consequence, in a nonefficient market, an active investment approach of responsible investing and environmental economics seems to be the best option (Shank and Shockey 2016).

Passive Strategy and Efficient Environmental Reporting

In an efficient market, a passive investment approach of responsible investing and environmental economics seems to be the best option, as no bargains can be available (North and Stevens 2015). As such, publicly traded firms identified as leaders by the Dow Jones Sustainability Indices (DJSI) may not outperform a portfolio of firms as a whole (e.g., all firms comprised in the Dow Jones Sustainability Indices), because environmental information for sustainable publicly traded firms is efficiently priced by capital markets (Shank and Shockey 2016).

Passive investment strategies are thus appropriate for capital markets where there is high level of information efficiency (Shank and Shockey 2016). Newer measures of sustainability, such as the Dow Jones Sustainability Indices (DJSI) (i.e., which utilize a number of nonfinancial performance data simultaneously) increase the information efficiency for sustainable publicly traded firms. Supporting this argument, academic research has compared the financial performance of firms in one of the major sustainability indices to benchmarks of broader markets (Shank and Shockey 2016). Consolandi et al. (2009) found that the Dow

Jones Sustainability Indices (DJSI), for sustainable European firms, slightly underperformed market benchmarks from 2001 to 2006. Xiao et al. (2013) investigated the role of corporate sustainability (i.e., using the Dow Jones Sustainability Indices) in asset pricing from 2001 to 2007 and found no significant impact of sustainability on equity returns.

Active Strategy and Inefficient Environmental Reporting

In a nonefficient market, however, an active investment approach of responsible investing and environmental economics seems to be the best option, as bargains can be available (North and Stevens 2015). As such, in limited situations (e.g., capital markets with low information efficiency for sustainability), firms identified as leaders by the Dow Jones Sustainability Indices (DJSI) may outperform a portfolio of sustainable firms as a whole (i.e., all firms comprising the Dow Jones Sustainability Indices), because information for sustainable firms stocks is not efficiently priced by capital markets (Shank and Shockey 2016).

Active investment strategies are thus appropriate for capital markets where there is low information efficiency (Shank and Shockey 2016). Orlitzky (2003) conducted a meta-analysis that examined 52 prior studies dating back to 1975 (i.e., lower level of information efficiency in the past) and found that accounting-based measures of financial performance better reflected Corporate Social Responsibility (CSR) efforts compared to market-based measures (Shank and Shockey 2016). Rodgers et al. (2013) found that a firm's Corporate Social Responsibility (CSR) commitment led to better financial performance on both accounting-based and market-based financial metrics in capital markets with low information efficiency (Shank and Shockey 2016). Support of the link between more sustainable firms and lower corporate risk was found in Lee (2009) and Ghoula (2011) in capital markets with low information efficiency (Shank and Shockey 2016).

As such, depending on the information efficiency level, responsible investing and environmental economics may or may not be aligned with the Efficient Market Hypothesis (EMH). This is based on the connection existing between the implementation of such sustainable business strategies, and the stabilization of future cash flows (e.g., no bursting of the carbon bubble) as well as the reduction of corporate risk exposure (e.g., no assets stranded because of climate change-related regulations).

Exercises in Practice

The World's Most Sustainable Companies

Pre-Class Preparation

The professor goes to the lists of all the best performing companies according to "Sustainalytics" (http://www.sustainalytics.com/sustainability-research-rankings/) for three categories of corporations:

- Information and Communication Technology (https://knowthechain.org/benchmarks/1/)
- Food and Beverage (https://knowthechain.org/benchmarks/2)
- Apparel and Footwear (https://knowthechain.org/benchmarks/3/)

In-Class Games

Game 1: Hang Man (or Growing Flower)
Hangman (i.e., the hangman may be replaced by a drawing of a growing flower if considered more appropriate) involves the students having to guess a particular corporation's name with a high "Know the Chain" ranking according to "Sustainalytics."

- The professor or a student thinks of one company with a high ranking according to "Sustainalytics," and draws a line of blank boxes on the board which indicates how many letters the name has.
- Students then ask for clues to the company and then add letters. For every letter students get wrong, a body part of the hangman (i.e., or a part of the flower) is drawn on the board. Once the picture is complete, the man is "hanged" (i.e., or the flower is "grown") and students lose. If they win, however, the entire word is spelled out on the board. In both cases, the details of the company are explored in class via the company website and/or in-class discussions about the company.

This is a fun way of promoting a better knowledge of companies with high "Know the Chain" rankings and getting the class involved.

Game 2: Five Questions
This game may get a few laughs from the students while increasing their understanding of "Know the Chain" rankings according to "Sustainalytics."

- Have a student sit in front of the board, facing the class, and write the name of a corporation with a high "Know the Chain" ranking above his/her head (i.e., without him/her being able to read the name; but with the class being able to read the name).
- The student then has to ask the class up to five questions about the corporation until he/she finds out which corporation it is.

Game 3: Class Survey
A nice way of helping students break the ice in their first class may be using the below ice-breaker.

- Get students to survey each other on a wide range of topics related to "Responsible Investing and Environmental Economics."
- You may use the "Chapter End Questions" as a guide and divide the class into groups of two or three students discussing each question separately.

- Then, each group may disclose their answers to the "Chapter End Questions" in front of the whole class.
- Students may get bonus marks related to their performance on this task.

Engaged Sustainability Lessons

This chapter aimed at discussing responsible investing and environmental economics within the context of climate change.

In November 2015, in Paris, Governments have shown their interest in stimulating change and the transition to a low-carbon economy through global regulations (i.e., for details see the United Nations Framework on Climate Change at: http://unfccc.int/paris_agreement/items/9485.php). In the context of such a transition, nonresponsible investors may face financial risks for their investments in terms of stranded assets and bursting of the carbon bubble. As such, there may be a strong relationship between responsible investing and environmental economics.

Regarding responsible investing, it was observed that economic performance of publicly traded firms which are applying the concept of sustainability in their operations and capital spending is improved, because investors are more and more interested in responsible investing.

Regarding environmental economics, such changes towards more sustainability mitigate the risk for unburnable carbon and the bursting of the carbon bubble on capital markets as demonstrated with the example of some Canadian crude oil extraction firms.

Through information asymmetry and/or moral hazard, however, some concerns arise that material amounts of publicly traded firms' actual environmental performance may be hidden from financial statements and annual reports (e.g., greenwashing).

As a consequence, the challenge for responsible investors is not only: (i) to evaluate the scale of actual environmental operational and capital investments (e.g., information efficiency) but also (ii) to evaluate publicly traded firms' environmental performance associated with the asset that may be locked into their reserves.

Chapter-End Reflection Questions

This chapter aims at discussing responsible investing and environmental economics within the context of climate change.

Questions to use with students may cover (i.e., but are not limited to) the following items:

Responsible Investing

- What is the impact of current responsible investing practices on climate change mitigation?
- What is the impact of responsible investing on fostering energy transition?

- What is the ethical versus financial relevance of responsible investing?
- What are emerging practices in green finance?
- Which socially responsible investment strategy has the greatest potential to bring out the "good organization"?

Environmental Economics

- Which are the key stakeholders for fostering environmental economics?
- What are the main drivers for environmental economics? e.g., regulatory pressure, stranded asset and carbon bubble risks, new social movements...
- What are the barriers for environmental economics?
- How can corporate governance be developed to enable the "good organization"?
- What are the societal implications of different discourses on corporate governance?

Cross-References

▶ Low-Carbon Economies (LCEs)
▶ Responsible Investing and Corporate Social Responsibility for Engaged Sustainability

References

Aglietta, M., & Rebérioux, A. (2005). *Corporate governance adrift: A critique of shareholder value*. Cheltenham: Edward Elgar.

Ameer, R., & Othman, R. (2012). Sustainability practices and corporate financial performance: A study based on the top global corporations. *Journal of Business Ethics, 108*, 61–79.

Ansar, A., Caldecott, B., & Tilbury, J. (2013). *Divestment campaign: What does divestment mean for the valuation of fossil fuel assets?* Working paper. U.K. University of Oxford.

Bakx, K. (2016a). Canada shouldn't lose resolve for a carbon tax, says Shell exec Oilpatch has worries about impact on competitiveness. Retrieved from http://www.cbc.ca/news/business/shell-canada-crothers-carbon-tax-oilpatch-1.3866261

Bakx, K. (2016b). Avoiding high emission taxes would make innovative alternatives more economical. Retrieved from http://www.cbc.ca/news/business/ccs-carbon-tax-shell-saskpower-1.3905724

Bakx, K. (2016c). Oilpatch friendly royalty system takes effect in Alberta Provincial government made only small changes to how much companies must pay in royalties. Retrieved from http://www.cbc.ca/news/business/alberta-royalty-oilpatch-oilsands-1.3905075

Campiglio, E. (2016). Beyond carbon pricing: The role of banking and monetary policy in financing the transition to a low-carbon economy. *Ecological Economics, 121*, 220–230.

Canadian Geographic. (2014). Energy rich. Exploring the top ressources powering our nation. Retrieved from http://www.canadiangeographic.ca/magazine/jun14/

Carbon Tracker. (2014). The carbon tracker initiative. Retrieved from http://www.carbontracker.org/

Chapple, L., Clarkson, P. M., & Gold, D. L. (2013). The cost of carbon: Capital market effects of the proposed emission trading scheme (ETS). *Abacus, 49*(1), 1–33.

Charpentier, A., Bergerson, J., & MacLean, H. (2009). Understanding the Canadian oil sands industry's greenhouse gas emissions. *Environmental Research Letters, 4*(14), 345–356.

Clarkson, P. M., Li, Y., Richardson, G. D., & Vasvari, F. P. (2008). Revisiting the relation between environmental performance and environmental disclosure: An empirical analysis. *Accounting, Organizations and Society, 33*(4), 303–327.

Considine, T. J., & Larson, D. F. (2009). *Substitution and technological change under carbon cap and trade: Lessons from Europe*. Policy Research working paper no. WPS 4957. Washington, DC: World Bank.

Consolandi, C., Jaiswal-Dale, A., Poggiani, E., & Vercelli, A. (2009). Global standards and ethical stock indexes: The case of the Dow Jones Sustainability Index. *Journal of Business Ethics, 87*, 185–197.

Davidson, A. (2015). A guide to sustainable investing. *Wall Street Journal*. Retrieved from https://www.wsj.com/articles/a-guide-to-sustainable-investing-1447038115

Davidson, A. (2016). Does socially responsible investing make financial sense? Some think it gives investors an edge, but critics point to poor results. Retrieved from https://www.wsj.com/articles/does-socially-responsible-investing-make-financial-sense-1456715888

Davis, G. F. (2009). *Managed by the markets: How finance reshaped America*. Oxford: Oxford University Press.

Deutsche Bank Report. (2017). Deutsche bank invests EUR 1 billion in green bond portfolio. Retrieved from http://www.m2.com

Dolan, K.A. (2016). *Bill gates launches $1 billion breakthrough energy investment fund*. Forbes. Retrieved from http://www.forbes.com/sites/kerryadolan/2016/12/12/bill-gates-launches-1-billion-breakthrough-energy-investment-fund/#498549c559a7

Drinkwater, R. (2016). Alberta minister says province 'still standing' on Day 2 of carbon tax Wildrose critic says 'science isn't settled' on climate change. Retrieved from http://www.cbc.ca/news/canada/edmonton/alberta-minister-province-still-standing-on-day-two-carbon-tax-1.3918863

Epstein, G. A. (2005). Introduction: Financialization and the world economy. In G. A. Epstein (Ed.), *Financialization and the world economy* (pp. 3–16). Cheltenham: Edward Elgar.

Fonda, J. (2017). Indigenous reconciliation will never flow from a pipeline. *The Globe and Mail*. Retrieved from http://www.theglobeandmail.com/opinion/indigenous-reconciliation-will-never-flow-from-a-pipeline/article33646470/

Ghoula, S. (2011). Does corporate social responsibility affect the cost of capital? *Journal of Banking & Finance, 35*, 2388–2408.

Giovannetti. (2016). Alberta bill would cap oil sands greenhouse gas emissions. *The Globe and Mail*. Retrieved from http://www.theglobeandmail.com/news/national/alberta-bill-would-cap-oil-sands-greenhouse-gas-emissions/article32638790

Griffin, P. A. (2013). Cap-and-trade emission allowances and US companies' balance sheets. *Sustainability Accounting Management and Policy Journal, 4*(1), 7–31.

Griffin, P. A., Lont, D. H., & Sun, Y. (2012). *The relevance to investors of greenhouse gas emission disclosures*. Working paper. U.C. Davis.

Hawley, J. (2015). Carbon risks and investment implications. Retrieved from https://www.insight360.io/carbon-risk-implications/

Heidari, N., & Pearce, J. (2016). A review of greenhouse gas emission liabilities as the value of renewable energy for mitigating lawsuits for climate change related damages. *Renewable and Sustainable Energy Reviews, 55*, 899–908.

Hill, R., Ainscough, T., Shank, T., & Manullang, D. (2007). Corporate social responsibility and socially responsible investing: A global perspective. *Journal of Business Ethics, 70*, 165–174.

Jackson, G. (2000). Comparative corporate governance: Sociological perspectives. In J. E. Parkinson, A. Gamble, & G. Kelly (Eds.), *The political economy of the company* (pp. 265–288). Oxford: Hart Publishing.

Johnson, C., & Boersma, T. (2015). The politics of energy security: Contrasts between the United States and the European Union. *WIREs Energy and Environment, 4*, 171–177.

Lee, D. (2009). Corporate sustainability performance and idiosyncratic risk: A global perspective. *Financial Review, 44*, 213–237.

Lopez, M., Garcia, A., & Rodriquez, L. (2007). Sustainable development and corporate performance: A study based on the Dow Jones Sustainability Index. *Journal of Business Ethics, 75*, 285–300.

Lounsbury, M., & Hirsch, P.M. (2010). *Markets on trial: The economic sociology of the U.S. financial crisis*. Research in the sociology of organizations book series, Vol. 30, Part A. Bingley: Emerald Group Publishing Limited.

Marti, E., & Scherer, A. G. (2016). Financial regulation and social welfare: The critical contribution of management theory. *Academy of Management Review, 41*(2), 298–323.

Matsumura, E. M., Prakash, R., & Vera-Muñoz, S. C. (2014). Firm-value effects of carbon emissions and carbon disclosures. *The Accounting Review, 89*(2), 695–724.

Mc Diamid, M. (2015). G20 countries spend $450B a year on fossil fuel subsidies, study says. *CBC News*. Retrieved from http://www.cbc.ca/news/politics/g20-fossil-fuel-subsidies-450b-1.3314291

McNulty, T., & Nordberg, D. (2015). Ownership, activism and engagement: Institutional investors as active owners. *Corporate Governance: An International Review, 24*(3), 346–358.

North, D., & Stevens, J. (2015). Investment performance of AAII stock screens over diverse markets. *Financial Services Review, 24*, 157–176.

Olson, H. (2016a). Exxon Mobil profit revenue slide again. *Wall Street Journal*. Retrieved from http://www.wsj.com/articles/exxon-mobil-profit-revenue-slide-again-1477657202

Olson, H. (2016b). Exxon accounting practices are investigated. *Wall Street Journal*. Retrieved from https://www.wsj.com/articles/exxons-accounting-practices-are-investigated-1474018381

Olson, B. (2016c). Exxon warns on reserves as it posts lower profit. *The Wall Street Journal*. Retrieved from http://www.wsj.com/articles/exxon-mobil-profit-revenue-slide-again-1477657202

Orlitzky, M. (2003). Corporate social and financial performance: A meta-analysis. *Organization Studies, 24*, 403–441.

Peritz, I. (2014). Alberta riled by Leonardo DiCaprio's position on oil sands. *The Globe and Mail*. Retrieved from http://www.theglobeandmail.com/news/national/alberta-riled-by-leonardo-dicaprios-position-on-oil-sands/artic le20187391/

Rodgers, W., Choy, H., & Guiral, A. (2013). Do investors value a firm's commitment to social activities? *Journal of Business Ethics, 114*, 607–623.

Sadler, A. (2017a). *Green bonds*. Investopedia. Retrieved from http://www.investopedia.com/articles/bonds/07/green-bonds.asp#ixzz4bu1SJ4VL

Sadler, A. (2017b). *Sustainability gives work and investing purpose*. Investopedia. Retrieved from http://www.investopedia.com/advisor-network/articles/022217/sustainability-gives-work-and-investing-purpose/#ixzz4bu1JecTT

Shank, T. M., & Shockey, B. (2016). Investment strategies when selecting sustainable firms. *Financial Services Review, 25*(2), 12–30.

Shank, T. M., Manullang, D., & Hill, R. (2005). Doing well while doing good revisited: A study of socially responsible firms' short-term versus long-term performance. *Managerial Finance, 30*, 33–45.

Velasquez-Manoff, M. (2016). Cashing in on climate change. *The New York Times*. Retrieved from http://www.nytimes.com/2016/12/03/opinion/sunday/cashing-in-on-climate-change.html?_r=0

Veldman, J., & Willmott, H. (2016). The cultural grammar of governance: The UK code of corporate governance, reflexivity, and the limits of 'soft' regulation. *Human Relations, 69*(3), 581–603.

Xiao, Y., Faff, R., Gharghori, P., & Lee, D. (2013). An empirical study of the world price of sustainability. *Journal of Business Ethics, 114*, 297–310.

Responsible Investing and Corporate Social Responsibility for Engaged Sustainability

Managing Pitfalls of Economics without Equity

Raghavan 'Ram' Ramanan

Contents

Introduction 1044
Evolution of Socially Responsible Investing 1045
Voluntary Responsible Investment Principles 1047
    Guidelines for Multinational Enterprises (OECD Principles) 1047
    Equator Principles 1048
    Principles for Responsible Investing (PRI) 1049
    The UN Global Compact 1050
Emerging Global Sustainability Governance Regulations 1050
    Sustainability Regulations: Evolving Globally 1050
    Sustainability Governance Regulations: Select Country Examples 1052
Impact Investing and Organization Structures 1054
    Impact Investing Primary Drivers 1054
    Impact Investment Organization Structures 1054
    Philanthropic Capitalism and Venture Philanthropy 1055
    Impact Investment Standards and Reporting Frameworks 1056
Responsible Investing and Private Public Partnership (PPP) 1057
Ethical Dimension of Resource Management 1058
    Ecosystem and Resource Management: Sociopolitical Choice 1058
    The New Social Contract and a Clarion Call for Ethical Leadership 1060
    Decision-Making and the Ethical Dimension 1061
Managing Pitfalls of Economics without Equity 1062
    Climate Change 1062
    Global Warming and Climate Change: The Issue and the Impact 1063
    Climate Change Paris Agreement, 2015 1064
    Ethical Considerations in Climate Change 1065
Chapter Summary and Management/Leadership Lessons 1066

This chapter is primarily based on Ramanan, "Introduction to Sustainability Analytics," ISBN-10: 1498777058 under publication, CRC Press 2018.

R. 'Ram' Ramanan (✉)
Desert Research Institute, Dallas, TX, USA
e-mail: Ram.Ramanan@fulbrightmail.org

© Springer International Publishing AG, part of Springer Nature 2018
S. Dhiman, J. Marques (eds.), *Handbook of Engaged Sustainability*,
https://doi.org/10.1007/978-3-319-71312-0_14

Chapter End Reflection Questions 1067
Cross-References 1067
References 1067

Abstract

Corporate social contract has morphed from Milton Friedman's "only social responsibility of business (is) to use its resource ...... to increase its profits .... within the rules of the game" on to the 'New Social Contract' defined as "business is one thread in the complex web of interwoven society... responsible for not just its inanimate inputs and outputs, but for all related human and environmental interactions." Today, corporate social responsibility encompasses investments that create positive social and environmental impacts beyond financial returns.

In this chapter, the author focuses on socially responsible investing by organizations, both corporate and government, within the context of sustainability, and expands on the value of impact investing and public-private partnership to preempt the disastrous pitfalls of economics without equity. The chapter highlights the emerging global regulations and the crucial roles of corporate social responsibility and public policy stewardship. It also presents the foundations of sustainability analytics and frameworks for ethical resource management and for managing pitfalls of climate change economics without ethics.

Keywords

New social contract · Socially responsible investing · Responsible investment principles · Sustainability governance regulations · Impact investing · Sustainability analytics

Introduction

Corporate social contract has morphed from Milton Friedman's (Friedman, Milton, "The Social Responsibility of Business is to increase its profits," New York Times Magazine, Sep. 13, 1970 available and accessed May 25, 2016 at http://deloitte.wsj.com/riskandcompliance/files/2013/04/scc_Drivers-of-Long_Term-Value.pdf) "only social responsibility of business (is) to use its resource ...... to increase its profits .... within the rules of the game" on to the 'New Social Contract' defined as "business is one thread in the complex web of interwoven society... responsible for not just its inanimate inputs and outputs, but for all related human and environmental interactions" (Taback and Ramanan, The New Social Contract' in "Environmental Ethics and Sustainability"). Today, corporate social responsibility encompasses investments that create positive social and environmental impacts beyond financial returns. Sustainability issues affect the various sectors of private and public finance and financial approaches, and integrating sustainability principles and practices into finance can be used to help business and governments become more efficient and effective, reduce risks, create opportunities, and develop competitive advantage. Sustainable development and ecosystem management commonly involve tough sociopolitical choices. Corporations and leaders have to manage corporate social responsibility and a public policy leader is often faced with balancing human needs and environmental considerations; the end goal in both cases is sustainability, to protect and preserve our only planet for future generations and to create positive social and environmental impacts beyond financial returns.

The next section introduces the evolution of socially responsible investing by organizations. The reader is taken thru a journey that started with the faith-based approach of the Quakers in the 1500s and carried on thru the current mission-driven impact investing by the Bill and Melinda Gates Foundation and the Clinton Global Initiative. The following two sections present a summary of the emerging global voluntary principles for responsible investments and governance regulations in select major countries, to ensure sustainable development, to protect investors, and to collect a fair share of taxes. The fifth section presents a discussion of the impact investing organization structures.

The sixth section introduces the concept and value of public-private partnerships for addressing select mega issues and highlights the crucial investment or resource allocation roles of corporate social responsibility (CSR) and public policy stewardship in sustainability. The final two sections present the foundations of sustainability analytics and frameworks for ethical resource management and for managing pitfalls of climate change economics without ethics.

Evolution of Socially Responsible Investing

Quakers in the 1500s and Churches in the 1920s used a negative screening and deliberately opted out of investing in gambling, tobacco, and alcohol. These pioneering socially responsible investors were faith or values based. In the 1970s, Global Sullivan Principles for social justice motivated others to selectively divest from South Africa to dissent apartheid, and the Vietnam War drove some investors to opt out of nuclear and military weapons production.

In the 1990s, driven by the Brundtland Commission's sustainability, corporate social responsibility (CSR) took into account social and environmental behavior; socially responsible investing continued on the path of social alignment by negative screening of unacceptable social and environmental conduct, building portfolios of assets that exclude companies deemed irresponsible or ones that are contrary to the mission or values of the investors. A further shift occurred toward incorporating environmental and social factors in investment decisions. However, explicitly seeking financial returns as well, nontraditional criteria, e.g., policies, were included in evaluating risk and return. The mantra was to do good for society but not do harm to financial returns. The key shift was the growth in active ownership or shareholder activism and inclusion of positive screening for best-in-class sustainability performance. These corporate social responsibility (CSR)-guided triple bottom-line investors and investments, using positive screening for best-in-class, were now able to aggregate the "triple bottom-line" economic, environmental, and social performance of organizations.

In addition to economic, environmental, and social factors, increasing emphasis on governance emerged with the passage of Sarbanes-Oxley in 2002. Institutional investors generally have investments that are diversified across asset classes, sectors, and geographies with long time horizons and closer ties to the markets and economies as a whole. These investors, also known as "universal owners" of private enterprise, alongside other mission-driven foundations and high-net-worth individuals, sought greater insight into the opportunities and risks in the nonfinancial performance of organizations. They engaged actively as shareholders with the organizations they invest in, rather than just mandate negative screening, and incorporated environmental, social and governance (ESG) factors into their investment process. Faith-based and CSR-guided investments that use ESG factors in a best-in-class approach evolved into ESG-integrated investments, and while early faith-based investors were driven by inherent value of the investor, today's responsible investors incorporate external realities.

Concomitant with making a positive societal impact, responsible investment strategy considers ESG criteria to achieve competitive and long-term financial return. Capturing the upside needs appropriate, often industry disruptive innovation strategy that in turn requires better understanding of the ESG advantage and leveraging the information arbitrage; the focus is on what ESG factors are "material."

In 2007, the Rockefeller Foundation coined "impact investing," "an umbrella term to describe investments that create positive social impact beyond financial returns" (Griffin 2013). Unlike the CSR-guided negative screening investors with exclusionary strategy, impact investors focus on inclusion, that is, positive screening for best-in-class social impact and the entity could be structured to serve different program or mission (e.g., agriculture, health) areas and use different legal entities (e. g., benefit corporations and community interest companies).

Investing in sustainability includes all the socially responsible investments that enhance one or more of the sustainability components or objectives, without significantly harming the other. For instance, a mission- or program-related investment may focus on eliminating toxics from chemicals that harm unborn children, which is clearly aligned with sustainability goals. Socially responsible investing covers a broad range of investments, faith or values based, CSR-guided negative screening, CSR-guided best-in-class triple bottom line, ESG-integrated, and program- or mission-related impact investing. The financial sector focused on socially responsible investment that has grown from $2.7 trillion in 2007 to $21.4 trillion in 2014 (Global Sustainable Investment Alliance, http://www.gsi-alliance.org/wp-content/uploads/2015/02/GSIA_Review_download.pdf accessed Mar 2017). Investors in this sector actively prefer to invest in corporations that have been vetted by and are high on the dominant sustainability indexes (Meg Voorhes et al., "Executive Summary – Fig. B: Growth of SRI $2.7 trillion in 2007 to $3.0 trillion in 2010," in 2010 Report on Socially Responsible Investing Trends in the United States, Social Investment Forum Foundation, accessed December 2012, available at http://ussif.org/resources/research/documents/2010TrendsES.pdf.).

Voluntary Responsible Investment Principles

Voluntary adoption of a set of principles to guide investment decisions helps direct companies and governments conduct their activities responsibly. Some of the established ones that cover large investments and investors, ranging from governmental development projects and multinational enterprise expansions to private equities and mission-driven charities, are highlighted below. These voluntary investment principles include Organisation for Economic Co-operation and Development Guidelines for Multinational Enterprises, Equator, UN Global Compact, and the Principles for Responsible Investing. Narrower range of investments and/or objectives are focused on by others, such as INSEAD's Global Private Equity Initiative for assimilating ESG in private equity and Impact Reporting and Investment Standards (IRIS), an initiative of the Global Impact Investing Network (Global Impact Investing Network (GIIN) https://iris.thegiin.org/about-iris accessed on Mar 22, 2017) with a goal to increase the scale and effectiveness of impact investing, Global Sustainable Investment Alliance (Global Sustainable Investment Alliance http://www.gsi-alliance.org/ accessed Mar 22, 2017) with a vision to integrate sustainable investment into financial systems, and CDC (CDC Investment Works, UK's Development Finance Institution (DFI) and wholly owned by the UK Government http://www.cdcgroup.com/Who-we-are/Key-Facts/ accessed Mar 21, 2017), the development fund arm of the UK with a focus on Africa and South Asia.

Guidelines for Multinational Enterprises (OECD Principles)

Adopted in 1976, the Organisation for Economic Co-operation and Development (OECD) (Organisation for Economic Co-Operation and Development http://www.oecd.org/corporate/mne/1922428.pdf accessed March 21, 2017; Organization for Economic Co-operation and Development (OECD) is an intergovernmental economic organization with 35-member countries.) Guidelines for Multinational Enterprises (MNE) (Organization for Economic Co-operation and Development, accessed March 20, 2017, available at http://www.oecd.org/investment/mne/38783873.pdf) establishes legally nonbinding principles and standards for responsible business conduct for multinational corporations. They cover such areas as human rights, disclosure of information, anti-corruption, taxation, labor relations, environment, competition, and consumer protection. Select components are highlighted below:

(a) Develop policies that consider country programs and other stakeholder views, respect human rights, and contribute to economic, social, and environmental progress for sustainable development. The policies should also promote human capital formation, capacity building, and good governance.
(b) Ensure disclosures regarding activities, structure, financial situation, and performance are timely, regular, reliable, and relevant. The disclosures should also be of high quality and cover financial and required nonfinancial information, including social and environmental performance.

(c) Employee relations practices should respect, within the framework of applicable law, the employee's right to form trade unions; abolish child labor and any forced labor; avoid discrimination in employment based on race, color, sex, religion, political opinion, national extraction, or social origin; and take adequate steps to ensure occupational health and safety.
(d) Environmental policies and practices should protect the environment, public health, and safety, and operations should be conducted in a manner that contributes to the wider goal of sustainable development. This component is amplified further, calling for actions as follows: collect and evaluate adequate and timely information on the environmental, health, and safety (EHS) impacts of enterprise activities and verify progress toward measurable goals; engage in timely communication and consultation with the public and employees directly affected by the EHS policies and activities of the enterprise; incorporate, in decision-making, the foreseeable EHS-related impacts associated with the processes, goods, and services and when needed, prepare an appropriate environmental impact assessment; maintain contingency plans for preventing, mitigating, and controlling serious events; not use scientific uncertainty to postpone cost-effective measures to mitigate damage; improve environmental performance by adoption of technologies and development of products or services with better EHS performance; provide adequate education and training to employees in safe handling of hazardous materials and the prevention of accidents; and help develop environmentally meaningful and economically efficient public policy.
(e) Issues such as combating bribery, protecting consumer interest, and building local science and technology capacity, fair competition, and timely payment of appropriate amount of taxes are addressed by other guidelines.

Equator Principles

At the dawn of this millennia, growing social expectations associated with the move from shareholder to stakeholder primacy put pressure on the financial investment sector to commit to sustainability, which called for measuring environmental and social impacts, continuous improvement of portfolios, proactively fostering sustainability, building capacity, and linking performance. The Collevecchio Declaration on Financial Institutions in 2002 was a move by over 100 NGOs to advocate environmentally responsible behavior in the financial sector (Collevecchio Declaration, BankTrack (Amsterdam: BankTrack, January 2003), accessed December 2012, http://www.banktrack.org/download/collevechio_declaration/030401_colleve cchio_declaration_with_signatories.pdf.), and it served as a precursor to the Equator Principles. The first principle, sustainability, calls for measurements of environmental and social impacts, continuous improvement of portfolios, and proactive fostering of sustainability, building capacity, and performance. The second principle is to "do no harm," which requires the creation of sustainability procedures and the adoption of international standards. The next three principles involve taking full responsibility for impacts, accountability for public consultation and stakeholder rights, and transparency through corporate sustainability reporting and information disclosure. The final principle is sustainable markets/governance, which covers public policies and regulations that recognize government's role and discourage unethical use of tax havens and currency speculation.

Around the same time, the World Bank and its project financing arm, International Finance Corporation (IFC), were sued by impacted parties and NGOs for not ensuring that their borrowers operate their project responsibly. This lawsuit led to the development of Equator Principles (Equator Principles, accessed October 2017, available at http://www.equator-principles.com/index.php/about) in 2003. It was an industry group voluntary initiative designed to manage environmental and social risk in project financing. Although it was led by IFC, later signatories include Goldman Sachs and Citigroup.

Equator Principles (2003) comprise of conducting environmental and social impact assessments (ESIA), compliance with all applicable social and environmental standards, covenants in financial documentation, public consultation and disclosure, grievance mechanisms, independent review, monitoring, and reporting. Furthermore, the public consultation and disclosure process requires conferring with all stakeholders for the development of the ESIA, disclosure of ESIA results to public and ongoing discussions during construction and operation. These communications and engagements must be conducted in local languages, showing respect for local traditions and ensuring that the groups involved are representative.

Principles for Responsible Investing (PRI)

Principles of Responsible Investing (PRI) was launched by the United Nations (UN) in 2006, following a finding that environmental, social, and governance (ESG) issues affect long-term shareholder value, which in some case could be profound (UN Principles of Responsible Investment, https://www.unpri.org/about accessed March 21, 2017). PRI is not associated with any government, and while supported by, it is not part of the United Nations. PRI is specifically designed for institutional investors and the financial sector and reflects the core values of large investors whose investment horizon is long, and portfolios are diversified. PRI has grown to over 1,700 signatories and US $62 trillion associated assets under management.

There are six principles for responsible investment for incorporating ESG factors into investment practice. Principles 1 and 2 seek incorporation of ESG issues into investment analysis and decision-making process and into ownership policies and practices. Principle 3 requires appropriate disclosure on ESG issues by the entities invested(ing) in. Principles 4–6 call on signatories to promote acceptance, enhance effectiveness, and report implementation progress on the principles.

The UN Global Compact

The UN Global Compact's principles are derived from the Universal Declaration of Human Rights, the International Labor Organization's Declaration on Fundamental Principles and Rights at Work, the Rio Declaration on Environment and Development, and the United Nations Convention against Corruption (UN Global Compact https://www.unglobalcompact.org/what-is-gc/mission/principles accessed Mar 22, 2017). There are ten principles in the areas of human rights, labor, environment, and anti-corruption, and signatories are required to provide annual communication on progress. Failure to do so can result in expulsion.

Principle 1 requires support and respects the protection of internationally proclaimed human rights; Principle 2 seeks to ensure that they are not, unwittingly or otherwise, complicit in human rights abuses; Principle 3 calls for upholding the freedom of association and the effective recognition of the right to collective bargaining; Principles 4, 5, and 6 support the effective abolition of child labor and the elimination of all forms of forced and compulsory labor and discrimination in respect of employment and occupation; Principles 7 and 8 support a precautionary approach to environmental challenges and promote initiatives for greater environmental responsibility; Principle 9 encourages the development and diffusion of environmentally friendly technologies; and Principle 10 urges work against corruption, including extortion and bribery.

Emerging Global Sustainability Governance Regulations

Today, corporate social responsibility encompasses investments that create positive social and environmental impacts beyond financial returns. Traditional financial reports do not adequately account for how corporate sustainability performance can enhance or impede both shareholder and stakeholder value. "Integrated corporate reports" that combine financial and sustainability reporting could close the gap by incorporating externalities and other intangible assets by capturing intrinsic values and enable investors to make better informed decisions. Nonfinancial information coming directly from company reports is more likely to be valued by investors (EY "Tomorrow's investment rules: global survey of institutional investors on nonfinancial performance," 2014).

Sustainability Regulations: Evolving Globally

Stakeholders relevant to sustainability are participants, influencers, and vulnerable groups. Participants are directly involved in the commercial exchange process of business and industry and include consumers, corporations, employees, financial institutions, shareholders, state-owned enterprises, and supply chains. Influencers are instrumental in the development of public opinion and policy and include local authorities and nongovernmental organizations (NGOs) such as regulatory agencies, industry trade groups, scientific communities, public interest activists, and the media. The third group comprises of the vulnerable sections of the society that require special protection from exploitation and include children, women, employees, and select socioeconomic groups.

While every group may benefit long term from sustainability regulations, near term the requirements and impact of regulations vary by group. Influencers formulate regulations and monitor compliance, participants comply, and the vulnerable are protected, and it is likely that the society at large benefits long term. Equity and institutional investors as well as corporate and public policy stewards clearly have a significant role in shaping sustainable development decisions and accomplishing the quadruple bottom line. Pension funds and institutional investors often file corporate shareholder resolutions seeking data on companies' risks and initiatives related to climate change, such as policy, emission levels, and mitigation plans. Equity investors are equally concerned about environmental and other sustainability risks to operations and the longevity of corporations.

Emerging sustainability regulations seek disclosures and emanate from government departments of environment, trade and commerce, and finance and treasury to ensure sustainable development, to protect investors, and to collect their fair share of taxes. Increasingly, lenders and institutional investors are required to disclose through integrated reporting how their investments are channeled into responsible operations from the perspectives of longevity, risk, and reward. Stock exchanges are recognizing the need for transparency on corporate sustainability strategy. The US Securities and Exchange Commission (SEC) and several stock exchanges across the developed world call for reporting material risks in their operations as part of their annual financial reports.

Governments have a dual role of leadership in sustainability reporting. State-owned enterprises have a natural stewardship role in progressing sustainability reporting. First, the state-owned agencies mandate and monitor sustainability reporting thru their public governance arm. Second, they serve as pilots and role models. Their development of metrics and measurement of sustainability helps advance government mandates for sustainability reporting across all sectors. Some European nations such as France, Spain, and Sweden as well as all the BRICS nations, Brazil, Russia, China, India, and South Africa, specifically target and mandate sustainability reporting from state-owned enterprises (Columbia University, http://spm.ei.columbia.edu/files/2015/06/SPM_Metrics_WhitePaper_2.pdf accessed Mar 2017).

Many governments and stock exchanges seek third-party verifications for assurance. Likewise, because of the growing linkage of sustainability impacts to financial performance, multinational companies and global investors want qualified and vetted third-party verification assurance. Global assurance standards available today include ISAE 3000 (International Federation of Accountants, Accessed Apr 2017 and available at https://www.ifac.org/publications-resources/international-standard-assurance-engagements-isae-3000-revised-assurance-enga) of International

Auditing and Assurance Standards Board of the International Federation of Accountants and ISO 14064–3 for GHG assertions (ISO Available at and accessed Apr 2017 https://www.iso.org/standard/66455.html).

Sustainability Governance Regulations: Select Country Examples

(a) Since 2001, companies listed on the stock exchange in France have been required to include social and environmental impacts in their annual reports. The 2010 Grenelle II Act of France expanded the mandate beyond environmental and social performance reporting and requires third-party verification (Institut RSE Management, "The Grenelle II Act in France: a milestone towards integrated reporting," 2012.). The assurance of verification related to the company's transparency obligations on social and environmental matters is mainly designed to comply with ISAE 3000 (International Federation of Accountants, Accessed Apr 2017 and available at https://www.ifac.org/publications-resources/international-standard-assurance-engagements-isae-3000-revised-assurance-enga) of International Auditing and Assurance Standards Board of the International Federation of Accountants and French professional standards.

(b) Starting 2004, the Australian Stock Exchange, ASX, requires listed companies to disclose material sustainability – economic, environmental, or social risk and mitigation plans. DR03422 General Guidelines on the Verification, Validation and Assurance of Environmental and Sustainability Reports 2003 was issued by Standards Australia.

(c) The 2007 Environmental Information Disclosure Act of China mandates public disclosure of compliance and serious releases. Incentives such as grant priority are offered for voluntary disclosure of environmental information on resource use, emission level, reduction targets, etc. The 2008 Green Securities Policy adopted in China requires several highly polluting industry sector companies listed on the Shenzhen and Shanghai stock exchanges to disclose environmental information to the public. The China Ministry of Finance issued China Certified Public Accountant Practicing Standard, CAS3101 which follows ISAE 3000, but requires sign-off by a certified practitioner.

(d) For over a decade, China Stock Exchanges Shanghai (Sustainable Stock Exchanges Initiative, "Notice on Strengthening Listed Companies' Assumption of Social Responsibility" http://www.sseinitiative.org/fact-sheet/sse/ accessed Mar 23, 2017) and Shenzhen ("Shenzhen Stock Exchange Social Responsibility Instructions to Listed Companies" http://www.szse.cn/main/en/rulseandregulations/sserules/2007060410636.shtml accessed Mar 2017) have required all companies listed on the stock exchange and all companies listed in the SSE Corporate Governance Index 240 to provide ESG reports (BSD Consulting, http://www.bsdconsulting.com/insights/article/sustainability-reporting-standards-in-china accessed, Mar 23, 2017). Hong Kong Stock Exchange's new HKEx ESG Reporting Guide (Hong Kong Exchange, "ESG Reporting Guide," accessed Mar 23, 2017 and available at http://www.hkex.com.hk/eng/rulesreg/listrules/listsptop/esg/guide_faq.htm) came into effect on 1 January 2016.

(e) The 2012 requirement of India's Securities and Exchange Board calls for business responsibility reports from the top 100 companies. A unique feature of the Companies Act 2013 of India is that it requires companies, beyond a certain size of operation, to set up a corporate social responsibility board committee to develop corporate social responsibility policies and to ensure allocation and application of "at least 2% of the average net profits of the company made during the three immediately preceding financial years" on "CSR" activities (Business for Social Responsibility (BSR) https://www.bsr.org/en/our-insights/blog-view/india-companies-act-2013-five-key-points-about-indias-csr-mandate accessed Mar 23, 2017) to implement those policies. If the company fails to spend this amount on CSR, the board must explain why, in its annual report. The act defines CSR as activities that promote poverty reduction, education, health, environmental sustainability, gender equality, and vocational skills development.

(f) Since 2012, the UK Department of Environment requires all companies listed on the London Stock Exchange to report their greenhouse emissions in their annual reports. One very interesting feature is the requirement to include at least one ratio that relates reported GHG emissions to company activity, such as carbon intensity. AA1000 Assurance Standard issued in 2008 by UK-based Account-Ability helps ensure that sustainability reporting and assurance meets stakeholder needs and expectations.

(g) The European Union adopted Directive 2014/95/EU (European Commission initiative for Mandatory Environmental, Social and Governance Disclosure in the European Union.) on disclosure of nonfinancial and diversity information by organizations with over 500 employees. They must include in their management report policies and main risks and outcomes on environmental, social, and employee aspects, human rights, anti-corruption and bribery, and diversity. The directive is under transposition to national law by EU member states.

(h) In the USA, the need for mandated corporate transparency is becoming acknowledged at a steady pace, with regulations on the rise. US investment banks are required to conduct due diligence for material risks, including environmental liabilities prior to preparing prospectus for any new initial public offering (IPO). The US financial reform holds banks responsible for their actions long past the date of transaction. US Dodd Frank (U.S. Securities and Exchange Commission, "Fact Sheet: Disclosing the Use of Conflict Minerals," 2014.) requires reporting of conflict minerals. US SEC, to protect equity investors investing in publicly listed stocks, has guided listed companies to manage climate risk like any other business risk. In 2010, US SEC created guidelines for companies to report on climate risks in their proxy statements, which accelerated the integration of ESG factors in mainstream financial risk disclosures for US companies. They suggest that "the climate risk mitigation may require internal capacity-building and stakeholder and community engagement and warn that uncertainty is not a reason for inaction" (Taback and Ramanan 2013).

Impact Investing and Organization Structures

Impact Investing Primary Drivers

In 2007, the Rockefeller Foundation coined "impact investing," "an umbrella term to describe investments that create positive social impact beyond financial returns" (Griffin 2013). Impact investment is emerging as a separate asset class, and the industry is estimated to grow to US $500 billion by 2020 (Monitor Institute 2009).

Impact investors invest with an intent to generate measurable social and/or environmental impact, while making financial returns. The idea is to align profit making with generating positive social impact. Also, known as social investing, they could be broadly categorized, based on primary motive as:

(i) Impact first to primarily maximize impact
(ii) Investment first to primarily get financial returns
(iii) Catalyst first to seed funds to collaborators to initiate or strengthen impacts

Impact investing includes program-related investments (PRI) which have been around since the 1970s and mission-related investments (MRI), (Rockefeller Philanthropy Advisors, "Mission Related Investing – a Policy and Implementation Guide for Foundation Trustees" available at and accessed Mar 2016 http://rockpa.org/document.doc?id=16) a term coined in the last decade. PRI is below market rate investment by foundations, deeply focused on impact and counting toward endowment payout requirements for foundations. MRI is a market rate investment by private foundation endowments that uses the tools of social investing, sometimes including shareholder advocacy and positive and negative screening (Monitor Institute, "The Future of Impact Investing," available at and accessed Mar 2016, http://monitorinstitute.com/downloads/what-we-think/impact-investing/Impact_Investing.pdf).

Impact Investment Organization Structures

Social enterprise is an impact investing business that reinvests profits directly to serve social needs. Unlike nonprofit entities, social enterprise does not seek support from government or philanthropists. Also, it is distinct from a socially responsible business that engages in CSR. The entity could be structured to serve different program or mission (e.g., agriculture, health) areas. One example is an energy savings mission supported by energy conservation consulting services.

Impact investing could use different forms of hybrid organizations (community interest companies in the UK). Examples of such legal entities include low-profit limited liability company, benefit corporation, and B corporation to meet the investor's specific legal, tax, and mission needs and achieve financial returns while prioritizing social benefit objectives.

Low-Profit Limited Liability Company

Low-profit limited liability company is a hybrid of for-profit and nonprofit. It limits liabilities and protects officers from shareholder lawsuits that question business choices that prioritize social or environmental returns over profits. It can also attract charitable donations or funds that accept below market return.

Benefit Corporation

Benefit corporation is a for-profit company that creates a material positive impact or public benefit. They cannot seek charitable contributions and must produce benefits and report to rigorous standards with third-party independent assessment that adhere to high transparency and accountability. Officers are not liable for damages if the public benefit is not achieved. However, they are required to consider broad array of stakeholders.

B Corporation

B corporations are organizations that are certified by B Lab, a nonprofit third-party entity, much like the Underwriters Lab, to ensure that the B corporation meets social and environmental transparency, accountability, and performance standards. Unlike benefit corporation, B corporation must be certified. Some examples of public benefits that B corporations provide are: buy from low-income communities and make donations to other nonprofit organizations.

Philanthropic Capitalism and Venture Philanthropy

Historically, commercial and social capitals have been clearly separated. Traditionally, the approach was to get rich using the commercial capital and then indulge in philanthropy. Starting with Rockefeller, Carnegie and, today, Bill and Melinda Gates and Warren Buffett are icons of business philanthropy. Over the years, corporate philanthropy became more professionalized but philanthropic capitalism – the business effort to do well by doing good – could not yield a superior model of capitalism. President Bill Clinton calls the Clinton Global Initiative a laboratory to test philanthropic capitalism ideas. He says "...the twenty-first century has given people with wealth, unprecedented opportunities...to advance public good......our interdependent world is too unequal, unstable and because of climate change, unsustainable. It failed to turn around our global environmental, social and ethical trends, and it may in fact be distracting us from true systemic sustainability and responsibility" (Bill Clinton in his Foreword in Mathew Bishop and Michael Green, Philathrocapitalism – How Giving Can Save the World, (New York: Bloomsbury Press, 2008).).

A more recent phenomenon is venture philanthropy – the idea that corporate foundations can improve effectiveness through monitoring where they invest, providing management support and staying long enough until those ventures become self-supporting. Other emerging models include traditional foundations practicing high-engagement grant-making, organizations funded by high-net-worth individuals but with all engagements done through professionals, and a partnership model where both the partner and individuals donate the financial capital and engage with the grantees.

Impact Investment Standards and Reporting Frameworks

The World Economic Forum defines impact investing, in the context of measurement, as "an investment approach that intentionally seeks to create both financial return and positive social or environmental impact that is actively measured." (World Economic Forum, Accessed Apr 2017 available at http://reports.weforum.org/impact-investment/ and http://www3.weforum.org/docs/WEF_Social_Investment_Manual_Final.pdf) Another study (Ebrahim, Alnoor et al., HBS, Accessed Apr 2017 and available at http://www.hbs.edu/socialenterprise/Documents/Measuring Impact.pdf) suggests that impact measurement efforts could be classified by measurement objectives as:

(i) Estimating impact – for due diligence prior to investment,
(ii) Planning impact – selecting metrics and data collection methods to monitor impact
(iii) Monitoring impact – to improve program
(iv) Evaluating impact – to prove social value

The same study (Ebrahim, Alnoor et al., HBS, Accessed Apr 2017 and available at http://www.hbs.edu/socialenterprise/Documents/MeasuringImpact.pdf) identifies four impact measurement methods:

(i) Expected return, one that takes into account the anticipated social benefits of an investment against its costs, discounted to the value of today's value
(ii) Logic model, a tool used to map a theory of change by outlining the linkage from input to activities, to output, to outcomes, and ultimately to impact,
(iii) Mission alignment method, which measures the social value criteria and scorecards to monitor and manage key performance metrics
(iv) Experimental and quasi-experimental

Impact Reporting and Investment Standards (IRIS)

Acumen Fund, B Lab, and the Rockefeller Foundation founded the Impact Reporting and Investment Standards (IRIS) to create a common framework for defining and reporting impact capital performance. The IRIS (Global Impact Investing Network (GIIN), Accessed Apr 2017 available at https://iris.thegiin.org/

guide/getting-started-guide) is a catalog of metrics for impact investors that measure the performance of an organization.

The key metrics of IRIS include:

(i) Financial performance, including standard financial reporting metrics
(ii) Operational performance, including metrics to assess investees' governance policies, employment practices, and the social and environmental impact of their business activities
(iii) Product performance, including metrics that describe and quantify the social and environmental benefits of the products, services, and processes offered by investees
(iv) Sector performance, including metrics that describe and quantify impact in particular social and environmental sectors, including agriculture, financial services, and healthcare
(v) Social and environmental objective performance, including metrics that quantify progress toward specific objectives such as employment generation or sustainable land use

B Impact Assessment (BIA)

B Impact Assessment (BIA) (Global Impact Investing Network (GIIN), Accessed Apr 2017 and available at https://iris.thegiin.org/b-impact-assessment-metrics) is another tool to assess a company's overall social and environmental performance. The impact of a business on all stakeholders is assessed through an online, easy-to-use platform. The BIA is a free, confidential service administered by the nonprofit organization B Lab. The BIA uses IRIS metrics in conjunction with additional criteria to come up with an overall company or fund-level rating, as well as targeted sub-ratings in the categories of governance, workers, community, environment, and socially and environmentally focused business models.

Responsible Investing and Private Public Partnership (PPP)

Public-private partnerships are typically between a government agency and a private sector entity to finance, build, and operate projects, such as public transportation networks, sustainable development of an underserved region of the world, or elimination of avoidable infant mortality (Investopedia, "Public Private Partnerships," http://www.investopedia.com/terms/p/public-private-partnerships.asp accessed Mar 2017). The government agency could be a federal, state, or municipal authority of a country or could even be funding agencies such as the United States Agency for International Aid (USAID) or International Finance Corporation (IFC), the financial arm of the World Bank, or United Nations Sustainable Development Program (UNDP). The private sector entity could be an entrepreneurial venture, a for-profit company, or one of several emerging responsible investment business models for sustainable development, such as philanthropic capitalism, venture philanthropy, mission-driven charitable foundations, and impact driven high-net-worth individuals with patient capital. Risks and returns are shared between the partners according to the ability and missions of each.

While financing could come from either or both partners, it requires repayments from the public sector and/or users over the project's lifetime. For instance, for a wastewater treatment plant, payment comes from fees collected from users. Toll-based bridges, tunnels, and highways have been following the public-private partnership models for over a century. However, because of the nature of sustainable development projects that calls for large investments, long-term patient capital and often a passion for social responsibility make the public-private partnership (PPP) model rather attractive and sustainable! One partner's authority to enforce long-term repayment by large captive set of users, coupled with the other partner's desire and drive to make an impact, is a powerful recipe. Private sector innovation could provide operational efficiency. However, they bear the burden of project delay, budget exceedance, and insufficient demand. In case of sustainable development projects for underserved regions, geopolitical risk may be overwhelming. The private and public entities could collude and siphon off major parts of the resources, as was seen during the Haiti earthquake; with over ten thousand NGOs misallocating global aid, Haiti has become the infamous "Republic of NGOs" (WorldPost, "Haiti's Multi-Billion Dollar Humanitarian Aid Problem," accessed Dec 2018 available at https://www.huffingtonpost.com/young-professionals-in-foreign-policy/haitis-multi-billion-doll_b_8207494.html).

Ethical Dimension of Resource Management

Ecosystem and Resource Management: Sociopolitical Choice

Ecosystem and resource management, especially optimal allocation, commonly involve tough sociopolitical choices. As a society progresses, it often faces a conflict between economic benefits and environmental degradation (This section is based largely on book chapter, Ram Ramanan and Hal Taback "Environmental Ethics and Corporate Social Responsibility" for the book titled "Spirituality and Sustainability: New Horizons and Exemplary Approaches," Springer 2016.). How much environmental protection is appropriate? What is the right balance between environmental protection and exploitation of natural resources for sustainable development? Both public (government) and private (corporate) stewards bear significant responsibility.

When the market exchange process between the seller and the buyer impact an external entity that has no say in setting the exchange price, a market externality is created. Economists continue to battle externality, which often challenges the traditional market efficiency principles and makes ecosystem development issues tough sociopolitical choices. This leads to highly charged debates of public opinion, and regulatory intervention becomes inevitable. In the environmental context, one could think of these as stakeholder demand for "right to no pollution" or the "right to compensation." Stakeholder engagement in cost-benefit analysis and ethical choice of effective regulatory mechanisms are the key drivers in resolving this sociopolitical conflict. In particular, when some of the negatively impacted entities are either not invited or not involved, there is significant opportunity for greed and corruption to creep in the policy development process.

Stakeholder preferences for comparing ecosystem development alternatives can be biocentric, anthropocentric, or centered on sustainability. With biological world as the center, biocentrism focuses on the intrinsic value of life and does not consider usefulness to human beings to be one of its core values. "The environment is there to provide material gratification to humans" is at the core of anthropocentrism. Sustainability strives to preserve the integrity of ecosystems.

Resource and ecosystem management alternatives for sustainable development are commonly evaluated and compared using cost-benefit analysis. The benefits and costs are quantified where possible but presented with a description of uncertainties. Stakeholders are engaged and their inputs considered. In most environmental policy decisions, this requires core assumptions regarding the social discount rate and the value of reducing risks of premature death and of health improvements.

However, decision-makers are not bound or limited by strict cost-benefit tests. In particular, because equity is a noneconomic factor, it is crucial to identify important distributional consequences to ensure that ethical choices are made. This is especially true in such areas as climate change and environmental justice – not unduly impacting people of lower socioeconomic strata because they tend to live closer to regions that are more vulnerable with no adequate plans for adaptation or to neighborhoods that face the brunt of highest emissions and discharges from manufacturing facilities that pollute the environment.

Governments commonly deploy one or more of the regulatory intervention mechanisms to manage ecosystem development; these are command and control, toxic torts (liability through law suits), and economic incentives such as emissions fees or tax savings and market-driven approaches such as emissions credit trading. Command and control is the most dominant form of regulation, in which regulation mandates specific pollution control equipment, technology, or emission limit for type of plant or specific pollutant(s). Noncompliance with the regulations carries significant financial and personal criminal liability/penalties. They are enforced through the legal framework and courts. The toxic tort or "liability" approach, where polluters are responsible for the consequences and pay for all damages, creates incentives for the polluter to take precaution, as some jury awards may be very significant, often, enough to eliminate an entire industry, for example, "asbestos."

The other two regulatory mechanisms offer more direct monetary incentives. "Emissions fees" calls for a charge per unit of pollution – it is in the polluter's interest to reduce pollution to lower the fees they must pay. "Tax savings" encourage investment in low-emission technologies. These approaches could achieve predefined environmental standards at a lower possible cost. But the control authorities often do not know the exact fee to charge or tax savings to offer in order to reach the optimum pollution for market efficiency. "Marketable permits/emissions credits trading" allows polluters and speculators to buy and sell rights to pollute; it separates who pays and who installs controls (Koutstaal 1996). A polluter may install excess controls at units that are more cost-effective at reducing emissions, and that yield emission reduction credits (ERC), and use the revenue from the sale of ERCs to pay for the controls elsewhere. Efficiency is achieved through the use of purchased or internally generated ERCs to avoid more expensive controls for operating facilities and equipment while achieving the same level of overall emission reduction. This preserves society's resources to achieve highest *bang for the buck*.

The New Social Contract and a Clarion Call for Ethical Leadership

The role of the public corporation and the nature of the "social contract" have been changing over the past two centuries but have changed at a faster pace in the recent decades. Capitalism in general and the American dream in particular interpret greed to be a healthy trait. Greed has become pervasive in business from executives, corporations, banks, and financial markets. This mantra, along with an obsession with the primacy of shareholder interests, has driven most early ventures to privatize gains and socialize costs. The role of business is transforming from one merely fulfilling a social contract to taking on social responsibility with the growing recognition that shareholders are only one of many stakeholders. A principal driver of this societal transformation is the recognition that business is no longer the sole property or interest of a very few. Notably, synchronous interactive connectivity among stakeholders has had a significant role in this change.

"The corporate (and corruption) scandals and implosions of the past decade, climaxing in the recent global financial crisis and environmental disasters have highlighted how critical ethically, environmentally, and socially responsible decision making and leadership are to the long-term survival and success of both individual businesses and society" (Ramanan and Ashton 2012a). It is not feasible to ignore the changing business ambiance and social contract under which corporations and public service organizations have to operate. In today's global environment, societal needs are defining markets, and both private and public leaders have to address a range of issues from poverty and hunger to sustainability and ethics. Ethical issues include bribery, fraud, greenwashing, inequity, and a culture of corruption. Corporations and leaders have to manage corporate social responsibility and integrate it into their global strategy, and a public policy leader is often faced with balancing human needs and environmental considerations; the end goal in both cases is sustainability – to protect and preserve our planet for future generations.

"With increasing focus on sustainability factors from the marketplace (regulators, investors, financiers and consumers), corporate sustainability reporting is shifting from voluntary to vital; and more recently becoming an integral part of annual financial reporting. Many stock exchanges are requiring corporations to provide citizenship or social responsibility reports prior to listing them. Advances in enterprise systems are making it feasible for corporations to track and transform sustainability performance. The materiality of these seemingly noneconomic impacts is the critical link between sustainability and business strategy. Leaders need insight into how to determine which sustainability metrics are material to them and relevant to their business" (Ramanan and Ashton 2012b). Long overdue, only now are ethics metrics being incorporated within sustainability reporting (G-4 56–58 Ethics and Integrity within Governance metrics of Global Reporting Initiative, https://g4.globalreporting.org/general-standard-disclosures/governance-and-ethics/ethics-and-integrity/Pages/default.aspx).

Decision-Making and the Ethical Dimension

Formation of Human Value System

Most humans are not naturally (information asymmetry apart) data centric, evidence driven, cold calculative, consciously thoroughly choosing rational robots (individuals). They are rather emotional spontaneous beings with sociocultural upbringing bias who decide and then rationalize their choices. Character, which is formed at an early age, defines the extent to which people will go to achieve an objective.

Ethical Decision-Making: Characteristics

Three qualities affect ethical decision-making: competence in identifying issues and evaluating consequences, self-confidence in seeking different opinions and deciding what is right, and willingness to make decisions when the issue has no clear solution. The development of these qualities in individuals depends on their intrinsic personality and their stage of moral development at the point of decision. Gandhi always held that a prerequisite to making ethical choices is to build a strong character and that requires one to always be cognizant that the means is as important as the end goal.

Other factors that contribute to individuals' ethical decision-making process are the moral intensity of the consequence of the action; the individual's empathy, knowledge, and intellectual and emotional ability to recognize the potential impacts on stakeholders; and the influence of the decision environment. For instance, sending a personal email from an office computer may not be seen as unethical at all.

Some individuals are more capable of understanding the broader impacts of an issue than others. For example, the natural attenuation remediation of a contaminated site may be better understood by someone with expertise in the area. They may also realize that in some situations, natural attenuation remediation is a good use of the community's resources and that by not moving contaminated materials, it reduces public health risks. However, this does not imply that such an individual is more ethical and will make more ethical decisions. For instance, one could be an environmental activist with a set agenda to make the company responsible for soil contamination pay more for the remediation, or one may be an environmental consultant who offers remediation contract service, and choosing this natural attenuation option may eliminate contract work. Under this circumstance, a greed-consumed consultant may choose the less ethical option of "dig, move, and treat the dirt" that provides him additional contract work and compensation.

Ethics Training for Leaders and Professionals

Building a culture of ethics is critical; and effective ethics training is crucial to overcome our inherent selfishness. When faced with a real-world ethical dilemma, a person cannot formulate an appropriate response from the hypothetical "two-on-a-raft" situation. Ethics training is valuable for everyone. It sensitizes one to ethical issues and prepares one to respond appropriately to ethically questionable situations, which are often unexpected. Frequent in-house ethics training and organizational ethical culture building supplemented with appropriate incentive/deterrent system helps develop ethical values and minimizes the temptations to cheat.

Training is not a one-time activity but an ongoing process. Leaders and professionals should participate in a planned series of participatory workshops that discuss real-world relevant dilemmas and help people learn how to do the right thing in a guided setting rather than leave it to their instinct. Workshops with significant number of participants debating opposing opinions promote deeper understanding, while those that include top management are especially effective because executives can share their values.

Managing Pitfalls of Economics without Equity

Climate Change

In 2004 when author Ramanan had lunch with Nobel Laureate Mario Molina, one who discovered the root cause of stratospheric ozone depletion, at one point, discussions turned to who parallels his discovery in the climate change arena. What surfaced quickly was the name of Nobel Laureate Svante Arrhenius; indeed, his paper of year 1896 describes how carbon dioxide could affect the temperature of the Earth. Recent NOAA (NOAA, "A Paleo Perspective on Global Warming," accessed December 2012 available http://www.ncdc.noaa.gov/paleo/globalwarming/paleolast.html) data shows a strong linkage between the surface temperature of the Earth and the carbon dioxide level of the atmosphere.

In 2010, at a professional luncheon event, Honorable Former Vice President of the USA, and a Nobel Laureate, Al Gore, in response to author's question on equity in global policy, highlighted the potential devastation climate change could unleash on the most vulnerable segment of our society. In 2013, Nobel Laureate Rajendra Pachauri, Chair of IPCC, at a dinner with the author Ramanan, said that scientific consensus among the thousand plus scientists was a tough task, not as much because of differences in scientific views, but more due to the political pressure of the interest groups they served.

In 2015, Pope Francis, leader of the Catholic faith with a following of over one billion people has drawn the world's attention to one of the mega issues of sustainability and said, "Climate change is a dire threat that humans have a moral responsibility to address" (Pope Francis encyclical on climate change, "On Care for Our Common Home" accessed 6/24/2015, available at http://w2.vatican.va/

content/francesco/en/encyclicals/documents/papa-francesco_20150524_enciclica-laudato-si.html). However, despite pleads from the Pope, President Trump has elected to withdraw the USA from the Paris Agreement; today the USA is the only country outside the Paris agreement.

This issue gets further exacerbated when a recent tweet from the President of the USA says, "In the East, it could be the COLDEST New Year's Eve on record. Perhaps we could use a little bit of that good old Global Warming that our Country, but not other countries, was going to pay TRILLIONS OF DOLLARS to protect against. Bundle up!" (CNBC "Climate scientists blast Trump's global warming tweet," accessed Dec 2017, available at https://www.cnbc.com/2017/12/29/climate-scientists-around-the-world-respond-to-trumps-global-warming-tweet.html). Unless humor was intended, the tweet demonstrates either callousness or a complete lack of understanding of the real concern. Climate change, "the two-degree classic," its complexity, truly tests how intergenerational equity and distributive justice is incorporated in making ethical choices.

Global Warming and Climate Change: The Issue and the Impact

Global warming and the resultant climate change is the issue and the way the change in environment affects lives is the consequential impact. One of the most complex applications of analytics in the sustainability arena is the NOAA (NOAA, "A Paleo Perspective on Global Warming," accessed December 2012 available http://www.ncdc.noaa.gov/paleo/globalwarming/paleolast.html) simulation demonstration of linkage between the surface temperature of the Earth and the carbon dioxide level of the atmosphere. Human activity is at least partially responsible for this increase in carbon dioxide level and the resultant warming of the planet. This has been shown by integrated assessment, modeling, and analysis and has the consensus of most scientists in the world. Although many gases contribute to global warming, the gases of most concern, based on their abundance and potential to impact global warming (over 99%), are carbon dioxide (CO_2), methane (CH_4), nitrous oxide (N_2O), and fluorinated gases. Anthropogenic carbon dioxide is by far the most dominant greenhouse gas (GHG) known to cause global warming. GHG emissions are expressed in terms of carbon dioxide equivalents, and the term "carbon" has been used to represent that here.

Conservation, or reduced usage, is often the norm to protect depletion of most resources. Carbon is a natural resource, but unlike other dwindling substances, its use has to be constrained to contain the generation of carbon dioxide. However, reducing the use of carbon is very challenging because of the near omnipresence of carbon in human life.

Complexity and inconsistency in the methodology of monetizing benefits, emergence of new materials and discovery of new adverse health effects, uncertainties in science, and political sensitivity are additional confounding factors. Furthermore, "with different countries likely to undertake different levels of climate-change mitigation, the concern arises that carbon intensive goods or production processes could shift to countries that do not regulate greenhouse gas emissions." (Jeffrey Frankel, "Global Environmental Policy and Global Trade Policy – Harvard Project on International Climate Agreements," accessed December 2012, available http://belfercenter.ksg.harvard.edu/publication/18647) When coupled with currency exchange rates and other geopolitical uncertainties, the problems compound and confound exponentially. Additional complexity in allocating resources to mitigate carbon use comes from who should bear the burden, reduce consumption, tolerate increase in cost of production, or pay more for the same exact functionality.

As stated earlier, climate change, "the two-degree classic," also truly tests how intergenerational equity and distributive justice is incorporated in making ethical choices. The impact of global warming and climate change and the benefits of averting this catastrophe are mirror images covered below. Traditional cost-benefit analysis uses discount rates to bring benefits over different, especially later years to a common base year. For instance, some of the benefits in climate change mitigation result in benefits that may be a generation or even two centuries away. Almost any rate of discount brings the present value to near zero. Distributive justice, an ethical mandate, requires that all human beings get equal share of public goods – the Earth's atmosphere. Absent purpose as a moderator, powerful stakeholders could skew the objective through the inherent bias of self-interest.

Climate Change Paris Agreement, 2015

The First World Climate Change Conference was held in Geneva, in 1979; and the First Assessment Report was produced by the Intergovernmental Panel on Climate Change (IPCC) in 1990. IPCC findings in their Fourth Assessment Report in 2007 received general scientific consensus, and the same year, they received the Nobel Peace Prize. The Paris Agreement 2015 on climate change leads to binding targets (European Capacity Building Initiative, "A Pocket Guide to the Paris Agreement," Accessed Apr 2017, available at http://www.eurocapacity.org/downloads/PocketGuide-Digital.pdf). The Paris Climate Change Agreement has to be ratified by individual countries. This year (2017), the USA has decided to withdraw from the Paris Agreement, the only country to do so.

Highlights of the agreement are presented below:

(a) Global Temperature Goal: The goal is to keep global temperature rise well below 2 °C above preindustrial temperatures while pursuing efforts to limit it to 1.5 °C, increase the ability to adapt, and make finance flows aligned toward low emissions and climate-resilient development.
(b) Mitigation Goal and Nationally Determined Contributions (NDC): The goal is to achieve a balance between anthropogenic greenhouse gas emissions by sources and removals by sinks of greenhouse gases in the second half of this century. All countries are encouraged to formulate and communicate low-emission development strategies and NDCs in 2020, and plans to strengthen them,

based on their national abilities. Parties communicate their NDCs when they join the Agreement.
(c) Adaptation Goal: The Agreement establishes a notional and aspirational "global goal on adaptation" to enhance adaptive capacity, strengthen resilience, and reduce vulnerability to climate change. Adaptation is recognized as a key component of the long-term global response to climate change and as an urgent need of developing country parties.
(d) Loss and Damage Basis: It incorporates the Warsaw International Mechanism for Loss and Damage and calls for its strengthening. Notably, the loss and damage text contain the cryptic words "does not involve or provide a basis for any liability or compensation," reflecting the concern by some that it could be construed as an admission of liability for climate change-related damage and could potentially result in claims for compensation.
(e) Compliance Mechanism: A compliance mechanism is established to facilitate implementation and promote compliance in a transparent and nonpunitive manner. Developing countries will receive support to implement transparency measures.
(f) Capacity-Building Support: It stipulates that developed countries will provide financial support to developing countries to assist them with capacity-building, which includes ability to implement adaptation efforts and take mitigation actions; develop, transfer, disseminate, and deploy mitigation technology; access climate finance; educate, train, and raise public awareness; and enable transparent, timely, and accurate communication of information.
(g) Broader Scope: The path to the common goal must reflect equity and differentiated capability-based national responsibility. It is not solely an environmental problem – it cuts across and affects all areas of society. Must respect and promote human rights: the right to health; the rights of indigenous peoples, local communities, migrants, children, persons with disabilities, and people in vulnerable situations; and the right to development, gender equality, the empowerment of women, and intergenerational equity.
(h) Exchange Mechanisms: Market-based as well as nonmarket-based mechanisms are established to allow parties to voluntarily cooperate in mitigation and adaptation to implement their NDCs. Nonmarket-based approaches promote mitigation and adaptation ambition and enhance public and private sector participation in implementing NDCs.
(i) Finance: The issue of differentiation in the finance was sorted by stating that developed countries "shall" provide climate finance for developing countries, while developing countries are "encouraged" to provide support voluntarily.

Ethical Considerations in Climate Change

The following will be significant in shaping the outcome in the ethical allocation of carbon share to balance ecological effectiveness against economic efficiency and equity:

(a) Avoiding Perceived Vulnerability: Many stakeholders call for action to protect climate to protect humans. However, the primary objective is clearly the protection of man from perceived vulnerability and not vice versa, protection of nature from man. Climate change may endanger human health, wealth, and ultimately survival, in particular, of the weaker sections of the world population.
(b) Optimizing Resource Use: Nature is considered a free and potentially infinite good. However, recognition of the value of resource use optimization results in egocentric ethics giving way to utilitarian ethics. The narrow pursuit of self-interest calls for collective rules instead and advocates the regulation of individual action in the name of the greater good of a greater number of people for a longer period of time. As a consequence, target selection is guided by aggregate benefits and costs rather than the individual actor's self-interest.
(c) Holding in Trust for the Future: "Justice across generations demands restraint today. The concept extends the principle of equity among the human community along the axis of time." It is indeed a question of ensuring intergenerational equity. The approach shifts from posterity, seen only as future beneficiaries of progress, to a possible victim of it. Being a beneficiary of the global commons today, therefore, also implies being their trustee. Protecting the climate system for the benefit of present and future generations suggests considering the well-being of future generations as one of the factors to be considered for decision-making in the present.
(d) Beyond Anthropocentric: People in general are anthropocentric and give humans a strong preference over other species. However, ethicists like Peter Singer (Cavalieri and Singer 1993) value wildlife and wild animals with equal status, and opine that nonhuman beings have rights as well.
Humans are not entitled to inflict climate change upon the communities of plants and animals, which – along with humans and inanimate matter – are not just instrumental but also have intrinsic value in the biosphere, for instance, biodiversity. An associated driver could be the motivation of humans to rejoice in creation.

Chapter Summary and Management/Leadership Lessons

Ecosystem and resource management, especially optimal allocation, commonly involve tough sociopolitical choices and often face a conflict between economic benefits and environmental degradation. This chapter traces the evolution of responsible investing to its current forms and demonstrates how corporate social responsibility (CSR); environmental, social, and governance (ESG); and mission or principle issues have become financially material and have a direct impact on risk exposure and goal accomplishment of public, private, and government investments. It also highlights some of the voluntary principles and provides an overview of globally emerging sustainability regulations.

Most humans are not naturally (information asymmetry apart) data centric, evidence driven, cold calculative, consciously thoroughly choosing rational robots (individuals).

"Aristotle is often cited to describe unethical behavior as when man's rationality is overcome by his desire. Most humans naturally draw their sense of values from multiple sources – reason (philosophical or secular), realization (spiritual) or religion (faith) – and often these have synergistic effect. Plato's rational charioteer could reign in the irrational passionate horses using the head (philosophical or secular), the heart (spiritual) or the heavenly (faith). These diverse inputs cement convictions about identifying the right thing and a commitment to doing the right thing. This inspires one to act more like a Centaur, where the horse and the rider are one – which, if steered correctly could effectively detoxify rampant materialism and preserve our only planet" (Ramanan and Taback 2016).

Chapter End Reflection Questions

1. As a leader would you rather be a rational charioteer or centaur?
2. What is the right balance between environmental protection and exploitation of natural resources for sustainable development?

Cross-References

▶ Collaboration for Regional Sustainable Circular Economy Innovation
▶ Ecopreneurship for Sustainable Development
▶ Environmental Intrapreneurship for Engaged Sustainability
▶ Ethical Decision-Making Under Social Uncertainty
▶ Expanding Sustainable Business Education Beyond Business Schools
▶ Low-Carbon Economies (LCEs)
▶ Moving Forward with Social Responsibility
▶ Responsible Investing and Environmental Economics
▶ Smart Cities
▶ Social Entrepreneurship
▶ Sustainable Decision-Making
▶ The Sustainability Summit

References

Cavalieri, P., & Singer, P. (Eds.). (1993). *The great ape project: Equality beyond humanity* (p. 152). New York: St. Martins Griffin.
Griffin, M. H. (2013). *Impact investing a guide for philanthropist and social investors*. Northern Trust.
Koutstaal, P. R. (1996). Tradeable CO2 emission permits in Europe: a study on the design and consequences of a system of tradeable permits for reducing CO2 emissions in the European

Union (p. 17). (PhD Diss., University of Groningen). http://www.unicreditanduniversities.eu/uploads/assets/CEE_BTA/Dora_Fazekas.pdf.

Ramanan, R., & Ashton, W. (September, 2012a) Green MBA and Integrating Sustainability in Business Education. Air and Waste Management Association's Environmental Manager (pp. 13–15); also accessed Dec 2012 available at http://stuart.iit.edu/about/faculty/pdf/green_mba.pdf.

Ramanan, R., & Ashton, W. (September, 2012b) Green MBA and Integrating Sustainability in Business Education. Air and Waste Management Association's Environmental Manager (pp. 13–15).

Ramanan, R., & Taback, H. (2016). Environmental ethics and corporate social responsibility. In *Spirituality and sustainability: New horizons and exemplary approaches*. Cham: Springer.

Taback, H., & Ramanan, R. (2013). *Environmental ethics and sustainability* (p. 56). Florida: CRC Press.

The LOHAS Lifestyle and Marketplace Behavior

Establishing Valid and Reliable Measurements

Sooyeon Choi and Richard A. Feinberg

Contents

Introduction	1070
LOHAS Is Born in the Natural Marketing Institute	1071
Conceptualizing LOHAS	1072
LOHAS: The Gospel or Science of Sustainability	1075
LOHAS Around the World	1077
The LOHAS Marketplace	1078
LOHAS as a Market Segment	1079
LOHAS as a Lifestyle Segment	1080
Measuring LOHAS	1081
Proposed Plans for Research	1082
LOHAS Measurement and Culture	1083
Cross-References	1084
References	1084

Abstract

Increasing interest in sustainable consumption/development suggests the need to define sustainable lifestyles. People have become concerned about personal and environmental health, social ethics, and morality, and they have incorporated these values into their daily practices. LOHAS (Lifestyle of Health and Sustainability) is an empirically defined lifestyle that reflects this shift. Saying that LOHAS exists does not make it "real" or important to research areas or useful to marketing. Although the LOHAS lifestyle defines a life in pursuit of balance between the individual, the environment, and the society, there is very little work empirically validating its measurement or its relationship to actual behaviors. This chapter will provide an overview of the extant research on the LOHAS

S. Choi (✉) · R. A. Feinberg
Department of Consumer Science, Purdue University, West Lafayette, IN, USA
e-mail: choi268@purdue.edu; xdj1@purdue.edu

© Springer International Publishing AG, part of Springer Nature 2018
S. Dhiman, J. Marques (eds.), *Handbook of Engaged Sustainability*,
https://doi.org/10.1007/978-3-319-71312-0_10

lifestyle. We will discuss the market place behaviors of the LOHAS consumer and our development of a reliable and valid measurement tool for LOHAS.

Keywords
LOHAS · Health and sustainability · Psychometric scale development · Consumer behavior

Introduction

The importance of sustainability is simply etched by the belief that the global future depends on it. Pick a topic...environment, energy, health, well-being, and clean water, and one could easily create an argument, a compelling argument, that sustainability is the broad overarching issue. Sustainability (sustainable development) is about how people consume and dispose of products and how people make choices to innovate and develop yet protect the environment and others lives. But sustainability is not simply about the environment; it is broader in that it is about illuminating how what we do affects health and well-being. Adams (2006) presented a well-developed history and definition of the intersecting world nature of sustainability and the challenge we face in becoming more sustainable in all areas of economic, social, and environmental health, and that discussion will guide this chapter.

Let's start with a definition of sorts. The issue of sustainability is not that we stop progressing or innovating but how do we live in harmony with our world protecting it from damage and guaranteeing that it will all survive (Liu and Diamond 2005). Why...because it is believed that societies collapse due to adoption of unsustainable practices. Whatever the issues under consideration, our world cannot endure if humans continue to use up, pollute, eat, and behave in unhealthy ways. It starts and ends by human action and inaction. A simple Google search (May 2017) using the search terms sustainability resulted in 184,000,000 web sites (3,870,000 scholarly works searching Google scholar). The search for "importance of sustainability turned up 96,800,000 web sites (2,710,000 scholarly books and papers). We use the Google search to highlight the fact that this is not a minor issue or an inconvenience...it is attracting scholarly and popular attention.

Over the last several decades, the increase of affluence and the growth of middle classes in countries have increased interest in quality of life (e.g., Senik 2014). High levels of economic welfare and material abundance have led people to take more of a philosophical and conscious approach in their lives (e.g., Mostafa 2013). In 13 years of survey research on Americans' values and lifestyles, Ray and Anderson (2000) revealed a newly emerging subculture differentiated from the traditionalists and modernists in the values of authenticity, harmony with surroundings, social conscience, and ecological and human welfare (Cortese 2003; Hoffman and Haigh 2010). These values are seen by some as an important new trend in our everyday life (e.g., Veenhoven 2010). The emergence of LOHAS reflects this.

LOHAS Is Born in the Natural Marketing Institute

The terminology that defined this emerging trend was called LOHAS (Lifestyle of Health and Sustainability) by the Natural Marketing Institute (NMI-http://www.nmisolutions.com). Originally, it was an attempt to integrate, describe, and define a cultural value and a growing body of research and thinking. It was an attempt to define a rapid growth of global trends (N.A 2016). LOHAS designates a group of people who make behavioral decisions based on the concern beyond their immediate environment (e.g., Cortese 2003). LOHASians (as we will call them) strive to live healthier and more sustainable lifestyles, and they consider how what they do and buy impact on the environment, the community, and the planet (e.g., Emerich 2011). The NMI believes that the LOHAS lifestyle represents nearly 25% of the US population and describes more than a $200-billion-dollar marketplace (Urh 2015).

While the NMI indicated that there is a clearly identifiable LOHAS segment, researchers from other countries have put an effort into investigating the existence of LOHAS in their own cultures (e.g., Cowan and Kinley 2014). Considerable writing and research have attempted to make LOHAS a brand (e.g., Wan and Toppinen 2016). However, research findings are quite mixed. Studies originating from the NMI have found behavioral patterns of the general US consumers to differ from those they identify as LOHASian. The LOHAS individuals, compared to the general population, were more inclined to purchase eco-friendly products such as organic foods or beverage. They tended to consider environmental impact of consumer products before they purchased (Natural Marketing Institute 2008). It is important to point out that this is NMI published and sponsored not independent and peer reviewed as we come to expect social research to be. They do not empirically validate the tool used to measure LOHAS. In addition, the NMI has a vested interest (profit) in becoming the place for LOHAS research. This may be why there are many independent studies that show that LOHAS differences do not reflect behavior differences.

Saying that LOHAS exists does not make it "real" or important to research areas or useful to marketing. Although appealing and instantly understandable when reading the background literature of the NMI, very little research has been completed on the nature and scope of this lifestyle. Indeed, it would be surprising if social researchers not connected to the NMI but interested in sustainability and environmental issues have ever heard of it. Indeed, a look at titles in the upcoming American Marketing Association annual conference (2017) does not have a single paper with LOHAS in the title although sustainability and environmental issues have a large research presence.

While not well known in the USA, LOHAS has attracted a loyal following in South Korea. To be specific, there is research defining consumption behaviors associated with or caused by the LOHAS lifestyle and the level of the LOHAS consciousness. Another stream of research is concerned with identifying and studying the antecedents which affect the LOHAS-oriented behaviors or the LOHAS consciousness level. And finally, there is a track of research that investigates the strategic application of LOHAS in various industries. Although these studies have triggered an interest in LOHAS, it has limitations in leading LOHAS theory advancement in the scholarly arena for several reasons. First, much of this literature is published in South Korean journals and is not attainable to US researchers to read and replicate. Second, most of the research papers stay at the level of investigating managerial implications for expanding the LOHAS market or assessing the LOHAS industry. In these studies, there is little consensus about the LOHAS measurement instrument. Each study used different scales, and no agreement or consensus exists with defining an operational definition of the LOHAS lifestyle.

Conceptualizing LOHAS

LOHAS is an acronym standing for Lifestyle of Health and Sustainability. LOHAS is based on the two words of health and sustainability. Although the health and sustainability are the core concepts of LOHAS, the lack of agreement on more operational definitions of the concept has led to a wide variety of conceptions and perceptions for each word making it impossible to compare research (Emerich 2000). But given a superficial understanding of the lifestyle, we can infer certain things that could ultimately be empirically tested.

LOHAS defines people who have interests in increasing personal and their family's health and well-being (Ray and Anderson 2000). They should purchase a wide variety of natural and organic products that enhance physical fitness and its ranges from the foods to the personal care products (e.g., Emerich 2000). When purchasing healthy foods, they should trust the information recommended by similarly LOHAS-centric friends or media (Yeh and Chen 2011). In terms of disease prevention and health management, LOHAS consumers would be attracted to integrative healthcare approaches that address human being as a balanced entity of mind, body, and spirit (Emerich 2011). They should be interested in seeking out information and services relevant to the alternative and complementary medical care such as Ayurveda and acupuncture (e.g., Kettemann and Marko 2012).

The second theme in the discussion of the LOHAS marketplace pertains to the personal development (Cohen 2010). Developing self in LOHAS refers to achieving full potential as human being. It is concerned with getting in touch with deeper self and cultivating one's spirituality, which is a recovery of one's true nature (e.g., Mróz and Sadowska 2015). LOHAS personal development is fulfilled by understanding one's connectedness with all life and matters and by healing the self in the process of recognizing the unity of mind, body, and spirit (Nieminen-Sundell 2011). The LOHAS-oriented person should be interested in spiritual products or practices such as meditation, yoga, Qigong, aromatherapy, or macrobiotics as a means of the self-fulfillment (French and Rogers 2010). It is estimated that the spiritual market forms an eighteen percentage of the LOHAS market in the USA (Westerlund and Rajala 2006).

The third theme in the literature includes the philosophical and psychological values inherent in the LOHAS population. The values embrace an optimistic future view, experience of new challenges, desire for peace, and relationship orientation (e.g., Mróz and Sadowska 2015). Yeh and Chen (2011) believe and found that the LOHAS-oriented people are open-minded and have a positive view on things and it enables them to deal with problems in a positive way. Liu and Wu (2014) also argued that they tend to be more optimistic for their future and are against pessimism and cynicism.

The fourth constant factor in the literature is ecological orientation. The environmental consciousness is the manifest characteristic that defines the LOHAS consumer. LOHAS orientation should lead to concern about the impact of a product on the environment. In other words, they should consider the manner in which a product is made, sold, consumed, and disposed and if the process is done without harm or depletion to the nature (Emerich 2000). According to Korhonen (2012), the LOHAS segment should value environmental value more than the functional, emotional, or instrumental value of product packaging such as recyclability and biodegradability. Zentner (2016) argued that the people committed to LOHAS tend to live vegetarians or vegans because the food products need less energy and less harmful by-products than meat or fish during the production process. LOHAS consumers should prefer the local food products or organic foods because of its mild farming techniques to the environment in addition to its health benefit (Essoussi and Zahaf 2008). Furthermore, Cowan and Kinley (2014) suggest that the LOHAS mindset would facilitate purchase intention toward the environmentally friendly apparel.

The LOHAS consumers should be less price sensitive and more value sensitive (French and Rogers 2006). In other words, they are not deterred by higher prices for sustainable products or services in order to translate their sustainable belief into their purchasing choices. Wan and Toppinen (2016) found a positive impact of LOHAS orientation on the willingness to pay a price premium for children's furniture made by sustainable materials. The NMI estimates that LOHAS consumers are willing to pay up to 20% more for the products made in a sustainable way (Natural Marketing Institute 2008).

Lastly, discussions of LOHAS market practices should include social responsibility issues (Bamossy and Englis 2010). A strong sense of social justice should impel the LOHAS consumer to choose products from the companies which share the values that they endorse (Mróz and Sadowska 2015). Corporate social responsibility (CSR) that pertains to human rights, workplace equality, and solicitude for the minority such as children and women should attract and influence the LOHAS consumer (Urh 2015). LOHAS-oriented people should be inclined to purchase Fair Trade certified products which indicate, for example, farmers receive higher than product price for their commodities, producers from economically less-developed countries can earn financial benefit, no child labor is exploited in the production process, or no discrimination was made in wages between men and women (e.g., Heim 2011). LOHAS consumers might also dominate boycotts of businesses which they perceive to be socially irresponsible (Natural Marketing Institute 2008). It is noteworthy that LOHASians should reward ethical actions (purchasing ethical products over alternatives) as well as punish unethical behaviors (refusing to choose unethical products), given that the two types of social actions are theoretically distinct concepts in that many who reject the products/services deemed to harm society do not necessarily choose ethical products over alternatives (Carrigan and Attalla 2001).

There are three issues that need reflection:

First, just what is the nature of the National Marketing Institute that first proclaimed that LOHAS exists and is measurable using "our" (our words) LOHAS scale. Although the NMI site is clearly correct in suggesting the importance of a generalized sustainable lifestyle, the NMI is clearly a business and has a vested interest in "selling" the LOHAS concept and anything they do that comes of it. NMI is a marketing company selling their proprietary services to commercial concerns...not that there is anything wrong with that to quote Seinfeld. Yet without peer-reviewed scholarly publications into the nature and scope and even the existence of LOHAS, we must hold off proclaiming that LOHAS is to be revered. This does not deny any usefulness of the LOHAS concert for marketing and commercialization or even for anecdotal and heuristic understanding of human social behavior. It just means that it does not completely smell right to us.

Second, how is the thing to be measured? Measurement is not a minor annoyance or inconvenience. Science progresses isomorphically with our ability to measure it (whatever it is). As empiricists, we have to measure it to see it; otherwise, we are just guessing. The LOHAS definition is clear, and we should be able to measure it and demonstrate reliability and validity...we should but haven't. A concept can only be a construct if it can be reliably and validly measured. It is this hurdle that researchers have appeared to skip or ignore. The NMI published a LOHAS measurement scale, but no research has been found that demonstrates that the scale is a reliable and valid measure of the construct. This is important for doing research in one culture but is particularly important for comparing research cross-culturally and internationally. One scale even if reliable and valid in one culture may not be reliable and valid in a second culture. For example, a reliable and valid scale in country a is done. The scale has been validated and shown acceptable reliability for country a. The findings are valid for country a (pending replication and extension). Now one takes the scale to country b and assumes it is reliable and valid for that country/culture. If the answer is yes, then comparison between the two countries is easy and meaningful. But what if the scale is not reliable or not valid in country b. Assuming that it is and doing the research leads to misleading findings...the findings in country b are completely wrong since the sale on which it is based is not reliable and valid for that country. So cross-cultural reliability and validation of measuring instruments are critical for LOHAS-type research. Although the LOHAS lifestyle defines a life in pursuit of balance between the individual, the environment, and the society, there is very little work empirically validating its measurement or its relationship to actual behaviors. We are currently doing research to develop a reliable and valid measurement tool for LOHAS. The development of a tool to measure differences in LOHAS allows then an examination of how differences in LOHAS cause and explain differences in behaviors. This really is the whole ballgame and must be done before findings can be used to direct marketing campaigns and gurus can proclaim this and that about how people are and are changing.

Third, until LOHAS research is done independent of the sponsorship and publishing arm of the NMI, it will always be suspect. This does not mean that the research from NMI is fabricated. We just do not know. This highlights the importance and usefulness of independent replication.

LOHAS: The Gospel or Science of Sustainability

The problem with LOHAS is that it has not attracted the type of scholarly attention needed to move LOHAS from a friendly concept to an important scientific scholarly issue.

In a recent national workshop, the National Academies of Science brought together the leading scientists in the field of sustainability. Four needs to fill the gap in the research of what they viewed as an important scientific and applied field were agreed upon (Brose et al. 2016. p 53):

1. Sustainability indicators
2. Models for supporting decisions
3. Frameworks for environmental decision-making
4. Efforts needed to address the range of needs and opportunities related to sustainability

Here is the important issue for the point of this chapter. In their discussions and review of the research in this area, not once was LOHAS mentioned. Not once was LOHAS cited. We expected that given the supposed importance of LOHAS, there had to be a scientist somewhere who talked about it as a scholarly principle.

In her recent book, *The Gospel of Sustainability*, Monica Emerich (2011) quotes from an article published (preface pg 1) in the LOHAS Journal, "Businesses the world over are leveraging LOHAS as a way to understand the consumption preferences of a growing number of people who care deeply about personal, community, and planetary health and well-being, and are willing to spend accordingly" (The LOHAS Journal is now extinct which should tell us something). The problem is that the hyperbole is outpacing any research support for these notions. Until research catches up, LOHAS is just a word not a scientific or marketing construct. Emerich introduces LOHAS as a movement, as a marketplace, as a new demographic, as social justice, and as a new business opportunity. Emerich also believes that LOHAS received its greatest "validation" (her words not ours when it was mentioned in a front page article in the New York Times (Cortese 2003)), not the social scientific validation that really is needed.

Emerich is not hiding her belief that LOHAS is more than an empirical construct worthy of study. The tone of her book is that we have gone way beyond having to do traditional social science and marketing research straight to LOHAS is a movement, almost a religion (our interpretation but given the title of her book *The Gospel of Sustainability: Media, Market, and LOHAS* a reasonable statement).

Emerich's book is a well-written document outlining what LOHAS is claimed to be (and its history). It is not an integrative and descriptive work on research and theory development. Let us just review her methods and data in this book (from the preface) – participant observation, work as a journalist, textual analysis (an analysis of "hundreds" of web sites, radio programs, promotional and marketing materials, etc. (no claim for any representativeness)), interviews (12 upper management from leading LOHAS media (most if not all extinct now), and personal experience. We would simply say that if we attempted to publish a scholarly paper based on this methodology, it would not get off the editors desk...rejected. We wouldn't even get it published or presented at the leading LOHAS Journal and conferences because they are both out of business.

There can be no real question that consumers are at least thinking about (if not acting on) the quest for health, wellness, and environmental sustainability. The issue with the LOHAS religion is that the fact that these consumers exist doesn't necessarily mean that there is a scale that measures a lifestyle that could be called LOHAS in a reliable manner and that differences in scaled scores on that scale reflect predictable differences in some behavior that is part of a defined LOHAS lifestyle (validity). The fact that researchers call something LOHAS and there is a commercial research organization (NMI) who say there is LOHAS and that there are other marketing organizations that subscribe to that organization's research (e.g., Nielsen) does not mean there is LOHAS.

The motivation and need to understand consumers' increasing motivation for health and sustainable lives are not the same thing as saying that LOHAS exists. There is a need for a general scale to measure a consumer's motivation for a health and sustainable lifestyle. It is needed for research purposes as well as marketing purposes. It is clearly needed if we want something called LOHAS to be an integral part of scholarly discussions.

Although the consumer interested in and motivated by and to buy health wellness and sustainability products and services is growing, until such point that *ALL* consumers have the same levels of motivation and interest means that consumer segmentation is crucial to understanding marketing strategies for health, wellness, and sustainability products and services. Effective and valid segmentation is not only important for marketing and selling goods and services in this area but for understanding how to educate consumers to promote health and well-being and environmental sustainability. This does not mean that there is LOHAS only that it is a consumer issue.

From the research perspective, there is absolutely no good way to evaluate the hundreds (probably thousands) of studies in health and wellness and environmental sustainability unless we have a common measuring instrument. The growth of science is really measured by the growth of the ability to measure what is being studied. LOHAS needs a simple and effective reliable and valid measurement tool.

The problem we are addressing is highlighted by a recent joint NMI/Nielsen research publication called "Health & Wellness in America August 2014" Neilson Perishables Group, Natural Marketing Institute (NMI) (2014) (available for download at http://tinyurl.com/y7cseb8f). The newer reports cost from 5000 to 10,000 dollars. For evaluative reasons, let's look at what they say in the report as an illustration of the issues we see in the current state of LOHAS research.

From page 21 of the report cited above:

> NMI Health and Wellness Segmentation: A segmentation model that divides the entire U.S. adult population into one of five mutually exclusive segments. Developed in 2001, the model derived through a combination of advanced statistics including exploratory and confirmatory factor analysis, convergent cluster analysis, and discriminant functions, among other techniques. It has been validated across multiple industries and global geographies. A tool (algorithm) is used to segment data sets, with an accuracy of 80+ percent.

Oh my...who would not be impressed with this methodology? Who would not think that these guys and gals got their act together? Who would not think that this has got to be the bible of segmentation and illuminating how consumers think, feel, and buy in this area of study and marketing? Yet there the outcome cannot be accepted until the inputs can be assessed. Was their measurement valid and reliable for the purposes of the study? If we are buying a report for $5000 and then making million dollar marketing decisions, we really want to know if the measuring tool works in the traditional psychometric sense. All good scales have that characteristic.

LOHAS Around the World

The research seems to indicate that the LOHAS phenomenon is more relevant in the USA than Germany or the US LOHAS consumers are more likely to be active in engaging in the pro-environmental behavior. However, this comparison may be invalid unless the researchers used an identical measuring tool for LOHAS in the two countries. Assuming the two different scales for each research, one presumable reason for an inconsistency in pro-environmental behavior might be attributed to the different psychometrics of the measurement scales (Auger and Devinney 2007). If each scale measured different aspects of LOHAS, a meaning of LOHAS in one country would be different from that of another country. Furthermore, if the US scale reflected behavior questions more as opposed to attitudes, beliefs, and intentions, predictability for the consumers' actual engagement in pro-environmental behavior would be greater than the German scale (Carrigan and Attalla 2001).

Despite recognition that a standard global scale that thoroughly measures LOHAS is necessary, efforts have been scarce to work on developing a valid and reliable scale. Although several measures are devised for the purpose of individual research across a broad range of science, including social, behavioral, and life science contexts, each of them is for one's domestic uses, and they are discipline specific (e.g., Wan et al. 2015). In addition, the scales in the research have not undergone rigorous development procedure nor extensive validation test because a scale development for a hypothetical construct was considered secondary to the substantive scientific issues. In result, the scales differ in terms of operationalization of LOHAS and methods of measurement across the studies. In sum, the currently available scales are insufficient to thoroughly measure and empirically estimate the LOHAS as a cultural phenomenon.

Absence of an accurate and valid measurement scale raises several issues. First, it depreciates the legitimacy of research as a scientific work. Although imperfect measure other than no measure might provide some pieces of information, they may be flawed or fragmented (DeVellis 2003). Indeed, problems with reliability and validity of the measures make it difficult to interpret the results and temper the conclusions accordingly (Schriesheim et al. 1993). In this sense, without a valid instrument, mass production of the substantive findings on LOHAS phenomenon may not be warranted.

Second, it hampers theoretical progress. In order for a theory to be verified and generalized, it is important for findings to be replicated by a number of research (Tsang and Kwan 1999). However, diverse incomplete scales make it difficult to have a strong convergent conclusion (Schmidt et al. 1985). Despite the LOHAS theory being introduced about a decade ago, it is still nascent in terms of a replication. Although some general laws are founded in the sustainable behavioral context such as LOHAS-minded people's likelihood to purchase environmentally friendly products, little is known about the general tendency in health behaviors. A sound scale that accurately explains LOHAS lifestyle will generate reliable findings, and it will increase replicability of the findings by the following research. In sum, it will further strengthen the advance of the LOHAS theory (Schriesheim et al. 1993). A sound and global LOHAS scale is imperative particularly to the consumer scientists in that it provides a new lens for understanding the behaviors of the rapidly growing consumer segment in a postmodern era.

The LOHAS Marketplace

The economic value of the market for goods and services for the lifestyle that revolves around a healthy and environmental sustainable lifestyle is estimated to be 300 billion + (Urh 2015). The fact that the personality used to describe this lifestyle is called LOHAS does not diminish or increase the value of the goods and services that define that lifestyle nor does it prove that the lifestyle exists. There is well-meaning and serious-sounding literature that goes into detail about the types of products and services in this category. This literature also talks passionately about these consumers and people. These literature talks of these things in terms that make even a skeptic believe that the evidence exists to "prove" that these consumers exist. The literature (e.g., Urh 2015) even claims that "23% of the population (about 50 million adults) in the United States, and 29% of the population in Japan (about 37 million)" (Urh 2015, p.172) are in this group. Urh continues, "The speed with which this group is growing is astounding...Countries around the world are showing interest – Japan, Taiwan, China, Australia, New Zealand, India, Germany, Holland, England, France, Canada, and more – all want to integrate LOHAS principles into their own culture" (Urh 2015).

The problem is that there is no clear segment we can call LOHAS (Urh 2015). Even though a proponent of the existence of LOHAS as an important consumer group, Urh then goes on to dismiss her own conclusion by saying, "although they are not a homogeneous group of consumers, LOHAS share some certain characteristics, for example, they mainly live in urban areas. They do not only think about their own benefits but also about the effects their lifestyles have on other people and the environment. Therefore, for example, LOHAS tend to buy organic products, consider ethical standards, fair trade and sustainability" (Heim 2011; Urh 2015). Simple research studies like this are not by themselves reliable and valid endorsements of the LOHAS paradigm. There is no clear evidence (scientifically validated) for the existence of a LOHAS marketplace or a LOHAS segment.

LOHAS as a Market Segment

This sharply etches the main issues of this chapter. LOHAS is a term that supposedly represents a reliable and valid segment of the consumer public. Nothing is wrong with the hypothesis. But is there evidence for this?

1. As an acronym to describe a set of products and services, we have no argument, and having acronyms is useful but in our opinion until now only as a heuristic.
2. However, the act of naming this group and defining it within the context of a worldwide tsunami of beliefs has created a set of disciples who believe that there is some truth to this lifestyle rather than something less. Is LOHAS a fact or a myth? That is our question. The LOHASIANS believe that this is a movement. The LOHASIANS have created a cult of LOHAS (*The Gospel of Sustainability* Emerich 2011). They seem to believe that LOHAS exists, can be measured, is predictive of behavior and beliefs, and maybe will become a political movement (so we will have Democrats, Republicans, and now LOHASians).
3. From a purely marketing perspective, there is nothing wrong with defining a consumer segment as LOHASIAN if...and herein lies the main argument of the chapter:
   (a) The segment can be clearly identified. It is homogeneous and unique from other groups.
   (b) The segment is measurable in reliable and valid ways.
   (c) The segment is accessible.
   (d) The segment size is worthwhile.

LOHAS can be clearly identified by common concerns and beliefs, but these concerns and beliefs are not unique to males or females, old or young, and rich or poor. The beliefs and values cut across demographic and socioeconomic groups. The naming of this potential segment has outpaced the research necessary to measure it in reliable and valid ways. Indeed, the speed with which the commercial entity (NMI) has pushed the existence of LOHAS has probably deterred more academic basic research into the existence of a true segment that they call LOHAS...such is the danger of a cult. The potential segment is accessible by specific media attracting people with these beliefs. The segment size has yet to be determined. There is certainly a LARGE market for goods and services that could interest the LOHASIAN, but that does not mean there is a specific segment of consumers who purchase sustainable and healthy products that makes them unique.

LOHAS as a Lifestyle Segment

LOHAS would be considered a psychographic lifestyle segment. This type of segment is defined by research on attitudes, beliefs, interests, and opinions that through sophisticated statistical treatment (factor analysis, discriminant analysis, multivariate analysis) of representative, populations show segments or groups of people sharing common lifestyle factors. Focus groups, depth interviews, ethnography, and other qualitative techniques cannot be reliable and validity used to define and prove the existence of lifestyles...although these are probably the most common techniques used. In addition, spewing 500 questions and getting an online group of consumers to respond doesn't count either since online surveys are notoriously unreliable and have little validity. The LOHAS lifestyle is built on a platform of totally unreliable data that lack validity. In order to keep everyone involved in getting Bill Clinton elected President in 1992 focused and on message, James Carville (lead strategist for the campaign) coined the phrase "The economy, stupid." The phrase and its variation "It's the economy, stupid" became the tag line for all that was important to communicate to potential votes in the election. For our purposes we would say It's the reliability, stupid, show us the reliability. And, it's the validity, stupid, show us the validity. Reliability and validity are what is missing in most LOHAS research and need to be clarified if we are to generalize LOHAS across applications and cultures.

This does not mean we believe that LOHAS as a lifestyle segment does not exist. It may. But at this point in 2017, the work that describes and supposedly "proves" its existence is weak at best and fraudulent at worst.

The method of data collection is critical. The questions asked are critical. The questionnaire length is critical. The spokesperson for the study is critical. Are the researchers' skills for some organization that has a vested interest in the existence of LOHAS or does the researcher have an independent record of research that demonstrates neutrality and methodological and statistical knowledge and sophistication. LOHAS is built on a platform of all the wrong sides of these questions.

There is tremendous upside to proper segmentation:

(a) Advertising messages can be more effective than chance.
(b) Guidelines for branding flow directly from segmentation.
(c) Position of brands and products are easily designed to differentiate them from competition.
(d) Segments can be blueprints for new products and services.

LOHAS lifestyle is nicely laid out in a paper by Barbara Urh (2015). This paper is unfortunately published in a low-level relatively unknown journal of almost no "citation index" (arguably a measure of research quality). There is no primary data collected. The definition is the definition according to Conscious Media in Broomfield (www.consciousmedia.com) who organizes an annual LOHAS forum event and publishes the LOHAS Journal (that does not exist anymore – postponed publication in 2008). As a humorous aside, if one goes to the Google and searches for LOHAS Journal, this is the article that comes up (six steps to design original bowling team short (May 2017)). It is difficult to take LOHAS as a theoretical and/or empirical paradigm in marketing seriously. This does not mean that it is impossible to do so...only that the influence of LOHAS has far exceeded its reliable and valid support.

So...there is no shortage of potential predictive validity measures for researchers studying LOHAS. However, the use of unreliable measurement scales...the use of poorly defined constructs, the use of anecdotal reports, the use of vague survey data, and the dependence on a company with a vested interest in promulgating the existence of LOHAS cause great concern to us. The fact that there are consumers who show interest in sustainable development by behaving in a certain way or purchasing specific products and services does not mean that there is a LOHAS lifestyle. Calling some assumed relationship LOHAS does not make it so.

Measuring LOHAS

The argument we are advancing is that the existence of a LOHAS lifestyle, the existence of LOHAS people/consumers, and related behaviors, attitudes, and choices do not rest with a simple declaration by a commercial marketing firm but with empirical research. The empirical research rests with a valid and reliable measurement tool. By analogy, if I look at you and say you look like you have temperature above normal and then do research, looking at symptoms and consequences cannot be the evidence of how body temperature affects behavior and health. The ability of a physician to use body temperature as a measurement of health rests with the ability to reliably measure such temperature and be able to have independent physicians measure it and get the same temperature temporally.

The NMI proclaimed that LOHAS was measured by 12 questions:

1. I protect the environment.
2. My purchase decisions are based on its effect on the world.
3. I choose environmentally friendly products.
4. I prefer products manufactured in sustainable ways.
5. When possible, I choose sustainable-source products.
6. I prefer products made of recycled materials.
7. I am socially conscious.
8. I teach the benefits of environmentally friendly products to family or friends.
9. I am willing to pay an additional 20% for sustainably manufactured products.
10. I choose renewable energy sources.
11. I prefer sustainable agriculture practices.
12. I buy products from companies with values like my own.

While these questions have face validity, the NMI or independent researchers have not asked two important questions. Do these 12 questions really measure what we could call LOHAS (reliability and validity)? Are these 12 questions all there is to really measure LOHAS (e.g., spirituality, health)? Do these 12 questions work in the USA, South Korea, or any other country where researchers are interested in a LOHAS lifestyle?

The purpose of this section is to review LOHAS scales in the previous literature in terms of the conceptual requirements and reliability/validity. Most of the LOHAS research thus far has been done for the market research using questionable scales and techniques. For example, Cowan and Kinley (2014) measured a level of LOHAS using general questions pertaining to the environmental consciousness, environmental knowledge, and pro-environmental attitude but how can these findings be compared to a study by Häyrinen et al. (2016) that used a different set of questions still calling it LOHAS. Furthermore, Fu et al. (2012) measured LOHAS from a different perspective, which regards the LOHAS as an optimal mental state achieved from the satisfaction in a wide range of life domains, but they did not include sustainability issues when conceptualizing the LOHAS. Koszewska (2011) limited the ethics of LOHAS to the purchase behavior which is a choice of products from the companies showing socially responsible practices, with little to measure other domains.

It seems that few previous research view LOHAS from the perspective described in the NMI view of LOHAS one which embraces the issues of health, environmental sustainability, and ethics. In other words, almost none has developed a research instrument that matches the broad perspective that reflects general descriptions of the LOHAS lifestyle.

Proposed Plans for Research

The previous sections outline the problems we faced as we tried to describe and integrate and guide research in LOHAS. Here is what we propose for research in this area.

First, develop the LOHAS scale that is reliable and valid for consumers.

Using the original 12 items used by the NMI add items suggested by the literature by never tested and then using well-developed and tested standards of psychometric scale development and testing determine the best reliable and valid set of questions to use in research pertaining to LOHAS.

Second, following scale development in the USA, look at cross-cultural reliability and validity using standard methods of language translation.

It may be hard for a researcher to use the current 42 Korean consumer-oriented questions as universal LOHAS scale items unless the researcher proves the equivalency of Korean and American consumers. The researcher will develop comprehensive LOHAS questions that capture features of both Korean and American consumers. LOHAS measuring items for American consumers will be developed, and it will be combined with the current 42 Korean items.

Third, extend and redefine LOHAS to include health dimensions.

Health and wellness are related but are not synonymous. Wellness is commonly conceptualized as a self-evaluation on whether the healthfulness is satisfied in multiple domains. The wellness is an abstract and a higher order concept that integrates several areas of health (Schuster et al. 2004). Therefore, it seems more appropriate to conceptualize health in LOHAS as health "behaviors" that form healthy lifestyles. Health behavior questions will be included by adopting the scales from previous research.

Fourth, determine whether the spiritual health is a separate dimension or if it is associated with the emotional or mental health.

The spirituality includes a belief in the existence of a higher power and interconnectedness among the spheres of life such as a self, the environment, as well as a sense of meaning and purpose of life (Adams et al. 2000). In some research, the spiritual wellness is regarded as an independent area that affects overall quality of life in harmony with other areas of wellness (Mackey 2000).

Fifth, determine if LOHAS is even a broader concept.

It seems more appropriate to define LOHAS more comprehensively to encompass healthy, pro-environmental, and ethical consumption. The social justice dimension will be incorporated in LOHAS, and a corresponding scale will be added to the current scale.

Sixth, conduct validity tests defined by Churchill (1979) *looking at the relationship between LOHAS level and predicted behaviors.*

It may need to see whether those who put high values on health and sustainability would carry these values through into relevant behaviors such as particular consumption choices.

LOHAS Measurement and Culture

Culture determines systematic differences in behavior (Tse et al. 1988). The values and emphases in the culture shape people's perceptions, dispositions, and behaviors (Markus and Kitayama 1991). Hofstede (1980) specifically argued that the culture is not only reflected in the persistent preference for affairs over others but also in the interpretations and responses to environmental cues. Culture can be operationalized and defined at various levels (Steenkamp 2001). For example, if culture is defined at a national level in cultural anthropology, it means unique practices, norms, and beliefs shared by an ethnic or national group (Edgar and Sedgwick 2005). In contrast, a micro- or subculture is defined by cultural homogeneity within the country. The subculture develops its own unique contents and patterns of propensity as well as containing the main patterns of national culture (Steenkamp 2001).

There are three reasons why LOHAS research needs to be defined within the context of national culture.

- First, LOHAS can be considered to be a type of subculture. It appears to be differentiated from other national subcultures in emphasizing health, well-being, and environmental sustainability. However, given that LOHAS forms within the national borders, LOHAS is expected to share some national cultural values.

Second, LOHAS is understood as a personal lifestyle at the individual level as an integrated system of attitudes, opinions, interests, and behaviors. Given that cultural factors affect the formation of personal value systems, it can be assumed that the LOHAS lifestyle is indirectly influenced by culture. In this respect, it may provide useful insights to include culture as an external variable in the nomological network of LOHASian's consumption behavior. Let's suppose that we are studying on the effect of personal value on the LOHAS consumer's purchase behavior across nations. If differences are found in the strength or the direction between the constructs in the hypothesized nomological network across countries, we may be able to explain the underlying systematic differences in LOHAS consumption behavior in terms of the moderate role of the national culture.

Third, the validity of the LOHAS as a theory and an empirical phenomenon has been developed in the USA but has not been examined in other cultural settings although LOHAS is defined as a global market segment. The generalizability of the LOHAS has not been studied. An important issue in extending the LOHAS theory to other countries is whether the instrument designed to measure LOHAS is invariant across different cultures or nations. If evidence supporting measurement invariance is found, it demonstrates that the LOHAS phenomenon has universal appeal. If the nature and scope of LOHAS differ between cultures and countries, the usefulness and scope of LOHAS must be questioned. The cross-cultural comparison approach to the LOHAS measurement instrument is crucial to validate the argument that LOHAS is a standardized global market segment. In addition, examining the similarities and differences in the LOHAS phenomenon across cultures could be a basis for developing an international marketing strategy for the global LOHAS market.

Cross-References

▶ From Environmental Awareness to Sustainable Practices
▶ Sustainable Living in the City
▶ To Eat or Not to Eat Meat
▶ Transformative Solutions for Sustainable Well-Being

References

Adams, W. M. (2006). The future of sustainability: Re-thinking environment and development in the twenty-first century. In *Report of the IUCN renowned thinkers meeting* (Vol. 29, p. 31).

Adams, T. B., Bezner, J. R., Drabbs, M. E., Zambarano, R. J., & Steinhardt, M. A. (2000). Conceptualization and measurement of the spiritual and psychological dimensions of wellness in a college population. *Journal of American College Health, 48*(4), 165–173.

Auger, P., & Devinney, T. M. (2007). Do what consumers say matter? The misalignment of preferences with unconstrained ethical intentions. *Journal of Business Ethics, 76*(4), 361–383.

Bamossy, G., & Englis, B. (2010). Talk the green talk, shop the green walk? In M. C. Campbell, J. Inman, & R. Pieters (Eds.), *NA – Advances in consumer research volume 37* (pp. 102–105). Duluth: Association for Consumer Research.

Brose, D., Romitti, Y., Anderson, R., & Macalady, A. (2016). Transitioning toward sustainability: Advancing the scientific foundation. In *Proceedings of the National Academies of Sciences* (pp. 1–71). https://doi.org/10.17226/23533.

Carrigan, M., & Attalla, A. (2001). The myth of the ethical consumer-do ethics matter in purchase behaviour? *Journal of Consumer Marketing, 18*(7), 560–578.

Churchill, G. A. (1979). A paradigm for developing better measures of marketing constructs. *Journal of Marketing Research, 16*(1), 64–73.

Cohen, M. (2010). Wellness and the thermodynamics of a Healthy lifestyle. *Asia-Pacific Journal of Health, Sport and Physical Education, 1*(2), 5–12.

Cortese, A. (2003). Business; they care about the world (and they shop, too). *The New York Times.* Retrieved from http://www.nytimes.com/2003/07/20/business/business-they-care-about-the-world-and-they-shop-too.html?pagewanted=all

Cowan, K., & Kinley, T. (2014). Green spirit: Consumer empathies for green apparel. *International Journal of Consumer Studies, 38*(5), 493–499.

DeVellis, R. F. (2003). *Scale development: Theory and applications.* Los Angeles: Sage.

Edgar, A., & Sedgwick, P. (Eds.). (2005). *Key concepts in cultural theory.* New York: Routledge.

Emerich, M. (2000). LOHAS means business. Resource document. https://tinyurl.com/ya4dmuxx Accessed 20 June 2016.

Emerich, M. M. (2011). *The gospel of sustainability: Media, market, and LOHAS.* Urbana: University of Illinois Press.

Essoussi, L., & Zahaf, M. (2008). Decision making process of community organic food consumers: an exploratory study. *Journal of Consumer Marketing, 25*(2), 95–104.

French, S., & Rogers, G. (2006). Understanding the LOHAS consumer: The rise of ethical consumerism. Resource document. http://www.lohas.com/Lohas-Consumer. Accessed 17 Mar 2015.

French, S., & Rogers, G. (2010). Understanding the LOHAS consumer: Lifestyles of Health and Sustainability (LOHAS). Resource document. http://www.lohas.com/Lohas-Consumer. Accessed 24 July 2016.

Fu, M. H., Lee, K. R., Pai, M. C., & Kuo, Y. H. (2012). Clinical measurement and verification of elderly LOHAS index in an elder suited TV-based home living space. *Journal of Ambient Intelligence and Humanized Computing, 3*(1), 73–81.

Häyrinen, L., Mattila, O., Berghäll, S., & Toppinen, A. (2016). Lifestyle of health and sustainability of forest owners as an indicator of multiple use of forests. Forest Policy and Economics, 67, 10–19.

Heim, J. (2011). LOHAS. or: The consumption of sustainability. Resource document. https://tinyurl.com/ydepb9ym

Hoffman, A. J., & Haigh, N. (2010). Positive deviance for a sustainable world: Linking sustainability and positive organizational scholarship (Working Paper No.1139). Resource document. http://deepblue.lib.umich.edu/bitstream/handle/2027.42/66462/1139_Hoffman.pdf?sequence=1&isAllowed=y

Hofstede, G. (1980). *Culture's consequences: International differences in work-related values.* Beverly Hills: Sage.

Kettemann, B., & Marko, G. (2012). The language of alternative lifestyles. A critical analysis of the discourses of Emos and LOHAS. *AAA-Arbeiten aus Anglistik und Amerikanistik, 37*(1), 69–93.

Korhonen, V. (2012). Package value for LOHAS consumers-results of a Finnish study. In *18th IAPRI World Packaging Conference* (p. 156). San Luis Obispo California.

Koszewska, M. (2011). The ecological and ethical consumption development prospects in Poland compared with the western European countries. *Comparative Economic Research, 14*(2), 101–123.

Liu, J., & Diamond, J. (2005). China's environment in a globalizing world. *Nature, 435*(7046), 1179–1186.

Liu, M., & Wu, Y. L. (2014). Agriculture reform in Taiwan from a leisure farm visitors' perspective. *Journal of Food, Agriculture and Environment, 12*(2), 423–426.

Mackey, S. (2000). Towards a definition of wellness. *The Australian Journal of Holistic Nursing, 7*(2), 34.

Markus, H. R., & Kitayama, S. (1991). Culture and the self: Implications for cognition, emotion, and motivation. *Psychological Review, 98*(2), 224.

Mostafa, M. M. (2013). Wealth, post-materialism and consumers' pro-environmental intentions: A multilevel analysis across 25 nations. *Sustainable Development, 21*(6), 385–399.

Mróz, B., & Sadowska, M. (2015). Global consumption trends and consumption of ecological food in Poland. *Konsumpcja i Rozwój, 1*(10), 17–32.

N.A. (2016). What is LOHAS?. Resource document. http://www.lohas.com.au/what-lohas. Accessed 23 June 2016.

Neilson Perishables Group, Natural Marketing Institute (NMI). (2014). Report of the Health & Wellness in America. Resource document, p. 8. http://tinyurl.com/y7cseb8f

Nieminen-Sundell, R. (2011). Beautiful scenery, but no jobs. Resource document. SITRA. https://tinyurl.com/y7twxtc2 Accessed 2 June 2017.

Natural Marketing Institute. (2008). Understanding the LOHAS market report. Resource document. Natural Marketing Institute. https://tinyurl.com/yb64g5od Accessed 2 June 2017.

Ray, P. H., & Anderson, S. R. (2000). *The cultural creatives: How 50 million people are changing the world*. New York: Harmony Books.

Schmidt, F. L., Hunter, J. E., Pearlman, K., & Hirsch, H. R. (1985). Forty questions about validity generalization and meta-analysis. *Personnel Psychology, 38*, 697–798.

Schriesheim, C. A., Powers, K. J., Scandura, T. A., Gardiner, C. C., & Lankau, M. J. (1993). Improving construct measurement in management research: Comments and a quantitative approach for assessing the theoretical content adequacy of paper-and-pencil survey-type instruments. *Journal of Management, 19*(2), 385–417.

Schuster, T. L., Dobson, M., Jauregui, M., & Blanks, R. H. (2004). Wellness lifestyles I: A theoretical framework linking wellness, health lifestyles, and complementary and alternative medicine. *The Journal of Alternative & Complementary Medicine, 10*(2), 349–356.

Senik, C. (2014). Wealth and happiness. *Oxford Review of Economic Policy, 30*(1), 92–108.

Steenkamp, J. B. E. M. (2001). The role of national culture in international marketing research. *International Marketing Review, 18*(1), 30–44.

Tsang, E. W., & Kwan, K. M. (1999). Replication and theory development in organizational science: A critical realist perspective. *Academy of Management Review, 24*(4), 759–780.

Tse, D. K., Lee, K. H., Vertinsky, I., & Wehrung, D. A. (1988). Does culture matter? A cross-cultural study of executives' choice, decisiveness, and risk adjustment in international marketing. *Journal of Marketing, 52*(4), 81–95.

Urh, B. (2015). Lifestyle of Health and sustainability -the importance of health consciousness impact on LOHAS market growth in ecotourism. *Quaestus, 6*(1), 167–177.

Veenhoven, R. (2010). Greater happiness for a greater number. *Journal of Happiness Studies, 11*(5), 605–629.

Wan, M., Chen, J., & Toppinen, A. (2015). Consumers' environmental perceptions of children's furniture in China. *Forest Products Journal, 65*(7), 395–405.

Wan, M., & Toppinen, A. (2016). Effects of perceived product quality and Lifestyles of Health and Sustainability (LOHAS) on consumer price preferences for children's furniture in China. *Journal of Forest Economics, 22*, 52–67.

Westerlund, M., & Rajala, R. (2006). Innovative business models and offerings based on inconclusive evidence. *Innovative Marketing, 2*(2), 8–19.

Yeh, N., & Chen, Y. (2011). On the everyday life information behavior of LOHAS consumers: A perspective of lifestyle. *Journal of Educational Media & Library Sciences, 48*(4), 489–510.

Zentner, M. (2016). Body, health and the universe–a polemic and critical review of youth health behaviour. In H. Williamson & A. Wulff (Eds.), *Perspectives on youth: Healthy Europe: Confidence and uncertainty for young people in contemporary Europe* (pp. 145–154). Jouve: Council of Europe.

To Be or Not to Be (Green)

Does Communication Experts' Environmental Sensitivity Affect their Marketing Communication Plans?

Ebru Belkıs Güzeloğlu and Elif Üstündağlı Erten

Contents

Introduction	1088
Human Focused Action on Environmental Worldview	1091
Green Strategy: Environmental Sustainability and Going Green	1093
High Score Group "77–83": Market Role as "the Innovators"	1103
Middle Score Group "71–76": Market Role as "the Investors"	1105
Lowest Score Group "65–70": Market Role as "the Propagator"	1106
Green Communication: Corporate Social Responsibility and Marketing Communication	1106
Innovators: Presentation of Green Identity and Ethics	1124
Investors: Presentation of Green Image	1125
Propagators: Presentation of Green Appearance	1127
Conclusion	1130
Cross-References	1132
References	1132

Abstract

The rise of environmental sensitivity stands out as one of the most thriving trends of the twenty-first century that has gradually been reflected in marketing communication strategies. In this research, we aim to find out to what extent and how communication experts reflect their environmental sensitivity to the marketing communication strategies they develop. By purposive sampling, we choose senior communication students as the unit of analysis since they represent the

---

E. B. Güzeloğlu (✉)
Faculty of Communication, Department of Public Relations and Publicity, Ege University, İzmir, Turkey
e-mail: ebru.guzeloglu@ege.edu.tr; guzelogluebru@gmail.com

E. Ü. Erten
Faculty of Economics and Administrative Sciences, Department of Business Administration, Ege University, İzmir, Turkey
e-mail: elif.ustundagli@ege.edu.tr; elif.ustundaglierten@gmail.com

© Springer International Publishing AG, part of Springer Nature 2018
S. Dhiman, J. Marques (eds.), *Handbook of Engaged Sustainability*,
https://doi.org/10.1007/978-3-319-71312-0_44

forthcoming communication experts. Using a mixed-method approach, we follow a three-stage data collection procedure: In order to evaluate environmental sensitivity at an individual level, we first carry on a survey based on the New Environmental Paradigm (NEP) scale. In the second stage, we ask participants to develop a green marketing communication plan for a brand that they choose. After analyzing the survey results statistically and the communication plans with content analysis, we move on to the last and the final stage in which we did in-depth interviews with representative participants from the first two stages in order to go in deeper in our discussion. We discuss our results on the basis of the similarities and differences between the strategies and the ideological preferences reflected in communication plans and interpret how ideology and practices rather converge or diverge from each other. Finally, it is found that there is significant convergence between experts' green ideologies and the communication strategies that they planned which reflects one of innovator, investor, and propagator roles and characteristics.

Keywords

Marketing communication · Green · Ideology · Communication experts · New Environmental Paradigm (NEP)

Introduction

The rise of environmental sensitivity has come to the forefront as one of the most important tendencies of the twenty-first century which has also been influential on marketing communication strategies. Development in communication technologies to retrieve information in the postmodern market has changed the agenda tendencies toward alternatives that consider the ecological environmental balance in lifestyles and consumption preferences such that the demand for this sensitivity turn into action causing businesses to review their strategies (Özkaya 2010: 251). To this effect, new social movements such as anticonsumerism, boycotts, and do-it-yourself activism have been on the rise since the 1970s. This sensitivity has become controversial and is now a foundation for criticism whereby the green transformation, led by businesses, is not trustworthy and/or is merely profit-oriented.

Comprehension and gradual acceptance of a green consumerism marketing approach, particularly in the 1980s, created two fundamental properties in the ascension of this concept through: an increase in green literature and academic research in green business development (Peattie and Crane 2005: 358). By the end of 1980s and at the beginning of the 1990s, the pace accelerated with a new movement especially in the Anglo-Saxon countries and in the rest of the Europe (Chamorro et al. 2009; Chang et al. 2011). Within this frame, the sustainability concept that enables us to discuss the postmodern market environment and its fundamental actors creates a functional purpose for businesses to embrace all these developments under a single roof as "transformations" in business. In the 1990s, discussion developed around the gap between interest shown in the green approach and that of a transformed buying behavior that pointed to the limited success of green products. Publishing on the topic yielded two different schools of thoughts. One argued that green business was a fad and temporary, while the other asserted that green business was a recent trend procuring the attention it deserved. This second school of thought indicated that green business was not only environment-friendly but also a "good business" practice that would provide long-term prosperity (Gilbert 2007). The most influential consequence of these discussions, both in green literature and green consumerism, was when green ideology and the green transformation were included in academic studies, particularly in management, marketing, and communication. Today, universities orient curricula toward the green transformation by including courses, projects, and discussion platforms. Change in demand, led by green consumerism, has influenced universities to include educational programs that reflect contemporary discussions and developments. Of course, the effect this education has on forming a demanding consumer mass should not be ignored as it points to a reciprocal loop whereby each transforms the other. On a microscale, the green revolution is made possible through such training programs that equip managers, workers, decision makers, and planners with the skills to provide, encourage, and implement the green transformation in businesses.

The strategic dimension in environmental issues (Menon et al. 1999) has gained importance not only in businesses but also in macro fields such as state policy (Aidt 1998; Button and Pearce 1989) and sustainability (Tuxworth 1996), including the foundation for micro practices such as organic nutrition (Lockie et al. 2002). Dunlap underscores (1998) discussion has reached new levels through focusing on how environmental problems are perceived. With this comes the recognition that consumers and businesses may foster perceptual differences towards environmental issues and businesses. Especially marketing communication strategies and consumer decisions and actions have gained importance as they may vary due to expectations and observed values. All are affected by these transformations in the market environment. Inevitably new approaches will be adopted where businesses persuade consumers they should support environmental sensitivity or by making the marketing mix sensitive to the environment to convince consumers businesses are, in fact, green.

The starting point for this study is our observation that some important differences exist between the work and project presentations of environmental subject projects of students enrolled in the marketing communications course in the Public Relations program (Faculty of Communication), who are known to be interested in environmental problems in their daily lives, and other students who are indifferent to and/or whose interest level to environmental problems are unknown. Although they received the same education, due to the development of different project content, the impact of their individual sensitivies have caused us to explore this difference further. We believe it is important to determine the direction of this discrepancy and present it in our analysis. Koçarslan (2015: 130) points out that green marketing, which differs from traditional marketing based on a profit-oriented mentality, also simultaneously adopts an approach of avoiding environment harm when consumer needs are satisfied. However, there is a distinction between an employer's adoption of the green transformation or green behavior while seemingly caring about the subject due to some necessities and opportunities. More importantly it is about how consumers perceive it. The role of communication planners, in terms of corporate and brand communication, emerges at this point. The difference observed among the students also pointed to this importance. Communication planners' certain communication tactics might mislead public convictions through making false promises believable or by defending the company against the crisis activists cause, perhaps for the single purpose of creating a reality that even they themselves do not believe in and thus represent examples that may be evaluated as the foundations for these expectations and realities. Examining how communication planners make a choice between reflecting upon a reality such as the carbon footprint of a business/brand over its profits by seemingly adopting environmental sustainability may expose key findings between expectations and realities that will allow us to strategically evaluate environmental sensitivity. With this aim, using data accumulated from communication faculty students with previous experience designing a green communication project, answers to the above question will be sought throughout this research.

Every practitioner is a consumer at the same time. In a market environment, competition is intense, especially as a practitioner. They must weigh over issues of environmental sensitivity in consumption preferences while under pressure to ensure business profitability and how such success may influence their expected income as practitioners. During their education, Public Relations students learn strategies that will make businesses profitable and enable them to sustain their assets in the market environment, and are trained as decision makers of the future. However, considering that the decision process is not independent of their "green" identity, it is essential by means of sustainability, to consider what kind of dynamics are involved in their decision processes when it comes to a green communication plan.

Projects prepared by student groups are the starting point of our research. Within this scope, 17 projects in groups of between one to five students were subjected to content analysis according to planning decisions such as the purpose of promotion, message of interest, target audience, partners, and media tools. Within the cosmetics, food, tobacco, apparel, and furniture sectors, projects are diversified in product groups such as perfumes, oils, cigarettes, apparel, coffee, hair dyes, furniture, bicycles, creams, razors, tooth brushes, shoes, baby diapers. The students determined the sector, the product group, and the brand to work with and were only informed about topics such as purpose and target group. In the second phase of the study, to determine environmental sensitivity and attitudes of the participants, questionnaires using a purposive sampling method on a voluntary basis were collected from 93 students who took the course in marketing communications and who had previously completed a green marketing communication project. To determine the environmental sensitivity of the participants, the New Environmental Paradigm (NEP) (Dunlap 2008) scale was used (Alnıaçık 2009 for scale adaptation in Turkish): I definitely agree – (5 points); I strongly disagree – (1 point) on the Likert scale. In addition to this scale, expressions of environmental attitude were included by the researchers in corollary with the purpose of the research. This scale was used to determine the individual environmental sensitivities of the participants.

However, due to the ambiguity of agreeing or disagreeing, some statements failed as an indicator of environmental sensitivity and instead they are oriented to measure attitudes. The environmental sensitivity score was determined through finding the total score on the scale by individuals indicating an environmental tendency by choosing the statement: "I definitely agree." Accordingly, the minimum score is 18 and the maximum score is 90. Students scored lowest with 54 points, while the highest scored 89. Other demographic characteristics are as follows:

- 59.1% (55) of the participants were females; 40.9% (38) were males.
- Participants were aged 19–29. The largest groups are as follows: 18.3% (17 participants aged 20); second group 38.7% (36 participants aged 21) and 25.8% (24 participants aged 22).
- 90.3% (84) of the participants plan to work in the field of communication after they graduate.
- 90.3% (84) of the participants identify themselves as "environmentally sensitive."
- Only 29% (27) of the participants voluntarily worked on environmental projects.

Project groups were analyzed in accordance with individual environmental sensitivity scores, explained in the second phase of the study, and individual environmental sensitivity average scores of the participants were generated to form environmental sensitivity scores. It is noteworthy that individuals with high environmental sensitivity scores tend to group together, while those with low environmental sensitivity tend to group together. Projects with high environmental sensitivity scores fully reflected the green ideology, while projects with low scores had more sales and persuasion-focused content. In other words, participants reflected their individual environmental sensitivities regarding the content by grouping with individuals who converged with them.

In the third part of the study, out of the participants whose individual environmental sensitivity scores were determined, in-depth interviews were made on a voluntary basis with 4 low-score and 5 high-score participants, thus in-depth knowledge was accumulated from projects students completed; how they handled green transformation in their working life; and their decision-making processes and partner relations. In-depth interview data were coded and themed as descriptive analysis in corollary with the study parameters based on content analysis made on the projects. The aim here is to highlight the ways in which they address the green strategy and green communication.

Human Focused Action on Environmental Worldview

Participants take an environmentally sided stance at the point where they believe that nature is fragile and that nobody has the right to design nature according to their own needs, and that if this continues we will soon encounter a major ecological disaster. But somehow, if the human mind understands how to overcome problems and learn how nature functions, it equally understands that to control and develop it, a view that plenty of resources are already present in nature tends to prevail. Therefore, environmental problems are not linked to human activities but to people's inability to understand how to manage the situation. In short, the problem lies in management failure. Another point worth highlighting is that the participants are unaware of the fact that plants and animals also have the right to exist in the world and that the ecological crisis is overrated. The idea that people do not have the right to organize nature in accordance with their own needs or that they were not created to dominate the part of nature other than themselves is, in fact, only ideologically supported. Acceptance of human supremacy over nature is contradictory to the power and skill of people to dominate, which leads to an attitude that shifts between these two extremes. The environmentally friendly attitude that participants demonstrate in terms of "understanding" environmental problems is transformed into a human-focused comprehension of "action." This is particularly the case with business-based behaviors and opinions. Environmental sensitivity is considered more important than profitability, but it is stated that businesses are not conscious. The incommensurate approach of "seeming to be sensitive to the environment" raises an environmentally suspicious perception of the activities of businesses with a view that supports environmental sensitivity as a trend to be followed, bringing an environment-skewed doubt towards the activities of businesses. This approach is also supported by the view that businesses use these activities as marketing communication activities. This unfavorable attitude towards businesses is transforming into a committed responsibility for employees in decision-making and action-taking. According to these findings, which can also be interpreted as an impasse against negative views, participants say that they will not give up their environmental sensitivity activities for profitability and that environmental sensitivity is duly a responsibility of marketing communicators. However, some participants were undecided about whether environmentally sensitive individuals would reflect these attitudes in their marketing communication strategies and admit that they may have to "ignore" anti-environmental activities. The difference between ideology and action is also observed here. From business perspective, companies pretend to be green applying convincing green communication strategies, they are accepted as "vocal" while the opposite (nonconvincing) is accepted as "silent" indicated by Delmas ve Burbano (2011). At this point, it is possible to make the following interpretation: Individuals who are idealistic may not put these ideologies into action (Table 1). The aim of this research is to determine the extent to which these ideologies are reflected into action. Findings from the in-depth interviews will shed light on this discrepancy gap.

Differences among the groups were examined in terms of gender; the degree of communication in their discipline upon graduation; identifying themselves as "environmentally sensitive"; and the participation of volunteer work in their field. These characteristics were observed not to create dissensus among the participants but rather to reveal that they generally replied to the statements in similar ways. However, it is possible to say that nongender factors have a positive integrating effect on environmental sensitivity. Because the participant group is homogenous and assembles under a mutual education focus, there is evidence to suggest that education constitutes an important factor in environmental sensitivity and transformation of this sensitivity into action and creates a sphere that homogenizes the group and neutralizes other demographic elements.

Table 1 Green strategic match between consumer and business perspectives

Business aspect Consumer aspect	Green business	Pretend to be green business
Consumer believing/persuaded in green action	Win-win situation Leverage effect on marketing communication	Latent green washing – Strong marketing communication
Consumer not believing/persuaded in green action	Weak or false marketing communication	Apparent green washing – Poor marketing communication or strategic silence

The research of Bodur and Sarigöllü (2005) found consumers who have high environmental sensitivity are educated and more affluent. These participants also exhibit a tendency to take pleasure in meeting new people. The group disinterested in the environment largely tend to focus on making a living and meeting household expenses due to limited financial means. In fact, "being green" is not just the problem of liberal, educated, and wealthy people, it is the problem of all individuals. However, the idea that the upper class have the money, thus the power to control the world, causes the perception that they already control everything (Grant 2007: 19). Therefore, interest in environmental issues is influenced by demographic characteristics, particularly factors affecting lifestyles, education, and economic possibilities in corollary with their decisions and behaviors (also see chapter ▶ "The LOHAS Lifestyle and Marketplace Behavior"). In this study, due to participants possessing common educational qualities, other demographic factors lessen in importance thus supporting the authors' emphasis on education.

In the role of marketing communication actors, participants' educational development, strategies, and reactions demonstrate differentiation in their consumer identity which consists of a wide range of factors such as the level of acceptance of green ethics; ideals in business culture; management expectations; belief in the green activities of an enterprise; and reaction to the practices of internal and external actors. So, when there is a chance to build and manage a green strategy and all the elements involved in this management, what constitutes the planners' individual environmental sensitivity and environmental worldview?

Green Strategy: Environmental Sustainability and Going Green

In today's business world, factors such as competition, internal or external pressures, legislation, influence from voluntary organizations, and consumer demand are forcing businesses into a responsible transformation (Koçarslan 2015: 103; Cinioğlu et al. 2016: 199; Özdemir and Topsümer 2016: 235). While the sustainable transformation of companies in the market necessitates the reorganization of business activities with more ecological input-output processes, it also raises the question of whether supply and consumer demands should be met and harmonized. John Elkington states that it is possible to produce sustainable resolutions when the following three perspectives are taken into consideration: economic welfare conservation (profit); environmental sustainability (planet); and social justice (people) (cited from Stanković and Strižak 2015: 16). For this reason, the concept of sustainability has a broader meaning than "green." However, the thought of using resources for sustainability and the green transformation without seeking profit means the loss of profit for companies born out of the liberal system (Hussain 1999). From this point of view, it is possible to say that planners of the future must approach this process in accordance with their level of environmental sensitivity. Planners with low environmental sensitivity are more likely to differentiate green strategies from business activities or demonstrate a profit-oriented approach. Participants with high environmental sensitivity emphasize with this environmental stance and take the risk of foregoing profit for the assurance of green sensitivity.

> Now, there is a capitalist order. I mean, at least there is an economic system. Consumer society exists and one perspective of influencing this society is this order. And I think that as a company, your goal is never to be sensitive to the environment, but if you are sensitive to the environment, you do not start a company nor work in one (Male, score 65).

> Actually, I will contradict myself if I say this. If I am marketing and selling a product, it would be ridiculous for me to tell you not to consume or to consume less, but I would have said consume necessarily or consume correctly (Female, score 81).

Corporate ethics are shaped by the creation of a moral value system, which itself, applies to the company. Sustainability, framed by corporate ethics, makes it possible to relinquish pragmatist targets for general well-being, particularly natural and human resources, to ensure long-term planning and consistency in implementation. Some companies may choose to create value for their world, society, and ultimately their own existence by making a sustainable transformation with a holistic view. Especially, planners with high scores offer a perspective that consists of a corporate culture that creates and fosters these values: "The company already has ethical values, there is a basic vision and mission. There is corporate culture, and sensitivity to the environment which must also be harmonized and taken as a value. I think that is how it should be (Female, score 77)." However, the thought of the debt to be paid to society and a part of the profit to be spent for society, nature, and the world drives the corporations into a contradictory situation between taking an ethical stance and profitability in their management decisions (Steinmann 2008). For planners with a high score, not only output matters. The quality of the journey that leads them to outcomes is as important as the outcomes themselves. This view is mostly due to the current situation in the market environment. Capitalist market environments and current technologies do not meet an environmental conclusion as the high score planners desire. For this reason, (good) intention of the business comes to the forefront.

> Green production is a type of company or production that harms the environment at a minimum level from input to output until it reaches the consumer. Of course, you will fail at some point, because if we consider it, the cost is high. But here **the mindset is important for me**, the way you begin (Female, score 70).

Another condition that reinforces good intentions is that the brand name should not be involved in a negative event or being purchased by a company with anti-environmentalist behavior. This condition creates a "staying clean" perception (Female, score 80) and separates the brand from other brands, thus giving the above-mentioned "on the journey" insight. This also suggests that high score planners are aware of the rational attitude toward failure of the system. For participants with low scores, the limit is profitability. Particularly, at points where interests are at stake, environmental activities can be stopped. One participant expressed this situation as follows:

> I think if the costs are influencing the work itself, you can only think of the environment to a certain extent. Otherwise, what will be left [in terms of gain] for the owner of the business if this condition occurs (Male, score 62).

To rationalize this controversial situation within the behavior of businesses, it is of utmost importance that a corporate ethical framework be transformed into a win-win relationship between the company and society. The rise of social demands for green has also opened the door to profit-oriented investments, resembling that of ethical behavior, and green product transformation has also become a trend that creates new market opportunities. It is possible to see companies that do not want to miss these market opportunities with products they position as green or with a single product in many sectors. From this perspective, it is possible to cultivate an ethical stance, including sustainability, into a two-axis structure between a holistic transformation point and a pragmatist point, where profitability goals are met.

In terms of planners, this stance is not at the point of profitability but within the dimension of "making ends meet" and "interests." Thus, the macroperspective of businesses reflects the planner in terms of individual expectations and anxieties. Planners are responsible for the decisions they make or not/apply or not, and as units, they (agents) are the ones to implement these actions. However, when their sense of responsibility conflicts with individual susceptibility, on the one hand, planners with high environmental sensitivity scores tend to persuade the business to "turn green" against antienvironmentalist action or "quit their job" with no compromise, while on the other, planners with low scores admit they would do what is required within the framework of their "job description" to secure interests.

> I will talk about the ambitions of the company as I am an ambitious person myself. As you will eventually be married at the age of 40 or over, there is the issue of leaving your children a livable world. I believe I would contradict myself. Because, people want to get promoted. **And you must do what the system requires to get promoted** (Male, score 62).

> It is a highly risky situation in my view. I would resist, but if they insist, I would have nothing else to do, because everybody has their own job. And our job is public relations which is related to persuading or fooling others. Fooling others is an offensive discourse, let's say, it is about persuading them or forcing them to think about something. So, we should get used to these kinds of things (Female, score 61).

Is it the opinion of the person or do we need to look at the company's own corporate culture? I am in the corporation. I look at it as a public relations expert, it is not my opinion. I am of the mindset that when a person should be seen differently when he/she comes to work and when he/she finishes work. Life gives us direction (Male, score 60).

I would look at what is in it for me, what do I have to gain or lose? If the signature takes 15 from me and gives only 10 back, I would definitely not sign the document. Again, it comes to the case of interest. What is my benefit here? If it is something that I can risk, I can also bear the consequences (Male, score 60).

On one side of the scale, I would put my job and life, and on the other, the environment and nature. How long am I going to work? What will I do after retirement? I also have dreams. I am 21 but I want to be in nature after 45. **I would not have accepted it otherwise and have quit the job immediately. I am serious about that. If they wanted me to manipulate people against the environment, I would take the risk of quitting my job at that moment** (Female, score77).

I am not a person that would cling to that much money. I would quit and start over again. I would not give up, because why would I support something that does not suit me? (Female, score 80).

Maybe if they do not support me [in green activities] there will be a chance to quit and leave the company. And of course, if I do support them, many more things can be done if I get support from the company (Female, score 81).

There is an issue of "surrender" in low-score planners that turns into an "inquiry" in high-score planners.

There is an internal customer concept. If you cannot make your employees happy, you cannot be happy. If management does not value my ideas, why do I work there if I do not see respect for my ethical values? I am environmentally friendly person. I have a sensitivity to nature and I want a project related to this. If you want to manipulate me on top of it, I'll give you a minute. But I should move on to a little convincing part, so I would have convinced you ... I will try everything... Try everything for everything. Certainly, I believe that I will convince my manager because we already have the power to turn around. We take it from nature and I absolutely believe it. There is a point that absolutely affects them (Female, score 77).

Recognition of their expertise as planners and the acceptance of their identities to the point of valuing their views is also important. In the decision process, the acceptance of individual sensitivities holds weight by transforming their field of expertise into practice. Life-care anxiety brings to the agenda a persuasion over executives but may also mean leaving the job when there is no possibility to persuade.

About "being green" we can generally define three or four categorical grades. Prakash (2002) identified three levels: value-addition processes, management systems, and/or product levels. Value-addition processes are aimed at reducing the environmental impact at all processes with new technologies or technology modifications. Management systems focus on the decision at all levels of the company to adopt green, while product level transformations center on making the generated outputs green with strategies such as reconditioning, remanufacturing, reusing, recycling, and reduction (p. 286). In terms of green strategy, the levels are defined by a similar perspective: strategic greening, quasi-strategic greening, and tactical greening. At the strategic greening level, a change in the philosophy of the enterprise is the aim for an integrated transformation, while at the quasi-strategic greening level, application-specific transformation occurs only in certain parts of the system. Tactical greening, on the other hand, refers to a transformation that is more on the level of promotional activities about environmental issues in relation to the product (Polonsky 2001; Polonsky and Rosenberger 2001; Polonsky 2005). In the "shades of green" approach, which focuses on a different segment that includes market conditions and stakeholders, green conversion is defined at four levels: light, market, stakeholder, and dark. The dark green level implies that sustainable green principles are accepted as business principles of the company; where planet reputation is preserved in the process of value creation; and at this level, investors, managers, employees, and customers are bound by and agree to these principles. The stakeholder green level defines a process focusing on the development of relationships, where the green transformation is adapted to all company processes and is shaped by the involvement of partners. In the market green level, market demands are better taken into consideration and the green performance of the products and services are emphasized. Light green levels indicate a transformation, where fulfilling legal obligations is the focus and where the goal is process-oriented through technological adjustment (Freeman et al. 1998: 345–347, 2008: 15).

Finally, as an expanded approach to marketing, Kotler et al. (2010: 153–159) distinguishes the role that companies can play in environmental sustainability into three categories: the innovator, the investor, and the propagator. The innovator is the company that develops products that are not environmentally harmful, as well as companies that make major investments in research activities for the development of these innovations. The investor companies do not see environmental sustainability as their main business, but drive green growth in corporate culture and continue their investments in order not to drift from the potential green trends in the future, and who continue these activities to relieve themselves from pressure or include additional income from a share of the green market. The role of the propagator involves contributing to green businesses by way of raising environmental consciousness and sensitivity, in addition to their job. This role has the mission of supporting green-sensitive groups and lobbyists by raising green sensitivity in the public. From this perspective, a company's integration into environmental sustainability, behavior, and communication is shaped in relation to the role and status of the company and its commitment to environmental sustainability. While the role of the innovator requires the highest risk, it also increases the level of participation in environmental problems that shapes the way these issues are handled in strategic management. Each adopted role brings different levels of acceptance and approaches to the agenda. The higher the level of integration in question is, the more this transformation penetrates the company's culture.

In conjunction with sustainability and external pressures, companies are making as much commitment as they can to meet green transformation obligations. According to, one dimension of this transformation is the degree to which the ecological ideals of the company culture are accepted (environmental orientation), while the other dimension is the degree to which environmental issues are reflected in the strategic planning process (environmental strategy) (cited in Neagu 2011: 211). Suggest an internal and external structure as an integrated management system in terms of sustainable development and green marketing. In accordance with this, the organization culture is shaped by human resources, managers, and employees. Company culture that is shaped by the practices of business and the market includes internal factors that involve understanding and perception along with technical counseling from the sector, and a management system structured in a way that it involves external factors constituting different public participants such as customers and partners (cited in Stanković and Strižak 2015: 9–10). In accordance with this realization, companies shape a new stance in the market through involving internal and external environment factors, which then strategically reflect on marketing activities, most of the time.

When we talk about green transformation and the reshaping of a corporations' stance in the market, the reorganization of specific elements and actions may be necessary. Peattie and Crane (2005: 366–367) mention some elements as recognized indicators:

- Redefinition of the "product"
- Desire to change markets
- Focus on forthcoming benefits of product usage over pleasure of product possession
- Marketing communication that aims to inform rather than impress
- Focus beyond present consumer needs
- Willingness to manage demand and expectation
- Focus on cost rather than the price
- Taking more responsibility

When a company's market stance is in question, it is vitally important that they be appreciated by the consumer through this transformation effort in bearing the costs of a green transformation. The messages to the stakeholders, and especially to the consumers, give an idea of the level at which the transformation takes place. Neagu (2011) underlines that how the message is declared in terms of the strategy will create a change in consumer attitudes. According to this, three types of effects are mentioned, namely, agreement, identification, and internalization (p. 211). Agreement is the result of relationships between parties with authority involved in the communication process. The attractiveness of the communication source depends on the way the recipient perceives the characteristics. If the buyer perceives the sources (physical, lifestyle) as familiar and similar, or enjoys some features, communication becomes interesting and an attitude change is achieved through self-identification. While identification includes messages oriented at defining and introducing the product or the process to the consumer, how the consumer should feel because of the buy supports him/her through the message. Internalization harmonizes with the consumer, including the message that aids in the development of the attitude. The consumer is an active part of the message and every new reception brings with it new meaning. The consumer's perception of the message content is internalized, and a link is established between the source and the receiver to play an important role in the cultivation of relations.

The degree of interest and attention to social and environmental problems whether they relate to the sector(s) in which the company operates can have a dramatic effect on the company's relations and transformation in the public perception. The basic issue here is, beyond the inner and outer periphery of the company's own existence, how sensitive it is to the environmental agenda. In relation to the green transformation strategy, the level of "green responsibility" a company fosters can differ. The reason for any companies' existence is its profitability. Economic profitability is stated as one of the fundamental dimensions of conventional corporate social responsibility (CSR) theories and related approaches (Carroll 1991). Over time, however, it can be said that CSR efforts have broadened to include a range of policy goals, stakeholder expectations, and approaches and theories that add to ethical concerns and values. Institutional social responsibility theories involve four main perspectives: economic, political, integrative, and ethical (Garriga and Mele 2004). Van de Ven (2008: 354) mentions the following three approaches: an ethical product differentiation approach; creation of a virtuous corporate brand; and a reputation management approach, of which all are shaped by CSR efforts. The reputation management strategy focuses on the basic requirements of running a responsible business to obtain permission and maintain activity from the community. This strategy is particularly well suited for companies that systematically address the needs of key stakeholders and respond to criticism from the wider external environment of the company. A venerable corporate branding strategy exceeds the reputation, protection, and remediation strategy, as it has a clear commitment to stakeholders and public that it is superior to CSR efforts. The product differentiation approach focuses on the ethical codes of the product and its production processes. Van de Ven underlines the need to base one or more of these strategies on a strategic argument that corresponds to the ideal identity of the company otherwise companies may face crisis.

Increasingly, segments of conscious consumers, spontaneous interest groups, or Nongovernmental Organizations (NGOs) unite around environmental issues, and despite they are noninstitutionalized initiatives, they wield a fair amount of pressure on shareholders of companies in such a way that they uproot economic balances and relations. These developments force companies into CSR efforts and to embrace ethical codes like sustainable transformation, human rights, and the common good. Carroll (1991: 42) mentions that charitability is the least important, whereas the most desired and rewarded is corporate social responsibility. This can be attributed to companies approaching CSR from a pragmatist and neoliberal perspective (McDonnell et al. 2015: 35). Pragmatist targets for corporate responsibility are even closer to the crisis that companies will create through forms of activism.

In terms of environmental sustainability, this can of course lead to wider CSR initiatives than what is volunteered or rather more than what is envisaged for enduring CSR expenditure. However, in the long run, it may be the case that the responsibility of public expectation combined with corporations' willingness to engage will also be shared with the public. Thus, the risk of reputational erosion that creates crises due to the actions of activist opposition might be reduced through relations that are constructed with trust, transparency, dialogue, and participation. On the other hand, it can also be said that forms of social movements enable a new approach for companies to discover. Today, a corporate activism approach has emerged which takes social problems out of the public domain and places them in the center of the business world, representing a public role in the business world that also leads to the advocacy of interest groups. Dodd's (2016) conception of the relationship between the business world and society is addressed with an approach that challenges the profit argument that Friedman put forth in the 1970s. This approach assumes business leadership will meet public expectations based on democratic rights rather than a choice of corporate responsibility. In this manner, corporate activism can be an important agent of change with an understanding that CSR activities are directed by consumer demand for transparency (Disparte and Gentry 2015).

As you see, there are many criteria in the literature that determine a company's green strategy and its stance in the business world. To analyze how a company pursues this strategic structuring in green transformation and how it is positions this strategy, the subject of green strategies can present road maps. In this respect, considering the distinctions and definitions in the literature, green strategy subjects are listed as the first set of analysis criterion and communication planner projects are examined accordingly. In our analysis, a total of 17 projects were found to diverge into three distinct clusters (Table 2). According to the cluster result:

- The first three projects with the highest scores (77–83 points) were developed to provide the product/brand with *Innovator* characteristics.
- Six projects with middle scores (71–76 points) were developed to give the product/brand *Investor* characteristics.
- The last eight projects with the lowest scores (65–70 points) were found to develop the product/brand with *Propagator* characteristics.

These three groups have been compared with the three roles (*Innovator, Investor,* or *Propagator*) (Kotler et al., 2010: 153–159) that companies can utilize for environmental sustainability, and in terms of other criteria, the distinctions become clearer in accordance with the literature. The sequencing shows that the projects developed are designed according to the level of individual environmental sensitivity when planners have the chance to design and manage a green strategy. This implies that communication planners with a high level of individual environmental sensitivity are more likely to position the brand and/or organization as an *Innovator* and have the tendency to attribute the role of *Investor* and *Propagator* to the emergence of low levels of individual environmental sensitivity. It can be argued that young communicators' individual environmental sensitivities have a significant influence in their decision-making on green strategy communication plans, as they are seen to have designed the most successful green communication plan for their chosen brand. Particular emphasis

Table 2 Strategic issues for green conversion by projects of three groups

Referring strategic role[a]	Product	Ethical perspective[b]	Adoption of ideals and strategies reflections degree[c]	Structuring[d]	Going green levels[e]	Shades of green[f]	Green marketing activities[g]	Reshaping indicators[h]	Expected message effect on consumer[i]	Business activism tendency[j]	CSR approach[k]
Innovators	Coffee 1 (83)	Holistic	Strategy and orientation	Both	1	Darker	1	1,2,3,4,5,6,7,8	3	Activism	2,3
	SportShoes (81)	Holistic	Strategy and orientation	Both	1	Dark	1		3	Leadership	3
	Furniture (77)	Holistic	Strategy and orientation	Both	1	Dark-stakeholder	1		2,3	Leadership	2,3
Investors	Apparel 1 (75.5)	Holistic and Pragmatist	Strategy and orientation	External	2	Stakeholder-market	2,3	1,2,3,4,6,8	1,2	–	1,2
	Apparel 2 (75)	Holistic and Pragmatist	Strategy and orientation	Both	2	Stakeholder-market	2	1,2,3,4,7,8	2	–	2
	Perfume (75)	Holistic and Pragmatist	Orientation	Both	2	Stakeholder-market	2	1,2,3,4,8	2	–	2
	Apparel 3 (74,25)	Holistic and Pragmatist	Orientation	Both	2	Stakeholder	2	1,3,5,6,7,8	1,2	–	1,2
	Bicycle (72.2)	Holistic and Pragmatist	Orientation	External	3	Market	2	1,2,3,4,6	1,2	–	1
	Vegetable oil (71)	Holistic and Pragmatist	Orientation	External	2	Stakeholder	2	1,2,3,4,7	1,2	–	1

(*continued*)

Table 2 (continued)

Referring strategic role[a]	Product	Ethical perspective[b]	Adoption of ideals and strategies reflections degree[c]	Structuring[d]	Going green levels[e]	Shades of green[f]	Green marketing activities[g]	Reshaping indicators[h]	Expected message effect on consumer[i]	Business activism tendency[j]	CSR approach[k]
Propagators	Apparel 4 (71)	Pragmatist	Orientation	External	3	Market	3	1,2,3,5,6	2	–	1
	Cream (70,2)	Pragmatist	Orientation	External	3	Market	3	1,3,5,7	1,2	–	1
	Hair dye (70)	Pragmatist	Orientation	External	3	Market	3	1,3,5,8	1,2	–	1
	Dental brush (69,25)	Pragmatist	Orientation	External	3	Market	2	1,3,4,5,7	2	–	1
	Razor blade (68,6)	Pragmatist	Orientation	External	3	Market	3	1,3,5,6	2	–	1
	Cigarette (68)	Pragmatist	Orientation	External	3	Light	3	1,2,3	2	–	–
	Coffee 2 (67)	Pragmatist	Orientation	External	3	Light	3	1,2	1	–	–
	Baby diaper (65)	Pragmatist	Orientation	External	3	Light	3	1,2	1	–	–

[a]Strategic position or role for green: The Innovator, the Investor, the Propagator (Kotler et al. 2010: 153–159).
[b]Ethical Perspective: (1) Holistic conversion shaped by ethical framework. (2) Pragmatist conversion rationalized by win-win principle.
[c]Adoption of Ideals and Strategies Reflections Degree: Environmental orientation and Environmental strategy (Banerjee 2002; ve Banerjee vd, 2003 cited in Neagu 2011: 211).
[d]Internal Structuring and External Structuring (cited in Stanković and Stržak 2015: 9–10).
[e]Going green levels: (1) value-addition processes (firm level), (2) Management systems (firm level), and/or (3) Products (product level) (Prakash 2002).
[f]Shades of green: Light, Market, Stakeholder, Dark (Freeman et al. 2008).
[g]Green marketing activities in three level: (1)Strategic greening, (2) Quasi-strategic greening, (3) Tactical greening (Polonsky and Rosenberger 2001; Polonsky 2005: 124).
[h]Thought and Practice Reshaping Indicators: (1) a redefinition of the "product," (2) a willingness to change markets ("alternative" forms of production and consumption), (3) an emphasis on benefits from product use, (4) aims to inform rather than just impress, (5) a focus beyond current consumer needs, (6) a willingness to manage demand and expectation, (7) an emphasis on cost instead of price, (8) taking more responsibility (Peattie and Crane 2005: 366–367).
[i]Expected Message Effect on Consumer: (1)Agreement, (2) Identification and (3) Internalization (Neagu 2011: 211).
[j]Business Activism and Business Leadership (Disparte and Gentry 2015; Dodd 2016).
[k]Corporate Social Responsibility Approach (1) Product differentiation approach, (2) Establish a virtuous corporate brand, (3) Reputation management approach (Van de Ven 2008).

should be given to the fact that participants are not informed about the role they play regarding the literature from Kotler et al. (2010) and that these groupings represent the clustering of strategies created by the participants: ***Taking into consideration how well planners think they create the green transformation strategy, the visibility of the strategies they choose are within the context of the roles*** Kotler et al. (2010) ***emphasize as a reflection of their individual environmental sensitivities.*** In other words, green strategies created by planners with low environmental sensitivity scores will not cause stakeholders in the market to perceive them as "investors" or "propagators." This creates a situation in which businesses converge "to look like green" (Kardeş 2011) and are sincerely questioned by the stakeholders. Moreover, low-scoring planners have already stated that this should be the case:

> The company will never be sincere, because it is a company. But the fact that the company is not sincere does not mean that you never do something like that because the company is not sincere but it seems to be profitable for some people who seem to be intimidated by such things(Male, 65).

> [...]**Because it comes to the point 'we're the only ones not doing it.'** If everybody is green and if we don't do so too, you feel alone but when you look at a company that goes [brand name] searching for oil in the poles and they know they're melting, no one says anything. Or on purpose they know this but get in on another activity there. So, humans enter the environment, even knowing the environment they get used to consuming it...they think the environment is designed for them and don't need to worry that it may finish. Probably, I once felt comfortable thinking this way too(Male, score 62).

The activities of well-known brands become a legitimizing tool for low-score planners. Rather than thinking about transforming the market environment, it is more likely they will become complicit in the partnership by cooperating with companies operating in this direction.

Planners with high environmental sensitivity scores seem to be more successful having drawn from their own consumer insight. For this reason, these activities can be described as "deceiving people." An important factor is that they cannot distinguish their green identities from the decision-making process, and therefore the activities they are doing – wrong in themselves – are turned into identity conflicts. For this reason, internally, this process is not as effective as low-score planners. This leads to the development of attitudes about businesses and their activities.

> It's nothing more than a swindle. Because you haven't thought about it you can't say anything to anyone else. That's when I feel bad [...] It's only like you're saying, 'look we're green too' so it's all smoke and mirrors to me(Female, score 70).

High Score Group "77–83": Market Role as "the Innovators"

In the green strategy analysis, it can be said that projects in this group aim at a complete transformation of the products and processes for a selected institution and/

or brand. Planners forecast the conversion efforts they initiate for any product group to be transformed along with other product groups over a short- and mid-term. Long-term targets are structured to ensure a complete "green transformation" by way of raw materials, technology, processes, and product outputs. Moreover, if deemed necessary for a robust green transformation, as exemplified in a project like furniture, the aim is to reduce production volumes and diversity to invest more in technological research and development activities thus providing simplified life habits by fulfilling multiple needs with additional functions added to products. For a coffee project (Coffee 1), the aim is to utilize the distribution network to be "cleaned" through advanced technology to serve other social services.

Another common feature of projects in the innovator group is that the green strategy is structured by both internal and external management system factors as Stanković and Strižak (2015: 9–10) point out. What is envisaged is that the internal axis management system will deeply penetrate green values and ethics into the company culture, especially the employees and management. However, governance refers to the role of protecting company values expressed in ethical codes, such as fairness, honesty, transparency, and being clean, at any cost, even before economic benefit. From the point of view of an externally oriented management system, it is evident that all the projects in this group include consumer motives, activists, initiatives, public institutions, independent research companies, and universities who operate under the motto of collaborative co-production. In the project with the highest score (Coffee 1), an activist role was embodied to transform a company, while in the other two projects environmentalists could reorganize other organizations and institutions like NGOs and suppliers to persuade nonclean companies to transform, even become leaders themselves to gain a competitive edge, to promote solidarity through creating and sharing new technological information. Business leadership approaches used in these project groups lead companies to "dark green" as Freeman et al. (2008) predicted. However, in the highest-scoring project (Coffee 1), stakeholders who do not engage in green transformation efforts were studied and policy decisions were taken to create a mechanism to pressure such pseudo-actors through exclusion from the chain of suppliers. This project was described as "darker" (see Table 2 and also chapter ▶ "Ecopreneurship for Sustainable Development") because Freeman et al. (2008: 14) included a step closer to the views of "dark green logic is not anti-business, though many people will believe that it is." Another noteworthy feature is that of the consumer desire to transform their lifestyle and shopping habits. Especially in one project (Furniture) need and consumption are redefined by equipping and streamlining a product with as many functions as possible. In all three projects, it appears that political actors and legislative mechanisms have been included in the process of transforming their sectors and other relevant sectors, with emphasis on ensuring that ethical codes are guaranteed by legal regulations disseminated across all sectors.

Another commonality among project groups is the premise that rather than businesses using corporate social responsibility for competitive advantage, it is pertinent they firmly establish a transformative stance. In the first two projects, this principle is emphasized. For this group, the stance about being green "is taken as necessary, it can't be any other way" and use this principle to turn companies into green entities. In these types of businesses, close relations are created with stakeholders who espouse the motto "collaborative production" to promote transparent governance as sharing partnerships underscored in their annual reports (Coffee 1).

Middle Score Group "71–76": Market Role as "the Investors"

In the projects of the second group, planners were positioned as companies investing to become green, attributing investor characteristics. In these companies, it is evident that the green transformation began with special series and products. However, it seems that only a product or a special series of the produced brand was emphasized. Production of existing products can continue when new products or series are brought to the fore. As for future planning, it was determined that short- and midterm targets were established for green transformations in these projects. However, for long-term goals and/or targets, no cleanup, improvement, or green transformation was mentioned in all the processes.

In terms of structuring these projects, it has been determined that internal and external as well as external management mechanisms are included in the process. In the internal structure, more employees are mentioned and some projects emphasize the sensitivity of employees regarding green transformation to encourage a cleaner life in terms of consumer lifestyles (Apparel 3). While the projects underscore visible improvements in some processes, in product production in general, more than one sustainable benefit is highlighted. In two projects, it is mentioned that ethical elements such as animal rights, employee rights, and fair trade should be adopted as company values besides green conversion (Apparel 2 and Perfume). It is noteworthy that although projects can be said to be green in most respects through the affirmation of sustainability, stressing that projects should be "green" under company values, only pragmatist targets are explicitly or implicitly stated.

In this group, some strategic decisions become apparent such as the use of existing government incentives (Apparel 3); preference for existing organic producers; farmers and natural fragrance suppliers (Apparel 1, Perfume); utilization of ongoing green consumer trends (Bicycle); and joining new business sectors with recycling processes (Vegetable oil). Unlike the high-scoring group, the strategies in this group involve interrogation, such as which points can be earned from the green conversion they make and which trends will provide a faster return for the company. For example, in a project, the green cycle of the production process is predicted at certain points, targeting bio-diesel fuel production with the waste conversion system mentioned for the mid- or long term (Vegetable oil). In another project, it is stated improvements will be administered with the addition of cleaning technology in the environment by taking advantage of this trend in corollary with an increase in consumer cycling habits (Bicycle).

Stakeholders and the market environment are considered important determinants in the project strategy. Unlike the projects in the first group, it is important to note that companies form strategies based on green values and pressure groups, differentiated from an original value set definition for a green policy. Finally, in terms of the corporate social responsibility approach, it has been observed that the product differentiation approach is dominant in general even though there are strategic plan steps to be a virtuous institutional brand in some projects. Furthermore, we are aware that the principles and practices related to corporate social responsibility are mentioned in the products and processes although not a clear transformation is planned (Apparel 3, Vegetable oil, Bicycle). No business activism and leadership trends were observed in any of the projects in this group.

Lowest Score Group "65–70": Market Role as "the Propagator"

Projects in the last group were observed as the lowest level of green conversion reflected in the scores. Yet it is possible to clearly distinguish green strategies within their business plans. In terms of planning for the future, it is not possible to arrive at any conversion goal. Although there are messages about environmental sensitivity, it is believed that projects in this group are attributed to propagators because of limited green performance when compared to the other two groups. The projects focused on a product that is generally produced by the company brand. The green strategy is associated only with the product and often with a stage in the production of the product (frequently used raw material or added natural substances) to emphasize green conversion.

It was observed that the projects were entirely focused on external management mechanisms and consisted of suppliers, consumers, and some civilian stakeholders. Topics such as sales promotion, image, and profitability are clearly declared in their targets (Cream, Shaving Blade, Baby Diapers), and these targets seem to be the main reason for creating an environmental tendency. In the plans, pragmatist targets appear in the foreground and the fruits of investments are expected to be accrued in the short term. As a remarkable feature, in some projects, the function required from product expectations superseded environmentalist tendencies (Cream, Shaving Blade). Environmentalist features appear as only small, marginal additions.

In most of the projects, a special event in which the product is highlighted and operates through a sponsorship is coded as corporate social responsibility. In some projects, nongovernmental organizations are articulated in CSR studies, but NGOs unrelated to the green transformation have also been selected as stakeholders (Apparel 4, Hair Color, Shaving Blade). Finally, none of the projects in this group had any tendency towards business activism and leadership. One participant described this situation by saying, "Of course, some of the public will ask questions (Female, score 79)."

Green Communication: Corporate Social Responsibility and Marketing Communication

The distinction between "being less harmful to the environment and being environmentally friendly" to which Polonsky (1994) refers is the topic of green marketing; essentially a compromise in environmental communication that forms an environmental consensus that will create actions to "restore and rehabilitate the harmed environment." provides two definitions of environmental communication at the formal and informal level. Environmental communication is defined as informal communication that affects both our perception of the environment and each other, and our relationship with nature. Formally, it is defined around environmental issues as well as an understanding of our relationship to nature and the environment as a symbolic means of negotiating different responses in society to these problems (cited in Jurin et al. 2010: 14). The International Environmental Communication Association (IECA) considers environmental communication as a communicative device in the context of two social functions. Environmental communication is primarily used to inform, persuade, educate, and alert others or to organize, discuss, reconcile, policy defend, raise sensitivity, change behavior, influence public opinion, cooperate, legislate, or challenge hypotheses. Effective communication requires attention because it influences results (Meisner 2015).

According to Van de Ven (2008) and Benoît-Moreau, and Parguel (2011), the product differentiation approach involves the discernment of a product or service based on environmental or social elements. Therefore, CSR studies and communication are inevitably at the center of brand positioning. A vigorous corporate branding approach also has an open promise to the public and stakeholders, and open communication is necessary as proof. In the publication on ethical codes, corporate communication tools such as websites and marketing communication tools like advertising, sponsorship, public relations, and promotion are used. However, this communication is viewed as risky and has received less attention in recent times. One reason being there is a risk that responsibilities and consumer expectations may surpass CSR commitments. According to the Cone Communications Corporate Social Return Trend Tracker analysis (2012, 2015), 84% of American CSR companies with respected identities, whose aims are to continue progressive activities, remain committed in their mission to create and report effectively their progressive agenda. If CSR results are not reported, 40% of the participants said they would not buy a company's products or services.

On the other hand, the concept of "green" reputation management converges with CSR approaches with an ethical focus; the company's philosophy has penetrated and aggrandized the green ethic code to foster a greater comprehension of CSR efforts. Stanković and Strižak (2015: 16) define this as eco-centric CSR independent of the benefits of companies, with an aim to protect and improve the environment and its potential, and who act in a manner that recognizes and ultimately respects the environmental dependence of existing and potential resources. From this perspective, it can be said that by supporting eco-centric CSR with reputation-oriented communication efforts will provide great benefits for the company. However, as Benoît-Moreau and Parguel (2011) also pointed out, there is no open communication around the climate-focused CSR approach. In fact, according to some research, it is possible to create risks whereby some companies become clear targets for activism. This can be better understood by looking at two factors (reputational halo effect and reputational liability effect) investigated by King and McDonnell (2012: 6) regarding a reputable CSR approach. According to their research, having overall positive publicity and pro-social activities can be argued as a means or rather duty of a company to stave off future targeting by activists. This situation is called a halo effect. An increase in the likelihood of activists to target companies with a positive reputation separates and protects them through pro-social actions described as a reputational liability effect. King and McDonnell (2012) found that CSR communication activities increased the perception of being targeted through public visibility in the media. In other words, from the moment activists choose a company to boycott, it is the responsibility of the halo effect to come to the fore whereby high-visibility, socially responsible companies will channel their efforts toward audiences with the highest expectations (p. 20).

Marketing communication activities include another type of strategic communication effort employed to emphasize "greenness." Unlike the reputation management approach, marketing communication elements and tools are often preferred options for virtuous corporate branding, or for companies that only choose product differentiation. Branding and product differentiation-focused positioning activities involve clear communication about the sharing of green promises with corporate environments through green marketing communications activities and the demonstration of necessary actions. Unlike profit-oriented institutions, nongovernmental organizations or organizations such as green initiatives can also develop marketing communication strategies. Talpos and Meltzer (2013: 38) address strategic decisions to minimize negative effects through such means as determining communication objectives, identifying stakeholders and target groups, message design, determining communication tools, and environments to stage an effective environmentally integrated marketing communication plan that makes a good impression on the public through its dissemination. However, if marketing communications efforts are being implemented by profit-oriented organizations, risks in consumer interaction may become more visible. For planners with high environmental sensitivity scores, it is possible to achieve both profitability and environmental objectives:

> I think that we can reach the targets of the institution by considering environmental sensitivity, I mean, both can be realized logically. The public relations specialist at that institution already takes on the whole task him/herself. Of course, one must work with the top management to create an image for the target group, but as I said, it must be sincere (Female, score 81).

The nature of marketing communication goals can be cognitive, informative, affective, and conative. At the cognitive level, organizations inform consumers so they can compare environmental properties indifferent products. The more information consumers have such that they adopt this approach, more "environmentally responsible" actions will be discovered. At the affective level, communication should trigger consumer emotions about environmental problems like destruction, pollution, not feeling, or taking responsibility for the future of the planet. Communication must be creative to illicit impress and persuade environmental sensitivity. Conversely, message receivers should be encouraged to act out their pro-environmental behavior. For all these levels to be effective, the source of communication needs to be strong, effective, and reliable (Palekhova et al. 2015). Especially for planners with high environmental sensitivity scores, this situation is conical, while the approach to conservation on an individual level is considered as embedded activities. Therefore, the green strategy for these planners is a feature that must already exist. When a participant specifically refers to the communication strategy, he expresses it as follows:

> **In the message part, yes, I am a green production, but I would have planned a message saying that this is not a privilege for me. So, everyone should be like this.** For me to do this for you shouldn't only be my privilege, I would have created a message that would make people think about changing their conscience, with that kind of a message I would reach my target audience (Female, score 80).

For low-scoring participants, promotion is important. While businesses that effectively use powerful green messages can help attract customers, the word of mouth effect is likely to provide a positive or negative sense of the business:

> There is a mass of the public that uses green, and when it appeals to the audience, when you introduce your green project more prevalently, when you say, 'we are a company sensitive to the green environment', the customer base will grow. For one thing, I think that the good is not bad. It sounds like all kinds of things. Even if you do not like the green project, you can listen to it in a way. It is a familiar thing that spreads from ear to the ear (Female, score 61).

Messages sent to consumers in communication plans involving the preservation of the environment positively shape the attitude of the consumer if the messages are convincing. Messages that are not perceived as "sincere" will arouse suspicion on behalf of the consumer and therefore will not have the expected positive impact on consumers' attitudes toward the brand (Kardeş 2011: 167). The level of environmental sensitivity affects the positive perception of environmental attitudes and environmentally friendly products (Yılmaz et al. 2009: 8). Alnıaçık (2009: 74–75) found that brands in some product categories had higher scores indicating positive attitudes over those in environmentally friendly communication studies. This positive difference in favor of studies involving environmental claims becomes even more evident in people who are sensitive to the environment.

> I think green consumption should be explained more accurately. I think it's inadequate. So, people must be conscious about green. It is necessary to show the harm people do to nature by taking nature into consideration. I think we can do this by showing the opposite or inverse dimension to people. I mean, we may create a message reflecting that people are not living in a green world, and how can they live in such a world, or if they have such a green world, they have to question how they can live in a non-green world to affect them more emotionally (Male, score 80).

In the target of every green communication plan, some themes arise in terms of environmentalist claims. Green message themes generally focus on paying attention to an area related to the environment (care), eliminating harmful contents (free), protecting an area related to environment (save), and/or focusing generally on the environment, nature or the ecosystem (friendly). These claims are often reinforced with a message of benefit that customizes the perception of the product or service in the minds of people.

In the literature, the value proposition presented with environmental claims is achieved to shape and strengthen the link established with the consumer through a creative strategy for a product or service that can be: (1) functional, (2) perceptual, or (3) symbolic (Erdem and Swait 1998: 136). Additional reinforcement includes eco-labels, environmental product declarations, cause marketing, and third-party supporters. While eco-labels are particularly effective in creating demand through marketing, environmental product declarations provide approved statements about the effects of products on a life cycle (Ottman and Mallen 2014).

Groups and stakeholders as the target of these claims comprise an important strategic aspect of the communication plan in terms of their qualifications. Yet perspectives and reactions to environmental problems are different (Jurin et al. 2010: 14). According to the International Environmental Communication Association (IECA), from the most passionate environmental advocates to the hardest opponents of ecological conservation, everyone who participates in these discussions is a part of environmental communication (Meisner 2015).

Heijungs et al. (1992: 19) refers to three target groups related to environmental efficiency in the product life cycle: consumers in need of information to make decisions regarding purchase and consumption; producers who need information for innovation; and the public sector that needs to be aware of new developments and regulations. Nowadays, consumers who have become stronger, thanks to the effect of social movements and NGOs' efforts, have also become the main actors in environmental issues. Interest in the consumer as a strategic target is an important leverage in explaining corporate social responsibility communications and practices (Benoit-Moreau and Parguel 2011).

In general, customers, suppliers, employees, competitors, consumer communities, institutional structures such as financial and commercial associations, international organizations, specialized NGOs, political groups, governmental agencies and local public administrations, and the national and local mass media are considered as key stakeholder groups (Polonsky 1995; Freeman et al. 2008; Ertuğrul 2008; Talpos and Meltzer 2013). In some green communication studies, the public are considered the broadest group (Talpos and Meltzer 2013). While the public are important stakeholders in terms of management strategies, they are in fact a vehicle of consumption and a common platform where all stakeholders come together in terms of action, behavior, and attitude. In other words, we can say that every stakeholder, whether institutionally or individually, comes together around a "consumer" identity, and consumers who are more related to the environment tend to engage with green products (Schuhwerk and Lefkoff-Hagius 1995).

Today, standards set by legal regulators, media publications, reference and pressure group demands, and sectoral recommendations and expectations of environmental groups (Stanković and Strižak 2015: 9–10) have become the driving force of green transformation.

They force green promises to become more visible and make green positioning an essential aspect of a brand. For this reason, marketing communication elements and decisions should be considered as decision areas that are shaped according to target groups and stakeholders. As mentioned in the literature, the dimensions of green marketing communication such as advertising, public relations, sponsorship, promotion, internet marketing, exhibition, lobbying, noncommercial activities, viral and guerrilla activities, as well as communication tools such as television, radio, outdoor and web profiles, accounts, social networks, and electronics are frequently utilized. The credibility of the media makes it an important communication platform as well as media preferences of target groups and the effectiveness of communication tools and formats. With the rise of computer and mobile communication tools, we can see more sustainable solutions and modern technologies such as electronic media instead of printed media in communication activities (Benoit-Moreau and Parguel 2011; Ottman and Mallen 2014; Palekhova et al. 2015).

As for the participants of this study, especially high-scorers, technological tools are viewed as an important factor in both reaching target groups and reducing the amount of environmental damage. Thus, activities are directed toward these platforms.

> [...] My study in public relations, my study in advertising, the size and dimension of deadly waste, if green consumption is hurting, my aim has always been to keep this damage to the minimum, the least amount, at least I tried alternate ways. For example, I remember we wrote a press release. If I should write a press release I would prefer to write it on the internet through alternative means (Female, score 80).

Both CSR-focused corporate communications and marketing communication plans are designed with phases such as communication objectives, message strategy, target audience and stakeholder selection, decisions regarding the mix of marketing communications, and choice of communication media tools. Based on these decisions, the expectation from the plans is to reflect a green-position and a green vision for the market. From this point of view, analysis of green communication plans will enable us to discover the green-positions that planners find appropriate for their brands. Thus far, green strategy projects have been separated into three groups whereby in the first part of the analysis we analyzed them according to the second criterion based on the decisions taken regarding each project, the communication strategies, and their applications (Table 3).

Planners share a common view that green products are expensive for their target market. In the role of consumers themselves, planners jointly recognize the obstacle for buying green products is because of current limited purchasing power. However, the point at which their common view diverges is that participants with low environmental sensitivity scores assert green products will not be purchased, while high-score participants believe there is an inclination that they will be. This situation effects how target markets are chosen. Planners with low-scores choose high-income groups as their target while high-score participants do not distinguish, seeing all with green tendencies as their potential target market, believing they can render green products accessible and consumable by all.

> The topic of green products is currently discussed on the internet by people who take a lead role because they read, foresee, research and it is through these people that I can reach my target market. In my other work, for example my public relations work, I can direct my efforts to those who are not yet green consumers (Female, score 80).

Table 3 Green communication issues by projects of three groups

Referring strategic role[a]	Product	Communication aims	The nature of marketing comm. goals[b]	Main message	Eco-message theme[c]	Value proposition[d]
Innovators	Coffee 1 (83)	Achieve measurable goals in all processes in green strategy Make a call co-production Emphasize that the company is ready to serve the world with its capabilities	4	The least you can do for nature, A coffee break to global warming, pollutions and waste	1,2,3,4	1,2,3
	Sport Shoes (81)	Achieve measurable goals in all processes, especially zero carbon in the green strategy The call to use technology for the benefit of the environment The call to ensure everyone's participation	3,4	Only leave real footprint	1,2,3	1,2,3

Third party supports[e]	Specific target groups	Stakeholders[f]	PromoMix[g]	Communication channels, media, and tools[h]
1,2	15–40 age old people Consumer associations Employees and families Employee unions Coffee and cream suppliers Vegan and green suppliers Universities Governmental organizations municipalities and city councils Eco-friendly, animal friendly, vegan NGOs - initiatives Mainstream and alternative media	1,2,3,4,5,6,7,8	1,3,4,5,7,8,9,12,13	(1) TV programs, news programs, magazines, newspapers, outdoor advertising, and guerrilla activities (2) Administrative and strategic meetings (with consumer and representatives, employees, trade unions, competitors, green suppliers, NGOs and initiatives, green company representatives, green idea leaders, universities and independent auditors, local government and councils) Training programs and seminars, press conferences and newsletters, lobbying for local government, councils and government agencies, POP activities. (3) Responsible corporate web (activity reports, bank reports, audit reports, meeting reports and news), blog, forum and social network profile management, alternative media and green initiative links, viral works
1,2	18–65 age old people Employees and families Competitors Green suppliers Green NGOs Green opinion leaders	2,4,6,7,8	1,2,3,4,5,7,8,9,10,11	(1) TV programs and advertisements, Radio programs, (2) Administrative meetings (with consumers, employees, competitors, green suppliers, green NGOs, green opinion leaders) Training programs and seminars, press conferences and newsletters, lobbying for local government,

(*continued*)

Table 3 (continued)

Referring strategic role[a]	Product	Communication aims	The nature of marketing comm. goals[b]	Main message	Eco-message theme[c]	Value proposition[d]
	Furniture (77)	To promote simple life with multifunctional products produced by recycling To raise awareness of green process outputs	3,4	Everything for your home, Everything for earth	1,4	1,2,3
Investors	Apparel 1 (75,5)	To raise awareness of natural product To publicity supporting green producers and production with governmental incentives and collaborations	2,3	Apparel 1 will keep the word, people will win by nature We are not as you know	4	1,2,3

Third party supports[e]	Specific target groups	Stakeholders[f]	PromoMix[g]	Communication channels, media, and tools[h]
				councils and government agencies, Sponsorship and special events. (3) Corporate web (activity reports, meeting reports and news), social network profile management, green blog links, viral works, web ads.
2	Society Employees Competitors Green suppliers Opinion leaders	1,2,3,4,5,7	1,2,3,4,5,10,11	(1) TV commercials, outdoor advertising, sponsorship and special event planning (2) Administrative and strategic meetings (with consumers, employees, competitors, green suppliers, NGOs, green opinion leaders, and green bloggers) Training programs and seminars, (3) Corporate web (activity reports, meeting reports), social network profile management, social media ads.
1,2	Middle and upper income group Those who give importance to the brand name	2,3,4,5,7,8	1,2,3,5,7,8,9,11	(1) TV commercials, outdoor advertising, news programs (2) Administrative and strategic meetings (with employees, green suppliers, NGOs, green opinion leaders, and governmental organizations) training programs and seminars, press conferences and newsletters. (3) Corporate web (news), social network profile management, social media ads.

(continued)

Table 3 (continued)

Referring strategic role[a]	Product	Communication aims	The nature of marketing comm. goals[b]	Main message	Eco-message theme[c]	Value proposition[d]
	Apparel 2 (75)	Promoting cyanide-free production with the use of 100% linen and cotton Awareness of improving worker conditions	3	Add your elegance to your nature Charm in green	1,3,4	1,2,3
	Perfume (75)	Awareness of natural products and emphasis on reducing consumption with longer durability To persuade investors and partners to support the green conversion	3	Live with nature scent	2	1,2,3
	Apparel 3 (74,25)	To show that 50% recycled clothes from wastes reduce the harm done to the environment Take a respectful brand position	2,3	What nature wants	1,2	1,2,3

Third party supports[e]	Specific target groups	Stakeholders[f]	PromoMix[g]	Communication channels, media, and tools[h]
1,2	18–45 age old people People who care about environment and have never had a Apparel 2 customer before	2,5,7	1,2,3,5,8,10,11	(1) TV commercials and programs, radio programs, outdoor advertising, Magazines (2) Administrative and strategic meetings (with employees, suppliers and producers, NGOs, green bloggers) training programs and seminars, (3) Corporate web (news), social network profile management
2,3	Young and middle-aged women and men Employees Green NGOs Investors and partners	2,4,6,7	1,2,3,4,5,8, 10,12	(1) TV programs, magazines, TV dramas (2) Administrative and partnership meetings (with employees, suppliers and producers, green NGOs, opinion leaders) (3) Corporate web (news), social network profile management, word of mouth marketing, and viral works with vlogger and bloggers
1,2	Loyal consumers 16–45+ age old consumers, Environmentalist	2,4,6,7,8	1,2,3,4,5,8,13	(1) TV commercials and programs, Newspapers Outdoor advertising, (2) Administrative and partnership meetings (with employees, green suppliers, producers, shareholders, NGOs, and opinion leaders) Sponsorship and special event planning (3) Corporate social network profile management, and viral marketing.

(*continued*)

Table 3 (continued)

Referring strategic role[a]	Product	Communication aims	The nature of marketing comm. goals[b]	Main message	Eco-message theme[c]	Value proposition[d]
	Bicycle (72,2)	To gain attention to the product that produces oxygen with the use of bicycles To create an eco-sensitive consumers To gain attention to the brand	3	Another pedal for nature	1	1,3
	Vegetable oil (71)	Producing biodiesel from waste oils Protecting nature To support healthy production and consumption	2,3	We produce, we consume, we transform we care...	1	1,2
Propagators	Apparel 4 (71)	To Sell 500 thousand organic pants in 3 months To Donate 200 thousand saplings in 3 months	1	Friendly with your body and environment Organic fashion	4	1,2,3
	Cream (70,2)	Increase awareness of our green production Replacing the image with nonparaben herbal products	1,2,3	Totally natural, friendly to your skin Stay young, shine	3,4	2,3

Third party supports[e]	Specific target groups	Stakeholders[f]	PromoMix[g]	Communication channels, media, and tools[h]
2	People interested in cycling, People who are willing to protect environment 18–45 age old people	7,8	1,3,5,8,9, 11	(1) TV commercials, sport programs, sport magazines, outdoor advertising, (2) Sponsorship and special event planning (3) Corporate social network profile management, and viral marketing.
1,2	Society	1,2,7	1,2,3,4,5,10	(1) TV commercials, outdoor advertising, and guerilla advertising (2) Administrative and partnership meetings (with consumers, NGOs, car manufacturers, local governmental agencies,) Sponsorship and special event planning (3) Corporate social network profile management, and viral marketing.
1,3	Young and middle-aged people People sensitive to the environment and those who care about their health Green hippies or organic product user "cool" peoples	4,5,7	1,2,3,5,8,9,10	(1) TV commercials, (2) Partnership meetings (with green NGOs, opinion leaders and investors), training programs and seminars, POP activities (3) Corporate web and social network profile management, viral marketing with vloggers
1	18–60 ages people Trend followers	1,7,8	1,2,3,5,9,10,11	(1) TV commercials, Woman magazines and ads (2) Partnership meetings (with consumers and representatives) special events, press conferences with celebrities and opinion leaders, POP Activities. (3) Corporate web and social network profile management, viral marketing with vloggers

(*continued*)

Table 3 (continued)

Referring strategic role[a]	Product	Communication aims	The nature of marketing comm. goals[b]	Main message	Eco-message theme[c]	Value proposition[d]
	Hair Dye (70)	To promote product with decontaminated contents for woman. To feel good about hormonal changes	1,2	The most colorful state of nature	4	1,2,3
	Dental Brush (69,25)	To promote 3B functional product	1,2	3B (Beneficial care, basic design, balanced ecology)	1,4	1,2,3
	Razor Blade (68,6)	To promote sharper, nonirritating and water-saving formula	1	Natural as water	1	1,3
	Cigarette (68)	To promote our environmentalism by placing seeds in canvases	1	We created the cigarette blue, now in the green	4	1,3

Third party supports[e]	Specific target groups	Stakeholders[f]	PromoMix[g]	Communication channels, media, and tools[h]
1,2	Pregnant women High income group Women who care their hair care and health	1,7,8	1,2,3,5,8	(1) TV commercials, magazine ads, product placement ads (2) Partnership meetings (with consumers, green NGOs), special events and Press conferences (3) Corporate web and social network profile management, viral marketing with bloggers
1,2	Loyal consumers Environmentalists Trend followers	1,2,7,8	1,2,3,4,5,8,10,11,12	(1) TV commercials, Magazine ads, product placement ads, Outdoor advertising, mobile ads, (2) TV and Radio programs, guerrilla activities Partnership meetings (with consumers, green NGOs, green suppliers, opinion leaders), Special events and press conferences (3) Corporate web and social network profile management, viral marketing with bloggers
1	Men All environmentalists over 14 ages	2,7,8	1,2,3,4,10,11	(1) TV commercials, Magazine ads, Newspaper ads, Outdoor advertising, TV dramas. (2) Partnership meetings (with NGOs and suppliers), sport events sponsorships, special events and Press conferences with opinion leaders and sportsmen
2	Smokers	7	3,4,9,10	(3) Corporate web and social network profile management, viral marketing

(continued)

Table 3 (continued)

Referring strategic role[a]	Product	Communication aims	The nature of marketing comm. goals[b]	Main message	Eco-message theme[c]	Value proposition[d]
	Coffee 2 (67)	To gain attention to the product	1	Naturalness in your coffee	4	3
	Baby Diaper (65)	To position the product as a green, To ensure that it is preferred, and thus to make consumers feel that they are helping nature	1	We brought quality and nature together for baby cloth	1,4	1,2,3

[a]Strategic position or role for green: The Innovator, the Investor, the Propagator (Kotler et al. 2010: 153–159).
[b]The nature of marketing communications goals: (1) Cognitive, (2) Informative, (3) Affective, (4) Conative (Neagu 2011; Palekhova et al. 2015).
[c]Ecological MessageTheme: (1) Care, (2) Save, (3) Free, (4) Friendly.
[d]Value proposition with product/service: (1) Functional, (2) Perceptual, (3) Symbolic (Erdem ve Swait 1998: 136).
[e]Third party supports: (1) Ecolabels, (2) Environmental product declarations, (3) Cause marketing (Ottman and Mallen 2014).
[f]Direct/Indirect-Internal/External Stakeholders: (1) Consumers, (2) Suppliers, (3) Competitors, (4) Investors, (5) Employees, (6) Shareholders, (7) Regulators and auditors, (8) Media (Polonsky 1995; Freeman et al. 2008: 13; Ertuğrul 2008: 206; Talpos and Meltzer 2013:39–40).
[g]Promotional Mix: (1) Advertising, (2) Public relations, (3) Sponsorship or cause related marketing, (4) Noncommercial activities, (5) Internet marketing, web and social media, (6) Exhibitions, (7) Lobbying, (8) Word of mouth advice/viral marketing, (9) POP activities, (10)Packing, (11) Sales promotions (12) Guerilla activities, (13) Internal events and trainings (Benoit-Moreau and Parguel 2011; Ottman and Mallen 2014; Palekhova et al. 2015).
[h]Communication channels, media, and tools: (1) Mass Communication channels, (2) Interpersonal communication channels, (3) Digital media and social networks (Benoit-Moreau and Parguel 2011; Ottman and Mallen 2014; Palekhova et al. 2015).

Third party supports[e]	Specific target groups	Stakeholders[f]	PromoMix[g]	Communication channels, media, and tools[h]
1	Environmentalist 18–40 age old coffee consumers	5,7,8	1,2,3,5,8,10,11,12	(1) TV and radio commercials, (2) Partnership meetings (with employees, green NGOs and governmental organizations), Special events and press conferences with artists. (3) Corporate web and social network profile management, viral marketing with bloggers
1,2	Baby-owned parents and pregnant women	7	1,2,3,5,8,10	(1) TV commercials, magazine ads, newspaper ads. (2) Partnership meetings (with opinion leaders) and press conferences (3) Corporate web and social network profile management, viral marketing with bloggers

The target market for green are from a bit higher income bracket, or rather, should I say exactly this category? It's like that.[...] Once you get into green, it seems you are free to set the price and so I have some apprehensions about these issues (Male, score 62).

Innovators: Presentation of Green Identity and Ethics

Looking at the communication objectives in the projects to which the Innovation role is attributed, it is seen that the aim of the transformative green strategy is emphasized among the first set of conditions. The objectives are to promote sensitivity in the target groups in relation to the dissemination of green technologies and the output of these green processes. Another issue that emerges in the purpose of the communication projects in this group is that making co-productions calls into action the actors in the marketplace such as consumers, competitors, and suppliers (Coffee 1) and allows for a provision of everyone's contribution (Sport shoes) to transform the sector. These projects also emphasize the use of technological know-how and other company possibilities for the general well-being of the environment and society. The nature of the marketing communication goals is predicated on participation of the internal and external circles in the process together, but it is also aimed at influencing behavior change (encouraging contribution, encouraging a simple lifestyle, etc.).

Messages are also structured in such a manner as to highlight a company's green responsibilities and to emphasize the role they can undertake. In the first two projects with the highest score, main- and submessages were directed to cover almost all dimensions while all projects in this group emphasized functional, perceptual, and symbolic value proposals. The green-labels for this group can be viewed as a natural consequence of the activities carried out. So ecological, environmental, and animal-friendly labels are not the goal of only being clean to be well received but were the actual result of possessed values. For this reason, the perception of marketing activities is observed as actions that cause an underestimation of what should be done.

> When it is said 'marketing', something like 'hanky-panky' stuff is coming to my mind, however when green marketing is asked to me as there is 'marketing' in it, it creates the impression of bringing [unavailable] something in being, I mean I think yes there is green there but that *'green'* word is neutralized [devaluing] by 'marketing' word. I mean it adds negative meaning in it. While one is positive, as the other is more negative, it becoming dull. I do not believe that it is sincere. (Female, score 80).

For planners with low environmental sensitivity scores, marketing and communication activities are viewed as mere tools. There is an attitude that green activities only pacify consumers. For this reason, there is a belief that messages must be sincere.

> I don't believe it but I think people generally believe. I think it's because people believe whatever you show them and it depends on how convincing you are as to whether they will believe. In this case it seems to me more like we are kind of **fooling** them (Male, score 62).

In the three projects in this group, a broad target group in the environmental market has been identified as the whole of society. For consumer groups that were categorized by age from a wide selection range, the justification for "everyone's participation" was made in the first two projects (Coffee 1, Sport shoes). The expectation was that suppliers, as stakeholders, comprise an institution to be targeted in all processes of the green transformation. Additionally, we understood that the suppliers, the state institutions, and the local governments are included in the stakeholders because the projects are aimed at an integrated transformation in all the related sectors. Besides environmental NGOs, another group of stakeholders that emerged in this group were green initiatives and activist organizations. Representatives of these structures have been included in strategic planning processes as opinion leaders.

In all three projects, almost all promotional mix elements are included in the communication plans. Predominantly billboard advertising and public relations were used as the communication mix in the projects. Advertising remained quite limited in the plan, although mainstream media like TV, magazines, and newspapers were used as news and program media as opposed to advertising mediums.

Moreover, project plans included administrative, strategic, and advisory meetings with consumers, employees, labor unions, competitors, suppliers, NGOs and initiatives, green company representatives, green idea leaders, universities, and independent auditors, local government and local councils in the process. In contrast to other groups, as a distinctive promise in the first project which differentiated them from other groups, stakeholders in this group are positioned and integrated as proprietors, as part of the organization and its activities, such that they be a "responsible corporate entity."

This planning reflects the communication pattern proposed by McDonnell et al. (2015). Researchers have shown that companies react to negative activism efforts with CSR effort. This effort has often been achieved by establishing CSR committees with the participation of stakeholders such as NGOs, initiatives, and philanthropic organizations within the company or by reporting CSR efforts (p.36–37).

Because cooperation is so important at an individual level rather than that of the institution, plans focus on interpersonal channels rather than mass communication. In these projects, the media is also assessed as a stakeholder and it is used effectively in the creation of pressure groups for lobbying activities. Another prominent feature of the projects in this group is the effective use of corporate web and social media profiles for activities, auditing, meeting reports, and news about promotional endeavors. Links to profiles related to green initiatives and activists have been added, including active participation and interaction in green blogs and forums as planned in these projects.

Investors: Presentation of Green Image

In the communication plans of companies in the role of the Investor, the prioritized goal is to raise awareness of products and the visibility of green investment. The objective of some projects is to encourage investors and to increase green production through the provision of cooperation in the marketplace (Apparel 1, Perfume, Bicycle, Vegetable oil). Marketing communication focuses more on influencing target groups to inform and change behavior with new products. Because most projects refer to the green consumer market, it is important to be able to appeal to green consumers, to have a competitive advantage over competitors, and to become a priority choice in the green product market. Functional, perceptual, and symbolic value messages are also found in this group. Unlike the first group, in general, all green conversion efforts are planned to achieve compliance conditions envisaged in green labels, even if not necessarily by way of sectoral or legal regulations. For example, "a brand known to be experiencing a major crisis due to activists" was emphasized as "cleaned up to get a green label" (Apparel 1). For another brand, "self-verification" was emphasized (Apparel 2). In other words, it appears special importance is given to both consumers and stakeholders in the market. In these projects, the green labels function as evidence of cleanliness, but the action of green conversion was only limited to having the message stipulated on the label.

As stakeholders, the focus is mostly on consumers, NGOs, and raw material suppliers and it seems no specific strategies have been developed for other stakeholders and processes. In most of these projects, it is strategically important to meet the expectations of green consumers and implement improvements they may anticipate. For example, one project highlighted that "it is important to get the opinion and approval from people and groups who give importance to green products and are as environmental as an expert" (Apparel 3).

In this group, fewer projects mention environmentally friendly raw material suppliers and generally work with suppliers who produce green raw materials as output, not those who provide green conversion through all their processes, as in the first group. In the projects, green consumers were evaluated both as stakeholders, encouraging target groups to buy, and as target markets themselves. Specialized groups such as green consumers (Coffee 2, Razor Blade), green hippies, or organic product "cool" user people (Apparel 4) are deemed niche target markets. Opinion leaders and influencers are assessed during strategy development and promotional mix activity. Their "green, environmentalist, or activist identities" play a central role when included in the communication plan.

These groups of projects, again as in the first group, utilize promotion mix elements. However, unlike the first group, advertising applications are given more importance. In these projects, mainstream mass media channels and interpersonal channels are given almost equal importance. Unlike the first group, the stakeholders in this group are often positioned as "collaborative" in structure both in the green conversion process and for promotional activities.

Regarding promotional efforts, stakeholders have collaborative involvement and play an active role in communication activities such as press conferences, special events, and sponsorships. In some projects, stakeholders even appear to reinsure and prove green appearance and performance. For example, in one project, green consumers may prefer or accept a brand more easily if the support of green stakeholders is provided (Apparel 2). Digital media tools are often preferred in this group. Communication activities are managed better through social networks and viral efforts with influential vloggers and bloggers. Additionally, special events and sponsorship activities help cultivate a responsible company profile.

Propagators: Presentation of Green Appearance

There generally appears to be clear sales-driven objectives inclusive of goals to increase awareness of green-invested products in communication plans of companies in the propagators role. Additionally, communication messages in these groups have a comparative advantage by stipulating which benefits (natural and healthy, high performance experience, good feeling) are offered in products presented as "green" (Cream, Hair Dye, Dental Brush, Baby Diapers). For this reason, marketing communication goals in this group tend to be mostly cognitive (Palekhova et al. 2015). In the lowest-scoring project (Baby Diapers), a campaign goal is presented as follows: "to position the product as green, to ensure that it is preferred, and thus to make consumers feel that they are helping nature." In these plans, green labels are used as a positioning tool without the goal of reaching any standard. Approaches such as using labels for a "clean image" (Dental Brush) and boasting through the label alone (Cream, Hair Dye) are striking. Environmental claims are emphasized by declaration without any proof of being green. It seems planners in this group consider "being green" as a "marketable" element.

> After all, we are also talking about playing a kind of devil's advocate, because we are studyingit, or **I am not a green person, but I can market green, after all, because there is a benefit for you or your company.** Here, work ethic and morality emerge(Male, score 62).

Projects in this group target green consumers in general terms. Additionally, target groups are directly related to special consumer groups in the product market (pregnant women, parents with babies). In this group, green producers are viewed as a new market to be targeted by planners. In some projects, it has been determined that selected opinion leaders and influencers have focused on their popularity without regard for any relationship with or closeness to the environment (Apparel 4, Cream, Dental Brush, Baby Diapers). Generally, popular artists, actors, and sportsmen are preferred as opinion leaders. However, environmental NGOs and green consumer groups are included in project plans at the promotional activity stage. A planner specifically describes the involvement of NGOs in the process as follows:

> [NGOs] Having these sanctioning powers already makes them powerful. No matter what happens, you're disclosed anyway. When you are disclosed by the group, your image is already shaken. It [NGOs] could be seen like a bridge, because if I were company manager and doing something about being green, I will definitely consult Greenpeace as it's not your area of expertise. Whether you're producing something for your business or providing a service in the service sector, you are hoping to get paid. You think about profit, loss, crisis management, you think about a ton of things, and think it's your project, or I could say, only your point of view. Promotions will definitely be better when done with help from NGOs(Male, score 65).

"Strong" NGOs have great importance in making businesses "professional" by inciting them to communicate "more convincing" messages to consumers. These planners try to improve an image through "dealing with them" whereby they negotiate face to face with NGOs, even in the case of a possible negative situation. Therefore, where a negative situation is concerned, they place personal and private views aside to incorporate the professional interests of the company. One participant commented that "The place I work is not so much about morality, my job is not to check its morality already. My job is to try to solve the problem of that company. Think of it as like being an attorney (Male, score 65)." According to him, environment-damaging activities of a company do not concern the planner. The low-score planner does not feel any individual discomfort when the company is insensitive to the environment. This seems to be a coping strategy for the planner.

Projects in this group have planners citing vegetable ingredients in product content and positioning them as "natural" without mentioning the other harmful substances in the product contents (Apparel 4, Cream, Hair Dye, Cigarettes). Among the five unsuccessful manifestations expressed by Peattie and Crane (2005: 360–364) and Kotler et al. (2010), such efforts remind us of green spinning and green selling strategies. According to the authors, green spinning is prevalent in petrol, chemical, pharmaceutical, and automobile companies which are essentially "dirty" sectors who approach the market in an aggressive manner using the green logic "whatever is green sells." When it comes to product development and research, they are not included, and instead, merely give the same products a green theme in promotional activities to entice green tendency consumers who have concerns about damaging the environment.

Most of the projects focused on providing a green appearance within elements of the promotional mix. The products' green features are often highlighted by special events and sponsorships, and such efforts are presented as corporate social responsibility. In some projects, NGOs did articulate CSR efforts yet little attention was paid where NGOs were concerned. For example, the Apparel 4 project CSR activity was planned to secure the support of an NGO who was not interested in the company's "green" agenda (Table 4).

The issue of activism is especially important in marketing communication activities. Low score planners view activists as a means of persuading consumers while high score planners treat them as people and organizations they can collaborate with.

> I was able to organize an event for NGOs in such a way that enabled them to work with genuine sincerity, toward green aims and to depend on them. I also had them investigate our product. For example, I could include them in the time and processes of production (Female, score 79).

Where activists and businesses are confronted, the reactions and responses of planners differ from each other. For planners with high environmental sensitivity, recognizing the negative activities of enterprises is considered a problem that needs to be taken seriously. If the activist group misunderstands the activities of a truly green effort, high-scoring planners focus on incorporating activist groups into the process by trying to locate the "communication gap." During this process, the planner judges him/herself and considers "where he/she was wrong" (Female, score 81). Another striking approach is to accept when "mistakes" are made and to include activists in the process by asking them to "reveal the truth." Therefore, the aim is to adopt a collaborative approach and develop activities in this direction:

> If I am confronted with activists, there must certainly be an error in communication. I ask myself what should be done again and again (Female, score 81).

Table 4 Attitudes towards operational activities and attitudes of planners with high and low green sensitivity

Business assumed to be employed in communication planner aspect	Business showing green activity	Business pretend to be green	Approach to activism
Believed/ persuaded to show green activity high-scoring communication planner	Effective work performance, strong persuasion and effective marketing – **Identity**	Effective work performance and marketing communication	While activists react mutually-collaboratively with plan activists in the green business, whether positively or negatively; trying to understand what activists are reacting to planners and activists against the negative reaction of activists to the business which pretend to be green and questioning itself
Believed/ persuaded to show green activity low-scoring communication planner	Effective work performance and marketing communication	Effective work performance and marketing communication	If the reactions of the activists are positive, activists are involved in the marketing communication process in a way that would create leverage effect; if the reaction of the activists is negative, it is either ignored or an effort is shown to persuade the consumer
Not Believed/ persuaded to show green activity high-scoring communication planner	Trying to persuade the business executives to show more effective green behavior; questioning the green efforts of the business	Negative attitude against the business and/or quitting the job	Persuading the business managers in line with the reactions of the activists
Not Believed/ persuaded to show green activity low-scoring communication planner	Sustaining the activities of marketing within the framework of the job description	Sustaining the activities of marketing communications within the framework of the job descriptions	Activists being ignored Communicator running away from responsibility and quitting work in an effort to save himself/herself

I can do things like 'admit to them there are missing things with the product. Don't do anything, I am mentioning this to you directly' and convince them of my honesty. Yes, we have some deficiencies but we are trying to manage them. Or like this... Yes, there are some deficiencies that the company is aware of and must correct and if the company flat out says no, then it isn't right. I can't commit to them; it's not something I can do. But if the company is a bit flexible I can tell them: 'This is the route we plan to take so please support us. These are our strong qualities, and these are our deficiencies and you can help us improve' for example and by including them in the process together we can convince them (Female, score 79).

In the case of an inability to convince actors that corrections should be made to operational activities, high-scoring participants tend to prefer "changing side" on the side of environmentalist organizations and even tend to act against the corporation.

> I swearI would expose the company. I would definitely do that. I do not care about the consequences (Female, score 80).

Participants with low environmental scores, on the other hand, adopt a more dialogical approach, especially when confronted with activist groups, and are more likely to develop a discourse on the part of the employer they work with. From the outset, activist groups get involved to find a "middle way" in communication by adopting the role of an "intermediary" and thus are involved in the process.

> I definitely try to be a bridge because if the company already thinks about going green and works with community organizations or activists there must be adisconnected communication. Otherwise they must serve both. It can't be otherwise. That's why I try to find a middle way (Male, score 65).

> The only thing I will practice, like in my life, is denying. Forever continuing without accepting anything in anyway. It's a situation where you have denied something and it takes a while to get over it. I don't think I'll intervene too much. [...] If they do not accept just continue to deny (Male, score 61).

Thus, planners with high environmental scores showed an attitude that did not oppose (activist) discourse, while those with low scores showed an attitude toward advocacy (corporate) discourse.

Conclusion

This study aims to reveal how environmental sensitivity levels are influenced by decision-making processes of students who are educated in the field of public relations and marketing communication. Student choices especially determine how they will incorporate environmental sensitivity into decision-making and strategy-building processes as decision makers of the near future. Although they do not have clear results because they are not yet in professional careers, they do present important findings regarding "tendencies" in the business environment due to the fact they prepare their projects within the scope of their education and have an authority to supervise them as they are at the same time, as if in a workplace. What is important here is to show how decision makers of the future address environmental sensitivity according to the current market environment so that businesses will have a segmented view point to work with.

The NEP (New Environmental Paradigm) scale (Dunlap 2008) was an effective means to measure our participants' environmental sensitivities in the study. Particularly in relation to content analysis and in-depth interviews, this scale shows very successful results. However, it would be more effective for the participants to support this scale with other data collection techniques to address how they approach to the issue. Strikingly, participants with high environmental sensitivity scores tend to collaborate when working on a project while participants who have low scores form groups among themselves and even lack the knowledge to do so. Uniting those who share similar individual characteristics in the work environment will have an accelerating effect and especially when combined with the objectives of businesses, successful strategies may be reflected in the market.

While low-score planners utilize their decision-making mechanisms in line with their job descriptions, they define acts opposed to business interests as "unethical." On the other hand, since high-scoring participants see marketing activities as deceiving people, they define being part of these activities as "unethical." Thus, we may conclude that decision-making processes that side more with businesses are observed in low-scoring planners, while high-scoring planners tend to shape their decisions and strategies taking environmental sensitivity into consideration both as a part of their identity and in projects they prepare. However, high-score planners, who are part of the capitalist system and operate in this dominant market environment, appear to be have more difficulty in complying with market demands and thus tend to experience more contradictions. Especially when businesses they work with harm the environment, low-score planners adapt easily, but high-score planners display a more interrogative and reactive attitude to the point of contradicting themselves. While this questioning attitude can be welcomed in terms of environmental sensitivity, it is unlikely we can state the same in terms of their willingness to embrace the market environment. This attitude leads high-score planners to leave their jobs when they experience potential antienvironmental attitudes by an employer. NGOs are stakeholders that should be emphasized. Because low-score projects treat NGOs as a part of the promotion mix increasing credibility, and therefore consumer interest, high-score projects become part of the process as a holistic green strategy. For high-score projects that aim to develop and transform processes and the market as well as for planners, it becomes a way of learning from NGOs and, more specifically, confirmation of implementing their identities. Low-score planners, in general, did not talk about NGOs during interviews. They either questioned who NGOs were or did not consider this group a stakeholder.

The existence of planners with high environmental sensitivity is a matter up for debate regarding the sustainability of the capitalist market environment. The existence of businesses that are profit-oriented organizations alters project content targeted by this group. For this reason, from a pragmatist perspective, middle-score participants are considered the most appropriate group in the current market environment. This group stands out as being both environmentally sensitive and as individuals who accept market implementation and do not suffer the tension that high-score planners experience by avoiding the pure market-oriented approach of low-score planners who are able to balance both market expectations with individual beliefs. The projects they created also shed light on this situation. The most desirable outcome of environmental idealism is that high-score planners and their projects are efficient and ensure visibility in the market. However, in realization of this idealism the market environment needs to be transformed. It is envisaged that the implementation of these strategies under the current conditions will become problematic for both individual and business practices.

The fact that this study was conducted within an educational environment and that discussions derive from the institution; this places emphasis on the importance of environmentalist educational. Institutions that provide education for planners help support students to overcome a sense of hopelessness prevalent in the capitalist system by making transformative approaches possible in the market as mentioned above. Predominance of a family element, especially in meetings, has also shown that out-of-school actors are also important in creating environmental sensitivity. Education provides an uplifting effect for every individual in terms of the green transformation. This effect is particularly influential on individuals that tend to be environmentally sensitive and even sensitizes individuals with low sensitivity. Environmental sensitivity is effective in the process of positive or negative decision-making and determines the strategies that will be applied.

This study presents varied perspectives regarding approaches that will likely appear in the market of the near future. Certainly, the future is unknown but knowing the fact that individuals will operate in the market with their individual sensitivities and identities shows that in terms of businesses and policy makers, there will not be any surprises when it comes to evaluating consequent activities and decisions.

Cross-References

▶ Ecopreneurship for Sustainable Development
▶ Sustainable Higher Education Teaching Approaches
▶ The LOHAS Lifestyle and Marketplace Behavior

References

Aidt, T. S. (1998). Political internalization of economic externalities and environmental policy. *Journal of Public Economics, 69*(1), 1–16.

Alnıaçık, Ü. (2009). Tüketicilerin çevreye duyarlılığı ve reklamlardaki çevreci iddialar. *Kocaeli Üniversitesi Sosyal Bilimler Enstitüsü Dergisi, 18*(2), 48–79.

Benoit-Moreau, F., & Parguel, B. (2011). Building brand equity with environmental communication: An empirical investigation in France. *EuroMed Journal of Business, 6*(1), 100–116.

Bodur, M., & Sarigöllü, E. (2005). Environmental sensitivity in a developing country: Consumer classification and implications. *Environment and Behavior, 37*(4), 487–510.

Button, K. J., & Pearce, D. W. (1989). Improving the urban environment: How to adjust national and local government policy for sustainable urban growth. *Progress in Planning, 32*, 135–184.

Carroll, A. B. (1991). The pyramid of corporate social responsibility: Toward the moral management of organizational stakeholders. *Business Horizons, 34*(4), 39–48.

Chamorro, A., Rubio, S., & Miranda, F. J. (2009). Characteristics of research on green marketing. *Business Strategy and the Environment, 18*(4), 223–239.

Chang, C. T., Lee, Y. K., Chen, T. T., & Wu, S. M. (2011). Are guilt appeals good in green marketing? The moderating roles of issue proximity and environmental involvement. *ACR European Advances in Consumer Research, 9*, 449–451.

Cinioğlu, H., Atay, L., & Korkmaz, H. (2016). Önlisans Öğrencilerinin Yeşil Reklama İlişkin Algılarının Belirlenmesine Yönelik bir Araştırma. *Yaşar Üniversitesi Dergisi, 43*(11), 198–210.

Cone Communications Corporate Social Return Trend Tracker. (2012). http://www.conecomm.com/research-blog/2012-cone-communications-corporate-social-return-trend-tracker. Accessed 17 Feb 2016.

Cone Communications/Ebiquity Global CSR Study. (2015). http://www.conecomm.com/2015-cone-communications-ebiquity-global-csr-study-pdf/. Accessed 17 Feb 2016.

Delmas, M. A., & Burbano, V. (2011). The drivers of green washing. *California Management Review, 54*(1), 64–87.

Disparte, D. A., & Gentry, T. H. (2015). Corporate activism is on the rise. https://intpolicydigest.org/2015/07/06/corporate-activism-is-on-the-rise/. Accessed 21 Sept 2015.

Dodd, M.D. (2016). Corporate activism: The new challenge for an age-old question. http://www.instituteforpr.org/corporate-activism-new-challenge-age-old-question/. Accessed 28 Apr 2016.

Dunlap, R. E. (1998). Lay perceptions of global risk: Public views of global warming in cross-national context. *International Sociology, 13*(4), 473–498.

Dunlap, R. E. (2008). The new environmental paradigm scale: From marginality to worldwide use. *The Journal of Environmental Education, 40*(1), 3–18.

Ertuğrul, F. (2008). Paydaş Teorisi ve İşletmelerin Paydaşları İle İlişkilerinin Yönetimi. *Erciyes Üniversitesi İktisadi ve İdari Bilimler Fakültesi Dergisi, Sayı, 31*, 199–223.

Freeman, R. E., Pierce, J., & Dodd, R. (1998). Shades of green: Business, ethics, and the environment. In L. Westra & P. H. Werhane (Eds.), *The business of consumption: Environmental ethics and the global economy* (pp. 339–354). Maryland: Rowman and Littlefield Publishers.

Freeman, R.E., York J. G., & Stewart, L. (2008). Environment, ethics, and business. Business Roundtable Institute for Corporate Ethics. http://www.corporate-ethics.org/pdf/environment_ethics.pdf. Accessed 5 July 2017.

Garriga, E., & Mele, D. (2004). Corporate social responsibility theories: Mapping the territory. *Journal of Business Ethics, 53*, 51–71.

Gilbert, A. J. (2007). The value of green marketing education at the University of Wisconsin-La Crosse. *UW-L Journal of Undergraduate Research, X*, 1–16.

Grant, J. (2007). *Green is a principle, not a proposition: A review of the green marketing manifesto*. Chichester: Wiley.

Heijungs, R., Guinée, J. B., Huppes, G., Lankreijer, R. M., Udo de Haes, H. A., & Sleeswijk, W. (1992). Environmental life cycle assessment of products: Guide and backgrounds (Part 1), https://openaccess.leidenuniv.nl/bitstream/handle/1887/8061/11_500_018.pdf?sequence=1. Accessed 14 May 2017.

Hussain, S. S. (1999). The ethics of 'going green': The corporate social responsibility debate. *Business Strategy and the Environment, 8*, 203–210.

Jurin, R. R., Roush, D., & Danter, K. J. (2010). *Environmental communication: Skills and principles for natural resource managers, scientist and engineers*. New York: Springer.

Kardeş, İ. (2011). Markaların Çevre Dostu Uygulamalarının Tüketicinin Marka Tercihi Üzerindeki Etkisi. *Ege Akademik Bakış Dergisi, 11*(1), 165–177.

King, B., & McDonnell, M. H. (2012). Good firms, good targets: The relationship between corporate social responsibility, reputation, and activist targeting. https://papers.ssrn.com/sol3/papers.cfm?abstract_id=2079227 . Accessed 20 May 2016.

Koçarslan, H. (2015). İşletmelerin Sosyal Sorumluluk Bilincinde Çevre Duyarlılığının Yeşil Pazarlama Üzerine Etkileri. Kilis 7 Aralık Üniversitesi Sosyal Bilimler Enstitüsü, Yayınlanmamış Doktora Tezi, Kilis. sbe.kilis.edu.tr/dosyalar/tezler/Hüseyin%20KOÇARSLAN.pdf. Accessed 20 May 2016.

Kotler, P., Kartajaya, H., & Setiawan, I. (2010). *Marketing 3.0: From products to customers to the human spirit*. Wiley, New Jersey.

Lockie, S., Lyons, K., Lawrence, G., & Mummery, K. (2002). Eating 'Green': Motivations behind organic food consumption in Australia. *SociologiaRuralis, 42*(1), 23–40.

McDonnell, M. H., King, B. G., & Soule, S. A. (2015). A dynamic process model of private politics: Activist targeting and corporate receptivity to social challenges. *American Sociological Review, 80*(3), 654–678.

Meisner M. (2015). Environmental communication: What it is and why it matters, one-planet talking point, The International Environmental Communication Association (IECA), https://theieca.org/sites/default/files/optp/%20OPTP%231-EC_What_and_Why.pdf. Accessed 8 Mar 2017.

Menon, A., Menon, A., Chowdhury, J., & Jankovich, J. (1999). Evolving paradigm for environmental sensitivity in marketing programs: A synthesis of theory and practice. *Journal of Marketing Theory and Practice.* https://doi.org/10.1080/10696679.1999.11501825.

Neagu, O. (2011). Influencing the environmental behavior through the green marketing. The case of Romania, Paper presented at the 2011 International Conference on Financial Management and Economics, Hong Kong China.

Ottman, J., & Mallen, D.G. (2014). Five green marketing strategies to earn consumer trust. Green Biz. http://www.greenbiz.com/blog/2014/01/14/five-strategies-avoid-taint-greenwash-your-business>. Accessed 10 May 2015.

Özdemir, E. K., & Topsümer, F. (2016). Kurumsal itibar yaratma sürecinde yeşil olgusu sosyal sorumluluk mu? Yeşil aklama mı? Volkswagen "think blue" kampanyasi. In A. Göztaş (Ed.), *İletişimde Serbest Yazılar.* Konya: Literatürk Academia.

Özkaya, B. (2010). İşletmelerin sosyal sorumluluk anlayışının uzantısı olarak yeşil pazarlama bağlamında yeşil reklamlar. *Öneri Dergisi, 9*(34), 247–258.

Palekhova, L., Ramanauskiene, J., & Tamuliene, V. (2015). Methods to stimulate sustainable consumption in the system of promotion products in industrial markets. *Management Theory and Studies for Rural Business and Infrastructure Development 37*(2):264–274

Peattie, K., & Crane, A. (2005). Green marketing: Legend, myth, farce or prophesy? *Qualitative Market Research: An International Journal, 8*(4), 357–370.

Polonsky, M. J. (1994). Green marketing regulation in the US and Australia: The Australian checklist. *Greener Management International, 5*(1), 44–53.

Polonsky, M. J. (1995). A stakeholder theory approach to designing environmental marketing strategy. *Journal of Business and Industrial Marketing, 10*(3), 29–46.

Polonsky, M. J. (2001). Green marketing. In M. Charter & U. Tischner (Eds.), *Sustainable solutions* (pp. 283–301). Eastbourne: Greenleaf Publishing Ltd.

Polonsky, M. J. (2005). Green marketing. In R. Staib (Ed.), *Environmental management and decision making for business* (pp. 124–135). USA: Palgrave Macmillan.

Polonsky, M. J., & Rosenberger, P. J. (2001). Reevaluating green marketing: A strategic approach. *Business Horizons, 44*(5), 21–30.

Prakash, A. (2002). Green marketing, public policy and managerial strategies. *Business Strategy and the Environment, 11*(5), 285–297.

Schuhwerk, M. E., & Lefkoff-Hagius, R. (1995). Green or non-green? Does type of appeal matter when advertising a green product? *Journal of Advertising, 24*(2), 45–54.

Stanković, J., & Strižak, M. (2015). The Danube region protection–challenges for green marketing and corporate social responsibility in Serbia. *Journal of Danubian Studies and Research, 5*(1), 7–18.

Steinmann, H. (2008). Towards a conceptual framework for corporate ethics: Problems of justification and implementation. *Society and Business Review, 3*(2), 133–148.

Talpos, M. F., & Meltzer, M. (2013). The internet and its potential for an efficient environmental communication. *Journal of Environmental Research and Protection, 37,* 28–36.

Tuxworth, B. (1996). From environment to sustainability: Surveys and analysis of local agenda 21 process development in UK local authorities. *Local Environment, 1*(3), 277–297.

Van de Ven, B. (2008). An ethical framework for the marketing of corporate social responsibility. *Journal of Business Ethics, 82*(2), 339–352.

Yılmaz, V., Çelik, E., & Yağızer, C. (2009). Çevresel duyarlılık ve çevresel davranışın ekolojik ürün satın alma davranışına etkilerinin yapısal eşitlik modeliyle araştırılması. *Anadolu Üniversitesi Sosyal Bilimler Dergisi, 9*(2), 1–14.

Banerjee, S. B. (2002). Corporate environmentalism: the construct and its measurement. Journal of Business Research 55(3):177–191

Banerjee, S. B. Iyer, E. S., & Kashyap, R. K. (2003). Corporate Environmentalism: Antecedents and Influence of Industry Type. *Journal of Marketing 67*(2):106–122

Erdem, T., & Swait, J. (1998). Brand Equity as a Signaling Phenomenon. *Journal of Consumer Psychology 7*(2):131-157

Agent-Based Change in Facilitating Sustainability Transitions

A Literature Review and a Call for Action

Katariina Koistinen, Satu Teerikangas, Mirja Mikkilä, and Lassi Linnanen

Contents

Introduction	1136
Methodology	1137
Challenges in the Study of Agency in Sustainability Transitions	1138
Challenge #1: Agency Is Neglected	1138
Challenge #2: Scattered Terminology in the Study of Agency	1140
What Is Known on Agency in Sustainability Transitions?	1142
Who Is an Agent?	1142
Individual Versus Collective Forms of Agency	1143
Agency Enabling Sustainability and Societal Transitions	1144
Future Research Directions	1147
Agency Formation	1147
Psychological Dimensions of Agency	1148
Diversity in Forms and Dynamics of Agency	1148
Representations of Agents Are Narrow	1149
Summary of the Research Trajectories	1150
Implications for Activists	1150
There Is Only One Planet: A Call for Action for Everyone	1151
Conclusion	1152
Cross-References	1153
References	1153

K. Koistinen (✉) · M. Mikkilä · L. Linnanen
School of Energy Systems, Lappeenranta University of Technology, Lappeenranta, Finland
e-mail: katariina.koistinen@lut.fi; mirja.mikkila@lut.fi; lassi.linnanen@lut.fi

S. Teerikangas
Turku School of Economics, Turku University, Turku, Finland

School of Construction and Project Management, Bartlett Faculty of the Built Environment, University College London, London, UK
e-mail: satu.teerikangas@utu.fi

© Springer International Publishing AG, part of Springer Nature 2018
S. Dhiman, J. Marques (eds.), *Handbook of Engaged Sustainability*,
https://doi.org/10.1007/978-3-319-71312-0_31

Abstract

How can sustainability transitions be enabled? The focus in the practice and research on sustainability transitions has traditionally been on large-scale institutional actors. However, in order to make a difference, also micro-led change, i.e., engaged individuals driving sustainable change, is needed. This chapter explores the role of engaged individuals in facilitating sustainability transitions. Who are they and what do they do? Despite much interest, this body of knowledge remains scattered across disciplines. In this chapter, the aim is to take stock of the subject matter by providing a review of the scholarly literature on agent-based change involved in sustainability transitions. The review is based on leading journals in environmental management and sustainability studies. Based on the review, challenges and key dynamics in the study of agent-based sustainability transitions are identified. Moreover, future research directions are provided. The chapter encourages individuals, starting with you and me, to awaken their agency and to start making a difference toward developing a better, more sustainable, tomorrow.

Keywords

Agency · Actor · Micro-led change · Sustainability transition · Multi-level perspective · Niche · Incumbent

Introduction

This chapter explores the role of actor-led change in enabling sustainability transitions. Actor-led change, or agency, refers to individuals or collectives, who take a proactive stance, acting toward a more sustainable planet. With the term sustainability, the focus is on environmental sustainability. Although the connections to social and economic sustainability are acknowledged, they are not actively discussed in this chapter.

Despite the potential of engaged actors in enabling sustainability transitions, a closer look at the sustainability transitions literature leads to observe that the role of agency tends to be underplayed at the expense of a focus on technological solutions and a policy orientation (Cashmore and Wejs 2014; Bakker 2014; Mercure et al. 2016). In other words, it appears that a macro-orientation has prevailed at the expense of a micro-level perspective. In this chapter, the aim is to take stock of the current knowledge in order to find out how agency has been studied in sustainability studies, with a particular focus on the sustainability transitions literature.

To our knowledge, this chapter is the first to undertake a systematic review of the literature on agent-based sustainability transitions. The chapter thus provides an overview of the hitherto scattered literature on agent-based change in the context of sustainability transitions. This is the chapter's main contribution. Within this contribution, the paper's findings are twofold. First, the chapter compiles together the hitherto scattered terminology on agents involved in sustainability transitions. Second, the chapter summarizes extant understanding on the means through which agents influence sustainability transitions. Given that the field is relatively young, the paper also identifies numerous research directions going forward to guide future research endeavors. Summing up, the review finds that agency, i.e., individual and collective action, bears the potential to accomplish major sustainability transitions. Hence, the role of agency should no longer be underplayed.

The chapter proceeds as follows. The first section presents how the literature review was carried out. The literature review enabled the identification of a number of challenges in the study of agency. These are presented in section three. The fourth section presents the main findings concerning the dynamics of agency in sustainability transitions. The final section concludes by summarizing the main observations, identifies future research directions, and suggests practical implications going forward be it for activists or ordinary citizens and consumers.

Methodology

As the aim of the chapter is to understand what is presently known on the role of individuals in facilitating sustainability transitions, a literature review was conducted. The review began by selecting journals from the fields of sustainability and environmental management. The aim was to achieve a representative sample of journals from the disciplines of environmental management and sustainability. The original sample included the following leading academic journals: *Journal of Environmental Economics and Management, Environmental Values, Business and Society, Sustainability Science, Global Environmental Change, Journal of Environmental Management, Journal of Environmental Planning and Management, Journal of Environment and Development, Sustainable Development, Social Research,* and *Environmental Innovation and Societal Transitions.*

The aim was to find out how agency has been studied in the sustainability literature. The selected journals were searched using the keyword "agency." As this initial keyword yielded only a small set of papers, the search was broadened to include multiple keywords. The aim of the journal search thus shifted from finding agency-termed papers to finding papers studying agency, however defined, in the sustainability literature. In so doing, the actually used keywords included the following set of terms: *agent, agency, actor, micro, change agent, social entrepreneurship, social movement, civil society, consumer, stakeholder, tempered radical, niche, enlightened user, early adopter, private actor, collective action, individual engagement, household engagement, grassroots initiative, activism, movement, bottom-up, pro-active, pro-active motivation, distributive leadership, sustainable innovation, positive social movement, and social inclusion.*

During the search for relevant articles, more journals were added to the original sample. In particular, in order to gain an appreciation of the field, reference and seminal papers were included into the literature review. This led to including journals, such as *Energy Policy, Research Policy and Business,* and *Strategy and the Environment* to the review.

The keyword search yielded a total set of over 60 journal articles published between 2002 and 2016. During the reading process, some articles were eliminated, as they were out of the scope of the paper. This led to a final sample of 25 journal articles (see Table 2) being selected for review. The analysis of these articles led to identifying two broad categories of findings: first, challenges in the study of agency in this literature; second, an overview of the means through which agents influence sustainability transitions. Going forward, the chapter proceeds using these headings.

Challenges in the Study of Agency in Sustainability Transitions

Challenge #1: Agency Is Neglected

The first observation arising from the analysis is the relative neglect of agency in the sustainability literature. All the while, the question of how transformation(s) towards more sustainable societal practices occur has gained much attention in recent years. The body of knowledge on sustainability transitions has increased, in particular, in policy research and in the social sciences (Markard et al. 2012). Four theoretical frameworks on sustainability transitions have come to dominate the discipline. These theoretical frameworks include (1) transition management, (2) strategic niche management, (3) the multi-level perspective on sociotechnical transitions, and (4) the technological innovation systems perspective (Markard et al. 2012). This chapter takes as its main focus agency in the multi-level perspective (MLP). Notwithstanding, some parts of the literature review, where agency is discussed in relation to sustainability transitions cover sustainability transitions from a more extensive theoretical perspective and adopts components from the other three theoretical frameworks discussed above.

The multi-level perspective (MLP) has been introduced and developed by Frank Geels (2002). This framework has gained prominence in the literature on sustainability transitions. This can largely be explained by the fact that the framework brings together the technical and social dimensions of sustainability transitions. Therefore, the MLP is often regarded as a theory of sustainability transition(s) especially in the frame of sociotechnical transitions. The MLP goes beyond studies of single technologies – such as wind turbines, biofuels, fuel cells, and electric vehicles – which previously dominated the literature on environmental innovation (Geels 2011). In MLP, transitions are instead described as complex, long-term processes involving multiple actors (Geels 2011). In sustainability transitions, the key question is solving how environmental innovations emerge and how these can challenge, replace, transform, and reconfigure existing, typically unsustainable, technologies and systems (Geels 2011).

The multi-level perspective is built on the assumption that structure exists at three different levels of analysis: the niche-level, the regime-level, and the landscape-level. Technological trajectories are situated in a sociotechnical landscape, consisting in a set of deep structural trends, such as economic growth or oil price (Geels 2002). The landscape is described as an external structure or as a context wherein the interactions of actors occur. Whereas regimes refer to rules that enable and constrain activities within communities, the sociotechnical landscape refers to wider factors beyond technology. Notwithstanding, the context of landscape is more difficult to change than the context of regimes. Landscapes eventually change, but slower than regimes. It is typical for regimes to generate incremental innovations. In contrast, radical innovations are generated in niches (Geels 2002). The status quo and hence the existing landscape and regimes are premised on actors continuing to conform to the current structure, as either they are seen as legitimate or their legitimacy is taken for granted (Cashmore and Wejs 2014). It is into this setting that hypothetically the role of agency can be set. If the perspective to agency is shifted from a passive conforming role (as above) to an active shaping role, then there is a space for individual actors to impact sustainability transitions. So what is it that extant literature states about active agency?

Several scholars have recognized that agency plays an essential role in sustainable transitions. Agency also forms an integral part of the multi-level perspective (Geels 2011; Upham et al. 2015). In the MLP, agency is considered in the form of bounded rationality, this including individuals' routines and interpretive activities. However, certain types of agency remain less developed, including rational choice, power struggles, and cultural discursive activities (Geels 2011).

In this literature, it is suggested that agency is crucial in order to create social change. Moreover, agency is necessary during particular episodes of a sustainability transition (Grin et al. 2011; King 2008; Wiek et al. 2012). Instead of one, in practice there are multiple agents seeking to influence the progress of a transition (Grin et al. 2011). Agency is critical, as agents possess the abilities, means, and power for deliberate action, and can thus contribute toward more sustainable societies (Wiek et al. 2012). Agency also bears an influence on the internal conception of sustainability and on individuals' interpretations of sustainability; in so doing, agency helps to further embed sustainability beyond themselves (Lehner 2015; Heijden et al. 2012).

Based on the literature, agency is recognized as an integral part of sustainability transitions. What is more, extant literature has found that agents are capable of contributing to sustainability transitions. Despite this acknowledgment, in practice the role of agency remains neglected in the multi-level perspective (Smith et al. 2005; Shove and Walker 2007; Genus and Coles 2008; Bakker 2014). Taking a critical stance, the MLP has been accused for downplaying the role of agency during transitions (Geels 2011).

Beyond the multi-level perspective, agency has been studied in the general literature exploring the interfaces on science, technology, and society. Here, a central debate concerns the recreation of governance and new environmental regimes. In this literature, the role of agency has been suggested to be significant. All the while, also in this literature, the role of agency tends to remain empirically neglected (Clapp 1998; Haufler 1998; Ehrlich 2006; Rothenberg and Levy 2012). If the focus of research on sustainability transitions relies on technological development and existing regime actors only, there is a risk that agency reinforces existing regimes rather than yield novel or radical innovations and forms a basis for transitions (Audet 2014). This explains the need for more agency-related research. Through an enhanced understanding of the role of agency in sustainability transitions, individuals can be empowered, and their agentic capabilities utilized to create better futures.

To summarize, despite the acknowledgment of agency in the literatures on sustainability transitions, it appears that the role of agency remains underplayed.

Challenge #2: Scattered Terminology in the Study of Agency

The second challenge in the study of agency in sustainability transitions relates to the difficulty of capturing the essence of agency. The analysis conducted for this review suggests that the terminology in the study of agency is scattered, definitions are loose or nonexistent, and the underpinning theoretical bases are thin. These issues are reviewed next, starting with a review of seminal theorizing on the concept of agency.

Agency is a central term in the social sciences. The term agency has its roots in sociology, where the nature of agency and its relation to the larger environment (or structure) is discussed. In sociology, agency tends to be interpreted as the human capability to make free choices and to have an impact on one's environment. In this chapter, agency is explored and defined using the lenses of Anthony Giddens and Margaret Archer, seminal sociologists whose theorizing has come to shape the contemporary debate on agency.

The interplay between individuals and their environment is an essential component in structuration theory, as elaborated initially by Anthony Giddens (1984) and later by, e.g., William Sewell (1992, 2001). In structuration theory, agency is considered as the bidirectional movement between individuals and their surroundings (Dean et al. 2016). Giddens emphasizes that agency and structures are ultimately inseparable (Giddens 1984). Agency determines structure, which consequently determines the possibilities for the expression of agency (Giddens 1984). This means that structure and agency are co-produced (Cashmore and Wejs 2014). Following Giddens, Margaret Archer (1995) offers another perspective to the debate on structure and agency. Archer sees structure and agency as appearing, instead of a dichotomy, as two separate functions in constant movement (Archer 1995). Agency constantly affects structure, yet agency is also constantly affected by structure (Archer 1995).

In addition to the structure-agency debate, there are numerous other definitions and perspectives to agency. Taking one example, John Law's (1992) Actor Network Theory (ANT) considers how the "social" can take forms beyond merely human ones. This bears implications for what is considered the remit of agency. Looked at from this perspective, agency can conceptually appear in social systems such as the economy or in ideas, in materials, or in other forms of life including animals (Law 1992). Recently, individualistic conceptions of subjectivity and human agency have come to be critiqued for failing to recognize the historical, political, and social conditions, and limitations of everyday life (Autio et al. 2009). Consequently, such considerations ought to be part of future research endeavors.

In contrast, the review of literature on agency in sustainability transitions reveals little connection to this seminal theorizing on agency. It thus appears that the term

Table 1 Authors of the final sample and if there is or if there is not a definition for agency

Authors who have defined agency in some way	Authors who have not defined agency
Audet 2014; Cashmore and Wejs 2014; Corry and Jørgensen 2015; Geels 2011; Grin et al. 2011; Larsen et al. 2011; Mercure et al. 2016; Wiek et al. 2012; Upham et al. 2015	Bakker 2014; Bergek et al. 2015; Bork et al. 2015; Fudge et al. 2016; Geels 2002, 2005, 2013, 2014; Kern 2015; King 2008; Klitkou et al. 2015; Rothenberg and Levy 2012; Seyfang and Longhurst 2013; Whetten and Mackey 2002; Wittmayer and Schäpke 2014; Åm 2015

agency is used "lightly" with little consideration of its sociological roots. In addition to the scattered terminology surrounding agency, we observed the definitions of agency to be lacking in numerous papers; see Table 1. When defined, the definitions in use varied. Some authors, such as, Audet (2014), Cashmore and Wejs (2014), and Upham et al. (2015), followed Giddens' (1984) definition for the agency, in which agency is co-produced within the larger environment. Other authors, such as, Corry and Jørgensen (2015) or Mercure et al. (2016), defined agency simply as the free will of actors.

What is more, the study of agency in sustainability transitions uses numerous terms. While the term agency or actor might be used in the MLP perspective, a closer look at the empirical work on agency shows a more scattered picture. Initially, the review and keyword searches were conducted using the term "*agency*." However, during the article search and literature review processes, it became apparent that this core term did not capture the richness and diversity of academic research on agency involved in sustainability transitions. This led to the observation that the notion of agency is implied using numerous terms. The term "*agency*" is thus not consistently used in the literature on sustainability transitions. A closer look at this terminology leads to the observation that the term "agency" appears most often in the journals sampled for this review. The second most frequent word in use is "actor." In addition to the terms "agency" and "actor," agency is referred to using terms such as "strategic agency," "collective action," "niche action," "stakeholder," "niche actor," "new entrant," "change agent," "private actor," "institutional entrepreneur," "niche," "civil society," "social actor," "social agent," "organizational agent," and "micro." This plethora of terms might, in part, explain why the study of agency in sustainability transitions is scattered and seems neglected. An overview of this scholarly literature, as provided in this chapter, has been missing. Table 2 summarizes this understanding of how agency appears in the core articles studied for this literature review.

In closing, in order to address the limitations of previous literature, in this chapter, the terms "*agency*" or "*actor*" are consistently used to refer to engaged individuals shaping sustainability transitions. Following Giddens and Archer, agency is considered as the individual's potential to act in one's surroundings and to affect structure (s). Moreover, this chapter acknowledges the ambivalent motivations and strategies for the agency, i.e., agency can act as a positive force encouraging change, or as a negative force resisting change. These dynamics are explored in greater detail in the next section.

Table 2 Agency-related terms in use in sustainability transitions literatures

Used terms	Authors
Agency	Bakker 2014; Cashmore and Wejs 2014; Corry and Jørgensen 2015; Geels 2002, 2005, 2011, 2013, 2014; Grin et al. 2011; Kern 2015; Larsen et al. 2011; Mercure et al. 2016; Upham et al. 2015; Åm 2015
Strategic agency	Grin et al. 2011
Actor	Audet 2014; Bakker 2014; Bergek et al. 2015; Bork et al. 2015; Geels 2002, 2005, 2011, 2013, 2014; Kern 2015; King 2008; Klitkou et al. 2015
Collective action	Fudge et al. 2016; King 2008
Niche action	Fudge et al. 2016
Stakeholder	King 2008; Wiek et al. 2012
Niche actor	Geels 2002, 2005, 2011, 2013, 2014
New entrant	Klitkou et al. 2015
Change agent	Wittmayer and Schäpke 2014
Private actor	Rothenberg and Levy 2012
Institutional entrepreneur	Rothenberg and Levy 2012
Niche	Seyfang and Longhurst 2013
Civil society	Seyfang and Longhurst 2013
Social actor	Whetten and Mackey 2002
Social agent	Wiek et al. 2012
Micro	Åm 2015

What Is Known on Agency in Sustainability Transitions?

In this part of the literature review, the focus shifts to appreciating and summarizing the main findings on agency in sustainability transitions. The findings are presented with respect to (1) who is an agent, (2) individual versus collective forms of agency, and (3) how agents affect sustainability transitions. In the latter category, the power struggle between active agents on the one hand and incumbent agents on the other hand takes center stage.

Who Is an Agent?

Extant literature considers that agency can be enacted at all systems levels in the multi-level perspective (Åm 2015). Geels (2011) argues that agency is apparent throughout the MLP. This effect is apparent in two ways. For one, trajectories and multi-level alignments are de facto constituted by social groups. This means that the pathways towards sustainability transition are ultimately dependent on social interactions, which occur throughout and across the different levels of the MLP. For another, different structural levels are continuously reproduced and adapted by agents. In other words, agency is implicitly and explicitly embedded into the MLP perspective.

All the while, the focus of agency research in sustainability transitions has largely retained a focus at the niche level (Geels 2011; Upham et al. 2015). Taking a look at exemplars of this research tradition, it can be observed that some have equated sustainability agency with resilience. Such agency is helpful in disaster risk reduction, e.g., when faced with societal sustainability threats (Larsen et al. 2011). Expressions of agency, such as green consumers or civil society, have also been related to the niche level. Thus, the question of sustainable consumption is widely acknowledged. Green consumers are described as goal-oriented agents and influential market actors, who use their purchasing power to bring forth social change by taking into account the environmental consequences of their consumption patterns (Moisander 2001; Autio et al. 2009).

In addition to proactive sustainability-driving agency, also incumbent agency exists. Incumbents refer to members representing the vested interests of the prevailing regime. Challenging the status-quo results in resistance from the current regime, i.e., the incumbent agents (Markard et al. 2012). At the level of governing systems, the existence and the role of stakeholder agency has also been acknowledged (Larsen et al. 2011). Furthermore, agency has been recognized as part of policy making. In this literature, the role of agency has been suggested to be significant (Clapp 1998; Haufler 1998; Ehrlich 2006; Rothenberg and Levy 2012), though in practice neglected in academic research and practice.

In summary, the role of agents in sustainability transitions is acknowledged, but an appreciation of their influence mechanisms remains limited. Moreover, the ways in which agents appear and act at different system levels remains understudied. Since the current knowledge related to agency mainly focuses on the niche level, the interplay between agents at different levels of MLP, and especially the role of regime and landscape agents, needs more attention.

Individual Versus Collective Forms of Agency

In addition to merely representing an individual's capability to act, agency can develop into collective action (Seyfang and Smith 2007). This literature emphasizes the agents' capability to influence one other and thus evolve from individual to collective action. This, by its sheer force, poses a greater threat to the current system than the micro-level action of lone individuals.

As an example, amidst the early green movements, broad-scale social change was seen as deriving from collective action (Gabriel and Lang 1995; Jamison 2001; Autio et al. 2009). Achieving sustainability benefits happens often through small community-based initiatives, since they typically utilize contextualized local knowledge (Seyfang and Smith 2007). The pursuit of sustainability transitions needs to be integrated with the contextual knowledge of consumers and local communities (Grin et al. 2011).

Change agency can support and empower local communities to shape a process that suits their purposes and allows them to place their own concerns on the sustainability agenda (Wittmayer and Schäpke 2014). The question of why and how some actors are able to facilitate change, and others not, deserves attention.

As agents are capable of influencing one another, the overall effect of agency needs to be considered within the remit of complex system dynamics, which themselves are formed by social interaction (Mercure et al. 2016). Such system social dynamics and their link to agency need further attention, in order to advance the understanding of emerging bottom-up sustainability transitions that grow from individual actions toward collective action.

Taking the opposite perspective, agency can also be derived from collective behavior. Thus, certain types of civic environmentalist group operate upon the collectivist assumption that the group/state is the agent of change/status quo (Corry and Jørgensen 2015). Agency is in this perspective considered as acts that are reflecting agents' values and norms (Hansen et al. 2015). Consequently, by finding other actors with similar values and norms, individual agency can develop into collective agency.

Instead of concentrating on one specific agent and the resulting "natural" limits to agency, the question of whether and how agents are able to connect needs more attention (Grin et al. 2011). Collective action is considered a promising, but underresearched, area in the realm of sustainability transitions (Seyfang and Smith 2007; Seyfang and Longhurst 2013). To conclude, agents' collective action bears potential in achieving large-scale sustainability transitions. However, the question of what kinds of processes drive collective action remains largely unknown.

Agency Enabling Sustainability and Societal Transitions

The potential of agents lies in their capability to shape and challenge the prevailing, currently unsustainable regime. Based on the literature review, it is observed that agents are crucial components of sustainability transitions. This section explores what is known on their role in this regard in extant scholarly literature.

Niche Versus Incumbent Actors

As noted earlier, the role of agency in sociotechnical transitions and its treatment in the literature have been widely debated. Several scholars have argued that the role of agency tends to be underrated in the multilevel perspective (Smith et al. 2005; Shove and Walker 2007; Genus and Coles 2008). During the recent years, researchers have attempted to fill this gap with analyses of actor strategies in transitions (Farla et al. 2012). These analyses attempt to cover both sides of the coin: agents supporting transitions (Markard and Truffer 2008; Musiolik and Markard 2011; Budde et al. 2012) as well as incumbents hindering transitions (Smink et al. 2015; Geels 2014).

Niche agency is considered essential for sustainability transitions, given that it bears the potential for systemic change and radical innovations (Geels 2011). Niche agents are often referred as engaged individuals driving societal innovation. The hope of niche agents is that the innovations they introduce will find their place in the existing regime, perhaps even replacing the latter (Geels 2011). However, incumbents bear a remarkable advantage over new agents, owing to being more widely acknowledged and enjoying legitimacy in the current system (Klitkou et al. 2015).

However, making a clear-cut divide and distinction between supporters and opponents of sustainability transitions is tricky. This distinction relies on the dichotomy between old systems and incumbent actors on the one hand, and newly emerging systems and proactive niche actors on the other (Bakker 2014). Yet, it needs to be remembered that different types of actors can enact many kinds of roles. For instance, incumbent regime actors may display ambivalent strategies (Bakker 2014). In practice, incumbent agents are constrained by parameters from the existing regime. Sustainability transitions enacted by incumbents have been found to follow trajectories set by the current regime, thereby evolving through incremental innovation (Geels and Schot 2007). It thus appears that there is potential for proactive agency also amid incumbent actors, though this agency is more incremental in nature in contrast with the more radical nature of niche agents. The present literature, however, seems not to have fully tapped into this perspective. The role of incumbent agents in contributing toward sustainability needs to be recognized.

Active Agent Strategies
A key conundrum in the literature relates to whether agency can influence the prevailing system (Bergek et al. 2015.) Several studies suggest that by empowering ordinary people and communities, agency might be the most effective means of creating sustainable futures (Walker et al. 2010; Fudge et al. 2016).

In the MLP paradigm, it is acknowledged that agents are able to introduce transitions outside of the prevailing regime. In particular, it has been observed that discursive activities at regime and niche levels result in cultural repertoires and changes at the landscape level (Geels and Schot 2007; Geels and Verhees 2011; Geels 2011). Agents have further been found to be capable of influencing the speed at which transitions occur. According to the sociotechnical approach, agency is considered to affect how and how fast a particular transition develops. In the complex systems approach, agents are able to utilize and create windows of opportunity for enabling transitions (Grin et al. 2011). Effective sustainable transitions depend on agency driving niche-level innovations, implementing regime-level changes, and connecting niche and regime levels (Grin et al. 2011). Essential for regime shaping agency is to understand the opportunities and limitations implied by the prevailing context, and the ability to expand one's span of agency by positioning oneself broader in space and time (Grin et al. 2011). The encouraging finding from extant research is that agency can make a difference.

The actions of agents are targeted toward replacing the prevailing regime. In order to challenge this regime, new innovations have to achieve legitimacy (Bork et al. 2015; Haxeltine and Seyfang 2009). Legitimacy is achieved by surpassing resistance to change. Understanding incumbent agents' but also niche agents' behavior, actions, and strategies is crucial to overcome the resistance toward large-scale transitions (Gazheli et al. 2012).

The current regime embodies power: the rules, resources, and actor configurations, which are part of the regime, privilege particular practices over others (Grin et al. 2011). Whereas the incumbent regime uses its power to create resistance towards change, regime changes eventually lead to changes in power relations (Grin et al.

2011). For regime-shaping agents, the challenge is to create mutually reinforcing transition and political dynamics that start to destabilize the prevailing power dynamics between incumbent and sustainable practices – this, over time, can bear a destabilizing effect on the established regime (Grin et al. 2011). Such a destabilization process may emerge through common visions or through the graduate, self-reinforcing structuring of (agentic) practices (Grin et al. 2011). All the while, the power balance and dynamics between agents and incumbents and how this leads to shifting power relations and new forms of legitimacy are not yet fully understood (Grin et al. 2011). One reason for this might be that agents are often considered merely as tools, instead of social beings. The social processes and dynamics of agentic activity might have considerable influence on the effectiveness of a transition. Presently, such processes remain underresearched.

Incumbent Strategies

Since agents bear the potential to transform the current regime, the existing regime is likely to react. Notwithstanding, the literature has acknowledged the confrontation between agents and incumbents. The typical question related to the early stages of institutionalization is thus the interplay between those who try to create change, agents, and those who oppose it, incumbents (Delbridge and Edwards 2008). As noted earlier, in transitions that also include broad systemic changes, there are multiple forms of agents that can act collectively (Farla et al. 2012). However, the involvement of several different kinds of agents may result in conflicts, since they might possess ambivalent interests and motivations (Coenen et al. 2012). Conflict can thus occur both at the level of active actors as well as between agents and incumbents.

As transitions challenge and ultimately aim to replace the existing regime, considerable resistance from incumbent agents can be expected (Geels 2005; Markard et al. 2012). Resistance may be passive, resulting from existing institutions that exclude the new, emerging system, or it may be active resistance from incumbent agents protecting their vested interests granted by the existing regime (Smink et al. 2013; Geels 2014; Bakker 2014). The strategies of incumbent agents are thus typically directed at delegitimizing the entering agency category (Geels and Verhees 2011; Bork et al. 2015).

Previous literature has primarily highlighted the incumbents' active resistance to change, but in reality the situation might be more complex (Bergek et al. 2015). Hence, instead of confrontation, the potential for synergies between agents and incumbents need consideration (Haley 2015; Bergek et al. 2015). For example, it has been observed that the opponents and proponents of transitions adjust their discursive framings to increase the salience of expression or discourse along five dimensions: actor credibility, empirical fit, centrality, experiential commensurability, and macro-cultural resonance (Geels and Verhees 2011; Geels 2011).

Many, often unsustainable, systems are rigid and filled with lock-in mechanisms (Geels 2011). A stable incumbent regime is the outcome of various lock-in processes, which become reinforced against novel innovations (Klitkou et al. 2015). Incumbents' institutional commitments, shared beliefs and discourses, power relations, and political lobbying reinforce the existing system (Unruh 2000; Geels 2011). Since lock-in mechanisms reinforce a certain pathway of economic, technological, industrial, and institutional development, the opportunity of upscaling a given niche depends on the characteristics of the regime in question (Klitkou et al. 2015).

At the regime level, incumbents use their power and control the status-quo. This leads niche agents to establish alliances with incumbents in order to conserve their innovations (Rothaermel 2001; Geels and Schot 2007; Bergek et al. 2015). This disparity between incumbents' and niche agents' power relations and available resources typically leads to transitions that are set by parameters of the current system. In order to achieve a truly sustainable system, cooperation between innovation-driving agency and incumbent agency is needed.

As in the MLP paradigm generally speaking, also in the agency-incumbent debate, agency is considered mainly at the niche level. However, it has been observed that incumbent actors may withhold strategies and motivations that overlap with those of niche agents. This finding bears potential for synergies between niche-level agents and regime, or landscape-level, agents. What is more, agency appears in multiple forms and in multiple levels – for this reason, a focus on the niche level is insufficient. More attention is needed to better understand the role of agents that are somewhat between active incumbents and active agents (Bakker 2014). In addition, the motivations of agents and how incumbents' situational experiences influence on their agency are not yet fully understood. It is noteworthy that there is an opportunity for incumbent agents to act differently. If they display ambivalent strategies, then they do withhold the potential to support and enable niche agents.

In closing, these findings suggest that the present understanding on agency is limited and especially the interplay between various agents is still underplayed in literature. In addition, the representation of agency is narrow and the forces driving agency are not yet well understood. To conclude, the most promising finding of the literature review related to agency in sustainability transitions is the agents' capability to make a difference toward transforming the world. However, many factors related to agency remain underresearched. There is a need to develop a holistic perspective on how agency influences sustainability transitions.

Future Research Directions

In addition to the afore-mentioned challenges and key findings, the literature review resulted in identifying four possible areas for future research. These are explored next.

Agency Formation

To begin with, more research is warranted to explore how agency is formed. While agency is known to play an important role in sustainability transitions, the means through which agency is formed remains rather unknown. Further, the processes through which individual agency and collective action emerge and the factors influencing the processes of agency emergence remain unresolved. It is thus critical to appreciate how agency shifts from individual-level action towards collective behavior (King 2008). Different forms of agency and the environments in which they appear further warrant investigation (Kern 2015).

In parallel, in order to understand agency that transforms the status quo, the drivers and motivations of agents need to be looked into (Cashmore and Wejs 2014). Individual circumstances and factors emerging from agents' personal interpretations will influence the transition process (Rothenberg and Levy 2012). Bakker (2014) suggests that more attention is warranted toward understanding agents' rationales to influence emerging sociotechnical systems, since agents likely have individual preferences and pathways toward achieving their desires. This leads to conclude that the formation processes and the motivations of agents, regardless of their type, are essential research pathways in order to develop a more profound understanding of sustainability transitions.

Psychological Dimensions of Agency

The second proposed research direction relates to studying the psychological and social dimensions of agency. Whereas the current literature acknowledges that agents play a crucial role in sustainability transitions, detailed knowledge and integration of the psychological processes involved in sustainability-related agency are largely missing (Geels 2013; Upham et al. 2015). Even though agency is part of the multi-level perspective and the processes by which agency is expressed are increasingly specified (Geels 2014), the psychological dimensions of agency in the MLP are largely unknown (Whitmarsh 2012). What is more, processes of agency in general tend to be neglected (Meadowcroft 2009). An appreciation of the life paths of agents is needed. Extant research thus does not acknowledge who agents actually are in a social or psychological sense.

Diversity in Forms and Dynamics of Agency

The third suggestion for future research is to extend the study of agency to the diversity of agent types and to examine the interplay between these agent types, whether at niche, regime, and landscape levels.

Even though there have been attempts to develop the MLP further and to pay attention to other types of agency, transition theory has focused on macro-level agency, including corporate, technology, and policy actors. Other types of agents, in particular consumers, representing the demand side, are largely neglected (Grin et al. 2011). By simultaneously addressing regime and niche agency, the MLP paradigm could benefit from incorporating insights from management studies, including organization theory and strategic management (Geels 2011). In this regard, the literature on strategic alliances offers relevant insights to the synergies between incumbents and agents driving sustainable transitions (Geels 2011).

Beyond the niche-incumbent continuum, variety with respect to forms and expressions of agency tends to be neglected in the sociotechnical transitions literature, including the multi-level perspective (Whitmarsh 2012). A diverse agent basis is crucial in the diffusion process of innovations, be it with respect to technological innovations or practice-based innovations (Knobloch and Mercure 2016). Nevertheless, the heterogeneity of agency has largely been neglected to date (Mercure et al. 2016). While the diversity of agency has recently been acknowledged, different forms of agency remain unexplored (Cashmore and Wejs 2014).

The variety of agency includes a diversity of representatives inside the current system, and representatives, for example, in the business environment, which could be interpreted as incumbent agents. Whereas their trajectory towards sustainability might differ from niche agents driving novel innovations, they nevertheless contribute to sustainability whether in passive or active roles. The drivers and motivations of incumbent agents are underresearched in the existing literature. By enabling interaction and cooperation between regime and niche agents, sustainability transitions could possibly emerge in a shorter time span. This further explains why an understanding of agency also at regime and landscape levels is needed.

Representations of Agents Are Narrow

The fourth research direction calls for research on the multiple representations of agency. Taking a critical stance, most of the reviewed literature focuses on the roles of agents from an outcome perspective, rather than from a motivation or power perspective. In the existing literature, agents are seen as components that have the potential to shape the status-quo and to enable sustainability transitions. Therefore, current representations of agents are relatively functionalist and narrow. Power dynamics and motivational factors of agents might bear a significant influence on sustainability transitions. Hence, they should no longer remain in the dark.

Even though the use of the term agency can sometimes be inconsistent and the knowledge about agency remains scattered across the discipline, the literature does not question the role of agency in sustainability transitions. Past literature has emphasized agency at the niche level. However, an appreciation of how agency appears across the different levels of the multi-level perspective remains absent. In addition, an understanding of how agents' motivation and power varies depending on the location within the system is lacking. Therefore, more research is warranted to address the representations, motivations, rationales, locations, and processes of agency across different locations of the multi-level perspective. In addition, beyond a focus on the multi-level framework, more research on agency is needed across the discipline of sustainability transitions. To conclude, holistic representations of agency are needed to develop a comprehensive understanding of sustainability transitions.

Table 3 What is known and what is not yet known

What is known: Who is an agent? – A tool for transition	What is not yet known: Who is an agent? – A social being
Agency exists	What are agents' rationales, life paths, psychological processes, and motivations?
Agency can grow from individual action to collective action	What are the processes behind the formation of agency at (1) the individual level and at (2) the collective level?
All levels of MLP have agency	Research concentrates on the niche level. Regime and landscape levels have agency too. What other kinds of agency is there? How could different agent types, particularly across the niche-regime-landscape levels, create synergy?
Agency drives or hinders sustainability transitions	The representation of agency is narrow; agents may have ambivalent strategies. What are the rationales of agents? What are the intertwined dynamics of proactive versus hindering agency? What kinds of power dynamics are at play?
Agents are seen as outcomes. They are known to be crucial parts of sustainability transition	Agents are critical toward sustainability transitions but what are their motivations? How agency is formed?

Summary of the Research Trajectories

Based on the literature review, it appears that the study of agency in sustainability transitions is a field calling for extensive research endeavors. Further, it seems that such research endeavors can support societal transitions toward better futures.

A number of questions remain unresolved. Table 3 provides a summative overview of the findings of the literature review in terms of what is known versus what is not yet known. In short, research is warranted to appreciate the processes, motivations, diversity, and representations of agency at the different levels of the multi-level perspective. Beyond a focus on the multi-level framework, more research on agency is needed across the discipline of sustainability studies and transitions. The motivations and rationales of agents were observed to be a theme cutting across the four identified research directions. This implies that in order to (1) to achieve a thorough understanding of agency and (2) sustainability transitions, it is critical to appreciate the motivations and rationales of agents. Are you an agent, and what motivates you?

Implications for Activists

In this section, the focus shifts onto the practical implications emerging from the literature review for environmental and sustainability activists.

The encouraging finding from the literature review is that agents, or individuals, matter. The literature emphasizes agents as the introducers and drivers of sustainability transitions. Active agents have been found to be able to exert pressure on the current system, thereby shaping the existing regime.

Sustainability activists may play a critical role as agents of sustainable change. Activists are typically deeply engaged to their agenda. Also, their resilience towards challenges is high. Hence the role of activists' in sustainability transitions is non-negligible, and might even emerge as system critical. Whereas sustainability activists are individually engaged toward the sustainability agenda, they are also capable of creating collective action. They inspire likeminded people to join them for the cause of creating a better future. As stated in the literature, collective action can be a solution for large-scale sustainability transitions. Activists that act as leaders of collective sustainability action can create large-scale leverage on the sustainability agenda. For example, many nongovernmental organizations (NGOs) act as sustainability intermediaries by exerting pressure on unsustainable business environments and governing systems. While activists and collective action may appear in several forms, every action towards sustainability matters. Such active agency is needed for several reasons: First, owing to the activists' overt role in acting toward a more sustainable future. Second, to inspire like-minded agents. Third, to awaken "sleeping agents," i.e., individuals that do not yet understand their potential for sustainable agency.

To conclude, different activists and activist groups can act as sustainability agents, in so doing facilitating the processes of sustainability transitions. The findings from this chapter encourage activists to pursue their actions and, where possible, to inspire others to join their agenda. No action is too small in contributing to sustainable futures. The role of activists as system catalyzing change agents is nonnegligible.

There Is Only One Planet: A Call for Action for Everyone

One of the underlying aims of this chapter was to make an impact on ordinary people via an appreciation of research-led findings on sustainable change agency. The assumption was that by communicating research findings, individuals can be empowered towards engaged sustainability and sustainable actions regardless of the role that they find themselves in, be it as citizens, consumers, commuters, and/or decision-makers. The final section of this chapter is thus a call for action for individuals, regardless of their location, profession, and identity. The time is now. There is only one planet.

Perhaps the singularly most inspiring message of this chapter is that every sustainable act, each sustainability movement, each consumption decision matters. Put bluntly, all forms and representations of agency matter. The characteristic of human nature is to bear agency. It is up to each individual to decide how this agency is enacted – whether by passively conforming to societal standards, including unsustainable or unethical ones, or by acting proactively toward creating a better future.

To begin with, everyone can be the agent of one's life. This encourages individuals to connect with and to cherish the agency within (Wood et al. 2018; Krogman 2018). This echoes the principles of personal development, as found in philosophical traditions across cultures. Further, individuals can recognize that each of their thoughts, actions, and behaviors has an impact on the surrounding systems, whether this action is explicit or implicit. Becoming aware and conscious of one's impact on the surrounding micro-environment is a first step in developing one's agency. Such impacts are visible in one's family environment, but also with one's peers be it at work or in social and societal situations at large (Schiele 2017; Zeiss 2018). The acts of agency may be, for example, moving towards paperless offices or to choosing vegetarian lunch (Dhiman 2018). Or just being nice to random colleagues or passers-by.

While ordinary citizens, you and me, might not change their lifestyles immediately, the continuously growing interest and demand for sustainable living environments and solutions is acting as a snowball effect that can shape the current regime toward greater degrees of sustainability. As an example to this end, vegan and healthy eating habits have increased considerably in the last years, resulting in leading retailers having to broaden their supply in order to cater for this customer base.

Individuals further have the potential to proactively express society-shaping agency, be it in professional or personal life, via voluntary work or peer conversations. In addition to individual agency, collective agency is also needed. Numerous social movements are active locally and internationally. There are a myriad of ways of connecting to like-minded peers be it via social media platforms or in person. Ordinary citizens can create greater leverage by joining together and generating new sustainable collective actions. The leverage offered by social media platforms is nonnegligible in this respect. These implications can also be used in educating future generations. Education might be one of the most potent forms of igniting sustainable agency.

What is your next move? Will you take hummus or steak for lunch today? Ride a bike or car to work tomorrow? Fly to long-haul Costa Rica or enjoy local holidays? Keep enjoying long showers or save water? Smile to your neighbor and an unknown passer-by? Reply to an emotional punch with empathy? The possibilities are unlimited. Such acts of agency can gradually develop into routines that may lead to sustainable practices. To conclude, there is perhaps no need to despair, once the role of everyday action and individuals become actively recognized. People do hold the key for a better future. What is your next move? How will you act on the sustainability landscape?

Conclusion

The aim of this chapter was to study how agency appears in the current literature of sustainability and environmental management. This chapter provided a multi-disciplinary literature review, which focused on the role of agency in sustainability transitions and especially on the multi-level perspective (MLP). The focus was on agency, since agency is a key component in successful sustainability transitions, and hence also in the pursuit of sustainable societies and environments.

This chapter contributes to the sustainability literature by bringing together the scattered knowledge related to agency. The most invigorating finding from this review of literature is that agency truly matters. Agents are acknowledged to introduce and drive sustainability transitions, and they have a crucial role in the trajectory towards sustainability. Agency is also known to evolve from individual to collective agency. In addition, as agency appears throughout the multi-level perspective, sustainability agents are positioned at every level of the society. The question then is, who is active and who is passive, in enacting their agency?

Secondly, this chapter contributed to the sustainability transitions literature by identifying several gaps in the current body of knowledge. The review revealed that, overall, agency remains neglected in the sustainability literature. Whereas agency is acknowledged, the representation of agency is narrow. Agents are seen as tools for sustainability transitions rather than individuals that possess individual rationales and aspirations. The life paths, motivations, and other psychological processes of agents remain currently underresearched. The knowledge on how agency is formed and how it evolves toward collective action is missing. Moreover, even though all levels of MLP have agency, the research has concentrated on niche agency. In order to further the understanding of sustainability transitions, the interplay between niche and regime, or landscape, agents needs to be addressed.

Based on the review, the chapter ended with important practical implications for activists on the one hand, and all citizens on the other hand. The message from the chapter is clear – agents, or individuals, matter. What kind of an agent are you, and how is it that you are contributing to the betterment of the planet through your daily actions? The future is here, and the choice to act is yours. As authors, we argue for the need to awaken and energize agency across the planet.

Cross-References

▶ From Environmental Awareness to Sustainable Practices
▶ Social License to Operate (SLO)
▶ Sustainable Higher Education Teaching Approaches
▶ To Eat or Not to Eat Meat
▶ Utilizing Gamification to Promote Sustainable Practices

References

Åm, H. (2015). The sun also rises in Norway: Solar scientists as transition actors. *Environmental Innovation and Societal Transitions, 16*, 142–153.
Archer, M. (1995). *Realist social theory: The morphogenetic approach*. Cambridge University Press, United Kingdom. 368 p. ISBN 0521484421.
Audet, R. (2014). The double hermeneutic of sustainability transitions. *Environmental Innovation and Societal Transitions, 11*, 46–49.
Autio, M., Heiskanen, E., & Heinonen, V. (2009). Narratives of 'green' consumers – The antihero, the environmental hero and the anarchist. *Journal of Consumer Behaviour, 8*, 40–53.
Bakker, S. (2014). Actor rationales in sustainability transitions – Interests and expectations regarding electric vehicle recharging. *Environmental Innovation and Societal Transitions, 13*, 60–74.

Bergek, A., Hekkert, M., Jacobsson, S., Markard, J., Sandén, B., & Truffer, B. (2015). Technological innovation systems in contexts: Conceptualizing contextual structures and interaction dynamics. *Environmental Innovation and Societal Transitions, 16*, 51–64.

Bork, S., Schoormansb, J., Silvester, S., & Joored, P. (2015). How actors can influence the legitimation of new consumer product categories: A theoretical framework. *Environmental Innovation and Societal Transitions, 16*, 36–50.

Budde, B., Alkemade, F., & Weber, K. (2012). Expectations as a key to understanding actor strategies in the field of fuel cell and hydrogen vehicles. *Technological Forecasting and Social Change, 79*, 1072–1083.

Cashmore, M., & Wejs, A. (2014). Constructing legitimacy for climate change planning: A study of local government in Denmark. *Global Environmental Change, 24*, 203–212.

Clapp, J. (1998). The privatization of global environmental governance: ISO 14000 and the developing world. *Global Governance, 4*, 295–316.

Coenen, L., Benneworth, P., & Truffer, B. (2012). Toward a spatial perspective on sustainability transitions. *Research Policy, 41*(6), 968–979.

Corry, O., & Jørgensen, D. (2015). Beyond 'deniers' and 'believers': Towards a map of the politics of climate change. *Global Environmental Change, 32*, 165–174.

David A. Whetten, Alison Mackey, (2002) A Social Actor Conception of Organizational Identity and Its Implications for the Study of Organizational Reputation. *Business & Society, 41*(4):393–414

Dean, W., Sharkey, J., & Johnson, C. (2016). The possibilities and limits of personal agency. *Food, Culture & Society, An International Journal of Multidisciplinary Research, 19*(1), 129–149.

Delbridge, R., & Edwards, T. (2008). Challenging conventions: Roles and processes during non-isomorphic institutional change. *Human Relations; Studies Towards the Integration of the Social Sciences, 61*(3), 299–325.

Dhiman, S. (2018). To eat or not to eat meat: Striking at the root of global warming, in: Dhiman, S. & Marques, J. (Eds.), "Handbook of Engaged Sustainability". Springer International Publishing, Switzerland. ISBN 978-3-319-71312-0.

Ehrlich, P. (2006). Environmental science input to public policy. *Social Research, 73*(3), 915–948.

Farla, J., Markard, J., Raven, R., & Coenen, L. (2012). Sustainability transitions in the making: A closer look at actors, strategies and resources. *Technological Forecasting and Social Change, 79*, 991–998.

Fudge, S., Peters, M., & Woodman, B. (2016). Local authorities as niche actors: The case of energy governance in the UK. *Environmental Innovation and Societal Transitions, 18*, 1–17.

Gabriel, Y., & Lang, T. (1995). *The unmanageable consumer, Contemporary consumption and its fragmentation*. London: Sage Publications.

Gazheli A., Antal, M. & Bergh, J. (2012). *Behavioral foundations of sustainability transitions*. WWW for Europe, European Commission.

Geels, F. (2002). Technological transitions as evolutionary reconfiguration processes: A multi-level perspective and a case-study. *Research Policy, 31*, 1257–1274.

Geels, F. (2005). Processes and patterns in transitions and system innovations: Refining the co-evolutionary multi-level perspective. *Technological Forecasting and Social Change, 72*(6), 681–696.

Geels, F. (2011). The multi-level perspective on sustainability transitions: Responses to seven criticisms. *Environmental Innovation and Societal Transitions, 1*, 24–40.

Geels, F. (2013). The impact of the financial–economic crisis on sustainability transitions: Financial investment, governance and public discourse. *Environmental Innovation and Societal Transitions, 6*(0), 67–95.

Geels, F. (2014). Regime resistance against low-carbon transitions: Introducing politics and power into the multi-level perspective. *Theory Culture & Society, 31*, 21–40.

Geels, F., & Schot, J. (2007). Typology of sociotechnical transition pathways. *Research Policy, 36*(3), 399–417.

Geels, F., & Verhees, B. (2011). Cultural legitimacy and framing struggles in innovation journeys: A cultural-performative perspective and a case study of Dutch nuclear energy (1945-1986). *Technological Forecasting and Social Change, 78*(6), 910–930.

Genus, A., & Coles, A. (2008). Rethinking the multi-level perspective of technological transitions. *Research Policy, 37*, 1436–1445.

Giddens, A. (1984). *The constitution of society: Outline of the Theory of Structuration*, illustrated reprint. Polity Press. 402 p. ISBN 9780745600062.

Grin, J., Rotmans, J., & Schot, J. (2011). On patterns and agency in transition dynamics: Some key insights from the KSI programme. *Environmental Innovation and Societal Transitions, 1*, 76–81.

Haley, B. (2015). Low-carbon innovation from a hydroelectric base: The case of electric vehicles in Québec. *Environmental Innovation and Societal Transitions, 14*, 5–25.

Hansen, M., Faran, T., & O'Byrne, D. (2015). The best laid plans: Using the capability approach to assess neoliberal conservation in South Africa—The Case of the iSimangaliso Wetland Park. *Journal of Environment & Development, 24*(4), 395–417.

Haufler, V. (1998). *Policy trade-offs and industry choice: Hedging your bets on global climate change*. Boston: American Political Science Association Annual Meeting.

Haxeltine, A. & Seyfang, G. (2009). *Transitions for the people: Theory and practice of 'transition' and 'resilience' in UK's Transition Movement*. Tyndall Centre for Climate Change Research, Norwich. Working Paper 134. Available at: http://www.tyndall.ac.uk/sites/default/files/twp134.pdf. Accessed 1 Mar 2017.

Heijden, A., van der Cramer, J., & Driessen, P. (2012). Change agent sensemaking for sustainability in a multinational subsidiary. *Journal of Organizational Change Management, 25*(4), 535–559.

Jamison, A. (2001). *The making of green knowledge*, Environmental politics and cultural transformation. Cambridge: Cambridge University Press.

Kern, F. (2015). Engaging with the politics, agency and structures in the technological innovation systems approach. *Environmental Innovation and Societal Transitions, 16*, 67–69.

King, B. (2008). A social movement perspective of stakeholder collective action and influence. *Business & Society, 47*(1), 21–49.

Klitkou, A., Bolwig, S., Hansem, T., & Wessberg, N. (2015). The role of lock-in mechanisms in transition processes: The case of energy for road transport. *Environmental Innovation and Societal Transitions, 16*, 22–37.

Knobloch, F., & Mercure, J.-F. (2016). The behavioural aspect of green technology investments: A general positive model in the context of heterogeneous agents. *Environmental Innovation and Societal Transitions, 21*, 39–55.

Krogman, N. (2018). Moving from the Individual to the collective approach to sustainability: Expanding and deepening sustainability curriculum across the disciplines in Higher Education, in: Dhiman, S. & Marques, J. (Eds.), "Handbook of Engaged Sustainability". Springer International Publishing, Switzerland. ISBN 978-3-319-71312-0.

Larsen, R., Calgaro, E., & Thomalla, F. (2011). Governing resilience building in Thailand's tourism-dependent coastal communities: Conceptualising stakeholder agency in social–ecological systems. *Global Environmental Change, 21*, 481–491.

Law, J. (1992). Notes on the theory of the actor-network: Ordering, strategy, and heterogeneity. *Systems Practice, 5*(4), 379–393.

Lehner, M. (2015). Translating sustainability: The role of the retail store. *International Journal of Retail & Distribution Management, 43*(4), 386–402.

Markard, J., & Truffer, B. (2008). Actor-oriented analysis of innovation systems: Exploring micro–meso level linkages in the case of stationary fuel cells. *Technology Analysis & Strategic Management, 20*, 443–464.

Markard, J., Raven, R., & Truffer, B. (2012). Sustainability transitions: An emerging field of research and its prospects. *Research Policy, 41*, 955–967.

Meadowcroft, J. (2009). What about the politics? Sustainable development, transition management, and long term energy transitions. *Policy Science, 42*, 323–340.

Mercure, J.-F., Pollitt, H., Bassi, A., Viñuales, J., & Edwards, N. (2016). Modelling complex systems of heterogeneous agents to better design sustainability transitions policy. *Global Environmental Change, 37*, 102–115.

Moisander, J. (2001). *Representation of green consumerism: A constructionist critique*. Helsinki: Helsinki School of Economics and Business Administration.

Musiolik, J., & Markard, J. (2011). Creating and shaping innovation systems: Formal networks in the innovation system for stationary fuel cells in Germany. *Energy Policy, 39*, 1909–1922.

Rothaermel, F. (2001). Complementary assets, strategic alliances, and the incumbent's advantage: An empirical study of industry and firm effects in the biopharmaceutical industry. *Research Policy, 30*, 1235–1251.

Rothenberg, S., & Levy, D. (2012). Corporate perceptions of climate science: The role of corporate environmental scientists. *Business & Society, 51*(1), 31–61.

Schiele, K. (2018). Utilizing gamification to promote sustainable practices: Making sustainability fun and rewarding, in: Dhiman, S. & Marques, J. (Eds.), "Handbook of Engaged Sustainability". Springer International Publishing, Switzerland. ISBN 978-3-319-71312-0.

Sewell, W., Jr. (2001). *Space in contentious politics*. In R. R. Aminzade (Ed.), *Silence and voice in the study of contentious politics* (pp. 51–88). Cambridge: Cambridge University Press.

Sewell, W., Jr. (1992). A theory of structure: Duality, agency, and transformation. *American Journal of Sociology, 98*, 1–29.

Seyfang, G., & Longhurst, N. (2013). Desperately seeking niches: Grassroots innovations and niche development in the community currency field. *Global Environmental Change, 23*, 881–891.

Seyfang, G., & Smith, A. (2007). Grassroots innovations for sustainable development. Towards a new research and policy agenda. *Environmental Politics, 16*(4), 584–603.

Shove, E., & Walker, G. (2007). CAUTION! Transitions ahead: Politics, practice, and sustainable transition management. *Environment and Planning A, 39*, 763–770.

Smink, M., Hekkert, M., & Negro, S. (2015). Keeping sustainable innovation on a leash? Exploring incumbents' institutional strategies. *Business Strategy and the Environment, 24(2)*, 86–101.

Smith, A., Stirling, A., & Berkhout, F. (2005). The governance of sustainable socio-technical transitions. *Research Policy, 34*, 1491–1510.

Unruh, G. (2000). Understanding carbon lock-in. *Energy Policy, 28*, 817–830.

Upham, P., Lis, A., Riesch, H., & Stankiewicz, P. (2015). Addressing social representations in socio-technical transitions with the case of shale gas. *Environmental Innovation and Societal Transitions, 16*, 120–141.

Walker, G., Devine-Wright, P., Hunter, S., High, H., & Evans, B. (2010). Trust and community: Exploring the meanings, contexts and dynamics of community renewable energy. *Energy Policy, 38*, 2655–2663.

Whitmarsh, L. (2012). How useful is the Multi-Level Perspective for transport and sustainability research? *Journal of Transport Geography, 24*(0), 483–487.

Wiek, A., Ness, B., Schweizer-Ries, P., Brand, F., & Farioli, F. (2012). From complex systems analysis to transformational change: A comparative appraisal of sustainability science projects. *Sustainability Science, 7*(1), 5–24.

Wittmayer, J., & Schäpke, N. (2014). Action, research and participation: Roles of researchers in sustainability transitions. *Sustainability Science, 9*, 483–496.

Wood, M., Carter, A., & Thistlethwaite, J. (2018). Sustainable solutions: Addressing collective action problems through Multi-Stakeholder negotiation, in: Dhiman, S. & Marques, J. (Eds.), "Handbook of Engaged Sustainability". Springer International Publishing, Switzerland. ISBN 978-3-319-71312-0.

Zeiss, R. (2018). The importance of routines for sustainable practices: A case of packaging free shopping, in: Dhiman, S. & Marques, J. (Eds.), "Handbook of Engaged Sustainability". Springer International Publishing, Switzerland. ISBN 978-3-319-71312-0.

Designing Sustainability Reporting Systems to Maximize Dynamic Stakeholder Agility

The Role of CSR

Stephanie Watts

Contents

Introduction	1158
Background: CSR Reporting and Organizational Strategy	1161
Stakeholder Engagement	1161
Stakeholder Agility	1163
Corporate Social Responsibility Reporting: Determinants and Impacts	1164
Dynamic Stakeholder Agility	1166
The Role of the Boundary-Spanning IT Artifact: CSR-Reporting Software	1167
Challenges to Successfully Utilizing the IT Artifact	1169
Electronic Integration of the CSR-Reporting Database with Other Enterprise Systems	1169
Inter-functional Coordination for Boundary Spanning	1171
Shared Transparency Beliefs and Boundary Spanning	1173
Conclusion	1177
Cross-References	1177
References	1178

Abstract

Most large companies file standardized corporate social responsibility (CSR) reports with organizations such as the Global Reporting Initiative and others. The information in these reports is in the public domain once filed, and it informs sustainability analysts' ratings of the reporting company; these ratings in turn affect assessment for inclusion of the company's shares in socially responsible investment funds. Inclusion in socially responsible investment funds generally lowers companies' borrowing costs and hence can increase profits. For this reason, companies' CSR reports are more than just advertising vehicles. The

---

S. Watts (✉)
Susilo Institute for Ethics in a Global Economy, Associate Professor of Information Systems, Questrom School of Business at Boston University, Boston, MA, USA
e-mail: swatts@bu.edu

© Springer International Publishing AG, part of Springer Nature 2018
S. Dhiman, J. Marques (eds.), *Handbook of Engaged Sustainability*,
https://doi.org/10.1007/978-3-319-71312-0_7

databases embedded in the software used to produce these reports are designed most effectively when the different subunits that provide the data and produce the reports share similar norms of organizational transparency. Also, electronic back-end integration of these databases with other company databases company can increase the timeliness and accuracy of the data released in CSR reports. This chapter explains the importance of CSR reporting for effective stakeholder engagement and for organizational agility – the capability of the company to respond most effectively to stakeholders' information needs. For companies to reap the benefits of their sustainability efforts, they need to communicate them effectively. Thus CSR reporting is integral to achieving organizational sustainability benefits, which is why it is important to understand the practices and benefits of effective CSR reporting.

Keywords
Sustainability reporting · CSR reporting · Organizational transparency · Database design · Electronic integration · Stakeholder engagement · Organizational agility · Norms

Introduction

In 2013, 93% of the 250 largest corporations in the Fortune Global 500 filed corporate social responsibility (CSR) reports with formal standards organizations, 82% of these specifically with the Global Reporting Initiative (GRI), and 56% with external auditor assurance (KPMG 2013). This is a significant increase from the 33% of large corporations that filed CSR (also called sustainability) reports in 2005 and which is increasing still (Deloitte 2015). CSR reporting has become the cost of doing business: both Coca-Cola and Pepsi file reports with the GRI, as do UPS and FedEx, Boeing and Lockheed Martin, AT&T and Verizon, Walmart and Target, and Microsoft and IBM. All major industries are represented in the public GRI database of CSR reports: ExxonMobil files GRI reports, as does Dow, Merck, Kellogg, Johnson and Johnson, McDonalds, Citibank, and General Motors, and thousands of others, drawing their competitors and suppliers along with them. The GRI voluntary reporting standard includes up to 91 specific indicators, many requiring quantitative data on operating resource usage, waste and emissions, employee statistics, product impacts, etc., as well as qualitative information on issues such as human rights, corruption, political influence, ecosystem impacts, etc., but reporting firms are not required to report on every indicator. Even so, these reports are very resource intensive to produce, reflecting months of data and information collection, usually by the firm's sustainability function. Importantly, the GRI rewards firms for their transparency, not for their CSR practices, and firms are able to achieve a higher ranking over time if they release successively more data and information annually. Due to the magnitude of this voluntary reporting endeavor, companies use software to aggregate and store their CSR data and information and to produce the actual reports. GRI reporting firms select this software from the list of GRI-approved vendors. This computer software application and its integrated database function as a boundary-spanning information technology (IT) artifact. The boundaries that it spans are those between the sustainability department and the other business units that supply the information needed to populate it on the input side and with the corporate communications department on the output side.

Notably, firms' CSR reports are in the public domain once they have been submitted to the GRI or other standards organizations. They correspond in varying degrees to the illustrated narrative versions of firms' sustainability reports displayed on their websites and produced by their corporate communications departments. Relative to these public relations reports, standardized CSR reports such as those in GRI format tend to be harder to locate online and are of sufficient detail to be daunting to comprehend except by those trained to do so. In many organizations, those in the corporate communications department serve as gatekeepers such that it is their responsibility to vet the standardized CSR report before it is publicly released. This reflects some marketing scholars' view that the marketing department should have a coordinating role in the production of CSR reports (Nikolaeva and Bicho 2011).

The GRI is the world's most widely used voluntary sustainability reporting framework (Levy et al. 2010; Manetti and Becatti 2009; Reynolds and Yuthas 2008) and has become the de facto standard. Because the GRI is designed to reward transparency, a company in a dirty industry, or a firm that is not sustainable yet but has committed to working toward being so, can get a high score from the GRI. In addition to the GRI, firms report to other standards organizations, such as the Carbon Disclosure Project, the United Nations Global Compact, the Dow Jones Sustainability Index, the Sustainability Accounting Standards Board, and others. The software applications used for reporting have multiple templates, making it easy for firms to report to multiple standards organizations once they have collected the necessary data. The widespread availability of CSR reports has created vast quantities of data, data that is easily searched and analyzed because it is submitted in a standardized tagging taxonomy called XBRL. Secondary markets for this data have emerged in the form of sustainability analyst reports and the databases that aggregate them for the SRI community, along with other aggregators serving consumers' and researchers' needs such as CorporateRegister.com, CSRHub.com, and GoodGuide.com.

CSR reporting is important because it is the principal mechanism by which companies' CSR practices are recognized and rewarded by their stakeholders. And the rewards can be significant: Over the long term, sustainability improves firm performance (Eccles et al. 2014; see Albertini (2013) and Orlitzky et al. (2003) for meta-analyses). Sustainability has also been found to increase customer loyalty (e.g., Du et al. 2007), lower firm-idiosyncratic risk (e.g., Luo and Bhattacharya 2009), and attract and retain talent (e.g., Greening and Turban 2000). Importantly, a stream of financial research finds that firms with high sustainability ratings have a lower cost of capital (Dhaliwal et al. 2011; El Ghoul et al. 2011): firms with high sustainability ratings are able to borrow funds for expansion, research and development, etc., at lower interest rates than firms with lower sustainability ratings. In sum, firms are engaging in sustainable practices both because their stakeholders expect it and because they acknowledge it can create organizational value (Klettner et al. 2014), value that depends on the ongoing production of CSR reports.

An important factor underlying the lower cost of capital for sustainable firms is the socially responsible investing (SRI) movement. By 2012, there were $13.6 trillion worth of sustainable investment assets worldwide (GSIA 2012). As of year-end 2011, 49% of all invested assets in the E.U. were invested along a responsible investment strategy, with Africa at 35.2%, Canada at 20.2%, Australia and Zealand at 18%, and the USA at 11.2% (Scholtens 2014). In the USA, $6.57 trillion was invested in SRI funds by the beginning of 2014 (USSIF 2014). Those that manage these funds use the information produced by sustainability analysts – in the form of ratings reports – to determine which companies' stocks to include in their SRI portfolios. And much of the information that sustainability analysts use to create these ratings reports comes from companies' own voluntary CSR reports, especially when they are GRI-formatted and have been audited. High GRI scores read favorably to CSR analysts that produce the sustainability reports that the socially responsible investment (SRI) community relies on to assess firms for inclusion in or exclusion from their financial products. Thus it is not just sustainability practices that affect firms' cost of capital, but sustainability reporting as well. CSR reporting can result in significant premiums in financial markets (Berthelot et al. 2012).

Clearly then, but only very recently, voluntary CSR reporting and the CSR practices they report on are beginning to have a major impact on firms, including resource allocation moves and new conceptions of risk and reputation management. This reflects a shift away from a shareholder primacy perspective toward a stakeholder view of business strategy (Klettner et al. 2014). Because the many benefits that CSR practices can confer to firms depend on the capability to communicate them to stakeholders, stakeholder communications is becoming a critical organizational capability. *Dynamic Stakeholder Agility* is the ability of the firm to *quickly and accurately* meet all stakeholders' information needs and to learn from doing so to increase the likelihood of being able to do so in the future. By optimizing their Dynamic Stakeholder Agility, companies can increase the likelihood of successfully engaging their stakeholders and of maximizing their return on investment in CSR practices and communications. But such agility is challenging to achieve, since the various organizational subunits involved in CSR reporting may differ in how transparent they believe the organization ought to be: different subunits often have different transparency norms. When subunits need to work together to produce information products for the public, differences in this norm can create conflicts that can suboptimize a firm's capability for Dynamic Stakeholder Agility, particularly at the boundary between the sustainability and corporate communications departments. At the same time, the boundary-spanning IT artifact can play an important role in optimizing Dynamic Stakeholder Agility, particularly when it is electronically integrated with other enterprise systems. However, such electronic integration will be difficult to achieve when there are conflicts around relational transparency norms.

Background: CSR Reporting and Organizational Strategy

Academics have two predominant ways of thinking about CSR behavior within the context of corporate strategy. Under the stakeholder view of the firm (Donaldson and Preston 1995; Freeman 1984), all the stakeholders of a company have needs that the company may fulfill when possible: the employees, the community, the environment, etc., not just the shareholders that have purchased company stock. This view stands counter to the nexus of contracts approach (Friedman 1970; Jensen 2002), in which shareholders should maintain financial primacy over other stakeholders, such as employees or the community. In this view, shareholder primacy prevents managers from misappropriating and misallocating funds for purposes of corporate social involvement (Margolis and Walsh 2003). However, as evidence accumulates that CSR practices improve firm performance over the long term and that CSR ratings affect the cost of capital in the short term, the inherent conflict between the stakeholder view and the nexus of contracts approach loses its potency. Further, legal scholars have found no legal precedence for requiring managers to maximize shareholder wealth relative to that of other stakeholders (Stout 2007, 2013), a conclusion recently confirmed by the US Supreme Court (Johnson and Millon 2015).

CSR reporting thus falls within the general framework of stakeholder theory (Freeman 1984; Donaldson and Preston 1995), where an activity is considered strategic to the extent that it is consequential for the strategic outcomes, directions, and survival and competitive advantage of the firm (Johnson et al. 2003). Thus strategy is not what an organization has, but what it does: strategy-as-practice, in this case the practice of agile stakeholder engagement. Despite their routinization, practices are diverse and variable, being combined and altered according to the uses to which they are put (Jarzabkowski et al. 2007). Paradox studies also apply here, since they explore how organizations can attend to competing demands simultaneously, as agile stakeholder engagement necessitates. Although choosing among competing tensions might aid short-term performance, a paradox perspective argues that long-term sustainability requires continuous efforts to meet multiple, divergent demands (Cameron 1986; Lewis 2000). The practices of cyclically responding to paradoxical tensions enable sustainable peak performance (Smith and Lewis 2011). Interacting with stakeholders requires constant learning, but these learnings need to be routinized by organizing. Organizational routines and capabilities seek stability, clarity, and efficiency while also enabling dynamic, flexible, and agile outcomes (e.g., Eisenhardt and Martin 2000; Teece et al. 1997).

Stakeholder Engagement

Stakeholder management is one of the three processes of corporate social responsiveness, along with environmental assessment and issues management (Wood 1991). Strong stakeholder engagement practices are associated with socially responsible companies (Sloan 2009). CSR itself can be conceptualized as corporate social responsiveness, since stakeholder engagement shows responsible commitment to stakeholders (Frederick 1994; Johansen and Nielsen 2012). Because organizations need to demonstrate to stakeholders that they are responding to their expectations, communication plays a central role in stakeholder engagement (Devin and Lane 2014). Stakeholder engagement should determine which information and data should be included in an organization's CSR reporting (Gray 2000; Hartman et al. 2007; Manetti 2011; McWilliams et al. 2006). And it should also indicate which new data and information needs to be collected in newly routinized practices. Most of the research on stakeholder engagement comes from the perspective of signaling theory and the need to create organizational legitimacy. Hence stakeholder engagement has been conceptualized as a process of organizational identity construction that involves dynamic and iterative processes (Scott and Lane 2000). It has also been conceptualized as a mechanism for accountability, for consent, for control, and for cooperation, as a method of enhancing trust, as a substitute for trust, as a form of employee involvement and participation, as a discourse to enhance fairness, and as a mechanism of corporate governance (Devin and Lane 2014; Greenwood 2007). Organizations seek to establish the legitimacy of the engagement process in order to legitimize the organizational decisions that result (Devin and Lane 2014). In practice, this involves engaging stakeholders' awareness of each other's competing and interdependent goals, identifying their shared problems, and motivating them to work collaboratively (Saravanamuthu and Lehman 2013). It can involve community engagement, social media engagement, employee engagement, and CSR engagement.

Historically, interacting with stakeholders has been the responsibility of those in the customer-facing function of the corporate communications department. However, the emergent phenomenon of widespread CSR reporting means that those in the sustainability function of large firms are now also in a customer-facing role when they produce publicly available CSR reports and may engage organizational stakeholders in the process of producing them.

The agility view of stakeholder engagement presented here encompasses practices at the edges of the firm – communicative interactions between stakeholders and the corporate communications department and between stakeholders and the sustainability department. Information learned from these communicative interactions become routinized when it is entered into the boundary-spanning IT artifact, where it then serves as an information source going forward, at the ready to support agile stakeholder responsiveness in the future. In this view stakeholder engagement does more than secure organizational legitimacy; it enables agile authentic response to stakeholders' needs and serves as an ongoing vehicle for organizational learning and innovation. It views stakeholder engagement not as a required response to external demands but as an opportunity to engage with the tensions and paradoxes that move the firm closer to sustainability. An agility perspective on stakeholder engagement suggests three aspects of engagement to be concerned with: the *speed* of response to stakeholder inquiries, the *accuracy* of the response, and, importantly, the *routinization* of new learnings to enable speedy, accurate future responses and to address potential information gatekeeping and bottlenecks.

Stakeholder Agility

Within the organization, the ability to detect changes in the environment and respond to them effectively with speed and surprise is called *agility* (Brown and Eisenhardt 1997; Christensen 1997; D'Aveni 1994; Goldman et al. 1995). The two components of organizational agility – sensing and responding – are key organizational capabilities that contribute to success in highly turbulent environments (Zaheer and Zaheer 1997). Agility underlies firms' success by continually enhancing and redefining their value creation through innovation. Because stakeholders are at the boundaries of the organization, interacting with them generates recombinative knowledge across different domains and practices (Van de Ven et al. 1999). Communication with stakeholders generates novel ideas, and so increases the invention phase of innovation (Garud et al. 2013). Such interaction also supports firm agility, since cross-boundary knowledge movement enables the organization to sense and adapt to rapidly changing environments (Brown and Eisenhardt 1997). Agility and innovativeness are thus interdependent and mediated by effective stakeholder interactions.

Organizations interact with numerous stakeholders that relate to the firm variously in power, legitimacy, and urgency (Mitchell et al. 1997). Setting aside the *shareholder* relations that historically have dominated stakeholder engagement, the two stakeholder groups that have received the most attention for contributing to organizational agility are customers and suppliers. Customer agility is the ability to sense and respond to customers. In addition to providing revenue, customers can be put to work cocreating firm value (Payne et al. 2008; Zwick et al. 2008). Such customer "coproduction" entails the direct involvement of customers as they participate in the design, delivery, and marketing of a firm's goods and services (Schultze and Bhappu 2007). It therefore implies a kind of partnership between the customer and the firm. Many companies have developed customer advisory panels to solicit ideas for new products, have built online customer communities to enable dialogue among customers, and have created toolkits that enable customers and engineers to codesign products (Schultze et al. 2007). However, such advisory panels and communities may not include customers whose participation goals do not align with those of the firm, for example, those representing the needs of the local community. Because the Internet and social media can give voice to disgruntled customers as easily as to supportive ones, customer agility needs to encompass all customers, not just those aligned with the agenda of the firm. By doing so, customer agility supports both risk mitigation and exploitation of customer learning opportunities.

In addition to customer agility, researchers have identified supply chain agility as the capability of the firm, both internally and with its key suppliers, to respond rapidly to marketplace changes and potential and actual supply chain disruptions (Braunscheidel and Suresh 2009). Like customers, suppliers can be engaged in interfirm collaboration for value cocreation (Cheung et al. 2010). Indeed, entire industries have been created around both customer relationship management and supply chain management systems for accomplishing these relationships and coordinating related interfirm processes. But as with customers, not all suppliers' goals are necessarily aligned with those of the firm – some may not engage in CSR practices as specified in their firm contract, or may seek undesirably exclusive contracts, or form coalitions to increase their power relative to the firm. And the Internet and social media increase the likelihood that unethical supplier practices may negatively impact firm reputation and sustainability ratings. Thus supplier agility must encompass the capability for both mitigating risks and exploiting supplier-based learning opportunities.

In addition to customers and suppliers, other stakeholders are coming to the fore and demanding engagement with the firm, both because the Internet and social media are giving them voice and because of generally higher CSR expectations: consumer and community activists, sustainability analysts, corporate coalitions, awards organizations, environmental groups, employees, regulators, and others. These stakeholder groups may provide additional opportunities for agility and innovation but may also present material and intangible risks. Firms need to respond to their requests for information and action in timely and effective ways and ensure that they are exploiting and exploring the learning opportunities that these groups offer; they need to go beyond customer and supplier agility and respond with agility to the full range of their stakeholders, using practices that together comprise Dynamic Stakeholder Agility.

Companies have been engaging their stakeholders for as long as public relations departments have existed. CSR reporting is a newcomer to firms' established routines for engaging their stakeholders.

Corporate Social Responsibility Reporting: Determinants and Impacts

CSR reporting can be conceived of as both an outcome and a part of reputation risk management processes (Hasseldine et al. 2005; Toms 2002). In general, large firms and publicly traded firms engage in more CSR reporting than smaller, privately held ones (Gallo and Christensen 2011), size and ownership being proxies for public pressure (Patten 1995). Similarly, high media exposure has been associated with CSR reporting (e.g., Haddock and Fraser 2008), as has firm financial performance and age of assets (e.g., Cochran and Wood 1984). Other studies have found that pressure from shareholders, regulators, and environmentalists leads to more reporting, with mixed results for the impact of pressure from the media (Neu et al. 1998). Internal determinants of reporting include systematic risk (e.g., Belkaoui and Karpik 1989) and the existence of a CSR committee (e.g., Cowen et al. 1987). Interested readers may want to see Fifka (2013) for a meta-analysis and Hahn and Kuehnen (2013) for a review of this literature. Private companies do CSR reporting to enhance brand value, reputation, and legitimacy, enable benchmarking, signal competitiveness, and motivate employees (Herzig and Schaltegger 2006). The reports themselves can provide early warnings about future mismanagement (Christofi et al. 2012; Bebbinton et al. 2008), be used to manage reputational risk

(Legendre and Coderre 2013; Bebbinton et al. 2008), and reinforce relations between firms and local communities (Alonso-Almeida et al. 2014).

While some studies have found CSR reporting to be associated with strong CSR performance (e.g., Brammer and Pavelin 2008; Eccles et al. 2014; Gelb and Strawser 2001), other have not (see Clarkson et al. 2008). This discrepancy has been attributed to differing assumptions between the two dominant theoretical frameworks used to frame CSR reporting; the economic theory of signaling versus the sociopolitical theory of legitimacy (Nikolaeva and Bicho 2011). According to the economics-based perspective, good CSR performers have more incentive to report their "good" practices, resulting in a positive association between CSR performance and reporting. In contrast, early social accounting researchers cited legitimacy theory to explain that companies that are weaker CSR performers face more legitimacy pressures and so tend to be more active reporters in order to influence stakeholders' perceptions (Brown and Deegan 1998; Clarkson et al. 2008; Deegan 2002; Deegan et al. 2002; Gray 2002; Milne and Patten 2002; O'Donovan 2002).

Because CSR reports are in the public domain, there is reason for those concerned about organizational legitimacy and reputation management to be concerned about their content. Green credentials can harm share prices (Fisher-Vanden and Thorburn 2011; Jacobs et al. 2010), and firms have incentives to exaggerate their environmental performance (i.e., to greenwash) (e.g., Kim and Lyon 2015). This reflects disclosure risk, whether it be because the issue revealed is immaterial to or misinterpreted by stakeholders or because the information released is inaccurate. On the other hand, CSR reporting and the CSR practices they report on are being assessed and rewarded separately, as distinct aspects of the socially responsible investing ecosystem (Watts 2015). Thus firms lagging in CSR practices can get a positive transparency assessment from the GRI, as long as reports include material CSR issues and plans to address them. Thus most large companies are now doing CSR reporting regardless of how mature their CSR practices are. Further, the design of the GRI has evolved to minimize opportunities for gaming it: Quantitative disclosure is rewarded over qualitative wherever possible; full disclosure is rewarded more than partial disclosure; and nondisclosure with explanation about plans for future disclosure is favored over nondisclosure alone. Assurance is rewarded as well. The recently released version of the GRI standard (4.0) places much emphasis on materiality: Corporations are now required to disclose their materiality priorities and assessment processes, so that if negative CSR behaviors are brought to light involving behaviors that have not been reported as being material, this is looked upon very unfavorably by CSR analysts. And because the Internet and social media have increased the likelihood that negative CSR behaviors will be revealed (O'Toole and Bennis 2009), it may be riskier for firms to withhold than to disclose information. Thus CSR-reporting standards have evolved to make it more difficult for weaker CSR performers to use CSR reporting to change stakeholders' perceptions about their CSR activities without actually improving their CSR behavior. The increasing use of external assurance further inhibits this. Thus the economic perspective, that CSR practices and CSR reporting are generally aligned, seems to be overtaking the legitimacy perspective.

Dynamic Stakeholder Agility

Widespread adoption of CSR-reporting practices is a new phenomenon. From the economic perspective, strong CSR *practices* are associated with strong CSR *reporting* (Brammer and Pavelin 2008; Eccles et al. 2014; Gelb and Strawser 2001). And strong CSR practices have recently been found to be positively associated with firm performance in the short (i.e., due to reduced cost of capital (Dhaliwal et al. 2011; El Ghoul et al. 2011)) and long terms (e.g., Eccles et al. 2014). Therefore strong CSR reporting should also be associated with firm performance. To the extent that the capability for Dynamic Stakeholder Agility can engender strong CSR reporting by increasing the speed, accuracy, and routinization of agile stakeholder engagement, this capability can enhance firm performance in the short and long terms. Figure 1 illustrates the logic of the above discussions: The remainder of this chapter describes CSR-reporting practices that occur within the organization, within the dotted lines, comprised of the reciprocal interactions between those doing the CSR reporting in the sustainability function and those doing stakeholder engagement in the corporate communications department. Two categories of marketplace response to *CSR practices* can engender positive returns to firm performance – short-term impacts on capital costs and the bundle of various positive CSR impacts that take longer to manifest. But these rewards to CSR practices cannot be realized without communicating those practices. Thus the practices of dynamic stakeholder agility *mediate* firms' CSR practices, marketplace recognition of these practices, and firm performance.

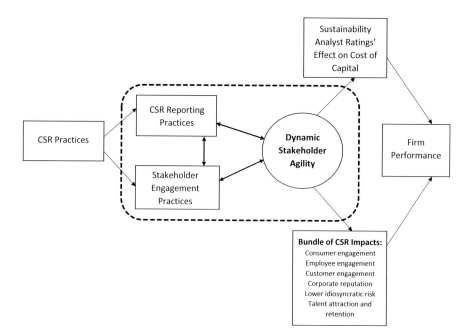

Fig. 1 The context and focus of the research (Watts 2016)

The Role of the Boundary-Spanning IT Artifact: CSR-Reporting Software

Within organizations, sustainability reporting software is used to gather, aggregate, and report the data and information needed to comply with formal reporting standards set by organizations such as the GRI. CSR-reporting software is not designed to *measure* nonfinancial information per se, as the technologies for doing so are still in a relatively nascent state of development (Simnett et al. 2009). They simply support the collection and storage of structured data and information from throughout the organization. For example, they are used to collect the (primarily) quantitative environmental measures of resource usage, waste and emissions from the firm's environmental management systems. They amass labor statistics from the company's human resource function, and supplier statistics from the supply chain managers, or suppliers can enter this data themselves over the Internet. Other data is collected from various other intraorganizational systems such as sustainability project management systems and others. Because the reporting indicators are broadly encompassing, ranging from environmental and labor practices to supply chain contracting and political engagement, the IT artifact supports data gathering from virtually all company functions.

CSR-reporting systems are designed for maximum flexibility, because the standards for reporting are evolving quickly and vary depending on the standards organization, and because newly reporting companies will store much less and data information in them than firms that are at higher levels of disclosure. Their flexible design means that companies do not have to use them to collect any data or information on indicators that they prefer not to. Companies can format fields for collecting data on new indicators that they are not currently reporting on, or on indicators that are not even included in standards' templates, essentially "making up" new indicators. Depending on the software product used, there may be a limit to how granularly the data can be collected: Some have an inherent temporal periodicity of 1 month, such that data for these fields cannot be stored more frequently. Others allow for greater update frequency. This may be due to the legacy of financial accounting or the fact that CSR reports are usually produced annually. But in general, these systems are aggregation systems, designed for responding to requests for trend information derived from quantitative data over time, for example, improvements in emissions or a more diverse workforce. In this way they are platforms of social and technical arrangements that, with the appropriate processes, structures and cultures, can function as generative memories (Garud et al. 2011).

These systems serve three general purposes. First, they store data and information in order to produce current and future CSR reports for standards organizations such as the GRI, usually annually. Second, they store data and information so that those in the corporate communications function have a repository of content to access during stakeholder engagement and to use to respond to stakeholder requests for information. This secondary, unintended function can support corporate communications specialists as they practice ongoing engagement with current and new stakeholders. Third, the system serves as a boundary object that enables knowledge transfer and transformation among all functions involved in producing content for stakeholders: the sustainability and corporate communications functions and also the executive leadership and risk/compliance officers. This reflects the pragmatic view that the materials created and produced from the content in this system, whether CSR report or for another stakeholder interaction, are themselves a new product. For each of the functional units at its boundaries, this IT artifact is localized, embedded, and invested and as such is susceptible to the negative consequences that often arise at problematic knowledge boundaries (Carlile 2002). This artifact functions as a boundary object to enable representation, learning and knowledge transformation to resolve these negative consequences. Because the artifact is also an information technology (IT), it facilitates sharing of information (Fuller et al. 2009), makes the process environment conducive for value cocreation (Aarika-Stenroos and Jaakkola 2012), and can play an important role in enabling organizational agility (Sambamurthy et al. 2003).

Lack of sophisticated information technology systems has been identified as an impediment to successful CSR reporting (Eccles et al. 2014), but the software industry is a very fast-evolving one and the newest CSR-reporting applications are relatively sophisticated. Currently, it is not lack of IT systems, but lack of integration of the CSR-reporting system with other enterprise systems, that is a source of frustration with the information technology used by those in the sustainability function.

Figure 2 illustrates the front office and back office components of CSR reporting and stakeholder engagement as they are mediated by the artifact of the CSR database and reporting system, with the top double-sided arrow indicating a two-way flow of

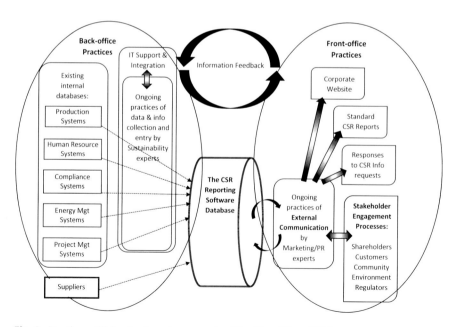

Fig. 2 Practices utilizing the boundary-spanning IT artifact (Watts 2016)

information providing feedback mechanisms between the two realms – front offices to back offices and forward again – that are necessary to move engagement from a legitimacy function at the front edge of the firm deeper into the organization in the form of routinized learning. By making CSR data and information readily available, this artifact can increase the speed and accuracy of the organization's stakeholder responsiveness, increasing Dynamic Stakeholder Agility. And as a boundary object, it can make tacit learnings explicit and overcome inter-functional conflict to support the routinization of new learnings that emerge from the practices of sensing and responding to stakeholders

Challenges to Successfully Utilizing the IT Artifact

The CSR-reporting software application can effectively support Dynamic Stakeholder Agility by increasing the speed and accuracy of responses to stakeholder requests for information. But to do so requires navigating the cross-functional nature of the CSR-report design process, so that communication interactions with stakeholders become routinized in future versions of the database. Part of this navigation process necessitates negotiations regarding transparency norms, norms that are likely to differ between subunits. The collective perceptions of those in a subunit regarding how transparent the organization ought to be (i.e., its transparency norms) will affect how well that subunit is able to work with other subunits at tasks that involve the release of information to the public, such as CSR reporting and stakeholder engagement. Before we discuss that challenge, an obvious means for increasing the speed and accuracy of responses to stakeholder information requests – electronic integration – bears attention.

Electronic Integration of the CSR-Reporting Database with Other Enterprise Systems

Enterprise systems are large-scale application software packages used by organizations to run their operations. No large firm operates without one or more. As the name suggests, the single enterprise software application integrates all or most of the organizational functions that comprise the enterprise, and even many that are external to it as well, such as customer and supplier support. In addition to providing the software to perform each function (e.g., accounting, purchasing, manufacturing, etc.), these systems support information flows between the functions via a single integrated database, which also enables reporting and data analytics. Within each organizational function, viewing and accessing the data in the central database is generally limited to that of the particular function. Enterprise systems enable a high degree of internal integration that allows internal organizational units to work in tandem, as a unified whole, and to be responsive to each other (Barki and Pinsonneault 2005). Both internal and external electronic integration can enhance a firm's agility (Nazir and Pinsonneault 2012).

Enterprise systems rely on a single underlying database in which to store enterprise-wide data collected by its various functional work modules. Standalone systems for each organizational function were common until the 1990s, before enterprise systems were invented to overcome the problems that resulted from storing data across the enterprise in multiple, separate databases. Now firms use enterprise systems to provide a single integrated view of their data, but these systems are not generally integrated with specialized and/or newly developed software applications such as CSR-reporting systems. And because CSR-reporting systems are not electronically integrated with their organization's enterprise system, the data they produce are not available in the central enterprise database, nor can they electronically source data from the central enterprise-wide database.

CSR-reporting software systems are not designed to be electronically integrated with the enterprise system and the other systems that provide the data it needs for reporting. Rather, internal data is gathered from around the firm and entered by hand or through manually controlled downloading and uploading of spreadsheets. This is because different systems from different venders have different data standards and periodicity, such that integration requires customization that is complex and expensive. But this is not to say that they cannot be electronically integrated with the databases that provide their source data. There are various technical options for securing such integration, of various degrees of expense and complexity. But to the extent that they are not electronically integrated, they are not able to realize several of the advantages of enterprise systems related to agility, such as being able to respond rapidly to the needs of those in other organizational units to work in tandem (Barki and Pinsonneault 2005). Where agility requires inter-functional coordination, as it does in this context, internal systems integration can achieve this to positively impact a firm's responsive capability (Roberts and Grover 2012). Automated information flows allow firms to quickly gain knowledge from their environments, and so sense impending changes. The close coordination enabled by internal integration, and access to up-to-date metrics, facilitates the firm's ability to proactively respond to these changes. Thus electronic data integration between back-end source systems and the CSR-reporting artifact can enhance dynamic stakeholder agility by *increasing the speed* by which data can be collected and delivered, with the potential for increasing its periodicity.

Such electronic back-end data integration also supports increased data quality, because data that is untouched by human hands is less prone to errors (Srinivasan et al. 1994) and hence more accurate. Also, by speeding data delivery, electronic data integration increases the likelihood that the data and the information derived from it will be timely (i.e., not arriving too late). It also lowers the cost of data collection, since automation reduces labor costs. In these ways electronic integration of the CSR-reporting IT artifact supports Dynamic Stakeholder Agility by serving data that is accurate and current.

However, not all the data and information required to produce a CSR report is amenable to back-end electronic integration. In the current (G4) GRI reporting standard, 72 of the 91 indicators have at least some quantitative component. Given the expense of electronic integration, it makes little sense to electronically integrate indicators that are primarily qualitative. Further, the periodicity and granularity of the data required for GRI reporting varies. Highly granular, detailed data points are generally aggregated during the integration process, enabling innovation by harnessing rather than reducing complexity (Brown and Eisenhardt 1997; Van de Ven et al. 1999). Integration makes sense for these data points, and for those that are frequently updated, having high periodicity. Quantitative data points that change infrequently can be inexpensively updated manually, although this can introduce errors. And obviously data for which there is no existing source-data information technology in place cannot be electronically integrated. Thus part of the practice of populating the CSR-reporting software artifact is one of prioritizing which data can and should be electronically integrated with back-end systems. Quantitative data that is available in existing organizational systems, that is highly granular, and that changes frequently are most appropriate candidates for back-end electronic integration with the CSR IT artifact.

The IT function is responsible for doing this appropriate data integration and also plays an important role in ensuring the accuracy and security of the data that moves electronically and automatically between source-data systems and the CSR IT artifact. Therefore the IT function can increase stakeholder agility by enabling back-end, appropriate electronic data integration. Since electronic integration can enhance stakeholder agility and the IT function performs the electronic integration, attempts to increase stakeholder agility without the involvement of the IT function are less likely to succeed than those that do involve the IT function.

Inter-functional Coordination for Boundary Spanning

In addition to the need for involving the IT function in the process of producing CSR reports, it is important that those in the corporate communications (CC) function are involved as well, as follows: The CSR-reporting IT artifact consists of functionality for data entry and for producing reports and also a database of organization-specific content reflecting ongoing data collection efforts. This database is a repository of data and information – both current and historical trends – that may be sought by stakeholders and/or helpful for engaging them. This repository can be utilized both by those employees trained in sustainability reporting *and* those in the CC function to communicate with stakeholders. In this way the practice of CSR reporting and the practice of stakeholder engagement come together as those in these two functions utilize this shared boundary object for communicating with stakeholders. Just as customer agility requires inter-functional coordination (Roberts and Grover 2012), so too does dynamic stakeholder agility, specifically between the CSR (or sustainability) function and the CC function.

The CSR-reporting IT artifact is a boundary object, with one boundary at its input side and another at its output side. At the input side of the CSR-reporting system, two design practices must occur. First, the initial data collection and entry processes need to be designed, along with specification of the processes for ongoing data collection and entry. This design practice encompasses what needs to be collected and where it is to be collected from (e.g., which systems, internal and or external, or processes) and how it is to be collected, for example, via electronic integration or a specified process of manual loading. It also needs to specify how often this data is to be updated, with controls to ensure its quality and timeliness. This design process takes place after senior management and risk management officers have done an assessment of the materiality of the risks posed by each of the six GRI aspects, for the particular industry sector. It necessarily involves the sustainability officer, since the result of this process will determine the content of the CSR reports that are subsequently produced and therefore affect the GRI level attained, which in turn affect analysts' sustainability ratings of the firm.

In addition to this initial design process, a second, meta-design activity is needed which specifies how the initial design will be *modified*, and how often, so that the content of the repository can evolve over time. Such ongoing modification enables routinization of emergent information needs in order to evolve sustainability disclosure by ensuring that historical data that will be needed in the future will be collected today. It also supports the future information needs of the CC function, information needs made apparent during current sense-and-respond stakeholder engagements. This is important because those in the CC function engage with stakeholders on an ongoing basis and so are well positioned to learn of current and future information needs that those in the sustainability function might not be aware of. In this way, those in the CC function play a critical role in designing for future information needs. By ensuring that representatives of the CC function are involved in these design practices, new learnings about stakeholders' information needs that emerge during stakeholder engagements can be routinized. In this way information for meeting future stakeholder needs will be readily available in the CSR-reporting IT artifact. Thus it is important to include those in the CC function in the design evolution of CSR data collection and entry practices.

Two similar practices also reside at the output side of the CSR boundary object. The first of these is the ongoing provision of information to stakeholders in active engagement. The second is the meta-design of future stakeholder engagement practices based on data and information that will be available in the CSR-reporting IT artifact in the future, if the organization begins collecting it now. At the *input* side, these two practices are generally under the purview of the sustainability function, while at the *output* side, the two practices are generally the responsibility of the CC function. The IT artifact serves as a boundary object that mediates the interactions of these two functions as they negotiate during the design of both the back-end, input-side practices necessary for routinizing new learnings and the front-end, output-side practices for engaging and communicating with stakeholders. The CSR-reporting IT artifact makes tacit knowledge explicit in this process of negotiation. By demanding both design and meta-design practices at both the input side of the repository and at the output side, the artifact supports organizational learning and consequent agility. Clearly the sustainability and CC functions need to work together to share this boundary object as they use it to design the future information needs of both functions. For this reason, Dynamic Stakeholder Agility requires ongoing design negotiations among those in the sustainability function and those in the CC function.

Certainly senior management and risk compliance officers need to be involved in these negotiations at a high level, and we have made the case above that the IT function should also participate. This is a more nuanced view of CSR reporting as the prerogative of marketing (Nikolaeva and Bicho 2011; Sweeney and Coughlan 2008), since it specifies the need for mutual, inter-functional design practices around the shared boundary object.

Shared Transparency Beliefs and Boundary Spanning

Shared vision is important for effective innovation (Pearce and Ensley 2004) and knowledge sharing (Abrams et al. 2003). Underlying the design practices described above are beliefs regarding how transparent the organization ought to be: To what extent *should* the company release information to the public versus keeping it within the organization? Since both the sustainability function and the CC function engage in innovative and routinized practices that release information to the public, divergent beliefs regarding *how transparent the organization ought to* be may create conflicts that can undermine effective knowledge sharing and innovativeness across this functional boundary.

Sustainability departments are a relatively new phenomenon in organizations, but firms have always released information to the public. In a practice sometimes called corporate social responsiveness, decision-makers work together anticipate, respond to, and manage the external impacts of organizational policies and practices (Epstein 1987; Strand 1983). Corporate social responsiveness has been characterized as a philosophy of response that can variously be reactive, proactive, or interactive (Carroll 1979) and is closely related to organizational "posture." Basu and Palazzo (2008) derive three dominant types of postures from the literature: defensive, tentative, and open. Defensive organizations accept no external feedback, presume their decisions are right, and insulate themselves from alternative sources of inputs. Such organizations might take a tentative posture toward their stakeholders due to inexperience with an issue or because of uncertain consequences. Open-posture organizations take a learning orientation based on willingness to listen and respond to alternative perspectives offered by others. Similarly, Spar and LaMure (2003) identify capitulation, resistance, and preemption as the three dominant ways that corporations react to external criticism. In these views, an organization's posture vis-à-vis stakeholders reflects a routinized approach to interacting with external critics (Basu and Palazzo 2008). However, the posture literature is limited for understanding corporate transparency today in two ways. First, posture is conceived of as a monolithic organizational construct reflecting the decision outcomes of senior management regarding the consistent face that the firm will present to external stakeholders. But, except in cases of a particularly large scandal, corporate transparency as it is practiced today involves the ongoing release of multitudes of small pieces of data and information, from both those in the sustainability function in the form of CSR reports, and from those in the CC function in the form of press releases, podcasts, social media interactions, etc. Second, because so much data and information are being released on a continual basis, and because speed of responsiveness is important for both stakeholder agility and to manage damage, it is unrealistic to conceptualize the process in terms of senior management decision-making. To engage senior management in the myriad of transparency, decisions that occur in large modern corporations would slow the process to the detriment of the organization's capability for stakeholder agility and learning. Thus the literature on organizational posture doesn't adequately characterize practices of organizational transparency in the context of widespread CSR reporting, practices that involve both the sustainability function and the CC function.

Transparency has traditionally been conceived of as a means for making governments accountable to their citizens. In the public sector, it enables public accountability and underlies recent recommendations for governments to regulate companies' transparency rather than their activities (Fung et al. 2008). Indeed, the GRI has evolved as a means for assessing firms' transparency, not their sustainability. Within for-profit organizations, transparency is one of the eight subdimensions of the measure of ethical organizational cultures called the Corporate Ethical Virtues Scale (Kapstein 2008). However, this measure conceives of organizational transparency as taking place within the company (i.e., either because the management is able to physically observe the behaviors of their employees or because the employees are able to observe each other's behaviors), not between the company and its stakeholders. This reflects the view of transparency as accurate observability of an organization's low-level activities, routines, behaviors, output, and performance (Bernstein 2012). But transparency as observability does not describe the transparency of CSR disclosure, which occurs *between* the organization and its external stakeholders. Similarly, Hofstede (1998) has identified open and closed systems within organizations at the function level of analysis. In open-system units, members consider both the organization and its people to be open to newcomers and outsiders. In closed-system units, the organization and its people are closed and secretive, even to insiders. But again, such open or closed systems lie within the organization and not at its boundaries with stakeholders. Or we might look to communication climate, a facet of the broader construct of psychological climate (James and Jones 1974), to shed light on the new transparency. Communication climate has subdimensions of openness and candor, participation in decision-making, and supportiveness (Guzley 1992; Poole 1985). But it too is conceived of as occurring within the organization. Other researchers have highlighted the importance of creating an internal corporate culture of candor (O'Toole and Bennis 2009), advising executives to set information free. But these studies do not address transparency as it is practiced at the organizational edges.

For individuals, relational transparency is one of the four dimensions of authentic leadership, along with internalized moral perspective, self-awareness, and balanced processing (Walumbwa et al. 2008). Relational transparency refers to presenting one's authentic self to others. Such behavior promotes trust through disclosures that involve openly sharing information and expressions of one's true thoughts and feelings (Kernis 2003). Also for individuals, some but not all measures of interpersonal trust include dimensions pertaining to transparency, such as truthfulness

(Gurtman 1992) and openness (Clarck and Payne 1997). And certainly transparency depends on trust, defined as the willingness to be vulnerable (Dean et al. 1998). When a firm trusts that it will not be punished for being transparent, it will be more likely to be transparent. And nondisclosure can be risky too, especially when the marketplace punishes obfuscation. Applying the relational transparency construct to the firm, it follows that firms that are relationally transparent with their stakeholders openly share authentic information with them, and this disclosure promotes trust with them.

Within the firm, some organizational subunits may believe that the organization ought to be very transparent with the public, while other subunits may believe that the organization ought to be less transparent with the public. These beliefs come into play during the design practices described above, such that for some subunits, the default stance will be that information should be freely available to stakeholders, but for others, the default position will be that there is need information gatekeepers to prevent information disclosure to the public. Subunits within the organization vary in the extent that they believe in the need for the firm to behave in relationally transparent ways with stakeholders.

When organizational subunits differ significantly in their beliefs about how relationally transparent their firm ought to be, they lack a shared vision in this regard, which can create conflicts between these subunits during design of organizational disclosure practices and policies such as those described above. In particular, beliefs about the firm's appropriate level of relational transparency held by those in the sustainability function may differ from those in the CC function: while some CC functions may believe in high relational transparency, following prescriptions to address tough issues head on and without greenwashing (Illia et al. 2013), this is a new prescription for public relations. According to agency theory (Eisenhardt 1989), executives have incentive to withhold information from shareholders and managers from employees, since this enables rent extraction and is a source of financial gain. Information obfuscation can raise profits by making customers less informed (Ellison and Ellison 2009). And increasingly, marketing and public relations practices are being called to task for presenting promotional information as independent, unbiased news and information (e.g., Stauber and Rampton 2002). Certainly there is a legitimate need for firms to protect their intellectual capital. Meanwhile, those in the sustainability function are often concerned about corporate externalities that harm the natural environment and society, and to the extent that CSR reporting can minimize these, they are likely to believe that more CSR disclosure is better than less. Their knowledge of the socially responsible investment industry may also make them more likely than those in the CC function to be aware that companies are being rewarded by the financial markets for being transparent as well as for being socially responsible. For these reasons, the sustainability subunit and the CC subunit may not share a vision of the firm's optimal level of relational transparency, making it a challenge for them to collaborate effectively during design of practices that release organizational information to the public.

To the extent that those in the sustainability function believe in higher levels of relational transparency than do those in the CC function, they may view those in a

CC function as barriers to appropriate information release, with consequent conflicts between these two subunits (Markus 1983). As it performs stakeholder engagement, the CC function serves as the ears of the firm when it hears about new stakeholder information needs. These new information needs can be effectively integrated into future design and meta-design of the information gathering processes that support the CSR-reporting IT artifact. However, where design practices are dominated by the sustainability function, and conflict between the two subunits exists, there may be a lack of feedback information from the CC function, impairing the capability for routinizing future stakeholder information needs into the CSR-reporting IT artifact.

At the same time, if these practices are dominated by the CC function, less information may be disclosed on CSR reports, to the potential detriment of the organization's sustainability ratings, ratings that may affect their cost of capital (Dhaliwal et al. 2011; El Ghoul et al. 2011).

Where design practices are dominated by the CC function, there may be a lower likelihood of achieving high sustainability ratings due to lower levels of information disclosure.

These effects may complicate efforts at back-end electronic integration between the source-data systems and the CSR-reporting IT artifact. On the one hand, doing so can increase Dynamic Stakeholder Agility by speeding the data update and delivery process, and by increasing its accuracy and timeliness. However, removing the

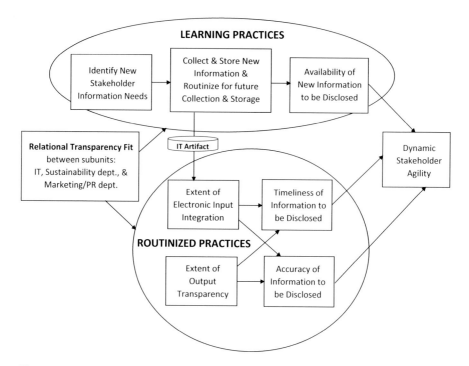

Fig. 3 Achieving dynamic stakeholder agility

human hand from the process of populating the CSR-reporting artifact also removes the option for information gatekeeping at the input side, which may present a threat to those who prefer low levels of firm relational transparency. This reflects the tension in the paradox between the positive effects of back-end electronic data integration and the lack of human control that it represents. Where design practices are dominated by the CC subunit, there may be a lower likelihood that effort will be undertaken to electronically integrate source-data systems with the CSR-reporting IT artifact. Further, the benefits of electronic integration come with the additional need to negotiate with those in the IT subunit. If those in the IT subunit believe that high levels of relational transparency are appropriate and thus share a vision closer to that of those in the sustainability function than the CC function, coalitions may form and have negative consequences for the negotiations necessary for building Dynamic Stakeholder Agility.

Overall then, a shared vision of the appropriate level of relational transparency for the firm should improve communication and reduce conflict between the functions working together to design and practice transparency-related activities such as CSR reporting and stakeholder engagement. Figure 3 depicts the role of shared *relational transparency vision* in achieving Dynamic Stakeholder Agility.

Conclusion

This explains the importance of effective design of CSR-reporting systems in the context of stakeholder engagement and organizational information disclosure practices. It extends the known advantages of organizational agility beyond customers and suppliers to the realm of all stakeholders. The CSR-reporting software application serves an important role as a boundary-spanning artifact, the effectiveness of which can be increased through appropriate back-end electronic integration. In today's companies, information technologies play a crucial role that should not be ignored by sustainability proponents. In designing the repositories embedded in these technologies, multiple organizational subunits need to work together to ensure they meet the information needs of current and future stakeholders. In doing so, interunit differences in transparency norms may impede the negotiations that these design practices entail and may reduce the likelihood of achieving the benefits of electronic integration specifically and CSR reporting more generally. Well-designed CSR-reporting systems can enhance dynamic stakeholder agility, which in turn can positively affect firm performance.

Cross-References

- ▶ Environmental Intrapreneurship for Engaged Sustainability
- ▶ Responsible Investing and Corporate Social Responsibility for Engaged Sustainability
- ▶ Responsible Investing and Environmental Economics
- ▶ Social License to Operate (SLO)

Acknowledgments The author would like to thank the Susilo Institute for Ethics in a Global Economy and its director, Laura Hartman, for the support and encouragement of this project.

References

Aarika-Stenroos, L., & Jaakkola, E. (2012). Value co-creation in knowledge intensive business services: A dyadic perspective on the joint problem solving process. *Industrial Marketing Management, 41*(1), 15–26.

Abrams, L. C., Cross, R., Lesser, E., & Levin, D. Z. (2003). Nurturing interpersonal trust in knowledge-sharing networks. *The Academy of Management Executive, 17*(4), 64–77.

Albertini, E. (2013). Does environmental management improve financial performance? A meta-analytical review. *Organization & Environment, 26*(4), 431–457.

Alonso-Almeida, M., Llach, J., & Marimon, F. (2014). A closer look at the 'Global reporting initiative' sustainability reporting as a tool to implement environmental and social policies: A worldwide sector analysis. *Corporate Social Responsibility and Environmental Management, 21*(6), 318–335.

Barki, H., & Pinsonneault, A. (2005). A model of organizational integration, implementation effort, and performance. *Organization Science, 16*(2), 165–179.

Basu, K., & Palazzo, G. (2008). Corporate social responsibility: A process model of sensemaking. *Academy of Management Review, 33*(1), 122–136.

Bebbington, J., Larrinaga, C., & Moneva, J. M. (2008). Corporate social reporting and reputation risk management. *Accounting, Auditing & Accountability Journal, 21*(3), 337–361.

Belkaoui, A., & Karpik, P. G. (1989). Determinants of the corporate decision to disclose social information. *Accounting, Auditing & Accountability Journal, 2*(1), 223–250.

Bernstein, E. S. (2012). The transparency paradox: A role for privacy in organizational learning and operational control. *Administrative Science Quarterly, 57*(2), 181–216.

Berthelot, A., Coulmont, M., & Serret, V. (2012). Do investors value sustainability reports? A Canadian study. *Corporate Social Responsibility and Environmental Management, 19*, 355–363.

Brammer, S., & Pavelin, S. (2008). Factors influencing the quality of corporate environmental disclosure. *Business Strategy and the Environment, 17*(2), 120–136.

Braunscheidel, M. J., & Suresh, N. C. (2009). The organizational antecedents of a firm's supply chain agility for risk mitigation and response. *Journal of Operations Management, 27*(2), 119–140.

Brown, N., & Deegan, C. (1998). The public disclosure of environmental performance information – A dual test of media agenda setting theory and LT. *Accounting and Business Research, 29*(1), 21–41.

Brown, S. L., & Eisenhardt, K. M. (1997). The art of continuous change: Linking complexity theory and time-paced evolution in relentlessly shifting organizations. *Administrative Science Quarterly, 42*(1), 1–34.

Cameron, K. S. (1986). Effectiveness as paradox – Consensus and conflict in conceptions of organizational-effectiveness. *Management Science, 32*(5), 539–553.

Carlile, P. R. (2002). A pragmatic view of knowledge and boundaries: Boundary objects in new product development. *Organization Science, 13*(4), 442–455.

Carroll, A. B. (1979). A three-dimensional conceptual model of corporate performance. *Academy of Management Review, 4*, 497–505.

Cheung, M.-S., Myers, M. B., & Mentzer, J. T. (2010). Does relationship learning lead to relationship value? A cross-national supply chain investigation. *Journal of Operations Management, 28*(6), 472–487.

Christensen, C. (1997). Patterns in the evolution of product competition. *European Management Journal, 15*(2), 117–127.

Christofi, A., Christofi, P., & Sisaye, S. (2012). Corporate sustainability: Historical development and reporting practices. *Management Research Review, 35*(2), 157–172.

Clarck, M. C., & Payne, R. L. (1997). The nature and structure of workers' trust in management. *Journal of Organizational Behavior, 18*(3), 205–224.

Clarkson, P. M., Li, Y., Richardson, G. D., & Vasvari, F. P. (2008). Revisiting the relation between environmental performance and environmental disclosure: An empirical analysis. *Accounting, Organizations and Society, 33*, 302–327.

Cochran, P. L., & Wood, R. A. (1984). Corporate social responsibility and financial performance. *Academy of management Journal, 27*(1), 42–56.

Cowen, S. S., Ferreri, L. B., & Parker, L. D. (1987). The impact of corporate characteristics on social responsibility disclosure: A typology and frequency-based analysis. *Accounting, Organizations and Society, 12*(2), 111–122.

D'Aveni, R. (1994). *Hyper-competition: Managing the dynamics of strategic maneuvering*. New York: Free Press.

Dean, J. W., Brandes, P., & Dharwadkar, R. (1998). Organizational cynicism. *Academy of Management Review, 23*(2), 341–352.

Deegan, C. (2002). The legitimating effect of social and environmental disclosures – A theoretical foundation. *Accounting, Auditing & Accountability Journal, 15*(3), 281–311.

Deegan, C., Rankin, M., & Tobin, J. (2002). An examination of the corporate social and environmental disclosures of BHP from 1983–1997: A test of legitimacy theory. *Accounting, Auditing & Accountability Journal, 15*(3), 312–343.

Deloitte. (2015). Sustainability practices 2015 – Key findings. Available at: https://www2.deloitte.com/content/dam/Deloitte/us/Documents/center-for-corporate-governance/us-aers-ccg-sustainability-practices-report-the-conference-board-050815.pdf

Devin, B. L., & Lane, A. B. (2014). Communicating engagement in corporate social responsibility: A meta-level construal of engagement. *Journal of Public Relations Research, 26*, 436–454.

Dhaliwal, D. S., Li, O. Z., Tsang, A., & Yang, Y. G. (2011). Voluntary nonfinancial disclosure and the cost of equity capital: The initiation of corporate social responsibility reporting. *The Accounting Review, 86*(1), 59–100.

Donaldson, T., & Preston, L. E. (1995). The stakeholder theory of the corporation: Concepts, evidence and implications. *Academy of Management Review, 20*(1), 65–91.

Du, S., Bhattacharya, C. B., & Sen, S. (2007). Reaping relational rewards from corporate social responsibility: The role of competitive positioning. *International Journal of Research in Marketing, 24*(3), 224–241.

Eccles, R. G., Ioannou, I., & Serafeim, G. (2014). The impact of corporate sustainability on organizational processes and performance. *Management Science, 60*(11), 2835–2857.

Eisenhardt, K. M. (1989). Agency theory: An assessment and review. *Academy of Management Review, 14*(1), 57–74.

Eisenhardt, K. M., & Martin, J. A. (2000). Dynamic capabilities: What are they? *Strategic Management Journal, 21*(10–11), 1105–1121.

El Ghoul, S., Guedhami, O., Kwok, C. C. Y., et al. (2011). Does corporate social responsibility affect the cost of capital? *Journal of Banking & Finance, 35*(9), 2388–2406.

Ellison, G., & Ellison, S. F. (2009). Search, obfuscation, and price Elasticities on the Internet vol. 77, 427-452.

Epstein, E. M. (1987). The corporate social policy process: Beyond business ethics, corporate social responsibility, and corporate social responsiveness. *California Management Review, 29*(3), 99–114.

Fifka, M. S. (2013). Corporate responsibility reporting and its determinants in comparative perspective – A review of the empirical literature and a meta-analysis. *Business Strategy and the Environment, 22*(1), 1–35.

Fisher-Vanden, K., & Thorburn, K. (2011). Voluntary corporate environmental initiatives and shareholder wealth. *Journal of Environment Economic Management, 62*(3), 430–445.

Frederick, W. C. (1994). From CSR1 to CSR 2: The maturing of business and society thought. *Business and Society, 3*(2), 150–164.

Freeman, R. E. (1984). *Strategic management: A stakeholder approach.* Boston: Pitman/Ballinge.
Friedman, M. (1970). The social responsibility of business is to increase its profits. *New York Times Magazine, 32–33*(122), 124–126.
Fuller, J., Mirhlbacher, H., Matzler, K., & Jawecki, G. (2009). Consumer empowerment through internet-based co-creation. *Journal of Management Information Systems, 26*(3), 71–102.
Fung, A., Graham, M., & Weil, D. (2008). *Full disclosure: The perils and promise of transparency.* Cambridge: Cambridge University Press.
Gallo, P. J., & Christensen, L. J. (2011). Firm size matters: An empirical investigation of organizational size and ownership on sustainability-related behaviors. *Business & Society, 50*(2), 315–349.
Garud, R., Gehman, J., & Kumaraswamy, A. (2011). Complexity arrangements for sustained innovation: Lessons from 3M Corporation. *Organization Studies, 32*(6), 737–767.
Garud, R., Tuertscher, P., & Van de Ven, A. H. (2013). Perspectives on innovation processes. *The Academy of Management Annals, 7*(1), 775–819.
Gelb, D. S., & Strawser, J. A. (2001). Corporate social responsibility and financial disclosures: An alternative explanation for increased disclosure. *Journal of Business Ethics, 33*(1), 1–13.
Goldman, S. L., Nagel, R. N., & Preiss, K. (1995). *Agile competitors and virtual organisations: Strategies for enriching the customer.* New York: Van Nostrand Reinhold.
Gray, R. H. (2000). Current developments and trends in social and environmental auditing, reporting and attestation: A review and comment. *International Journal of Auditing, 4*, 247–268.
Gray, R. H. (2002). The social accounting project and accounting, organisations and society: Privileging engagement, imaginations, new accountings and pragmatism over critique? *Accounting, Organizations and Society, 27*(7), 687–708.
Greening, D. W., & Turban, D. B. (2000). Corporate social performance as a competitive advantage in attracting a quality workforce. *Business and Society, 39*(3), 254–280.
Greenwood, M. (2007). Stakeholder engagement: Beyond the myth of corporate responsibility. *Journal of Business Ethics, 74*(4), 315–327.
GSIA. (2012). Global sustainable investment review 2012. Available at: http://gsiareview2012.gsi-alliance.org/pubData/source/Global%20Sustainable%20Investement%20Alliance.pdf
Gurtman, M. B. (1992). Trust, distrust, and interpersonal problems: A circumplex analysis. *Journal of Personality and Social Psychology, 66*(6), 989–1002.
Guzley, R. M. (1992). Organizational climate and communication climate: Predictors of commitment to the organization. *Management Communication Quarterly, 5*, 379–402.
Hahn, R., & Kuehnen, M. (2013). Determinants of sustainability reporting: A review of results, trends, theory, and opportunities in an expanding field of research. *Journal of Cleaner Production, 59*, 5–21.
Haddock-Fraser, J., & Fraser, I. (2008). Assessing corporate environmental reporting motivations: Differences between "close-to-market" and "business-to-business" companies. *Corporate Social Responsibility and Environmental Management, 15*(3), 140–155.
Hartman, L. P., Rubin, R. S., & Dhanda, K. K. (2007). The communication of corporate social responsibility: United State and European Union multinational corporations. *Journal of Business Ethics, 74*(4), 373–389.
Hasseldine, J., Salama, A. I., & Toms, J. S. (2005). Quantity versus quality: The impact of environmental disclosures on the reputations of UK Plcs. *The British Accounting Review, 37*(2), 231–248.
Herzig, C., et al. (2006). Understanding and supporting management decision-making. South East Asian case studies on environmental management accounting. In *Sustainability accounting and reporting* (pp. 491–507). Springer Netherlands.
Hofstede, G. (1998). Attitudes, values and organizational culture: Disentangling the concepts. *Organization Studies, 19*(3), 477–493.

Illia, L., Zyglidopoulos, S. C., Romentl, S., Rodriguez-canovas, B., & Gonzalez del Valle Brena, A. (2013). Communicating corporate social responsibility to a cynical public. *Sloan Management Review, 54*(3), 2–4.
Jacobs, B. W., Singhal, V. R., & Subramanian, R. (2010). An empirical investigation of environmental performance and the market value of the firm. *Journal of Operations Management, 28*(5), 430–441.
James, L. R., & Jones, A. P. (1974). Organizational climate: A review of theory and research. *Psychological Bulletin, 81*, 1096–1112.
Jarzabkowski, P., Balogun, J., & Seidl, D. (2007). Strategizing: The challenges of a practice perspective. *Human Relations, 60*(1), 5–27.
Jensen, M. (2002). Value maximization, stakeholder theory, and the corporate objective function. *Business Ethics Quarterly, 12*, 235–256.
Johansen, T. S., & Nielsen, A. E. (2012). CSR in corporate self-storying- legitimacy as a question of differentiation and conformity. *Corporate Communication: An International Journal, 17*(4), 434–448.
Johnson, L. P. Q., & Millon, D. K. (2015). *Corporate law after Hobby Lobby.* Washington & Lee University School of Law Scholarly Commons, Lexington, VA.
Johnson, G., Melin, L., & Whittington, R. (2003). Micro strategy and strategizing: Towards an activity-based view? *Journal of Management Studies, 40*(1), 3–22.
Kapstein, M. (2008). Developing and testing a measure for the ethical culture of organizations: The corporate ethical virtues model. *Journal of Organizational Behavior, 29*(7), 923–947.
Kernis, M. H. (2003). Toward a conceptualization of optimal self-esteem. *Psychological Inquiry, 14*(1), 1–26.
Kim, E.-H., & Lyon, T. P. (2015). Greenwash vs. brownwash: Exaggeration and undue modesty in corporate sustainability disclosure. *Organization Science, 26*(3), 705–723.
Klettner, A., Clarke, T., & Boersma, M. (2014). The governance of corporate sustainability: Empirical insights into the development, leadership and implementation of responsible business strategy. *Business Ethics, 122*, 145–165.
KPMG. (2013). KPMG survey of CSR reporting 2013. Available from: http://www.kpmg.com/Global/en/IssuesAndInsights/ArticlesPublications/corporate-responsibility/Documents/kpmg-survey-of-corporate-responsibility-reporting-2013.pdf
Legendre, S., & Coderre, F. (2013). Determinants of GRI G3 application levels: The case of the fortune global 500. *Corporate Social Responsibility and Environmental Management, 20*(3), 182–192.
Levy, D. L., Brown, H. S., & de Jong, M. (2010). The contested politics of corporate governance: The case of the global reporting initiative. *Business & Society, 49*(1), 88–115.
Lewis, M. W. (2000). Exploring paradox: Toward a more comprehensive guide. *Academy of Management Review, 25*(4), 760–776.
Luo, X., & Bhattacharya, C. B. (2009). The debate over doing good: Corporate social performance, strategic marketing levers, and firm-idiosyncratic risk. *Journal of Marketing, 73*(6), 198–213.
Manetti, G. (2011). The quality of stakeholder engagement in sustainability reporting: Empirical evidence and critical points. *Corporate Social Responsibility and Environmental Management, 18*, 110–122.
Manetti, G., & Becatti, L. (2009). Assurance services for sustainability reports: Standards and empirical evidence. *Journal of Business Ethics, 87*, 289–298.
Margolis, J. D., & Walsh, J. P. (2003). Misery loves companies: Rethinking social initiatives by business. *Administrative Science Quarterly, 48*(2), 268–305.
Markus, M. L. (1983). Power, politics, and MIS implementation. *Communications of the ACM, 26*(6), 430–444.
McWilliams, A., Siegel, D. S., & Wright, P. M. (2006). Corporate social responsibility: Strategic implications. *Journal of Management Studies, 43*(1), 1–18.

Milne, M. J., & Patten, D. M. (2002). Securing organizational legitimacy: An experimental decision case examining the impact of environmental disclosures. *Accounting, Auditing & Accountability Journal, 15*(3), 372–405.

Mitchell, R. K., Agle, B. R., & Wood, D. J. (1997). Toward a theory of stakeholder identification and salience: Defining the principle of who and what really counts. *Academy of Management Review, 22*(4), 853–886.

Nazir, S., & Pinsonneault, A. (2012). IT and firm agility: An electronic integration perspective. *Journal of the AIS, 13*(3), 150–171.

Neu, D., Warsame, H., & Pedwell, K. (1998). Managing public impressions: Environmental disclosures in annual reports. *Accounting, Organizations and Society, 23*(3), 265–282.

Nikolaeva, R., & Bicho, M. (2011). The role of institutional and reputational factors in the voluntary adoption of corporate social responsibility reporting standards. *Journal of the Academy of Marketing Science, 39*, 136–157.

O'Donovan, G. (2002). Environmental disclosures in the annual report: Extending the applicability and predictive power of legitimacy theory. *Accounting, Auditing & Accountability Journal, 15*(3), 344–371.

O'Toole, J., & Bennis, W. (2009). What's needed next: A culture of candor. *Harvard Business Review, 87*(6), 54–61.

Orlitzky, M., Schmidt, F. L., & Rynes, S. L. (2003). Corporate social and financial performance: A meta-analysis. *Organization Studies, 24*(3), 403–441.

Patten, D. (1995). Variability in social disclosure: A legitimacy-based analysis. *Advances in Public Interest Accounting, 6*, 273–285.

Payne, A. F., Storbacka, K., & Frow, P. (2008). Managing the co-creation of value. *Journal of the Academy of Marketing Science, 36*(1), 83–96.

Pearce, C. L., & Ensley, M. D. (2004). A reciprocal and longitudinal investigation of the innovation process: The central role of shared vision in product and process innovation teams. *Journal of Organizational Behavior, 25*(2), 259–278.

Poole, M. S. (1985). Communication and organizational climates: Review, critique, and a new perspective. In R. D. McPhee & P. K. Tompkins (Eds.), *Organizational Communication* (pp. 79–108). Beverly Hills: Sage.

Reynolds, M. A., & Yuthas, K. (2008). Moral discourse and corporate social responsibility reporting. *Journal of Business Ethics, 78*, 47–64.

Roberts, N., & Grover, V. (2012). Leveraging information technology infrastructure to facilitate a firm's customer agility and competitive activity: An empirical investigation. *Journal of Management Information Systems, 28*(4), 231–269.

Sambamurthy, V., Bharadwaj, A., & Grover, V. (2003). Shaping agility through digital options: Reconceptualizing the role of information technology in contemporary firms. *MIS Quarterly, 27*(2), 237–263.

Saravanamuthu, K., & Lehman, C. (2013). Enhancing stakeholder interaction through environmental risk accounts. *Critical Perspectives on Accounting, 24*(6), 410–437.

Scholtens, B. (2014). Indicators of responsible investing. *Ecological Indicators, 36*, 382–385.

Schultze, U., & Bhappu, A. D. (2007). Internet-based customer collaboration: Dyadic and community-based modes of co-production. In N. Kock (Ed.), *Emerging E-collaboration concepts and applications* (pp. 166–192). Hershey: CyberTech Publishing.

Schultze, U., Prandelli, E., Salonen, P. I., & Van Alstyne, M. (2007). Internet-enabled co-production: Partnering or competing with customers? *Communications of the AIS, 19*, 294–234.

Scott, S. G., & Lane, V. R. (2000). A stakeholder approach to organizational identity. *Academy of Management Review, 25*(1), 43–62.

Simnett, R., Vanstraelen, A., & Chua, W. F. (2009). Assurance on sustainability reports: An international comparison. *Accounting Review, 84*(3), 937–967.

Sloan, P. (2009). Redefining stakeholder engagement: From control to collaboration. *Journal of Corporate Citizenship, 36*, 25–40.

Smith, W. K., & Lewis, M. W. (2011). Toward a theory of paradox: A dynamic equilibrium theory of organizing. *Academy of Management Review, 36*(2), 381–403.

Spar, D. L., & La Mure, L. T. (2003). The power of activism: Assessing the impact of NGOs on global business. *California Management Review, 45*(3), 78–101.

Srinivasan, K., Kekre, S., & Mukhopadhyay, T. (1994). Impact of electronic data interchange technology on JIT shipments. *Management Science, 40*(10), 1291–1304.

Stout, L. (2007). Why we should stop teaching Dodge v. Ford. Research paper No. 07-11, Law & Economics Research Paper Series, UCLA School of Law.

Stout, L. A. (2013). The shareholder value myth. *European Financial Review*, April. http://www.europeanfinancialreview.com/?p=883

Strand, R. (1983). A systems paradigm of organizational adaptations to the social environment. *Academy of Management Review, 8*, 90–96.

Sweeney, L., & Coughlan, J. (2008). Do different industries report corporate social responsibility differently? An investigation through the lens of stakeholder theory. *Journal of Marketing Communications, 14*(2), pp. 113–124.

Teece, D. J., Pisano, G., & Shuen, A. (1997). Dynamic capabilities and strategic management. *Strategic Management Journal, 18*, 509–533.

Toms, J. S. (2002). Firm resources, quality signals and the determinants of corporate environmental reputation: Some UK evidence. *The British Accounting Review, 34*(3), 257–282.

USSIF. (2014). Report on U.S. sustainable, responsible, and impact investing trends 2014. Available at: http://www.ussif.org/files/publications/SIF_Trends_14.F.ES.pdf

Van de Ven, A. H., Polley, D. E., Garud, R., & Venkataraman, S. (1999). *The innovation journey*. Oxford, UK: University Press.

Walumbwa, F. O., Avolio, B. J., Gardner, W. L., Wernsing, T. S., & Peterson, S. J. (2008). Authentic leadership: Development and validation of a theory-based measure. *Journal of Management, 34*(1), 89–126.

Watts, S. (2015). Corporate social responsibility reporting platforms: Enabling transparency for accountability. *Information Technology & Management, 16*(1) SI, 19–35.

Watts, S. (2016, August 11–13). Electronic Integration of CSR reporting systems for stakeholder responsiveness. In *Proceedings of AMCIS 2016*, San Diego.

Wood, D. J. (1991). Corporate social performance revisited. *Academy of Management Review, 16*, 691–718.

Zaheer, A., & Zaheer, S. (1997). Catching the wave: Alertness, responsiveness and market influence in global electronic networks. *Management Science, 43*(11), 1493–1509.

Zwick, D., Bonsu, S. K., & Darmody, A. (2008). Putting consumers to work: 'Co-creation' and new marketing governmentality. *Journal of Consumer Culture, 8*(2), 163–196.

Index

A
Abolition of child labor, 1050
Academic circles, 229
Acceptance, social license as, 584
Accidental ethical decision-making failures, 236
Action-based learning, 436
Actions, 223, 225, 228, 1091–1093, 1095, 1100, 1108, 1113, 1115
Active investment approach/strategies, 1037
Activism, 298, 303, 305
Activists, 1090, 1100, 1104, 1108, 1114, 1115, 1119, 1120
Actor, 1140–1141
  See also Agency
Adi Śaṅkara, 38
Advertising, 1107, 1111, 1114, 1115
Affluenza, 414
Agency
  dynamics, 1148–1149
  formation, 1147–1148
  individual *vs.* collective forms, 1143–1144
  neglect of, 1138–1140
  psychological dimensions, 1148
  related terms in sustainability transitions, 1142
  sustainability and societal transitions, 1144–1147
  sustainability transitions, 1140–1141
Agency model, 1035
Agility, 1160, 1163, 1166, 1169
  customer, 1163, 1171
  electronic integration, 1171
  enterprise systems, 1170
  and innovativeness, 1163
  inter-functional coordination, 1170
  organizational, 1163, 1168, 1177
  stakeholder engagement, 1162
Agribusiness, 655

Ahiṁsā, 8, 13, 39, 57, 59
Akron Children's Hospital (ACH), 784, 786, 789, 790, 794, 799–805
Albuquerque, 864
Alexander, C., 860–862
Algal-dominated systems, 392
Allen, W., 854
Alternative and complementary medical care, 1072
Altruism, 18–19, 222, 224, 237, 410
Altruism:The Power of Compassion, 18
Ambience, 561
American Association of Community Colleges (AACC), 487
American dream, 1060
Amsterdam, 961
Anchoring, 224
Anderson, R., 27, 73
Andrews, A. D., 48
Anglo American, 132, 133
AngloGold Ashanti, 132
Animal-based diets, 159
Animal rights movement, 161
Animal suffering, 160
Antarctica, 485
Anthropocene, 557
Anthropocentrism, 203, 207, 215, 1059
Anthroposene, 21
Apartheid, 1023
Apathy, 5
Appadurai, 568
Appreciative inquiry (AI), 360
  anticipatory principle, 365
  City of Cleveland, 372–377, 379
  constructionist principle, 364
  definition, 362
  design phase, 367
  destiny phase, 367
  discover phase, 366–367

Appreciative inquiry (AI) (cont.)
  dream phase, 367
  Fairmount Minerals, 368, 379
  poetic principle, 364
  positive principle, 365
  principle of simultaneity, 364
  stakeholder engagement, 362–364
Approval for project, 585
Aquinas College, 479
Arctic Ocean, 485
Aristotle, 13
Arizona State University (ASU), 712
Article 89, 864
Asian Pacific Network of Oregon
    (APANO), 287
Aspiration, 385
Assets, 1029
Associate in Applied Science (AAS), 481, 483
Associate in Arts (AA), 481–483, 495
Associate in Science (AS), 481, 483, 487
Association for the Advancement of
    Sustainability in Higher Education
    (AASHE), 448, 476, 487
Association to Advance Collegiate Schools of
    Business (AACSB), 475, 476
Assumptions, 151
Atmosphere, 392, 556
Atmospheric CO_2, concentration of, 392
Attitudes, 154, 157, 161
Auction market, 643
Australia, 480
Automatic response, 387
Autonomy, 155
Availability, 224
Average ticket price, 560
Avoidable food waste, 687
Awareness, 5, 14, 20
  eco-friendliness, 399
  empathy, 390
  engaged sustainability, 399

B
Babylon, 300
Back-end electronic integration, 1170, 1171
Bailey, R., 46
Bakelite, 569
Barcelona, 695, 946, 959
Barlow, Z., 397, 399, 400
Barnard, N. D., 51, 54
Barthes, R., 551, 568
Batson, D., 18
Baudrillard, J., 550

Bauman, Z., 551
B Corporations, 1055
Beef, 39, 40, 43, 45–47, 51
Behavior, 157, 158, 557
Behavioral economics, 224, 235
Behavioral law, 235
Benefit Corporation, 1055
Bennett, L., 397, 399, 400
Berry, W., 399, 841, 842, 853
Bezos, J., 95
Bhagavad Gītā, 12, 14, 16, 43
Bible, 57, 298, 301, 303, 306
Bicycles, 953
Big Brother, 947, 963
Big data, 948, 963
B impact assessment (BIA), 1057
Biocentrism, 408, 1059
Biochemical cycles, 390
Biodegradable, 558
Biodiversity, 65, 392, 668
  conservation, 210
  credits, 670
  in urban spaces, 919
Bio-economic education, 316
Bio-economic theory, 317
Bio-economy, 310
  circular economy, 315
  conceptualization of, 311–312
  as ecological economics, 313
  global ethics, 322–323
  as green economy, 312–313
  holistic approach, 316
  innovation culture and capacity
      development, 317–319
  knowledge–based economy, 319
  new economic epistemology paradigm,
      314–315
  social capital and culture of peace, 323–324
  as strategy, 324–328
  sustainable development, 315–316
  transdisciplinary approach, 316–317
Biomimicry, 707
Bio-pool, 1005
Biosphere, 386, 391
Biospherical systems, 392, 402
B Lab, 350
Blake, W., 303
Bloomington Urban Forestry Research Group
    (BUFRG), 282
Bluesign® standard, 343
Bodhi, B., 21, 23
Bohm, D., 10
Bonds, 1021

Bonnett, M., 385
Boston, 864
Bottom-line, 668
Bottom-up approach, 735
Boundary object, 1167, 1169, 1171, 1172
Boundary-spanning information technology (IT) artifact
    electronic integration, CSR-reporting database, 1169
    inter-functional coordination, 1171–1173
    role of, 1167
    shared transparency beliefs, 1173
Bourdieu, P., 551
Brasilia, 555
Brazil, 550, 634, 643, 1051
Brazlândia, 559
Breaching experiment, 744, 747–749
Breakdown of family, 415
Bṛhadāraṇyaka Upaniṣad, 43
Bricolage, 994
    by Chitra, 1007
    forms of, 1002
    by Sanjay, 1011
    by Vikram, 1001
Bricoleur, 994
    ecopreneurship, 1012
BRICS nations, 1051
British Columbia, 638
Brown, B., 400
Browning, D., 853
Brundtland Report, 7, 247
Buddha, 42, 57
Buddhism, 39
Buddhist psychology, 6
Business
    models, 970, 981
    plan, 978
    sustainable decision making
        business jargon, 118–121
        language and intent of business-as-usual, Millennials disagree with, 121
        language implementation, 132–135
        logic of appropriateness and sustained motivation, 130–131
        poetry and arts, holistic frame of, 122
        sustainable organizations, poetry, 124–130
    sustainability, 473, 480, 484, 487, 492
Business ethics, 93
Business model canvas, 981
Business Responsibility Reports, 1053
Business Youth For Sustainable Development (BY4SD), 517–520

Buy in, 350
Buying, 1111
BY4SD, *see* Business Youth for Sustainable Development (BY4SD)

C

Campbell, E., 54, 55
Campbell, T., 54, 55
Campus labs, 976
Canada, 1030
Canadian crude oil
    extraction firms, 1030
    in 2011, 587
Canons, 845–848
Canons of Sustainable Architecture and Urbanism, 845
Cap and trade, 639
Capital, 227, 229
Capitalism, 474
Capitalist, 474, 476, 490, 491
    marketing, 157
Capital knowledge, 553
Capital markets, 1020, 1021, 1026, 1029
    already reflected in, 1034
Carbonate ion, 392
Carbon bubble, 1020, 1030, 1033
    bursting of, 1031, 1032
Carbon dioxide equivalents, 1063
Carbon emissions, 638, 639, 641, 642, 664
Carbon footprint, 25, 26, 31, 44, 45, 47
Carbon pricing, 632
Carbon sequestration, 893
Carbon tax, 637
Cardboard tents, 234
Care-givers, 387
Cargill, 132
Carlyle, T., 24
Carnegie Commission on Higher Education, 482
Cascio, J., 44
Catastrophic consequences, 391
Central European countries, 859
CERES, 114
Certificates, 477, 479–481, 483, 486–488, 491, 493
Challenge, 1039
Challenging conventions, 744
Champion, 349
Chāndōgya Upaniṣad, 13, 17
Changed mindset, 133
Change management, 491, 493
Chapin, F. S., 386, 390

Charity giving, 236
Charter, 842–845
Checkland, P., 389
Chicago, 947, 963
China, 641, 1051
Choice architecture, 224
Circular economy (CE), 315, 685, 704, 810–818, 822, 827, 828
　evolution of, 706–711
　in Phoenix, 714
Citizen stewardship, 275, 285
City/cities, 298
　ecovillages (*see* Ecovillages)
City of Cleveland, 372, 379
City of Phoenix, 711
Civic engagement, 505
Clark, M., 45
CleanMetrics, 45
Clean technology, 24
Climate, 670
　patterns, 392
Climate change, 19–22, 114, 116, 117, 121, 122, 124, 126, 128, 131, 135, 159, 161, 164, 236, 436, 439, 440, 474, 485, 492, 550, 635, 638, 839, 851, 863
　NGP, 591
Closed loop production, 811, 818
CO_2 emissions, 39, 42
Cognition, 228
Cognitive, 385
　empathy, 387
　legitimacy, 759
Cognitive strategy, 223
Collaboration/collaborative, 153, 339
Collaboration systems, 787
Collaborative co-production, 1104
Collaborative learning, 460
Collapse, 299
Collective actions, 761
Collectively shared goals, 227
Collective outcomes, 385
Collector, 574
Collectors and Recyclers Association, 572
Colorado, 857
Columbia Sportswear, 71
Comfort, 751
Command and control, 1059
Commercial entrepreneurs, 95, 96
Commitment systems, 787
Common goals, 228, 231, 236
Commons and polycentricity, 453
Common threads initiative, 344

Communication experts, marketing communication, *see* Marketing communication
Communication planners, 1090, 1100
Community, 236, 391, 667, 758
　college system, 484
　　community aspect, 484–485
　　entry-level occupational training, 485
　　sustainable business and sustainability curriculum, 486
　　US (*see* US community college system)
　green spaces, 890
Community-based conservation (CBC), 205
Community-based entrepreneurship, 757, 764
Community college system, 484
Community development
　sustainable management for, 927
　urban green spaces in, 922
Community green spaces, 890
Community involvement, 67, 72, 78, 79, 81, 83
　in urban green spaces, 898–904
Community resilience, 274, 278, 291
Community service-learning (CSL), 461
Community Watershed Stewardship Program, 281
Companies, 1094, 1095, 1097–1100, 1103–1105, 1107, 1108, 1114, 1116, 1119
Compassion, 6, 7, 13, 14, 18, 23–24, 28–31, 38, 39, 41, 50, 56–57, 59
Compassionate, 385
　empathy, 387
　perspective, 157
Competence, 155
Competitive, 337
　advantage, 1032, 1104, 1115
　disadvantage., 1032
　individualism, 157
Complex adaptive systems, 150
Complexity, 1030
Complexity science, 149–151
Compliance, 227, 228, 231
Compliance mechanism, 1065
Compliance systems, 787
Compromise, 746
Conflict-focused case study learning, 459
Congress of New Urbanism, 840–842
Coniferous trees, 891
Concientización, 185
Conscious consumption, 558
Conscious decision, 385
Consciousness, 395, 402
Conservación Patagonica, 345

Conservation alliance, 343
Conservation and environmentalism, *see* Environmentalism
Conspicuous consumption, 552
Constrained regulatory framework, 1031
Construction industry, 993
Consultative Group to Assist the Poor (CGAP), 605
Consumerism, 19, 28, 157, 872, 1088
Consumers, 25, 234, 554, 734
    issues, 67, 77, 83
    segmentation, 1076
Consumer society, 552, 558
Consumption, 550
    sociology, 565
Consumption behavior, 1071, 1084
Containers, 737
Convenience matters, 743
Cooling
    capacity of green spaces, 891
    effects, 891, 893
Cooperrider, D. L., 362, 365, 368, 370, 378
Co-ownership, 585
Coral reefs, 392
Corell, R. W., 386, 390
Corporate branding, 1099, 1107, 1108
Corporate governance, 1036
Corporate greed, 415
Corporate shareholder resolutions, 1051
Corporate social contract, 1044
Corporate social responsibility (CSR), 65, 339, 475, 505, 1027, 1035, 1036, 1045, 1073, 1099, 1100, 1106–1120
    reporting, 1158
        boundary-spanning IT artifact, CSR-reporting software (*see* Boundary-spanning information technology (IT) artifact)
        CSR disclosure, 1174
        database, 1169–1171
        determinants and impacts, 1164–1165
        GRI database of, 1158
        stakeholder agility, 1162–1166
        stakeholder engagement, 1161
Corruption in water and sanitation, 419
Cosmic order, 12
Costa, L., 555
Costanza, R., 386, 390
Costa Rica, 645
Cost of urban living, 947
Cradle to Cradle^{TM} (C2C), 685, 708, 810, 813, 821, 829
Creativity, 750
    and experimentation, 161
Credibility, 585
Credit, 225
Critical pedagogy, 186
Critical theory, 489
Critical thinking, 971
Crude oil
    benchmark prices, 588
    extraction firms in Canada, 1029
Crutzen, P., 386, 390, 557
CSR, *see* Corporate social responsibility (CSR)
CSR-reporting systems, *see* Corporate social responsibility (CSR), reporting
Cultural capital, 229
Cultural Environmental Services, 286
Cultural services, 660
Culture, 100, 102, 110, 492, 1074, 1078, 1083
Curricula/curriculum, 474–480, 485–492, 494, 496
Cybercriminals, 963
Cynefin framework, 150

D

Dairy, 44–46, 49
Darwin, 18
Database design, 1170
Databases, 1158, 1159, 1168, 1169
Data collection, 231, 234
da Vinci, L., 54
Dayananda Saraswati, 58, 59
Day-to-day environmental protection, 236
De-centralization, 842, 863, 867
Decision making, 222, 223, 726
    autonomy, 336
    business (*see* Business, sustainable decision making)
Deep ecology, 24, 202, 203, 207, 209, 212, 214
Defense mechanism, 155
Deloitte 2016 Millennial Survey, 121
Denmark, 646
Dense green space, 888
Denver, 857
Destruction, 384
De Wit, C. A., 386, 390
Dharma, 11–12
Diamond, J., 299
Diamonds, 301
Diary, 744
Dietary change, 159–162
Diet-related risk factors, 160
Digital technology, 430

Dimension, 348
Discarded, 573
Discomfort, 386, 749
Disruptive innovators, 995
Dissolution of coral reefs, 392
Dissolved oxygen, 392
District parks, 920
Diverse, 486, 1027
Divestment, 1022
Dormitory, 229, 231
Duany, A., 849–853, 856–857
Dynamic stakeholder agility, *see* Agility

E

Earth's life-support systems, 402
Earth Summit, 478, 501
Eating habits, 746
Eco-bag, 567
Ecocentrism, 489
Ecocide, 213
Ecocritic, 567
Eco-entrepreneurs, 979
Ecoliterate, 400–401
Ecology, 45
 connection, 400
 cost of food, 44–47
 dimension, 656
 disaster, 313
 economics, bio-economy as, 313
 footprints, 414
 imperative, 575
 intelligence, empathy, 390–391
 justice, 203, 207, 209, 211, 213, 214
 modernization, 489
 services, 864
 sustainability, 393
Economics, 1033
 development, 839
 dimension, 656
 growth, 415
 performance, 670
 profitability, 1099
 prosperity, 684
 sustainability, 406
 and transactional costs, 793
Ecopreneurs, 1012
Ecopreneurship, 997
 bricoleur, 1012
Ecosphere, 398–400
Ecosystems
 based approach, 672, 673, 677
 based risks, 667

 definition, 268
 marine, 255, 256
 services, 257, 268, 654–656, 668
 and wine industry, 656
Ecovillages, 872
 bio system challenge, 879
 built-environment challenge, 879
 consumerism and materialism, 872
 definition, 871–872
 economic challenge, 879
 glue challenge, 880
 governance challenge, 880
 LAEV (*see* Los Angeles Eco Village (LAEV))
 principles of permaculture, 872
 whole-system challenge, 880
Education, 227, 1089, 1090, 1093, 1121, 1122
Educational sector, 229
Education for sustainable development (ESD), 478
Education in human values (EHV), 408
Efficiency
 high level of information, 1036
 not always perfectly, 1034
Efficient market hypothesis (EMH), 1033
Ehrlich, A. H., 43
Einstein, A., 10, 53
Ekman, P., 385, 387
Electronic integration, CSR-reporting database
 back-end data, 1170, 1171
 external, 1169
 internal, 1169, 1170
Elevation prospects, 224
Elite education, 227, 229
Ellen MacArthur Foundation (EMF), 708
Ellul, J., 298
Elmsworth, 54
Embedded nature, 352
Emergence, 152, 153
Emission reduction credits (ERC), 1060
Emissions Database for Global Atmospheric Research (EDGAR), 20
Emissions
 fees, 1059
 trading, 639–640
Emotions(al), 224, 227, 228
 compensation, 223
 condition, 386
 empathy, 387
 experiences, 228
 self-awareness, 389
 self-management, 389
 whim, 236

Emotional intelligence (EQ), 178
  self-awareness, 388
  self-management, 389
Emotional loss compensation act, 237
Empathetic muscle, 402
Empathetic nature, 399
Empathetic skills, 388
Empathy, 386
  cognitive, 387
  emotional, 387
  EQ, 388–390
  knowing and feeling, 388
Endowment, 222, 225, 226
Enduring solution, 385
Energy
  auctions, 643
  consumption, 223, 230, 633, 982
  light consumption conscientiousness, 234, 235
  meter data, 234
Energy-efficient, 955
Engaged sustainability, 25, 752
Engagement, 361, 366, 368, 378, 379
Engage the mind, 385
Enterprise, 391
Enterprise thinking, 491
Entrepreneur, 337
Entrepreneurship, 226, 476, 479, 488
Entry-level, 493
  employee, 488
  occupational training, 485–486
  technical work, 482, 494
  trade, 472, 484
Enviro internships, 345
Environment(al), 72, 73, 98, 107, 229, 231, 387
  activist, 1061
  awareness, 742
  conscientious energy consumption, 232
  consciousness, 1073, 1082
  crisis, 564
  degradation, 5
  economics, 1020, 1029, 1030, 1033, 1034, 1036, 1037, 1039
  education, 810, 811, 815
  entrepreneurship, 994
  ethicality, 236
  friendly products, 1081
  friendly technologies, 1050
  grants program, 345
  health, 342, 1070
  impact, 982
  intrapreneurship, 346, 349
  issues, 656
  justice, 203, 204, 209, 211, 214, 288
  management, 655, 657
  movement, 161
  performance, 670
  policies, 752
  problems, 93
  protection, 232, 236
Environmental and social impact assessments (ESIA), 1049
Environmental Information Disclosure Act of China, 1052
Environmentalism, 204, 224
  biodiversity, 210
  CBC, 205
  conservationist industry, 202
  deep ecology, 203, 207
  ecocentric view, 207
  ENGO, 202, 204
  environmental justice, 211
  environmental racism, 211
  HEP, 209
  indigenous rights, 205, 206
  instrumental motivation, 203
  mainstream environmentalists, 202, 204, 208, 212
  nature and origins of, 214
  nature-culture dichotomy, 209
  NCS, 205
  neoliberal conservation, 202
  population pressure, 210
  radical environmentalists, 208
  reconciliatory approaches, 212
  social justice, 202, 211
Environmental nongovernmental organizations (ENGO), 202, 204
Environmental Protection Agency (EPA), 474, 484
Environmental quality, 684
Environmental, social, and governance (ESG), 1021, 1024
  indexes, 1026
  integrated investments, 1046
Environmental stewardship, 274, 275, 290
  definition, 276–278
  environmental problems, social constructions of, 275–276
  historic socio-economic inequities, 278–279
  motivations, 280
  Portland's Tree Inventory Project, 285
Environmental sustainability, 25, 116, 244, 246, 406, 660, 1076, 1081, 1083
  climate change impacts, 259
  and economic growth, 252

Environmental sustainability (*cont.*)
    language game of, 244
    linguistic economy, 264
    (*see also* Sustainable, development)
Environmental values, 278, 281, 285, 1073
Environmental Working Group (EWG), 30, 45, 47
EPA, *see* Environmental Protection Agency (EPA)
Ephesus, 300
Equality, 531, 546
Equator principles, 1048–1049
Equity investors, 1051
Espigoladors, 695–697
Ethicality, 224, 227, 229
Ethics, 12–13
    causes, 228
    codes, 1099, 1104, 1107
    consumption, 236, 1083
    dilemma, 1062
    downfalls, 236
    metrics, 1061
    of plant-based diet
        alternative for sustainable future, 49–50
        vs. animal-based, 43
        carbon footprint, 45
        compassion, 56
        environmental perspective, 50
        global warming, 41–42
        greenhouse gas emissions, 45, 46
        harmlessness, 57
        humans, 47–49
        livestock raising, cost, 44
        meat atlas, 55
        meat-based diet drawbacks, 39–40
        reduce climatic and environmental impacts, 47
        vegetarianism, 53
    products, 1073, 1074
Eudaimonia, 154
EU Plastic Bags Directive, 732
Europe, 480, 1025
European Union, 639
European Union Emissions Trading System, 639
Everyday life, 734
Evolutionary-based natural law, 228
Evolutionary theory, 758–762
Exchange traded funds (ETFs), 1021, 1026
Exercise, 744
    in lived sustainability, 752
    of packaging-free shopping, 751
Existing practices, 742, 743

Experience, 385
    lived sustainability, 752
    packaging-free shopping, 744
Exploitative relationships, 413
Exposure, 234
Extinction, 21
Extrinsic aspirations, 155
Extrinsic materialistic values, 412
Extrinsic values, 407

F
Fabry, V. J., 386, 390
Factories, 384
Factory farming, 53, 160
Fairmount Minerals, 368–372, 379
Fair operating practices, 67, 72, 75
Falkenmark, M., 386, 390, 391
Farmers market, 862–864
Farming, 686
Farmland, 850, 864
Favorable group membership, 225
Featherstone, M., 563
Feedback loops, 148
Feijoada, 566
Field experiment, 229, 233
Field observation, 231
Fifth Discipline, 394
Financial investors, 228
Financial performance, 1028
Financial resources, 688
Fixed-income instruments, 1025
Flexible cohesion, 784, 789, 796–798
Flourishing, 141–154, 948
Flourish-smart, 951
Foley, J., 386, 390
Folke, C., 386, 390
FoodLoop, 693
Food
    management, 686
    packaging, 751
    security, 857
    waste, 686
        avoidable, 687
        implications, 690
        national, 699
Footprint chronicles®, 342
Forbes, S., 42
For-profit college, 485
Fossil fuel
    companies, 1023
    divestment, direct impact of, 1022
    equity exposure ratio, 1022

Framing
  language of business (*see* Business, sustainable decision making)
  role of, 116–118
Free progress' school, 1011
Froggatt, A., 46
Fulfillment, qualities of, 385
Function optimally, 387
Future investigations, 237
Future-oriented cognitive strategy, 237
Future planned research, 237
Future studies, 223

G

Gains, 222, 223
  preferences, 226
  prospect, 228
Gamification
  applications, 431–432
  description, 428
  fit for green, 434
  greenify, 434
  history, 429
  meaningful action, 438–440
  popularity, 428
  positive peer pressure on sustainability issues, 437–438
  powerAgent, 434
  powerhouse, 435
  sustainability, 432–434
  working principles, 429–431
Gandhi, Mohandas Karamchand, 42, 54
Garbage, 573
  collection, 953
Garcia-Canclini, N., 552
Garden city, 849
Garden of Eden, 301
Gardens, 860–862, 864
Garrard, G., 567
Geertz, C., 557
Gehl, J., 306
General Agreement on Tariffs and Trade (GATT), 643
General Electric, 476
Generational shift, 1023
Generation Z, 486, 492
Genetically modified organisms (GMO), 27
Geographic information system (GIS), 935
Geo-political uncertainties, 1064
Georgescu-Roegen (GR), 317
Germany, 859
GI Bill, 482

Gigafactory, 647
Gītā, *see* Bhagavad Gītā
Glacial freshwater supplies, 392
Gleaner, 696
Global climate regulation, 666
Global Ecovillage Network (GEN), 871, 872, 879
Global footprint, 64
Globalization
  benefits of, 506
  dangers of, 509
  economic issues, 507
  negatives/consequences, 511
Globalized goods, 552
Global market segment, 1084
Global movement, 64–66
Global Reporting Initiative (GRI), 114, 133, 350, 361
Global Sullivan Principles, 1045
Global warming, 39–44, 648, 1063
Golden Rule, 408, 409
Goleman, D., 385–388, 390, 391, 397–400
Goodwill/relational benefits, 793
Gore, A., 384, 1062
Governance tool, 734
Governmental policy control, 236
Governmental regulation, 229
Governments, 1039
Grameen Bank, 99–101, 108
Granite State Farm to Plate, 851
Grassroots innovations, 756–758
Greed, 6
Green, 552
  barriers, 909
  bonds, 1024
  building, 957
  business, 1088
  consumerism, 1088
  dark green level, 1097
  economy, bio-economy as, 312
  energy, 436
  environmental sustainability, 1093–1106
  ethics, 1093
  finance, 1021
  funds, 1026
  ideology, 1089, 1091
  light green level, 1097
  luxury, 24
  management systems, 1096
  market green level, 1097
  marketing communication (*see* Marketing communication)
  product level, 1096

Green (*cont.*)
 quasi-strategic greening level, 1097
 revolution, 1089
 stakeholder green level, 1097
 strategic greening level, 1097
 tactical greening, 1097
 transformation, 1088–1090
 value-addition processes, 1096
 walls, 893
 washing, 349, 1035, 1036
 wave, 555
Greenhouse gas (GHG), 22, 636, 1063
 emission, 42, 45, 46, 55, 159
  cost of, 1034
  reporting, 1034
Green Mountain College, 479
Green Organics System Design project, 723
Green roof, 892, 893
Green space, 918
 audits, 938
 biological quality of, 925
 economic value of, 929
 linear, 921
 origin, 918
 scale, 891
 types, 921 (*see also* Urban green spaces)
Greger, M., 51, 52, 54
Grenelle II Act of France, 1052
GRI, *see* Global Reporting Initiative (GRI)
Grocery shopping, 731
Gross domestic product (GDP), 588
Grounded theory, 765
Group membership, 223, 224
Group polarization, 227

H
Habit of cooking, 746
Hannover Principles, 707
Hansen, J., 390
Happiness, 6, 10, 13, 27, 226, 948, 950
Hard skills, 971
Hardwired, 408
Harmlessness, 13, 39, 53, 57
Harmony, 14
Harper, W. R., 481
Haves and have-nots, 507
Health
 behaviors, 1078, 1083
 care costs, 160
 hazards, meat-based diet, 49, 56
 management, 1072

 and obesity epidemic, 161
 and sustainability (*see* Lifestyle of Health
  and Sustainability (LOHAS))
 wellness, 1076, 1083
Healthy eating, 700
Healthy foods, 1072
Healthy products, 1080
Hedonic approaches, 154
Hedonic happiness, 157
Heidrich, R. E., 49, 54
Helping behavior, 787–790, 792, 796
Herbivores, 47, 48
Herd behavior, 224
Heroic imagination, 228
Heterodox economy, 313
Heuristics, 317
High complexity, 142
Higher Education for Sustainable Development
 (HESD), 478, 494
Higher education institutions, 447
Higher income social, 574
High-net-worth individuals, 1057
Hinduism, 39, 57
Hinterland, 838
Holistic health movement, 161
Holistic management, 122–124
Holistic substantive awareness, 386
Hope, 300–302, 304–306
Howard, E., 849
Hughes, T., 386, 390
Human
 decision-making, 223
 health
  and ecosystems, 392
  hydrocarbon on, 591
 potential, 157
 survival
  life-support systems, 390
  sustainable for, 391
Human action, 387
Human capital, 658, 666
Human communities, 391
Human exemptionalism paradigm (HEP), 209
Humanity, 402, 556
Human rights, 67
Hungry Harvest, 695
Hybrid non-profit, 97
Hybrid organizations, 1054
Hydrocarbon, on human health, 591
Hydroelectric power, 643
Hygiene, 742
Hyperloop, 99, 105, 108, 110

I

"I"-centeredness, 411
Ideology, 1089, 1091, 1092
IKEA, 68
Illness, 386
Impact investing, 1045, 1046, 1054–1057
Impact Reporting & Investment Standards (IRIS), 1056–1057
Inconsistent criteria, 1027
Inconspicuous consumption, 552
Inconvenience, considerations of, 742
Incremental changes, 1029, 1033
Incumbent
   actors, 1145
   strategies, 1146–1147
India, 480, 1051
Indigenous sovereignty, NGP, 592
Individual behavior, 223
Individual reference point, 222, 229
Inductive case, 339
Industrial Ecology, 707
Industrial Revolution, 567
Industrial symbiosis, 707
Industry disruptive innovation, 1046
Industry leaders, 581
Information asymmetries, 689, 1035, 1039
Information technology (IT) artifact, CSR-reporting software
   electronic integration, 1169
   inter-functional coordination, 1171
   role of, 1167–1169
   shared transparency beliefs, 1173
Inhabitants, 889
Inherent skill, 387
Initial public offering (IPO), 1053
Innovation, 92, 94, 97, 103, 105, 685, 715, 726
Innovator, 1097, 1100, 1103–1105, 1113–1114
Institute of Leisure and Amenity Management (ILAM), 920
Institute of Sathya Sai Education, 418
Institutional entrepreneur, 996
Institutional entrepreneurship, 761
Institutional theory, 758
Institutions, 227
Instrumental attitudes, 156
Instrumental solutions, 140
Integrated Corporate Reports, 1050
Integrated marketing communication, 1108
Integrated project delivery (IPD) method, 784, 801

Integration
   programs and structural barriers, 529, 539–543
   societal, 532–539
Integrative healthcare approaches, 1072
Intellectual intelligence (IQ), 178
Intensive rotational grazing, 859
Intentional communities, 871, 872, 877
Inter-American Development Bank (IDB), 605
Interbeing principle, 11
Interconnectedness, 397
Intercultural business, 608, 614–616, 618–620, 623, 625, 626
Intercultural education, 605, 614
Interdisciplinary approach, 490, 493
Interface, 27, 73
Intergovernmental Panel on Climate Change (IPCC), 556, 1064
Internal legitimacy, 760
Internal locus of control, 94
International Auditing and Assurance Standards Board of the International Federation of Accountants, 1052
International Finance Corporation (IFC), 793, 1049
Internet-of-Things (IoT), 946
Interpersonal interaction, 394
Interpersonal relationships, 412
Intrinsic, 144
   human values, 413
   motivation, 131
   values, 155, 407
Intrinsic goals, 162
Investing, 1021, 1033
Investment tax credit, 642
Investor, 1097, 1100, 1105–1106, 1114–1115
Invisible plague, 384
Involuntary, 335
Ironman triathlons, 49, 54
Irreversible changes, 392
Isaiah, 299
ISO 26000, 67
Iśvara, 15
IT artifact, CSR-reporting software, *see* Information technology (IT) artifact, CSR-reporting software
Ivy League, 227

J

Jacobs, J., 306
Jainism, 39, 57

Jeremiah, 300
Johnson & Johnson, 476
Joliet Community College, 481
Judgment, 223, 226, 228
Julie Ann Wrigley Global Institute of
    Sustainability, 713

K
Kafka, F., 54
Karlberg, L., 386, 390
Kasser, T., 18, 19
Kawasaki, G., 108
KBBE, *see* Knowledge-Based Bio-Economy
    (KBBE)
Keep Indianapolis Beautiful (KIB), 281, 284
Kenan-Flagler Business School, 479
Kenya, 645
Kilikili, 1005
Ki-Moon, B., 475
Kleiner, A., 396, 398, 400, 401
Knowledge-based bio-economy (KBBE), 312,
    319–322
Kunstler, J. H., 854

L
Labels, 740
Laboratory experiments, 223, 225
Labor practices, 70–71
Lack of sensitivity, 384
Lag in response, 1030
Lakin, J. L., 386
Lama Tsering Gyaltsen, *see* Tsering Gyaltsen
Lambin, E., 386, 390
Language of business, *see* Business, sustainable
    decision making
Large group planning, 368
Last Minute Sotto Casa, 693
Laudato Si:On Care for Our Common Home, 841
Leaders, 229
Leadership, 229
Leadership in Energy and Environmental
    Design (LEED), 957
Learning college, 482
Learning communities, 400–401
Learning organization, 393
Legal codifications, 228
Legal gaps, 228
Legal requirements, 228
Legitimacy, 582
    defined, 582
    of firm, 584

Lenient regulatory framework, 1031
Lenton, T. M., 386, 390
Leveler, 124
Leveraged non-profit, 96
Leverage points, 145–147
Liability approach, 1059
Liberal arts, 475
Libertarian paternalism, 224
Library, 233
    visitors, 234
Life and non-life forms, 402
Life cycle analysis (LCA), 491
Life-cycle sustainability, 782
Life cycle thinking (LCT), 491
Lifestyle of Health and Sustainability
    (LOHAS)
    consciousness, 1071
    consumers, 1072, 1073, 1077
    cross cultural reliability and validation of
        measuring instruments, 1074
    environmental consciousness, 1073–1082
    environmental value, 1073
    health and sustainable lives, 1076
    health dimensions, 1082
    industry, 1072
    as lifestyle segment, 1080–1081
    market, 1072, 1084
    marketplace, 1072, 1078, 1079
    as market segment, 1079–1080
    measurement
        and culture, 1083–1084
        instrument, 1084
        invariance, 1084
        scale, 1074
        tool, 1074, 1076, 1081
        items, 1082
    mind, body and spirit, 1072
    mindset, 1073
    new challenges, 1072
    NMI, 1071
    operational definition, 1072
    operationalization of, 1078
    optimistic future view, 1072
    orientation, 1073
    peace, 1072
    personal, community, and planetary health
        and well-being, 1075
    personal development, 1072
    phenomenon, 1077, 1078, 1084
    philosophical and psychological values,
        1072
    pro-environmental behavior, 1077
    questions, 1082

relationships orientation, 1072
scale, 1074, 1078, 1082
segment, 1073, 1078
self-fulfillment, 1072
social responsibility, 1073
social scientific validation, 1075
spiritual products, 1072
theory, 1072, 1078, 1084
vegetarians/vegans, 1073
Lifestyles, 1088, 1093, 1104, 1105
Life-threatening, 385
Light consumption, 230
Limits to growth, 863
Linear green space, 921
Linear thinking, 141
Lipton, B., 385
Lithium triangle, 641
Lived sustainability, 734
Livelihood, 384
Liverman, D., 386, 390
Livestock, 40, 42, 44, 46, 55
  production, 159
Living, 976
Living Building Challenge criteria, 715
Living environment, 394
Local food products, 1073
Local park, 921
Location, 736
Locus of control (LOC), 94
Logic of appropriateness, 130
LOHASians, 1071, 1073, 1079, 1084
London Array, 647
Long-term economic viability, NGP, 592
Long-term inflation-adjusted price of oil, 1029
Los Angeles Eco Village (LAEV), 873–875
  ecological dimensions, 875–876
  economic dimensions, 876–877
  socio-cultural dimensions, 877–879
Loss aversion, 226
Loss of biodiversity, 392
Love, 407
Low-carbon economy (LCE)
  carbon tax, 637–639
  definition, 635
  emissions trading, 639
  energy auctions, 643
  government subsidies and tax credits, 642–643
  Paris Agreement, 635–637
  renewable energy storage technology, 641
  transactive energy systems, 640–641
  transition to, 632
  WTO, 643–644

Low-emission development strategies (LEDS), 635
Low information efficiency, 1037
Low-meat diets, 159
Low profit limited liability company, 1055

M

Machiavellianism, 414
Mahābhārata, 8, 16
Malthusian theory of population, 502
Management, 22, 31
  of emotions, 390
  of environmental issues, 667
  intensive grazing, 858
  and policy outcomes, 667
Marine life, 392
Marital dissatisfaction, 415
Market, 228
  implementation, 1122
  as innovators, 1103
  as investors, 1105
  as propagator, 1106
  target, 1111, 1115
Marketable permits/emissions credits trading, 1059
Market-based instruments, 671
Market-based paradigm, 157
Market-driven approaches, 1059
Marketing communication, 1089, 1092
  and corporate social responsibility, 1106
  goals, 1108, 1113, 1116
  integrated, 1108
  role of, 1093
  strategies, 1089, 1092
Marx's powerful project, 254
Maslow, A. H., 228
Massachusetts Commercial Food Waste Ban, 864
Materialism, 414
  extrinsic goal, 156
Mathematical bio-economics, 315
Maya communities, 611, 612, 622
Mayor Greg Stanton, 711
McDougall, J. A., 54
Meadows, D., 146
Measurement, LOHAS, see Lifestyle of Health and Sustainability (LOHAS)
Meat Atlas, 55–56
Meat-based diets, 47, 49, 56
Meat consumption, 160, 161, 164
Media, 1030
Mediterranean diets, 46

Membership groups, 227
Mental models, 396–398
Metabolic rift, 452
Methane, 42, 46
Microcredit, 99, 101
Micro-led change, 1136
Microsoft, 128
Migration and Me, 286, 290
Millennials, 121–122, 131, 486, 492
Millennium Ecosystem Assessment, 286
Mills, M. R., 48
Mindfulness, 158, 402
Mirror neurons, 409
Misshapen food, 694
Mission related investments (MRI), 1054
Mitigate, 1029, 1033
Mixed community green space, 887
Modern Yucatec Maya communities, trade and commerce in, 607–614
Moisture feedback, 392
Monetary gains, 229
Montreal Protocol, 571
Moore, C., 569
Moral and spiritual impoverishment, 5, 8, 9
Moral awareness, 23
Moral dimension of social life, 236
Moral hazard, 1035, 1039
Morality, 1073, 1082
More-ism, 5
Morocco, 645
Mother Theresa, 228
Motivations, 225, 275, 278–285
Multiculturalism, 213
Multi-level perspective (MLP), 1138–1139, 1149
Multinational corporations, 65
Multi-problem-solvers, 141, 143
Muni Narayana Prasad, *see* Narayana Prasad
Musk, E., 104, 647
Mutual Funds (MFs), 1021, 1026
Myth, 563

N
Nadeau, R. L., 385
Nadella, S., 128, 129
Narayana Prasad, 58
NASA, 20
Nascent entrepreneurs, 758
National culture, 1083, 1084
National Environmental Education Act, 474

Nationally Determined Contributions (NDC), 1064
National Science Foundation (NSF), 488
Natural air-conditioning, air-flow, 1000
Natural attenuation remediation, 1061
Natural behavioral law, 235
Natural capital, 654, 657, 658, 666, 669, 671, 673, 677
Natural disasters, 397
Natural law, 228
Natural Marketing Institute (NMI), 1071–1073, 1076, 1079, 1081, 1082
Natural resources, 236, 655
Negative screening, 1045
Neighborhood park, 920
Neoclassical economics, 269, 489
    deficiencies of, 313–314
    model of, 315
Neoliberal paradigm, 157
Neuroplasticity, 411
Neuroscience, 123
New conservation science (NCS), 205
New Environmental Paradigm (NEP), 1090, 1121
New Hampshire, 851
New Mexico, 863
New social contract, 1044, 1060–1061
New Urbanism, 840
New World, 570
New York City, 838
New Zealand, 480
NGOs, *see* Non-governmental organizations (NGOs)
NGP, *see* Northern Gateway Pipeline (NGP)
Nicaragua, 645
Niche, 756, 1144
Niemeyer, O., 555
NMI, *see* Natural Marketing Institute (NMI)
Noble act, 224
Nobleness, 229
Non-exploitative relationships, 413
Non-governmental organizations (NGOs), 1051, 1099, 1104, 1106, 1110, 1114–1116, 1119, 1121
Non-institutionalized initiatives, 1099
Nonlinearity, 150
Non-place, 553
Non-profit, 96, 106
    organizations, 399
Non-vegetarian, 38, 51, 59
Non-violence, 6–8, 12, 39, 57, 407

Noone, K., 386, 390
Norms, 1160, 1169, 1177
North American campus, 231
Northern Gateway Pipeline (NGP), 580, 586–588, 594–597
  economic benefits, 588–590
  environmental risks, 590
  social risks, 591–593
Nudge theory, 224
Nykvist, B., 386, 390

O

Ocean basin scales, 392
Ocean beds, 392
Oil
  extraction firms, 1029
  spills, NGP, 590
Olio, 693
1% for the planet®, 343
Oneness, 409
Open space, 918
Operational environmental performance, scale of, 1027
Opportunities, 222, 224, 753
Orage, A. R., 8
Organic, 553, 558
  foods, 1071, 1073
  products, 740, 1072, 1079
  supermarkets, 736–742
Organization
  agility, 1163, 1168, 1177
  behavior, 123
  change, 362
  communities, 399
  Governance, 80
  initiatives, 347
  learning, 393, 758
  non-profit, 399
  transparency, 1174
Organization development (OD) process, 364
Organization for Economic Co-operation and Development (OECD) Guidelines for Multi-National Enterprises (MNE), 1047–1048
Outdoor apparel, 341
Out of sight, out of mind, 385
Outperform(ing), 1037
  of public policy, 228
Outreach, 725
Over-consumption, 23, 406, 414, 452

Owens, M., 853
Ozone, 571
Ozone layer, 392

P

Pacific region, 480
Packaging, 566
  benefits of, 730
Packaging-free lifestyle, 746
Packaging-free shop, 747
Packaging free shopping, 751
Packaging-free supermarkets, 734–743
Packaging-free transport, 743
Pão de Açúcar, 559
Paradigm shift, 163
Paris Agreement, 557, 637
  Climate Change, 1064–1065
Parks type, 920
Passive investment approach/strategies, 1036
Patagonia Inc., 79, 339
A Pattern Language, 860–862
Payments, environmental services, 671
Peace, 5, 13, 14, 411, 1072
Peak oil, 863
Pedagogic, 348
People, 970, 978
  at risk of exclusion, 983
Perceived social value, 795
Perceptions of water, 420
Permaculture, 142, 302
Perseverance, 94, 99
Personal development, 1072
Personal mastery, 401
Personal social responsibility (PSR), 505
Persson, Å. 386, 390
Pescetarian diets, 46
Pesticides, 30
Philanthropy, 228
  capitalism, 1055
Phosphorus, 392
Physical circulation systems, 390
Physical fitness, 1072
Physiocrats, 313
Pinker, S., 386, 390
Pitfalls, 350
Place-based experiential learning, 457–458
Planet, 394, 563, 970, 978
Planned obsolescence, 562
Plant-based diets, 161, 163
  alternative for sustainable future, 49

Plant-based diets (*cont.*)
   vs. animal-based, 43
   environmental perspective, 50
   humans, 47
   vs. meat based diet, 28–30
   reasons to change, 38, 57
Plastic, 568, 730, 751
Plato, 54
Pleasant surprises, 750
Poetry and business, 122–130
Polar ice caps, 392
Policy measures, 732
Policy recommendations, 228
Pollution, 982
Polycentricity, 453–455
Poly-crisis, 142
Pope Francis, 841
Populations, 758
Positive, 363, 365, 368, 379
   experiences, 750
   psychology, 154–158
   societal change, 223
Poster, 230
Poverty, 406
Powerpack, 647
Practices, 731, 734
Pra Você, 559
Predictability, 1077
Predictive dimensions, relational climate, 788
Predictive validity, 1081
Pre-Hispanic Maya, trade and commerce in, 606–607
Price, 737
Principal/city/metropolitan parks, 920
Principles for responsible management education (PRME), 475, 476
Principles of responsible investing (PRI), 1049
Prisoner's dilemma, 116
Private sector, 236
Private social entrepreneurship, 97
Problem-based learning, 458, 975
Problem-solving, 971
Product-service system (PSS), 813
Pro-environmental behavior, 1077
Profit, 970, 978
Profit-seeking entrepreneur, 996
Program related investments (PRI), 1054
Project approval, 585
Project-based activity, 980
Project-based learning, 458, 975
Project Prithvi, 283, 284
Project teams, 782–787, 789, 790, 792–797, 799, 801–803

Promotion, 1090, 1107, 1109, 1111, 1115
Propagator, 1097, 1100, 1106, 1115–1120
Prophets, 299, 300, 306
Pro-social behavior, 224, 227, 229, 231, 413
Pro-social decision-making, 235
Prospect theory, 222, 226
Provisioning services, 660
PSR, *see* Personal social responsibility (PSR)
Psychology, 224
   health, 155
   ownership, 236
Psychometric scale development, 1082
Public, 1108
   consultation and disclosure, 1049
   policy, 228, 236
   and private sectors, 230
   relations, 1107, 1111, 1114
   transit, 841, 843, 844, 847, 848
Public-private partnerships, 1057–1058
Public-sector social entrepreneurship, 98
Pythagoras, 54

Q
QR code, 566
Quadrants, 347
Qualitative data, 764
Quality
   of food, 738
   of life capital approach, 924
   management, 489, 493
Questions, 1021
Quintana Roo (Mexico), higher education in, 614–617

R
Radical transparency, 385
Ramana Maharshi, 43
Rāmāyaṇa, 16
Rational market calculus, 228
Rational profit maximization calculus, 237
Rational strategy, 223
Rawls, J., 54
Reactive, 335
Real-world relevant sustainability, 236
Rebound effect, 828
Reciprocity, 546
Recyclable packages, 558
Recycled disposals weight, 232
Recycled weight, 231
Recycle plastic pollution, 732
Recycling, 223, 230, 231

bins, 231
compliance, 232
Reductionist problem-solving mode, 141
Reference point, 225
Refugees, 529
    crisis, 546
    societal integration, 532
Regenerative City, 304
Regular supermarket, 736–742
Regulating services, 660
Regulations, 228, 335, 732
    present and/or future, 1035
Reimagine Phoenix, Turning Trash into Resources, 711
Relatedness, 155
Relational approach to project delivery, 784
Relational context, 792, 797
Relational contracting, 784, 789, 796
Relational team approach, 782–784
    relational climate (*see* Relational team climate)
    relational context, 792–796
    relational contracting, 784–787
Relational team climate
    ACH team, 789, 790
    attitudes and behaviors, 789
    collaborative climate, 788
    communal, 788
    helping behavior, 787, 788
    justice norms, 791
    motivation, 791
    opportunistic category, 788
    predictive dimensions, 788
    project team, 789, 790
    risk, 791–792
    social science research, 787
    team assessment matrix, 789
    trust between parties, 792
Relationship management, 391
Relationships orientation, 1072
Relative comparison, 226
Reliability and validity, 1074, 1078, 1082
Renewable energy, 640, 641, 646
Re-Nuble, 695
Reporting system, CSR, *see* Corporate social responsibility (CSR)
Representativeness, 224
Reputation, 229
Reserves, 1027
Resilience, 455
Resource-based theory, 758
Resource Innovation and Solutions Network (RISN), 712

Resource Innovation Campus (RIC), 715
Resource-oriented entrepreneur, 997
Respect, 222, 224, 227, 228
Responsible investing, 1033, 1034, 1036, 1037, 1039
Responsible investors, 1021, 1034, 1039
Responsible Research and Innovation (RRI), 733
Retirement, 224
Ricard, M., 18, 41–42
Richardson, K., 386, 390
Right conduct, 411
Right to compensation, 1058
Right to no pollution, 1058
Risks, 226
    aversion, 225, 226
    shifts, 227
Roberts, C., 396, 398, 400, 401
Rockström, J., 386, 390, 391
Rodhe, H., 386, 390
Rogers, J., 125
Romanticism, 303
Ross, R. B., 396, 398, 400, 401
Rotational grazing, 858
Routines, 745
RRI, *see* Responsible Research and Innovation (RRI)
Rubbish, 572
Russia, 1051

S

Sabaté, J., 49
Sadness, 386
St. Petersburg College, 487
Salecl, R., 571
Santa Fe Farmers Market Institute, 863
SAS Institute, 75, 76
Sathya Sai Baba, 410
Sathya Sai Education in Human Values (SSEHV), 417
Satisfaction, 750
Sāttvic, 43
Scale, 348, 1074, 1076, 1077, 1082
    development, 1077
    measurement, 1078, 1081
    valid and reliable, 1077, 1081
Scheffer, M., 386, 390
Schellnhuber, H., 386, 390
Schema, 347
Schumacher, E. F., 842, 863, 865
Schweitzer, A., 54
Scott, D., 49, 54

SDGs, *see* Sustainable development goals (SDGs)
Self-actualization, 155
Self-efficacy, 94
Self-enhancement values, 407
Self-esteem, 227, 229
Selfishness, 5, 41, 42
Self-management, 389
Self-organization, 152, 153
Self-transcendent values, 407
Self-worth, 227, 228
Seminole State College of Florida, 488
Senge, P. M., 386, 393, 394, 396, 398, 400, 401
Senses, 740
Sensitivity, marketing communication, *see* Marketing communication
Separateness, 409
Service business, 980
Service-learning, 505
Serviceman's Readjustment Act of 1944, 482
Shame, 228
Shared nature, 228
Shared vision, 398–400
Sharpener, 124
Shaw, G. B., 8, 54
Shield, 230
Shopping, 730
 time, 745
Siemens AG, 647
Simmel, G., 551
Skills, 971
Slippery slope, 228
SLO, *see* Social license to operate (SLO)
Small and medium-sized enterprises (SMEs), 318
Small is Beautiful, Economics As If People Mattered, 863
Small-wins strategy, 132, 134
Smart citizens, 950, 951, 956
Smart city, 950
 Canons, 845
 characteristics, 840
 Charter, 842
 definition, 838
 enabling, 946
 hinterland, 858–860
 services, 953
 sustainability, 839
 transect, 849–852
 and urban planning, 841
Smart connectivity, 953, 958
Smart education, 953, 954
Smart energy, 953, 955

Smart governance, 953, 959
Smart health, 953
Smart home, 955
Smart information and communication technology (ICT) infrastructure, 952
Smart lighting, 955
Smart parking, 956
Smart recycling, 958
Smart sanitation, 953, 957
Smart services framework, 952
Smart spaces, 953, 956
Smart street lighting, 955
Smart toll, 956
Smart traffic management, 956
Smart transportation, 953, 956
Smart water, 953, 958
Smart work, 953, 958
Smith, B. J., 396, 398, 400, 401
Smog screen, 384
Snyder, P. K., 386, 390
Social awareness, 391
Social business venture, 97
Social capital, 229, 531, 537, 658, 666
Social class, 227, 229
Social cognition, 116
Social comparisons, 223, 226
Social compound, 223, 228
Social conscience, 1070
Social conscientiousness, 225, 227, 229, 232
Social contract, 228
Social cost, 415
Social dimension, 656
Social ecological economic development studies (SEEDS), 458
Social empowerment, 529
Social entrepreneurs, 92
 and commercial entrepreneurs, 96
 innovation, 94
 in Israel, 95
 Musk, Elon, 104–108
 perseverance, 94
 self-efficacy, 94
 Yunus, M., 99–104, 108–109
Social entrepreneurship, 108–110
 definition, 92
 hybrid non-profit, 97
 legitimacy of, 93
 leveraged non-profit, 96
 private, 97
 public-sector, 98
 social business venture, 97
 sustainability, 98
Social environments, 236

Social equity, 684
Social ethics, 1073
Social forces, 223, 228, 232, 236
Social groups, 227
    membership, 236
Social identity, 223, 227
Social incentives, 228
Social intelligence, 390
Social justice, 202, 206, 207, 209, 211, 213, 290, 1073, 1075, 1083
Social learning, 460, 461
Social license, as acceptance, 584
Social license to operate (SLO), 579–597
    concept, 580, 581
    defined, 581
    gaining and sustaining, 582–583
    identify, 583
    levels and boundaries of, 583–586
Socially responsible investing (SRI), 1021, 1025, 1045–1046, 1160
    investors, 1028
    lack of uniform information, 1027
    undervalue, 1028
Social mission, 983
Social mobility, 227
Social modeling, 412
Social norms, 228, 232, 748
    compliance, 228, 232
    cues, 228
    indirect change in, 1022
    insignia, 231
Social performance, 670
Social problems, 99, 105
Social responsibility, 92, 222, 227, 228, 1073
Social scientific validation, 1075
Social skills, 229
Social standing, 229
Social status
    attributions, 236
    drops, 226
    elevation opportunities, 222
    elevation prospects, 229
    endowments, 225
    enhancement, 223, 224
    gains, 222, 223, 232
    losses, 223, 224, 236
    outlooks, 226
    pedestal, 229
    prospects, 222, 226, 228, 229
    striving, 222
    symbol, 229
Social stratification, 227
Social sustainability, 119, 406
Social uncertainty, 225, 226
Social upward mobility, 227
Social workers, 387
Societal contribution, 222
Societal norms, 744
Societal transformation, 140
Societal welfare, 236
Society, 224, 227, 228
Socio-economics, 223
Sociology, 225, 235
Socio-political choices, 1045
Socio-political legitimacy, 760
Socio-psychological motives, 227
Sociotechnical transitions theory, 757
Socratic method, 184–187
Soft, 971
Soft-wired, 411
Soil
    carbon, 665
    health, 664
Solar City, 105
Solar energy, 955
Solar photovoltaics (PV), 642
Solidarity, 228
Soret, S., 49
Sörlin, S., 386, 390
Soros, G., 228
Sources of inefficiency, 688
South Africa, 1051
South Pacific, 485
Space X, 105
Spain, 695
Spiritual and cultural transformation, 5
Spiritual intelligence (SQ), 178
Spirituality, 5, 17, 26, 174, 177, 189, 192, 194, 385
Spiritual products, 1072
Sponsorship, 1106, 1107, 1111, 1115
Sri Ramana Maharshi, *see* Ramana Maharshi
Stakeholders, 334, 361, 368, 369, 373, 379, 582, 583, 585, 1061, 1097–1099, 1103–1106, 1108, 1110, 1111, 1114, 1115, 1121
    engagement, 361–364, 708, 1161–1162, 1166, 1169, 1171, 1172, 1176
    social license from, 582
    theory, 581, 1161
Stapleton, 857
Starbucks Company, 80, 81, 476
State-owned enterprises, 1051
Status, 223, 224, 227, 228
    losses, 225
    positions, 224

Status (cont.)
 quo, 224
 quo bias, 225, 226
Steffen, W., 386, 390
Stewardship, 274
 barriers to, 285–289
 Canopy Story, 284
 citizen, 274
 community resilience, 274
 moral and ethical dimensions of, 275
 for pragmatism, 281–282
 Project Prithvi, 283
 United States Forest Stewardship Program, 278
Stimulus, 387
Stockholm Conference on the Human Environment, 473
Stock market
 collapse, 1029
 index, 1021
Stocks of corporation, 1021
Stories, 304
Stranded, 1031
 assets, 1020, 1029
Strategy, 124
Stratosphere, 392
Stress, 156
Stringent legal enforcement, 235
Students of test dormitory, 231
Subconscious emotional compensation, 237
Subculture, 1070, 1083
Summer school, 230
Supermarkets, 554, 730
Supermodernity, 553
Supporting services, 660
Sustainability, 6, 7, 25, 92, 98–99, 114, 115, 119, 122, 124, 132, 133, 135, 136, 181, 209, 214, 223, 229, 230, 236, 298, 304, 306, 360, 361, 363, 365, 367, 368, 370, 373, 377–379, 384, 406, 520, 655, 710, 871–873, 875, 876, 880, 992–995, 1158, 1162, 1166, 1168, 1171, 1177
 analytics, 1045
 definition, 171, 783, 839
 economic, 119
 eco-pedagogy approach to, 184–187
 education, 447, 448, 450, 456, 457, 460, 462, 463, 473, 475, 478, 486–487, 495
 engaged, 24–25
 entrepreneurship and, 996
 environmental, 25, 116

 examples, 840
 exercises, 732
 flexible cohesion, 797
 fun and rewarding experience, 435–437
 gamification, 432
 governance regulations, 1050–1053
 in higher education, 971–977
 interdisciplinary concepts, 449, 450, 456, 458
 LOHAS (*see* Lifestyle of Health and Sustainability (LOHAS))
 metrics, 1061
 narrative, 154, 158, 163
 pedagogical approaches to teaching, 462
 positive peer pressure on issues, 437
 reporting, CSR (*see* Corporate social responsibility (CSR))
 science, 145
  failing systems, 146
  feedback loops, 148
  leverage points, 145, 147
  multi-problem-solvers, 148
  potential risks and unintended consequences, 149
  virtuous cycles, 148
 social, 119
 Solutions Festival, 720
 student engagement, 456
 summit, 367
 transformative learning, 187–189
 transitions, 1140
  agency-related terms in, 1142
  and societal transitions, 1144
 Vikram's definition of, 999
Sustainability Matters, 24
Sustainability-oriented habits, 973
Sustainable, 979
 agriculture, 686
 bio-economy, 317, 319, 323, 324
 business, 810, 811, 813–816
 business education
  challenges of, 476–477
  community colleges (*see* Community college system)
  critical theory, 489
  curriculum development and best practices and challenges, 491
  ecocentrism, 489
  ecological modernization, 489
  entrepreneurship, 476
  ESD, 478
  Europe, Asia and Pacific region, institutions in, 480

generational and cultural role and
 influence, 492–493
HESD, 478
interdisciplinary approach, 490
justification for, 474–475
neoclassical economics, 489
quality management, 489
support for, 475–476
teaching, methods and tools for,
 490–491
transdisciplinary approach, 489
US, institutions in, 478–480
US community college system
 (*see* US community college system)
business practices, 1028
circular economy, 705
community development, 530
consumption, 810, 813
decision making
development, 115, 135, 245, 476, 478–480,
 490, 501, 505, 604, 605, 613–615,
 617, 621, 622, 624, 626, 732–734,
 783, 793, 993, 1081
application, 252
bio-economy, 315
and Brundtland Report, 247–251
conceptual model, 251
CSD, 473
definition, 704
economic, 504
environment, 502–504
ESD, 478
goals, 705
HESD, 478
social, 503
social and political construct, 252
WSSD, 473
diets, 43–44, 49, 687
entrepreneurs, 970, 977, 979
entrepreneurship, 476, 479, 488, 977, 978
education, 978
food systems, 654, 656, 671, 677
future, 397
game plan, 391
language of business (*see* Business,
 sustainable decision making)
lifestyles, 156
management
for community development, 927–933
practices in urban green spaces, 932
mankind, 237
power, 434
production, 810, 811, 813

prosperity, 236
role of framing, 116
transformation, 148, 157
well-being, 154–158
Sustainable development goals (SDGs), 305,
 448, 504–505, 512–517, 557, 871
Svedin, U., 386, 390
Swami Dayananda Saraswati, *see* Dayananda
 Saraswati
Swami Tattvabodhananda,
 see Tattvabodhananda
Sweden, 644
Synchronous interactive connectivity, 1060
"System 1" and "System 2" modes, 130
System dynamics, 146
Systemic balance, 392
System of signs, 561
Systems thinking, 394–396

T

Taittirīya Upaniṣad, 38, 58, 59
Take-away venues, 748
Tao Te Ching, 13
Target market, 1111, 1115
Taste, 161
Tattvabodhananda, 59
Team cohesion, 783, 796, 797
Team learning, 400–401
Tech-smart, 951
Tesla, N., 54
Tesla Motors, Inc., 105, 647
The First World Ocean Assessment, 255
The God Theory, 409
The Greening of Detroit (TGD), 282, 285, 287,
 288
The Natural Step, 708
Theology, 298, 306
The Road Not Taken, 125
Thessaloniki Declaration, 478
The Walt Disney Company, 476
Think local, 399
Thriveability, 163
Tibilisi Declaration, 478
Thich Nhat Hanh, 11
Tilman, G. D., 45, 46
Time banking, 531–532
Time-consuming, 750
Tobacco, 1023
Toll-based bridges, 1058
Tolstoy, L., 50, 51, 54
Toxic emotions, 24
Toxic tort, 1059

Trade-offs, 747
Trader Joe's, 77–78
Traditional investors, 1028
Transactive energy, 640
Transdisciplinary approach, 489, 493
Transdisciplinary teaching, 976
Transect, 842, 849
Transformational change, 146
Transformative learning, 189, 191
  sustainability, 187
Transition, 1024
  management, 734
Transparency, 1159
  beliefs and boundary spanning, 1173–1177
  norms, 1160, 1169
Transportation mode, 742
Transversal skill, 971
Trash picker, 573
Trigger social norm compliance, 228
Triple bottom line, 115, 116, 118, 122, 131, 136, 176–179, 335, 350, 394
Trujillo, J., 711
Trump, D., 474, 493
Trust, 585, 740
Truth, 411
Tsering Gyaltsen, 53
Tzu, L., 13

U

Überethicality, 222, 228
UHI effect, 891
UIMQRoo, 608, 614, 615, 618, 623
Uncertainties, 223, 656
Uncomfortable feeling, 749
Unconscious wish, 236
Unconventional crude oil, 1030
Unethicality, 228
Unhappiness, 225
Unintended consequences, 398
United Nations (UN), 448, 504–505, 512–517, 556
  Commission on Sustainable Development (CSD), 473
  Educational, Scientific and Cultural Organization (UNESCO), 555
  Environment Programme (UNEP), 478, 635
  Food and Agriculture Organization (FAO), 46
  Global Compact, 1050
  Global Compact Summit, 363
  Human Settlements Programme (UN–Habitat), 870
  International Environmental Education Programme, 473
  Secretary General's Special Advocate for Inclusive Finance for Development (UNSGSA), 605
  Sustainable Development Program (UNSDP), 1057
United States (U.S.), 473, 478, 480–482, 1025
  Advanced Energy Manufacturing Tax Credit, 642
  Agency for International Aid (USAID), 1057
  community colleges (see US community college system)
  culture, 492
  Department of Commerce, 43
  Department of Energy Pacific Northwest Demonstration Project, 640
  Environmental Protection Agency (EPA), 7, 40
  Forest Stewardship Program, 277, 278
  Government Accountability Office, 258
  Securities and Exchange Commission (SEC), 1051
Universal Consciousness, 408
Universal owners of private enterprise, 1046
University/ies, 970
  of Birmingham, 387
  of Chicago, 481
  logo, 231
  settings, 229
Unpleasant past social status losses, 237
Upaniṣads, 12, 13, 15–17, 38, 57
Urban Agrarianism, 852
Urban agriculture, 841, 855
Urban ecology, 24
Urban green spaces, 886
  aesthetics benefits of, 926
  assessment, 929
  audits, 929
  benefits of, 922
  biodiversity in, 919
  challenges and opportunities, 904–906
  community development, 922–927
  community involvement in, 898
  components of, 887–888
  cooling capacity of, 891
  declining quality of, 897
  density, 888
  designing and planning of, 907
  development of, 938
  distribution inequities, 896
  economic benefits of, 926

economic value of, 901
ecosystems functions and services of, 888–893
educational benefits of, 926
environmental and ecological benefits of, 922–923
functions of, 919
health and psychological benefits of, 924–925
management and maintenance of, 929
nature of, 918
negative economic impact of, 897
network, 921
non-users of, 898
public initiatives and actions, 906–908
quality of, 894
reasons of visiting, 889
role and behavior of, 892
self-management of, 931
social benefits of, 924
strategic management innovation, 931, 933–940
sustainable management practices in, 932
types, 892
typology of, 919–921
unequal distribution of, 895
users of, 894–898, 921
Urbanization, 870, 871, 880
Urban poor, 419
Uruguay, 645
US community college system
basic commitments, 481
characteristics, 482–483
history, 481–482
student characteristics, 483–484
UV-B radiation, 392

V

Value-based management approaches, 422
Value-based water education (VBWE), 410, 416
Value-belief-norm theory, 280
Values, 1070
and beliefs, 1079
cultural, 1071, 1083
economic, 1078
environmental, 1073
personal, 1084
psychological, 1072
van der Leeuw, S., 386, 390
Vedānta psychology, 6

Vedas, 16
Veda Vyāsa, 9
Vegan, 39, 45, 54, 55, 58
  See also Vegetarian
Vegetable gastronomy, 161
Vegetarian, 39, 43, 46, 48, 49, 1073
  diet, 160
Vegetarianism, 50–54, 161
Vegetation density, 890
Venture philanthropy, 1056
Vermont, 863
Vertical garden
  close-up, 1002
  irrigation, 1001
  set-up, 1001
Vicious cycles, 148
Vietnam, 567
Violence, 6, 7, 28
Virtues, 228
Virtuous cycles, 148
Vocational training, 481, 483, 485
Voluntary service, 228

W

Walberg, H. J., 387
Walkability, 840, 842, 848
Walker, B., 386, 390
Wang, M. C., 387
Warm glow, 224
Warren Buffett, 228
Warsaw International Mechanism for Loss and Damage, 1065
Waste production, 735
Water
  education, 420
  kiosks, 419
  resources, 668
  use efficiency, 664
  waste, 668
Watts, A., 54
WCED, see World Commission of Environment and Development (WCED)
Wealth distribution, 226
Weissberg, R. P., 387
Well-being, 224, 1070, 1072, 1075, 1076, 1083, 1094, 1113
Wellesley, L., 46
Wellness, 1076, 1083
Western Canada Select (WCS) prices, 587
Western spiritual traditions, 301
West Texas Intermediate (WTI) price, 587

Weyerhaeuser Productos, 132
Whole and organic food, 161
WHO's International Agency for Research, 52, 53
Wicked problems, 142, 150
Wildlife habitat, NGP, 591
Williams, R., 563
Winery waste management, 668
Wisdom, 6, 14, 23, 33
WISErg, 695
Wispe, L., 385
Woodstock, 863
Workplace, 387, 391
World Bank, 634
World Commission of Environment and Development (WCED), 501
World Economic Forum, 1056
World Intellectual Property Organization's (WIPO) Green, 644
World Summit on Sustainable Development (WSSD), 473
World Trade Organization (WTO), 643–644
Worldviews, 143, 156
  plurality of, 145

Y

Yajñas, 15–16
Yourofsky, G., 58
Yucatec Maya, 605, 608–610, 614, 617, 621, 623–625
Yunus, M., 99, 109

Z

Zeitgeist, 28
Zero-waste, 750
  economy, 685
Zika virus, 634
Zimbabwe, 641
Zins, J. E., 387
Zion, 300

PGSTL